Springer Series in Molecular Biology

Series Editor: Alexander Rich

Springer Series in Molecular Biology

Series Editor: Alexander Rich

Yeast Genetics
Fundamental and Applied Aspects
J.F.T. Spencer, Dorothy M. Spencer, A.R.W. Smith, eds.

Myxobacteria
Development and Cell Interactions
Eugene Rosenberg, ed.

DNA Methylation
Biochemistry and Biological Significance
Aharon Razin, Howard Cedar, Arthur D. Riggs, eds.

Cooperativity Theory in Biochemistry
Steady-State and Equilibrium Systems
Terrell L. Hill

Terrell L. Hill

Cooperativity Theory in Biochemistry

Steady-State and Equilibrium Systems

With 230 Figures

Springer-Verlag
New York Berlin Heidelberg Tokyo

Terrell L. Hill
Department of Health & Human Services
National Institutes of Health
Bethesda, Maryland 20205 U.S.A.

Series Editor:
Alexander Rich
Department of Biology
Massachusetts Institute of Technology
Cambridge, Massachusetts 02139 U.S.A.

Library of Congress Cataloging in Publication Data
Hill, Terrell L.
Cooperativity theory in biochemistry.
(Springer series in molecular biology)
Bibliography: p.
Includes index.
1. Cooperative binding (Biochemistry) I. Title.
II. Series.
QP517.C66H55 1985 574.19'282 85-2716

Softcover reprint of the hardcover 1st edition 1985

Typeset by Interactive Composition Corporation, Pleasant Hill, California.

9 8 7 6 5 4 3 2 1

ISBN-13: 978-1-4612-9555-6 e-ISBN-13: 978-1-4612-5082-1
DOI: 10.1007/978-1-4612-5082-1

To my friend and colleague Yi-der Chen
for his expert and gracious aid when
needed, over many years

Series Preface

During the past few decades we have witnessed an era of remarkable growth in the field of molecular biology. In 1950 very little was known of the chemical constitution of biological systems, the manner in which information was transmitted from one organism to another, or the extent to which the chemical basis of life is unified. The picture today is dramatically different. We have an almost bewildering variety of information detailing many different aspects of life at the molecular level. These great advances have brought with them some breath-taking insights into the molecular mechanisms used by nature for replicating, distributing and modifying biological information. We have learned a great deal about the chemical and physical nature of the macromolecular nucleic acids and proteins, and the manner in which carbohydrates, lipids and smaller molecules work together to provide the molecular setting of living systems. It might be said that these few decades have replaced a near vacuum of information with a very large surplus.

It is in the context of this flood of information that this series of monographs on molecular biology has been organized. The idea is to bring together in one place, between the covers of one book, a concise assessment of the state of the subject in a well-defined field. This will enable the reader to get a sense of historical perspective—what is known about the field today—and a description of the frontiers of research where our knowledge is increasing steadily. These monographs are designed to educate, perhaps to entertain, certainly to provide perspective on the growth and development of a field of science which has now come to occupy a central place in all biological studies.

The information in this series has value in several perspectives. It provides for a growth in our fundamental understanding of nature and the manner in which living processes utilize chemical materials to carry out a variety of activities. This information is also used in more applied areas. It promises to have a significant impact in the biomedical field where an understanding of disease processes at the molecular level may be the capstone which ultimately holds together the arch of clinical research and medical therapy. More recently in the field of biotechnology, there is another type of growth in which this science can be used with many practical consequences and benefit in a variety of fields ranging from agriculture and chemical manufacture to the production of scarce biological compounds for a variety of applications.

This field of science is young in years, but it has already become a mature discipline. These monographs are meant to clarify segments of this field for the readers.

Cambridge, Massachusetts

Alexander Rich
Series Editor

Preface

The primary purpose of this book is to present, by means of a very large number of examples, the *methods* that should be used to treat cooperativity in steady-state and equilibrium biochemical systems from a fundamental molecular, rather than empirical or thermodynamic, starting point. The main motivation has been to bring to the attention of a wider audience the treatment of cooperativity in steady-state systems that was developed by the author and his collaborators in 15 papers between 1977 and 1982. Collaborators, on some of the papers, were Y. Chen, L. Stein, and A. Levitzki. Equilibrium systems have been included here for completeness and as a background for the steady-state analysis. In the steady-state problems, intermolecular interaction effects are introduced at the outset, as they should be, into the rate constants of the relevant kinetic diagram of states. In the equilibrium problems, intermolecular interaction effects appear explicitly, in a natural way, in the terms of the grand partition function, which is the proper starting point for a molecular theory. The older, conventional methods used by biochemists to treat cooperativity in equilibrium and steady-state systems have been appropriate and extremely productive, but it seems inevitable that, with the growing availability of molecular details, young biochemists of the future will want to take a more fundamental molecular approach. It is to such readers that the book is principally addressed.

In some ways, the present book can be viewed as a continuation of the author's graduate-level textbook,[1] *Introduction to Statistical Thermodynamics,* which provides the molecular interpretation of equilibrium macroscopic (thermodynamic) properties.

The grand partition function has been applied to biochemical problems with cooperativity, of the sort considered here, beginning[2–5] in the early 1950s, by a few practitioners of statistical mechanics (for reviews see Hill,[1] Poland,[6] and Hess and Szabo[7]). But the recognition of this method as the method of choice for the treatment of cooperative equilibrium systems from a molecular, rather than thermodynamic, starting point remains almost non-existent among practicing biochemists.

Only a few examples of real systems with real data are discussed. This is because my objective is not to survey applications that have been made of the methods used here, but rather to emphasize the methods themselves so that readers can learn how to apply these methods to their own problems. Correspondingly, the treatment is meant to be pedagogical rather than sophisticated or advanced; this emphasis has been promoted primarily by the analysis of a very large number of different, relatively simple, molecular models. No explicit problem sets are included, but the serious reader will encounter innumerable places where missing details of proofs, further special cases, alternative methods, etc., furnish implicit problems.

Transients leading to final equilibrium or steady states are hardly mentioned, although the kinetic formalism developed can be applied in a direct and obvious way to transients. This was an arbitrary decision based on considerations of length. For the same reason, and because of the existence of the excellent book by Poland and Scheraga,[8] helix-coil theory in proteins and polynucleotides has been omitted. Also, to save space and to avoid overemphasis, the Bragg-Williams or mean-field approximation has been considered in the last chapter only. Other approximations have hardly been mentioned.

Theoretical physicists may be particularly interested in Chapters 9–11. From the point of view of statistical physics, these chapters extend the well-known Ising problem[1] from equilibrium to steady states arbitrarily far from equilibrium. In effect, the Ising problem acquires, at steady state, a new dimension or axis, namely, the thermodynamic force X driving the individual, interacting units of the lattice. In the usual Ising problem, $X = 0$.

After the first, introductory, chapter, the chapters alternate in covering essentially the same topic, first at equilibrium and then at steady state. Thus Chapters 2, 4, 6, and 8 deal with equilibrium systems, whereas Chapters 3, 5, 7, and 9 treat corresponding or related steady-state systems. Chapters 10 and 11 treat both equilibrium and steady-state problems, especially the latter.

Much of the material included here has not been published previously, but no attempt has been made to distinguish these topics.

Numbered references appear at the end of each chapter and of this preface. Additional general references, related to the material of the chapter, are also included.

Some readers may find it useful to have a complete list of the 15 papers, referred to above, on steady-state interaction problems. These papers contain many details and cases not included in the book. The list follows:

1. Theoretical study of the effect of enzyme–enzyme interactions on steady-state enzyme kinetics. *Proc. Natl. Acad. Sci. USA* **74,** 3632–3636 (1977).

2. Further study of the effect of enzyme–enzyme interactions on steady-state enzyme kinetics. *Proc. Natl. Acad. Sci. USA* **74,** 4111–4115 (1977).
3. "Virial" expansion of enzyme flux and use of quasi-chemical approximation for two-state enzymes with enzyme–enzyme interactions. *Proc. Natl. Acad. Sci. USA* **74,** 5227–5230 (1977).
4. Effect of enzyme–enzyme interactions on steady-state enzyme kinetics. IV. "Strictly steady-state" examples. *J. Theor. Biol.* **75,** 391–416 (1978).
5. Unsymmetrical and concerted examples of the effect of enzyme–enzyme interactions on steady-state enzyme kinetics. *Proc. Natl. Acad. Sci. USA* **75,** 1101–1105 (1978).
6. Monte Carlo calculations on critical behavior in two-state, steady-state Ising systems. *J. Chem. Phys.* **69,** 1126–1138 (1978). With Y. Chen.
7. Critical behavior of two-state, steady-state Ising systems, according to the Bragg–Williams approximation. *J. Chem. Phys.* **69,** 1139–1150 (1978). With L. Stein.
8. Interacting enzyme systems at steady state: location of the phase transition in approximations of the mean field type. *Proc. Natl. Acad. Sci. USA* **75,** 3015–3018 (1978). With Y. Chen.
9. Interacting enzyme systems at steady state: further Monte Carlo calculations on two-state molecules. *Proc. Natl. Acad. Sci. USA* **75,** 5260–5263 (1978). With Y. Chen.
10. Steady-state phase or cooperative transitions between biochemical cycles. *Proc. Natl. Acad. Sci. USA* **76,** 714–716 (1979).
11. Approximate steady-state properties of lattices of interacting three-state enzyme molecules: a novel phase transition. *Proc. Natl. Acad. Sci. USA* **76,** 1014–1017 (1979). with L. Stein.
12. Properties of some three-state, steady-state Ising systems, according to the Bragg–Williams approximation. *Proc. Natl. Acad. Sci. USA* **77,** 693–697 (1980). With L. Stein.
13. Subunit neighbor interactions in enzyme kinetics: half-of-the-sites reactivity in a dimer. *Proc. Natl. Acad. Sci. USA* **77,** 5741–5745 (1980). With A. Levitzki.
14. Three-state, steady-state Ising systems: Monte Carlo and Bragg–Williams treatments. *Proc. Natl. Acad. Sci. USA* **78,** 4–8 (1981). With Y. Chen.
15. On the one-dimensional steady-state Ising problem. *J. Chem. Phys.* **76,** 1122–1127 (1982).

References

1. Hill, T. L. (1960) *Introduction to Statistical Thermodynamics* (Addison-Wesley, Reading, MA).
2. Hill, T. L. (1950) J. Chem. Phys. **18**, 988.
3. Hill, T. L. (1953) J. Phys. Chem. **57**, 324.
4. Hill, T. L. (1955) J. Chem. Phys. **23**, 623.
5. Hill, T. L. (1956) J. Amer. Chem. Soc. **78**, 3330.
6. Poland, D. (1978) *Cooperative Equilibria in Physical Biochemistry* (Clarendon Press, Oxford).
7. Hess, V. L. and Szabo, A. (1979) J. Chem. Educ. **56**, 289.
8. Poland, D. and Scheraga, H. A. (1970) *Theory of Helix-Coil Transitions in Biopolymers* (Academic Press, New York).

Contents

9. One-Dimensional Array of Interacting Two-State Units at Steady State

10. Monte Carlo Study of Equilibrium and Steady-State Two-Dimensional Lattices

11. The Bragg–Williams or Mean-Field Approximation in Steady-State Systems

I

Systems without Explicit Interactive Cooperativity

1

Preliminary Background Material

This chapter covers several topics that are prerequisite—for the thorough reader—to the rest of the book, but are not really part of the main argument. Hence this chapter could be viewed as a collection of appendices, which may be omitted to begin with and referred to later as needed. On the other hand, for some readers, these fundamental topics will have considerable intrinsic interest.

The motivation, aim, and plan of the book have been set forth in the Preface. This material will not be repeated here. We turn immediately to the business at hand.

1. Canonical and Grand Partition Functions

In this section we review a few properties of the two most important partition functions for a one-component system. Generalizations will be required later, but these will be introduced as needed.

Canonical Partition Function

We consider an equilibrium thermodynamic system with independent variables N (number of molecules), V (volume), and T (temperature). The temperature is fixed by a surrounding heat bath. We are interested in the connection between the macroscopic thermodynamic properties of this system, on the one hand, and its (quantum mechanical) molecular properties, on the other. Statistical mechanics provides the desired formal connection, which can be made explicit for systems that are not too complicated.

If $E_j(N, V)$ is the energy (eigenvalue) of the jth energy eigenstate of the system, then the canonical partition function is

$$Q(N, V, T) = \sum_j e^{-E_j(N,V)/kT}, \qquad (1.1)$$

where the sum is over all states j, and k is the Boltzmann constant. Degenerate energy levels are represented by several equal terms in the sum. In practice, in solution physical chemistry (where there is more than one component), usually the internal vibrational partition function of each molecule is separated out, and the remaining coordinates (translation, external and internal rotation) are taken care of by a classical phase integral over $e^{-H/kT}$, where H is the Hamiltonian function of these coordinates. Details are given in Ref. 1, pp. 262–264.

The connection with thermodynamics is made via the Helmholtz free energy:

$$A(N, V, T) = -kT \ln Q(N, V, T) \tag{1.2}$$

and

$$dA = -SdT - pdV + \mu dN \tag{1.3}$$

or

$$d(-A/kT) = -\bar{E}d(1/kT) + (p/kT)dV - (\mu/kT)dN, \tag{1.4}$$

where S is the entropy, p the pressure, μ the chemical potential, and $\bar{E}$ the mean energy (E fluctuates). Differentiation of $-kT \ln Q$, Eq. (1.3), or $\ln Q$, Eq. (1.4), gives various thermodynamic properties of interest.

An example, from Eq. (1.4), is

$$\bar{E} = -\left(\frac{\partial \ln Q}{\partial\, 1/kT}\right)_{V,N}. \tag{1.5}$$

From Eq. (1.1) we then find

$$\bar{E} = \frac{\sum_j E_j e^{-E_j/kT}}{\sum_j e^{-E_j/kT}}. \tag{1.6}$$

This shows that the probability that the system, at equilibrium, will be observed in state j is $e^{-E_j/kT}/Q$. This is the famous Boltzmann probability distribution.

Grand Partition Function[2]

Here the one-component system is open with respect to N: the system exchanges molecules with a reservoir of the same molecules at chemical potential μ and temperature T. Consequently, N fluctuates. The independent thermodynamic variables are now μ, V, T rather than N, V, T. This approach is the natural one to use in binding problems. The bound molecules are the system, or part of the system, and the free molecules are the reservoir.

The corresponding partition function, introduced by Gibbs, is the grand partition function:

$$\Xi(\mu, V, T) = \sum_{N\geq 0} Q(N, V, T)\, e^{N\mu/kT} \tag{1.7}$$

$$= \sum_{N\geq 0} Q(N, V, T)\lambda^N, \tag{1.8}$$

where λ (the absolute activity) $\equiv e^{\mu/kT}$, and the sum goes over all possible values of N. Incidentally, when $N = 0$, $Q(0) = 1$ (unless the system includes an always present "background"). The term "grand" refers to the second, or higher-order, sum over N; the first sum is over j for each N, as in Eq. (1.1). The connection with thermodynamics is

$$pV = kT \ln \Xi(\mu, V, T), \tag{1.9}$$

$$d(pV) = SdT + pdV + \bar{N}d\mu. \tag{1.10}$$

Of course Eqs. (1.3), (1.4), and (1.10) are all equivalent thermodynamic relations, rearranged for convenience (according to the choice of independent variables).

From Eqs. (1.7), (1.8), and (1.10), we find the important relations

$$\bar{N} = kT\left(\frac{\partial \ln \Xi}{\partial \mu}\right)_{T,V} = \lambda\left(\frac{\partial \ln \Xi}{\partial \lambda}\right)_{T,V} \tag{1.11}$$

$$= \frac{\sum_{N\geq 0} NQ(N, V, T)\lambda^N}{\sum_{N\geq 0} Q(N, V, T)\lambda^N}. \tag{1.12}$$

Equation (1.12) shows that

$$P_N = \frac{Q(N, V, T)\lambda^N}{\Xi} \tag{1.13}$$

is the probability that this open system contains exactly N molecules. If $Q(0) = 1$, the probability that the system is empty ($N = 0$) is $1/\Xi$. Equation (1.13) is the analogue of the Boltzmann probability distribution for a closed system (see above). In other words, just as the separate terms in the canonical partition function, Eq. (1.1), give the relative probabilities that the closed system has various energy values E_j, so the separate terms in the grand partition function, Eq. (1.8), give the relative probabilities that the open system contains various numbers of molecules N.

For example, in a system with binding on several sites, the probabilities of various degrees of binding, at equilibrium, follow immediately from the corresponding terms in the grand partition function. Furthermore, the overall amount of binding is given by Eq. (1.12) (λ is proportional to the concentration activity or concentration). These are extremely useful properties that are very easy to exploit. Many examples will appear in later chapters.

A further differentiation of Eq. (1.12),

$$\bar{N}\Xi = \sum_{N\geq 0} NQ(N, V, T)\lambda^N,$$

leads to the variance in N:

$$\overline{N^2} - \bar{N}^2 = \overline{(N - \bar{N})^2} = \lambda\left(\frac{\partial \bar{N}}{\partial \lambda}\right)_{T,V}. \tag{1.14}$$

In summary: the first differentiation of $\ln \Xi$ with respect to λ gives $\bar{N}$, Eq. (1.11), and the second gives $\overline{N^2} - \bar{N}^2$, Eq. (1.14).

2. Effective Partition Functions for Solute Molecules in a Solvent

In the next section, we discuss the equilibrium constant for binding one molecule on another in solution. Here we introduce the effective solute partition functions needed for the formulation of this binding constant.

We shall consider, below, a binary solution with N_1 molecules of solvent and N_2 molecules of solute. The canonical partition function for such a solution is denoted by $Q_{N_1N_2}(V, T)$. In practice, the solvent might have several components (e.g., H_2O, NaCl, etc.); in this case N_1 would stand for a *set* of solvent numbers (N_a, N_b, etc.).

For reference, we start with a (possibly hypothetical—if the solute is not volatile) system of very dilute gaseous solute molecules in the absence of solvent ($N_1 = 0$). The grand partition function for this system is

$$\begin{aligned} \Xi = e^{pV/kT} &= \sum_{N_2=0}^{\infty} Q_{0N_2}\lambda_2^{N_2} \\ &= 1 + Q_{01}\lambda_2 + \cdots, \end{aligned} \tag{2.1}$$

where $\lambda_2 = e^{\mu_2/kT}$ and μ_2 is the chemical potential of solute in the gas. For our special purposes here, $\lambda_2 \to 0$, we need only the first two terms in Eq. (2.1). Q_{01} is the partition function for one molecule of solute in V at T. Q_{01} is proportional to V, and includes translational, rotational, and vibrational factors. From Eqs. (1.11) and (2.1),

$$\begin{aligned} \overline{N}_2 &= \lambda_2\left(\frac{\partial \ln \Xi}{\partial \lambda_2}\right)_{T,V} \\ &= Q_{01}\lambda_2 + \cdots. \end{aligned} \tag{2.2}$$

Only the first term is needed for a very dilute gas. The second term, in λ_2^2, is related to the second virial coefficient of the gas. If we divide by V,

$$c_2 \equiv \overline{N}_2/V = (Q_{01}/V)\lambda_2. \tag{2.3}$$

Here c_2 is the number concentration and Q_{01}/V is a function of T only because Q_{01} is proportional to V. This simple proportionality between c_2 and λ_2 holds only in the dilute limit $c_2 \to 0$. Then, on taking logarithms,

$$\mu_2 = \mu_2^{\circ}(T) + kT \ln c_2, \tag{2.4}$$

$$\mu_2^{\circ}(T) = -kT \ln(Q_{01}/V). \tag{2.5}$$

Equation (2.4) is the conventional thermodynamic relation between μ_2 and c_2 for a dilute gas, and Eq. (2.5) shows that the standard chemical potential μ_2° is determined by the partition function Q_{01}. The above procedure, starting with Eq. (2.1), is not necessary in order to derive Eq. (2.4), but this kind of approach *is* necessary to obtain, in a simple way, the corresponding result for a solute in a solvent.

We digress very briefly to examine an alternative derivation of Eq. (2.2).

For an ideal gas of N_2 molecules in V, that is, for a gas with no intermolecular interactions,

$$Q_{0N_2} = \frac{1}{N_2!} Q_{01}^{N_2}. \tag{2.6}$$

Then

$$\Xi = \sum_{N_2 \geq 0} Q_{0N_2} \lambda_2^{N_2} = e^{Q_{01}\lambda_2} \tag{2.7}$$

and

$$\overline{N}_2 = \lambda_2 \left(\frac{\partial \ln \Xi}{\partial \lambda_2} \right)_{T,V} = Q_{01} \lambda_2. \tag{2.8}$$

This is essentially the same result as in Eq. (2.2) for a very dilute real gas. Equation (2.8) was derived for an *ideal* gas, which is not quite the same thing.

We now consider the same dilute solute molecules, component 2, to be immersed in a solvent, component 1. Our object is to derive the equivalent of Eq. (2.4) for use in the next section. Actually, there we shall be interested in a solution that is very dilute with respect to *several* different solutes. Because of this diluteness, however, it suffices here to treat only a single arbitrary solute in the solvent. That is, the limiting ($c_2 \to 0$) expression analogous to Eq. (2.4) is unaffected by the presence of other solutes that are also very dilute.

As already mentioned, we use $Q_{N_1N_2}(V, T)$ to represent the canonical partition function of the binary solution. The generalizations of Eqs. (1.8) and (1.10) for a two-component system are

$$\Xi(\mu_1, \mu_2, V, T) = \sum_{N_1, N_2 \geq 0} Q_{N_1N_2}(V, T) \lambda_1^{N_1} \lambda_2^{N_2}, \tag{2.9}$$

$$d(pV) = SdT + pdV + \overline{N}_1 d\mu_1 + \overline{N}_2 d\mu_2. \tag{2.10}$$

Equation (1.9) is unchanged. Incidentally, if the solvent is itself multicomponent ($N_1 \to N_a, N_b, \ldots$), $\lambda_1^{N_1}$ would be replaced by $\lambda_a^{N_a} \lambda_b^{N_b} \ldots$ but the argument below would be unaffected.

Just as in Eq. (2.1), we need consider only the terms with $N_2 = 0$ and $N_2 = 1$ (because the solution is very dilute with respect to solute). However, the dense liquid solvent cannot be handled in this way. Thus we rewrite Eq. (2.9) as

$$\Xi = \left(\sum_{N_1} Q_{N_10} \lambda_1^{N_1} \right) + \left(\sum_{N_1} Q_{N_11} \lambda_1^{N_1} \right) \lambda_2 + \cdots \tag{2.11}$$

$$= \Psi_0 + \Psi_1 \lambda_2 + \cdots, \tag{2.12}$$

where

$$\Psi_0(\mu_1, V, T) \equiv \sum_{N_1} Q_{N_10} \lambda_1^{N_1}, \tag{2.13}$$

$$\Psi_1(\mu_1, V, T) \equiv \sum_{N_1} Q_{N_11} \lambda_1^{N_1}. \tag{2.14}$$

Ψ_0 is the grand partition function for the pure solvent; Ψ_1 is a grand partition function for the solvent with exactly one solute molecule always present. Since only one solute molecule is present, any location of this molecule in the volume V is equivalent to any other location, despite the presence of solvent. Hence $Q_{N_1 1}$ will, for any value of N_1, always contain one extra factor of V compared to the corresponding $Q_{N_1 0}$. This factor of V arises on carrying out the integration, in the classical phase integral part of $Q_{N_1 1}$, over $dx\,dy\,dz$ for the location within V of the center of mass of the one solute molecule. Thus the quotient Ψ_1/Ψ_0 is proportional to V. This property is significant below.

A more detailed treatment of Eq. (2.12), with other applications, is given in Ref. 2, pp. 341–350.

From Eqs. (2.10) and (2.12), in the limit $\lambda_2 \to 0$,

$$\overline{N}_2 = \lambda_2\left(\frac{\partial \ln \Xi}{\partial \lambda_2}\right)_{T,V,\lambda_1} = \left(\frac{\Psi_1}{\Psi_0}\right)\lambda_2. \tag{2.15}$$

On dividing by V,

$$c_2 = \overline{N}_2/V = (q_2/V)\lambda_2, \tag{2.16}$$

$$q_2 \equiv \Psi_1/\Psi_0, \tag{2.17}$$

$$\mu_2 = \mu_2^\circ + kT \ln c_2, \tag{2.18}$$

$$\mu_2^\circ = -kT \ln (q_2/V). \tag{2.19}$$

The quotient q_2/V is a function of μ_1 and T or, equivalently, of p and T. These relations are formally the same as in Eqs. (2.3)–(2.5), but $q_2 = \Psi_1/\Psi_0$, the effective partition function of a single solute molecule in a solvent, replaces Q_{01} for a solute molecule in the absence of a solvent.

We examine q_2 more carefully below, but we note a formal result here in passing: if there is an equilibrium between solute molecules in the dilute gas phase and solute molecules in dilute solution, the relative concentrations in the two phases are (because the chemical potentials, μ_2, or absolute activities, λ_2, must be equal)

$$\frac{c_2(\text{solution})}{c_2(\text{gas})} = \frac{(q_2/V)}{(Q_{01}/V)}. \tag{2.20}$$

We return now to q_2 itself. Although fluctuations in N_1 are taken into account in Ψ_0 and Ψ_1, Eqs. (2.13) and (2.14), the magnitude of the fluctuations relative to the mean values is extremely small:

$$\frac{\text{standard deviation}}{\text{mean value}} = O\left(\frac{1}{\text{mean value}}\right)^{1/2}. \tag{2.21}$$

Hence, when useful, one can work with only the maximum terms in the sums, rather than with the full sums. This is in fact true of any partition function for a macroscopic system (except near critical points and phase transitions) and is the reason why different partition functions give identical thermodynamic results for the same system. Consequently, in the ratio Ψ_1/Ψ_0, we keep only the maximum terms with $N_1 = \bar{N}_1$ in Ψ_0 and $N_1 = \bar{N}_1'$ in Ψ_1. These values of N_1 are the *mean* values of the distributions because, for such extremely sharp distributions, the maximum and mean coincide. We generally expect (though the opposite is possible) $\bar{N}_1 > \bar{N}_1'$ because, when one solute molecule is present, a few solvent molecules will be excluded from the open system in order to keep the volume constant at V.

This situation is shown schematically in Fig. 1-1(a), where the one solute molecule on the right is exaggerated in size for clarity. On the left, there is solvent only. We then have

$$q_2 = \frac{\Psi_1}{\Psi_0} = \frac{Q_{\bar{N}_1'1}(V)}{Q_{\bar{N}_10}(V)\,\lambda^{\bar{N}_1-\bar{N}_1'}}. \tag{2.22}$$

This expression is a little awkward to interpret because two different solvent numbers, $\bar{N}_1$ and $\bar{N}_1'$, appear. Therefore we derive an alternative, but equivalent, expression for q_2 starting with a different partition function that has considerable utility for work with solutions, quite aside from the present application. In the new partition function, Γ, p and N_1 are held constant (V fluctuates) rather than V and μ_1 (N_1 fluctuates). Γ is discussed in more detail, with other applications, in Ref. 2, pp. 362–368.

In place of Eq. (2.9), we use

$$\Gamma(N_1, \mu_2, p, T) = \sum_{N_2, V} Q_{N_1N_2}(V, T)e^{-pV/kT}\lambda_2^{N_2}. \tag{2.23}$$

This refers to a system at constant p and T, open with respect to component 2 (solute) but closed with respect to component 1 (solvent). The summation over V is symbolic; one would generally use an appropriate integration over V. The relation of Γ to thermodynamics is

$$N_1\mu_1 = -kT \ln \Gamma(N_1, \mu_2, p, T), \tag{2.24}$$

$$d(N_1\mu_1) = -SdT + Vdp + \mu_1 dN_1 - \bar{N}_2 d\mu_2, \tag{2.25}$$

$$\bar{N}_2 = \lambda_2\left(\frac{\partial \ln \Gamma}{\partial \lambda_2}\right)_{T,p,N_1}. \tag{2.26}$$

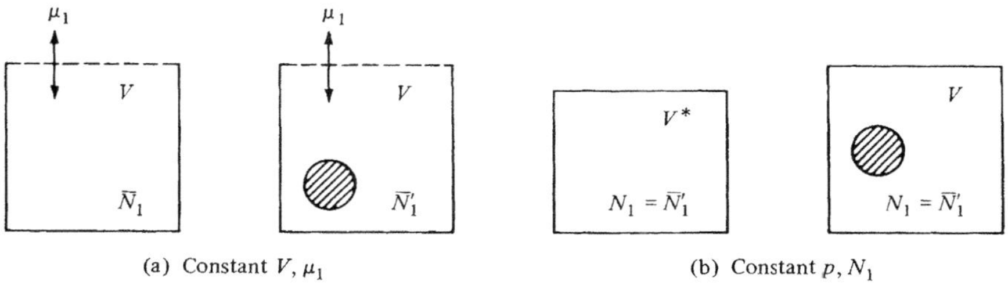

Fig. 1-1. (a) Volume V open with respect to solvent (component 1) at μ_1. On the right, one molecule of solute is present, which alters $\bar{N}_1$ to $\bar{N}_1'$. (b) Closed system at p and $N_1 = \bar{N}_1'$. On the right, one molecule of solute is present, which alters V^* to V.

As above, Eq. (2.9), we need only the $N_2 = 0$ and $N_2 = 1$ terms in the sum over N_2 in Eq. (2.23). Then

$$\Gamma = \Delta_0 + \Delta_1\lambda_2 + \cdots, \tag{2.27}$$

where

$$\Delta_0(N_1, p, T) = \sum_V Q_{N_1 0}(V)e^{-pV/kT}, \tag{2.28}$$

$$\Delta_1(N_1, p, T) = \sum_V Q_{N_1 1}(V)e^{-pV/kT}. \tag{2.29}$$

From Eqs. (2.26) and (2.27), in the limit $\lambda_2 \to 0$,

$$\overline{N}_2 = (\Delta_1/\Delta_0)\lambda_2. \tag{2.30}$$

This is as far as we need to go because comparison of Eq. (2.30) with Eq.(2.15) shows that

$$q_2 = \Psi_1/\Psi_0 = \Delta_1/\Delta_0. \tag{2.31}$$

This is another example of the equivalence of partition functions, referred to above.

We now simplify Δ_1 and Δ_0 by the maximum term method, as above. In Δ_1, with one solute molecule present [Fig. 1-1(b), right-hand side], we choose N_1 to have the value $\overline{N}_1'$ and $\overline{V}$ (maximum term) to have the value V, both as in Fig. 1-1(a) (right). That is, the right-hand systems in Figs. 1-1(a) and 1-1(b) are thermodynamically identical. This is accomplished by choosing p in Fig. 1-1(b) the same as p in Fig. 1-1(a) (right).

For Δ_0 [Fig. 1-1(b), left], which refers to pure solvent, N_1 has the same value as for Δ_1 but the maximum term V, denoted by V^*, will generally be slightly smaller than V (for Δ_1) because of the absence of the single solute molecule. This volume difference is exaggerated in Fig. 1-1(b) for clarity. In fact, the connection between $V - V^*$ [Fig. 1-1(b)] and $\overline{N}_1 - \overline{N}_1'$ [Fig. 1-1(a)] is

$$V - V^* = (\overline{N}_1 - \overline{N}_1')v_1, \tag{2.32}$$

where v_1 is the volume per molecule of the pure solvent at μ_1 and T (or p and T). Thermodynamically, $V - V^*$ is the partial molecular volume of the solute at infinite dilution.

The advantage of using Δ_1 and Δ_0 [Fig. 1-1(b)] is that N_1 has the same value in both; this is not true of Ψ_1 and Ψ_0 [Fig. 1-1(a)]. Thus, upon using maximum terms only, we have

$$q_2 = \frac{\Delta_1}{\Delta_0} = \frac{Q_{N_1 1}(V)}{Q_{N_1 0}(V^*)e^{p(V-V^*)/kT}}. \tag{2.33}$$

It is easy to show explicity that the right-hand sides of Eqs. (2.22) and (2.33) are identical, but this is left as an exercise for the interested reader. One has to use $Q = e^{-A/kT}$ and $G = N\mu = A + pV$, where G is the Gibbs free energy.

In Eq. (2.33), $Q_{N_1 0}(V^*)$ is the partition function of N_1 molecules of pure solvent. On the other hand, $Q_{N_1 1}(V)$ refers to the same N_1 molecules of solvent but with one molecule of solute moving in it. When the quotient of these Qs is taken in Eq. (2.33), what remains after cancellations are contributions of the following types. (a) Translation and external rotation of the single solute molecule. These contributions will be identical to those in Q_{01} (solute in gas) because the translational and external rotational kinetic energy of the solute is separable in the Hamiltonian H even when solvent is present and because intermolecular interactions between the solute and all the solvent molecules are identical for every location (within V) and for every external rotational orientation of the solute molecule. The proportionality of q_2 to V arises here (see pp. 262–264 of Ref. 1). (b) Internal rotation and vibration within the solute molecules, including any perturbations caused by the solvent molecules surrounding the solute. (c) Intermolecular interactions between the solute molecule and the surrounding solvent molecules. (d) Perturbation of all the degrees of freedom and of the related intermolecular interactions of those solvent molecules in the immediate neighborhood of the solute molecule as it moves about the volume V.

It is important to note that the only contributions to $Q_{N_1 1}(V)/Q_{N_1 0}(V^*)$ arise from the solute molecule itself and from those solvent molecules within the range of influence of the one solute molecule. All contributions from the unperturbed bulk of solvent molecules cancel in the quotient.

Thus $Q_{N_1 1}(V)/Q_{N_1 0}(V^*)$ strongly resembles Q_{01} for a solute molecule in a gas, but there are also very significant differences. Contribution (a) above is the same as in Q_{01}; (b) is also the same except for perturbations arising from the solvent, which may be significant, especially for internal rotational degrees of freedom in large solute molecules; but (c) and (d) are new for a solute molecule in a solvent (they do not occur at all in Q_{01}) and are of major importance. Clearly, (c) and (d) are the main effects in Eq. (2.20), with possibly some contribution from (b). There is no contribution from (a) because of cancellation.

The factor $e^{p(V-V^*)/kT}$ in Eq. (2.33) also has no counterpart in Q_{01}. However, because $V - V^*$ is small, of the order of the solute molecular volume, this factor is generally negligible for practical purposes. For example, if we take $V - V^*$ as a sphere of diameter 15 Å, $p = 1$ atm and $T = 300$ °K, $e^{p(V-V^*)/kT} = 1.044$. Furthermore, our main use for q_2 is in equilibrium binding constants where a quotient of these qs occurs (see Section 3) in such a way that the separate $e^{p(V-V^*)/kT}$ factors exactly cancel if the partial molecular volumes at infinite dilution of the participants are exactly additive in the binding process. One would not expect exact cancellation, but there would be approximate cancellation. Thus we shall use

$$q_2 \cong Q_{N_1 1}(V)/Q_{N_1 0}(V^*). \tag{2.34}$$

However, it is interesting in principle that Eq. (2.33), or Eq. (2.22), has the form of a *Gibbs* free energy change ($Q = e^{-A/kT}$, $G = A + pV$) whereas Q_{01} is related to the Helmholtz free energy. This difference arises, obviously, because in the former case room has to be made for a solute molecule in a solvent that is under a pressure p. In Q_{01}, the "solvent" is a vacuum.

Small Solute Bound on Large Solute

In a solution dilute with respect to all solutes, suppose L binds to P to form LP. Then the effective partition function of each of these three participants is given by Eq. (2.34); we denote these by q_L, q_P, and q_{LP}. Because Eq. (2.34) applies to a solute of any size, there is no necessary restriction on the relative sizes of L and P. However, in most binding problems in biochemistry, a relatively small molecule (L, a "ligand") binds on a site on a relatively large molecule (P; for example, a protein molecule). Because of its dominant importance, we consider this case explicitly here. It should be emphasized that although L might be, it need not be, a small molecule (e.g., ATP); the *relative* sizes are what is important. For example, L might be the S1 fragment of myosin and P a long aggregate of actin monomers (i.e., F-actin); or L might be a protein molecule that binds on DNA (P in this case).

If P is very much larger than L, the important cancellation feature implicit in Eq. (2.34) can be carried one step further. That is, almost all of the N_1 solvent molecules cancel out of Eq. (2.34); only those solvent molecules near the solute molecule need be taken into account. Similarly, if we form the quotient q_{LP}/q_P, because P is very large compared to L, all P contributions in the quotient will cancel (in particular, translation and external rotation of P) except for perturbations of P caused by the presence of L in the binding site on P. If there are no perturbations of P, cancellation is exact. These perturbations will generally be local (portions of P in and near the binding site for L), but they might include, as well, long-range effects such as an immediate conformational change in another part of P induced by the binding of L.

If we denote the above quotient by

$$q_L(P) \equiv q_{LP}/q_P, \tag{2.35}$$

then $q_L(P)$ is the effective partition function of L bound on P in the solvent. The motion of L in the site on P is what is left after the partial or complete P cancellation just referred to. This partition function is not to be confused with q_L, which is the effective partition function of an unbound, that is, of a *free* ligand molecule L in the solvent.

The partition function $q_L(P)$ includes several contributions. (a) The free energy of interaction between L and P that arises when L is brought through the solvent, from a distance, and placed in its most stable position and configuration in the binding site. This interaction includes any resulting perturbations of P or L, and also all alterations in the number, intermolecular arrangement, and motion of the solvent molecules that surround P and L. The solvent contribution could be of major importance. (b) Vibrational motion of the center of mass of L about the most stable position in the site. This corresponds to the translational motion of the free ligand, which is included in q_L. (c) Rocking of the rotational axes of L about their equilibrium (most stable) orientations. This replaces the free external rotational motion of the free ligand. (d) Internal vibrational and rotational motions of the atoms and atomic groups of bound L. Any perturbations here, owing to binding, are already included in (a). Hence this contribution would be the same as in q_L.

A limiting case, included here, is one in which P is immobile because it is part of some large structure. For example, P might be an actin filament in the myofilament structure of muscle. If P is much larger than L, it is immaterial (for the binding of L) whether P is mobile or not.

3. The Equilibrium Constant for Binding

In the first part of this section, we show how the equilibrium constant K for binding is related to the effective solute partition functions introduced in the preceding section. In the second part, we present an explicit example of the *a priori* calculation of K from partition functions.

For guidance, we begin with a very dilute gas mixture and consider the equilibrium state for the reaction A + B $\leftrightarrows$ AB. The condition for equilibrium is

$$\mu_A + \mu_B = \mu_{AB}. \tag{3.1}$$

Using Eq. (2.4) for each species, we then have, at equilibrium,

$$\ln K = \ln\left(\frac{c_{AB}}{c_A c_B}\right) = -\frac{(\mu^\circ_{AB} - \mu^\circ_A - \mu^\circ_B)}{kT}, \tag{3.2}$$

where K is the equilibrium constant. Let us use the notation Q_{0A}, Q_{0B}, Q_{0AB} for Q_{01} in Eq. (2.5). Then the relation between K and the Qs is

$$K(T) = \frac{(Q_{0AB}/V)}{(Q_{0A}/V)(Q_{0B}/V)}. \tag{3.3}$$

Equations (3.2) and (3.3) are standard results in thermodynamics and statistical mechanics, respectively. Equation (3.3) relates the thermodynamic quantity K explicitly to the molecular quantities Q. K is a function of T only because each Q is proportional to V and otherwise a function of T only. If the gas is not dilute, the concentrations c_A, c_B, c_{AB} in Eq. (3.2) have to be replaced by concentration activities. However, the relation of K to μ°s or to the Qs is not affected by gas imperfections; these are general results. Our use of a very dilute gas in order to derive these relations was simply a matter of convenience.

Turning now to the analogous very dilute solution case, we study the binding of L on P to form LP, with binding constant K. At the outset of this discussion, the molecules L and P may have arbitrary relative sizes. For simplicity, we assume that P has only one site for binding L (or, vice versa, L has only one site for P). Actually, if P has more than one site for L, our treatment here still applies to any one particular site, provided that binding on each site is independent of binding on the others. Following the same procedure as for a gas, using Eqs. (2.18) and (2.19), we have:

$$\mu_L + \mu_P = \mu_{LP}, \tag{3.4}$$

$$\ln K = \ln\left(\frac{c_{LP}}{c_L c_P}\right) = -\frac{(\mu^o_{LP} - \mu^o_L - \mu^o_P)}{kT}, \tag{3.5}$$

$$K(p, T) = \frac{(q_{LP}/V)}{(q_L/V)(q_P/V)}, \tag{3.6}$$

where the qs are given by Eq. (2.34). Each q is proportional to V and otherwise a function of p and T. As anticipated in Section 2, the factors $e^{p(V-V*)/kT}$ discussed there, which in any case are separately near unity, would cancel or nearly cancel in Eq. (3.6). If the solution is not very dilute with respect to L, P, and LP, the concentration quotient in Eq. (3.5) must be changed to a concentration activity quotient (reflecting intermolecular forces among L, P, and LP), but otherwise Eqs. (3.5) and (3.6) are unaltered.

For a solution dilute in L, P, and LP, let θ be the fraction of P sites occupied by L. Then $c_{LP} \propto \theta$ and $c_P \propto 1 - \theta$. Hence, from Eq. (3.5),

$$Kc_L = \frac{\theta}{1-\theta} \quad \text{or} \quad \theta = \frac{Kc_L}{1 + Kc_L}. \tag{3.7}$$

These are very familiar relations. The second of Eqs. (3.7) is the Langmuir binding isotherm. Incidentally, if θ' is the fraction of L sites occupied by P, Eq. (3.5) also leads to

$$Kc_P = \frac{\theta'}{1-\theta'} \quad \text{or} \quad \theta' = \frac{Kc_P}{1 + Kc_P}. \tag{3.8}$$

Thus, in general, there is a certain symmetry between L and P. The concentrations c_L and c_P are of course the concentrations of free (unbound) L and P. If the *total* concentration of L and P are c^o_L, c^o_P, then the three relations

$$c^o_L = c_L + c_{LP}, \qquad c^o_P = c_P + c_{LP},$$

$$K = \frac{c_{LP}}{c_L c_P} \tag{3.9}$$

determine c_L, c_P, and c_{LP} as functions of c^o_L, c^o_P, and K via a quadratic equation. The details are left to the interested reader.

Small Solute Bound on Large Solute

If P is much larger than L, two modifications of Eqs. (3.5) and (3.6) are significant. First, if activities are needed in Eq. (3.5) (e.g., because c_L is too large), we would have

$$Ka_L = \frac{\theta}{1-\theta} \qquad \text{or} \qquad \theta = \frac{Ka_L}{1+Ka_L}. \tag{3.10}$$

That is, the quotient a_{LP}/a_P can still be written c_{LP}/c_P or $\theta/(1-\theta)$ because the activity coefficients of LP and P would virtually cancel (because L is so small compared to P). Ordinarily we can set $a_L = c_L$ as in Eq. (3.7).

The second point is that Eq. (2.35) can now be used in Eq. (3.6), which leads to a conceptual simplification. Thus

$$K = \frac{\theta}{c_L(1-\theta)} = \frac{q_{LP}/q_P}{q_L/V} = \frac{q_L(P)}{q_L/V}. \tag{3.11}$$

Here $q_L(P)$ is the effective partition function of L bound on a site on P whereas q_L/V refers to free L in the solution. That is, the binding equilibrium can be viewed as a phase equilibrium:

$$\text{L(in solution)} \leftrightarrows L\text{(bound on P)}, \tag{3.12}$$

$$\mu_L\text{(solution)} = \mu_L^b\text{(bound)}. \tag{3.13}$$

If we replace μ_L in Eq. (3.13) by [see Eqs. (2.18) and (2.19)]

$$\mu_L = -kT \ln(q_L/V) + kT \ln c_L,$$

and then substitute, from Eq. (3.11),

$$\frac{\theta}{1-\theta} \cdot \frac{q_L/V}{q_L(P)}$$

for c_L, we obtain

$$\mu_L^b = -kT \ln q_L(P) + kT \ln\left(\frac{\theta}{1-\theta}\right) \tag{3.14}$$

for the chemical potential of bound L. It is easy to verify that this is just $\mu_{LP} - \mu_P$, as we should expect from Eq. (3.4). Equation (3.14) will be derived from a more direct statistical argument in Section 6. The term $-kT \ln q_L(P)$ is a standard chemical potential for bound L with standard state $\theta = \frac{1}{2}$ (half-saturation).

Explicit Calculation of K

The last form of Eq. (3.11) can be used to calculate K from first principles if an explicit molecular model is adopted. We pursue this topic in a rather general way at first, and then treat a specific example, approximately.

The partition function for a free L in the solution has the form

$$q_L = \left(\frac{2\pi mkT}{h^2}\right)^{3/2} V \cdot q_{rot} q_{vib}, \tag{3.15}$$

where m is the mass of L. The factors are from, in order, translation, external rotation, and internal vibration and rotation. These are contributions (a) and

(b) in the discussion following Eq. (2.33). Contributions (c) and (d) would contribute an additional factor $e^{-A_{0L}/kT}$, where A_{0L} is a free energy, but we choose A_{0L} as our zero of free energy; hence, it does not appear in Eq. (3.15). This choice of zero is appropriate because it is only the free energy *change* upon moving L from solution into the site on P that is physically significant for binding (see below).

If L is a spherical molecule with moment of inertia I_A about a diameter, then

$$q_{rot} = \pi^{1/2}\left(\frac{8\pi^2 I_A kT}{h^2}\right)^{3/2}. \tag{3.16}$$

For a rigid body of arbitrary shape, with principal moments of inertia I_A, I_B, and I_C, $I_A^{3/2}$ in Eq. (3.16) is replaced by $(I_A I_B I_C)^{1/2}$.

The partition function q_{vib} takes care, generally, of a very large number of degrees of freedom, $3n - 6$, where n is the number of atoms in L. However, we need not concern ourselves with q_{vib} because it also appears in $q_L(P)$ and cancels in K (see below).

We turn now to $q_L(P)$. We begin with the free energy change W that occurs when L is moved from a distant position in the solvent toward the binding site of P. W will be a function of the coordinates x, y, z of the center of mass of L relative to the binding site and of three rotational or orientational angles (θ, ϕ, ψ; defined below). So long as L is far from the site, $W = 0$ for any values of the six coordinates. For the most stable position and orientation of L in the binding site, $W = U$, a negative quantity, usually large. We choose this most stable position as the origin, $x = y = z = 0$. The z axis is chosen as normal to the surface of P at the site. A plot of $W(z)$ along the z axis, holding the most stable orientation constant, would have the appearance (schematic) shown in Fig. 1-2. As the molecule moves away from the site, z increases and $W(z)$ approaches zero. For $z < 0$, $W(z)$ increases very rapidly owing to van der Waals repulsions, etc. $-U$ is the depth of the binding free energy well. The function W is a well-defined statistical mechanical quantity known as a potential of mean force. The "mean" refers to statistical averaging over all coordinates of all solvent molecules for each set of values of the six coordinates of L relative to the site.

The translational motion (x, y, z) of the center of mass of L in q_L becomes three vibrational degrees of freedom of the center of mass about $x = y = z = 0$ in $q_L(P)$. The corresponding partition function is

$$q_{CMv} = \frac{kT}{h\nu_x} \cdot \frac{kT}{h\nu_y} \cdot \frac{kT}{h\nu_z}, \tag{3.17}$$

where ν_x, ν_y, ν_z are the vibrational frequencies of L moving in the neighborhood of the minimum in W (the x and y axes are located in such a way as to make x and y "normal" coordinates). The classical rather than quantal form is used for $kT/h\nu_x$, etc., because m and T are large. Similarly, the free external rotational motion of L in q_L becomes three orientational-vibrational degrees of freedom in $q_L(P)$ (coordinates θ, ϕ, ψ), with a partition function

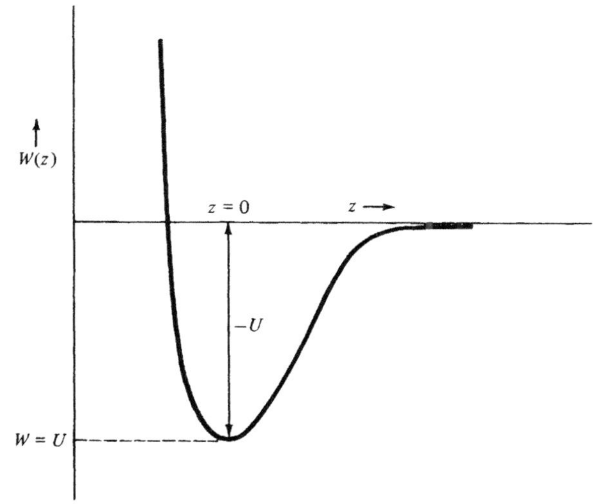

Fig. 1-2. Free energy change $W(z)$ as a ligand L approaches the binding site on P, holding constant the most stable orientation of L relative to the site on P.

we call q_{rock} because this is a rocking motion of the entire molecule L in the neighborhood of its most stable orientation at the minimum in W.

As already mentioned in the discussion of $q_{\text{L}}(\text{P})$ following Eq. (2.35), U would include all perturbations of P, L, or solvent molecules that occur when L is brought into the site on P. Hence q_{vib} is the same in $q_{\text{L}}(\text{P})$ as in q_{L}.

We then have, for comparison with Eq. (3.15),

$$q_{\text{L}}(\text{P}) = q_{\text{CMv}} q_{\text{rock}} q_{\text{vib}} e^{-U/kT}. \tag{3.18}$$

Hence, after cancelling q_{vib},

$$K = \frac{q_{\text{L}}(\text{P})}{q_{\text{L}}/V} = \frac{q_{\text{CMv}} q_{\text{rock}} e^{-U/kT}}{(2\pi mkT/h^2)^{3/2} q_{\text{rot}}}. \tag{3.19}$$

One can see immediately from Eq. (3.19) that the looser the motion of L in the site (the larger q_{CMv} and q_{rock}) and the greater the depth of the binding well $(-U)$, the larger K. If we use c.g.s. units in $(\ \ )^{3/2}$ (the qs are dimensionless), K has units of cm^3 molecule^{-1}. The conventional units for a binding constant are l mole^{-1} or M^{-1} (M is the molarity). Thus,

$$K(\text{M}^{-1}) = \frac{q_{\text{CMv}} q_{\text{rock}} e^{-U/kT}(N_{\text{o}}/1000)}{(2\pi mkT/h^2)^{3/2} q_{\text{rot}}}, \tag{3.20}$$

where N_{o} is Avogadro's number and $(\ \ )^{3/2}$ is still in c.g.s. units (cm^{-3}).

Because it seems easier to estimate the *range* of oscillation rather than the frequency in q_{CMv} or q_{rock}, we make the following modifications of Eq. (3.20). Consider first the factor $kT/h\nu_x$ in Eq. (3.17). This is the classical partition function for a one-dimensional harmonic oscillator of mass m with potential

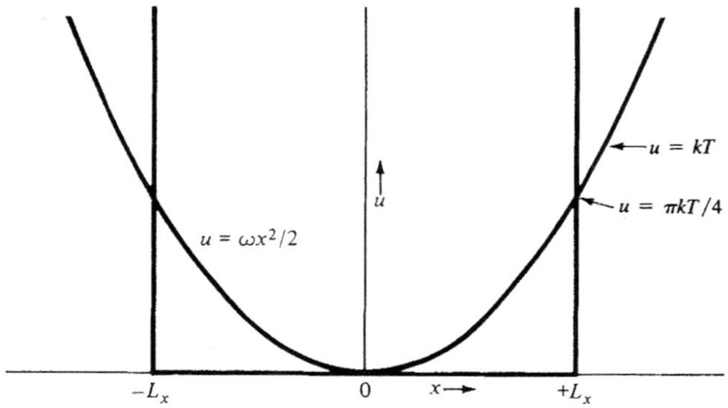

Fig. 1-3. Replacement of vibrational motion in $u(x)$ by motion in an equivalent square-well potential. See text for details.

energy $u = \omega x^2/2$ (Fig. 1-3) and frequency ν_x determined by

$$4\pi^2 \nu_x^2 = \omega/m. \tag{3.21}$$

The force constant is ω. Of course u here is just $W(x) - U$ near $x = 0$. We want to replace this motion by the motion of a particle of mass m in an equivalent square-well potential of width $2L_x$ (Fig. 1-3). This is free translational motion, but over a very restricted distance. By "equivalent" we mean that L_x is chosen to make the partition function of the latter system equal to that of the former, $kT/h\nu_x$. That is,

$$\left(\frac{2\pi mkT}{h^2}\right)^{1/2} \cdot 2L_x = \frac{kT}{h\nu_x}. \tag{3.22}$$

The expression on the left is a one-dimensional example of the translational partition function included in Eq. (3.15). Upon using Eq. (3.21) for ν_x in Eq. (3.22), we find that L_x is determined by the equation

$$\omega L_x^2/2 = \pi kT/4. \tag{3.23}$$

Not surprisingly, L_x is close to the turn-around point $x = x_0$ for an oscillator with the mean thermal energy kT: $\omega x_0^2/2 = kT$.

If we repeat the above procedure for the y and z motion in q_{CMv}, and use the corresponding square-well partition functions, as in Eq. (3.22), in place of vibrational partition functions, we see that

$$q_{\text{CMv}}/(2\pi mkT/h^2)^{3/2}$$

in Eq. (3.20) simplifies to

$$v \equiv 2L_x \cdot 2L_y \cdot 2L_z, \tag{3.24}$$

where v is the volume of a small three-dimensional box, determined by the ranges in x, y, z oscillation [Eq. (3.23)] of the ligand in the binding site.

The rocking motion is also vibrational. In essentially the same way as above, we replace the actual rocking motion in a three-dimensional (angular) parabolic potential (W) by equivalent free rotation over restricted angular ranges. The three angular coordinates we use are the Eulerian angles θ, ϕ, ψ. In Fig. 1-4, X, Y, Z are axes fixed with respect to P with origin chosen at the center of mass of L when L is in its most stable position in the binding site. The coordinates x, y, z locate the center of mass relative to this origin (as already mentioned). The axes $X'Y'Z'$ are fixed in the ligand itself with origin at its center of mass. That is, these axes move with the ligand. Then θ and ϕ (Fig. 1-4) establish the orientation of Z' relative to Z, and ψ relates to rotation of the ligand about Z' itself. The angular ranges are 0 to π for θ and 0 to 2π for ϕ and ψ. When L is in its most stable position and orientation, $x = y = z = 0$, $\theta = 0$, $\psi = 0$, and the axes $X'Y'Z'$ coincide with XYZ.

In completely free rotation, i.e., in q_{rot}, integration over $\sin\theta\, d\theta\, d\phi\, d\psi$ produces a factor 2 from $\sin\theta\, d\theta$, 2π from $d\phi$, and 2π from $d\psi$. This is the origin of the factor $8\pi^2$ in Eq. (3.16). The same integration for the "equivalent free rotation" of bound L, referred to above, over limited portions of the total solid angle 4π (for θ, ϕ) and of the total angle 2π (for ψ), near $\theta = 0$, $\psi = 0$, leads to $4\pi g_{\theta\phi}$ and $2\pi g_{\psi}$, respectively, where $g_{\theta\phi}$ and g_{ψ} are fractions representing the amount of rotational motion left available to the ligand when it rocks in its binding site. An example will be presented below. Thus, in place of the factor $8\pi^2$ in q_{rot}, q_{rock} has a factor $8\pi^2 g_{\theta\phi} g_{\psi}$. Hence, $q_{\text{rock}}/q_{\text{rot}}$ in Eq. (3.20) is simply equal to $g_{\theta\phi} g_{\psi}$.

On using this result, along with Eq. (3.24), Eqs. (3.19) and (3.20) become

$$K = v g_{\theta\phi} g_{\psi} e^{-U/kT}, \tag{3.25}$$

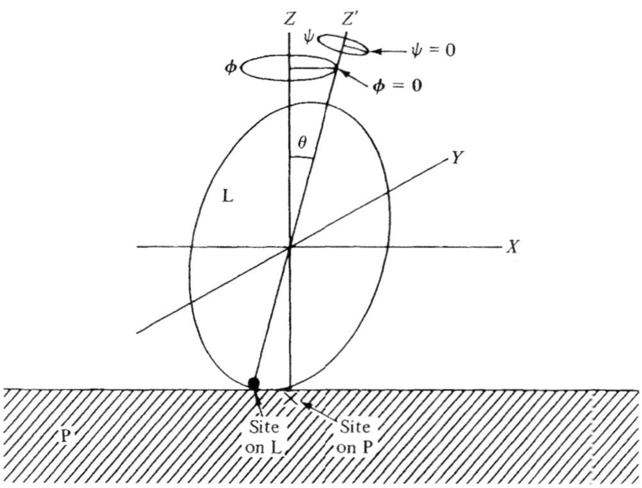

Fig. 1-4. Angular orientation of ligand L relative to binding site. The angle ϕ locates the axis Z' around Z.

$$K(\mathrm{M}^{-1}) = \left(\frac{v}{1000/N_o}\right) g_{\theta\phi} g_\psi e^{-U/kT}. \tag{3.26}$$

Formally speaking, this is not an approximation to Eq. (3.20) but rather an alternative and equivalent representation in terms of v and $g_{\theta\phi}\, g_\psi$, defined as

$$v \equiv \frac{q_{\mathrm{CMv}}}{(2\pi mkT/h^2)^{3/2}}, \tag{3.27}$$

$$g_{\theta\phi} g_\psi \equiv q_{\mathrm{rock}}/q_{\mathrm{rot}}. \tag{3.28}$$

However, the usefulness of Eq. (3.26) arises from the fact that it is relatively easy to make reasonable estimates of v and $g_{\theta\phi}\, g_\psi$. An example is given in the next subsection.

In the special case of a monatomic ligand (e.g., Ca^{2+}), q_{rock} and q_{rot} drop out of the problem, and $g_{\theta\phi}\, g_\psi$ is omitted from Eqs. (3.25) and (3.26). Then K has a particularly simple form, $ve^{-U/kT}$.

The relations

$$K = vg_{\theta\phi} g_\psi e^{-U/kT} = \frac{\theta}{c_{\mathrm{L}}(1 - \theta)}$$

lead to an interesting interpretation of the binding equilibrium. There is a certain balance in thermodynamic stability between free and bound ligand molecules when $\theta = \frac{1}{2}$; that is, when $Kc_{\mathrm{L}} = 1$. We denote this value of c_{L} by $c_{1/2}$ and its reciprocal by $v_{1/2}$ (the volume per molecule of ligand in solution at $c_{\mathrm{L}} = c_{1/2}$). Thus, $Kc_{1/2} = 1$ and $K = v_{1/2}$. Consequently,

$$\frac{v_{1/2}}{vg_{\theta\phi} g_\psi} = e^{-U/kT}. \tag{3.29}$$

At half-saturation ($\theta = \frac{1}{2}$), the relative freedom of motion of the free compared to the bound ligand [left-hand side of Eq. (3.29)] is balanced by the strength of binding of the bound ligand [right-hand side of Eq. (3.29)]. For a monatomic ligand (omit $g_{\theta\phi}\, g_\psi$), Eq. (3.29) is especially transparent.

In Eq. (3.26), $1000/N_o$ is the volume (in cm^3) occupied per molecule in a solution at a concentration of one mole per liter. This factor arises as a consequence of the particular choice of M^{-1} as the units for K. Also, v is in cm^3. However, other units of volume may be used so long as both $1000/N_o$ and v are expressed in the same units. For example, if v is in $Å^3$, which is very convenient, then $1000/N_o$ is 1660 $Å^3$.

Thus we see that K in Eq. (3.26) is a product of four factors, all of which have a simple physical significance.

Numerical Example for *K*

We suppose, as an illustration, that the ligand L is a spherical protein molecule of radius $a = 30$ Å and that the equivalent free *linear* range of motion, for bound L, for either vibration or rocking, is 6 Å, i.e., 3 Å in either direction. Thus, for vibration, we take

$$L_x = L_y = L_z = 3 \text{ Å} \qquad \text{or} \qquad v = 216 \text{ Å}^3, \tag{3.30}$$

from Eq. (3.24).

To translate a linear range of ± 3 Å into allowed angles, in the rocking motion, requires some further discussion. First we consider movement in θ and ϕ, and hold ψ constant. We assume in this example, as is plausible, that there is no restraint at all on ϕ but that the maximum possible range in θ is from $\theta = 0$ to θ^* such that the linear distance moved by the site on L, when $\theta = 0 \rightarrow \theta = \theta^*$, is $d = 3$ Å [Fig. 1-5(a)]. (The site on L would move a distance d in the opposite direction if $\phi = \pi$ instead of $\phi = 0$.) That is,

$$\sin(\theta^*/2) = d/2a. \tag{3.31}$$

From the definition of $g_{\theta\phi}$ above, we have

$$4\pi g_{\theta\phi} = \int_0^{2\pi} d\phi \int_0^{\theta^*} \sin\theta \, d\theta = 2\pi(1 - \cos\theta^*). \tag{3.32}$$

On combining Eqs. (3.31) and (3.32),

$$g_{\theta\phi} = \sin^2(\theta^*/2) = d^2/4a^2. \tag{3.33}$$

The significant result here is that $g_{\theta\phi} \propto (d/a)^2$. In our example with $a = 30$ Å and $d = 3$ Å, $g_{\theta\phi} = \frac{1}{400}$. This is a very small fraction of the total solid angle 4π.

In Fig. 1-5(a), the site on L moves a distance d when θ increases from $\theta = 0$ to $\theta = \theta^*$, holding ψ constant at $\psi = 0$. In Fig. 1-5(b), as a subsequent step, we hold θ constant at θ^* and allow a rotation $\pm\ \pi g_\psi$ in ψ about $\psi = 0$, such that the *original* position of the site on L moves a distance d. The site on L, itself, remains fixed. In Fig. 1-5(b), ψ moves in the positive

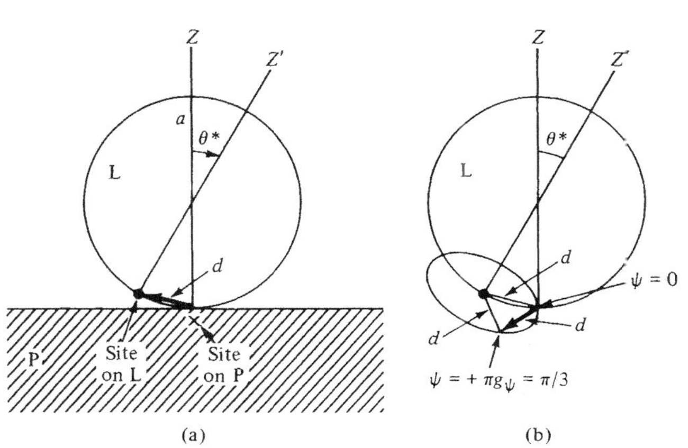

Fig. 1-5. Orientation of a spherical ligand L relative to its binding site on P. See text for details.

direction to $\psi = +\pi g_\psi$. Because the triangle in Fig. 1-5(b) is equilateral, with each side of length d, $\pi g_\psi = \pi/3$. Hence $g_\psi = \frac{1}{3}$ and

$$g_{\theta\phi} g_\psi = d^2/12a^2. \tag{3.34}$$

This relation would be relevant for any ligand represented by a sphere. In our numerical example, $g_{\theta\phi}\, g_\psi = \frac{1}{1200}$. The restraint, $g_\psi = \frac{1}{3}$, on the angle ψ is not very severe because the prior restraint on θ already introduces the necessary small-scale distances.

Returning now to Eq. (3.26) for K, in our numerical example,

$$\begin{aligned} K(\text{M}^{-1}) &= \left(\frac{216\ \text{Å}^3}{1660\ \text{Å}^3}\right)\left(\frac{1}{1200}\right)e^{-U/kT} \\ &= 1.084 \times 10^{-4} e^{-U/kT}. \end{aligned} \tag{3.35}$$

Thus, if $U = -12$ kcal mole^{-1} and $T = 300$ °K,

$$e^{-U/kT} = 5.49 \times 10^8, \qquad K = 5.95 \times 10^4\ \text{M}^{-1}. \tag{3.36}$$

The binding site on P will be half-saturated by L, $\theta = \frac{1}{2}$, when $Kc_L = 1$, that is, in this example, when the ligand concentration is $c_L = c_{1/2} = 1.68 \times 10^{-5}$ M.

In summary, the freedom of translational and rotational motion of the ligand in solution is greatly inhibited when it is bound in the site on P, but this can be much more than compensated for by the deep binding well ($-U$) in Fig. 1-2. Hence K in Eq. (3.26) can be large. See the discussion of Eq. (3.29) in this connection.

Related examples will be found in Refs. 3 and 4.

4. Rate Constants for Binding and Escape

It is a relatively simple matter to give an approximate *a priori* treatment of the diffusion-controlled rate constant for binding of a ligand, and for the inverse process, escape or release of the ligand from the binding site. We do this here as a continuation of the preceding section. To simplify the discussion, we maintain the assumption that P is much larger than L, or is some fixed structure. The general strategy is to derive the binding rate constant and then to use the already known K to find the release constant.

A feature of considerable general interest that arises naturally in the following analysis is that the familiar frequency factor kT/h in the Eyring rate theory must be replaced by a new frequency factor ($D/R\Lambda$; see below) for diffusion processes.

The discussion is presented in three related stages: binding of a monatomic ligand from a dilute gas; binding of a monatomic ligand from solution; and binding of an arbitrary ligand, for example, a protein, from solution.

Monatomic Ligand in a Gas

First we consider ligand atoms from a dilute gas phase that bind on sites on larger molecules P, or on sites on a surface. In either case, we can use the approach and notation in Eqs. (3.11)–(3.14). The partition functions are [compare Eqs. (3.15) and (3.18)]

$$q_{\mathrm{L}} = V/\Lambda^3, \tag{4.1}$$

$$q_{\mathrm{L}}(\mathrm{P}) = q_{\mathrm{CMv}} e^{-U/kT}, \tag{4.2}$$

where it is convenient at this point to introduce the so-called "thermal" de Broglie wavelength

$$\Lambda = h/(2\pi mkT)^{1/2}, \tag{4.3}$$

which has dimensions of length (e.g., cm in c.g.s. units). U now has a simpler interpretation than in Section 3: it is the value of the intermolecular interaction potential W between the atom L and all the atoms in and around the binding site on P, when L is at its most stable position in the site. Any perturbation of P as a result of the binding is also included in U. The binding constant K is then

$$K = \frac{q_{\mathrm{L}}(\mathrm{P})}{q_{\mathrm{L}}/V} = q_{\mathrm{CMv}} \Lambda^3 e^{-U/kT}. \tag{4.4}$$

Having introduced Λ above, it should be noted that v in Eq. (3.27) is equal to $q_{\mathrm{CMv}}\Lambda^3$.

To obtain the rate constant for binding, we use a well-known result from the kinetic theory of gases. We suppose that there is a small area, or "window," S, very near an empty site, on which $W = 0$, and with the property that whenever a ligand atom crosses S (coming from the gas), the atom enters the attractive potential field of P (Fig. 1-2) and is trapped (bound) in the potential well. From kinetic theory, the number of molecules hitting a surface per unit area per unit time is

$$\frac{p_{\mathrm{L}}}{(2\pi mkT)^{1/2}} = \frac{c_{\mathrm{L}} kT}{(2\pi mkT)^{1/2}}, \tag{4.5}$$

where p_{L} is the partial pressure of L in the gas. Then the probability per unit time that a ligand will cross S and be bound on the previously empty site is

$$\alpha = \frac{c_{\mathrm{L}} kTS}{(2\pi mkT)^{1/2}}. \tag{4.6}$$

This is a pseudo-first-order rate constant with units, say, s^{-1}. The corresponding second-order rate constant is α^*, related to α by $\alpha = \alpha^* c_{\mathrm{L}}$, with units determined by those of α and c_{L}. Thus

$$\alpha^* = \frac{kTS}{(2\pi mkT)^{1/2}}. \tag{4.7}$$

If β is the first-order rate constant, say, in s^{-1}, for escape or release of a bound L from a site, then α and β are related to K by

$$K = \alpha^*/\beta. \tag{4.8}$$

This follows from Eq. (3.11) and detailed balance at equilibrium:

$$\text{rate on} = \text{rate off},$$

$$\alpha(1 - \theta) = \alpha^* c_{\mathrm{L}}(1 - \theta) = \beta\theta. \tag{4.9}$$

From our expressions for K and α^* in Eqs. (4.4) and (4.7), we can obtain β:

$$\beta = \frac{\alpha^*}{K} = \frac{kTSe^{+U/kT}}{(2\pi mkT)^{1/2} q_{\mathrm{CMv}} \Lambda^3}. \tag{4.10}$$

We now wish to show that the main results above, Eqs. (4.7) and (4.10), also follow from Eyring's rate theory. According to this theory, we have

$$\alpha^* = \frac{kT}{h} \cdot \frac{q_{\mathrm{L}}^{\ddagger}}{(q_{\mathrm{L}}/V)}, \tag{4.11}$$

$$\beta = \frac{kT}{h} \cdot \frac{q_{\mathrm{L}}^{\ddagger}}{q_{\mathrm{L}}(\mathrm{P})}, \tag{4.12}$$

where $q_{\mathrm{L}}^{\ddagger}$ is the partition function for L in the transition state, except that the reaction coordinate (z in this case; Fig. 1-2) is not included in $q_{\mathrm{L}}^{\ddagger}$. The factor kT/h is the so-called frequency factor, with units, say, s^{-1}. The partition function ratios in Eqs. (4.11) and (4.12) formally resemble the similar ratio in Eq. (4.4) for the equilibrium constant. The motion of L in the transition state, in this example, is free two-dimensional translation within the small area S. Hence, by analogy with Eq. (4.1),

$$q_{\mathrm{L}}^{\ddagger} = S/\Lambda^2. \tag{4.13}$$

If we substitute this expression in Eqs. (4.11) and (4.12), we find

$$\alpha^* = \frac{kT}{h} \cdot S\Lambda, \tag{4.14}$$

$$\beta = \frac{kT}{h} \cdot \frac{Se^{U/kT}}{q_{\mathrm{CMv}}\Lambda^2}. \tag{4.15}$$

These are easily seen to agree with Eqs. (4.7) and (4.10). It is interesting that, in this case, Eyring's quantum-mechanical approach gives correctly the classical mechanical result, Eq. (4.7), by cancellation of h.

The rate constant for escape, β, is dominated by the small factor $e^{U/kT}$ (U is negative): $-U$ is the potential barrier that must be surpassed in order for L to escape from the site. There is no potential barrier in the attachment process, only the barrier of restricted geometry; the incoming L must cross the small area S.

Monatomic Ligand in Solution

Here L and P are in solution; L finds the site on P by Brownian motion, which we handle as a (well-known) diffusion problem. To simplify the mathematics, we assume spherical symmetry, as in Fig. 1-6. The ligand L is shown in the figure at its most stable position in the site, $r = 0$ in spherical coordinates. We suppose that the capture window of area S is a hemisphere of radius R around $r = 0$. Whenever a ligand diffuses from solution across S, it is captured in the site. The size of the area S is $2\pi R^2$.

The boundary conditions in the diffusion problem are $c = c_L$ at $r = \infty$ (i.e., far from the site) and $c = 0$ at $r = R$ ("absorption" on the surface $r = R$). At steady state, $\nabla^2 c = 0$. Because of the spherical symmetry, this simplifies to

$$\frac{d}{dr}\left(r^2 \frac{dc}{dr}\right) = 0$$

with solution

$$c(r) = c_L\left(1 - \frac{R}{r}\right). \tag{4.16}$$

The ligand flux is $J = -D\nabla c$, where D is the ligand diffusion coefficient. The number of ligand molecules reaching unit area of the absorbing surface $r = R$ per unit time [the analogue of Eq. (4.5)] is then

$$-J_{r=R} = D\left(\frac{dc}{dr}\right)_{r=R} = \frac{Dc_L}{R}.$$

Hence, corresponding to Eqs. (4.6) and (4.7),

$$\alpha = \frac{Dc_L S}{R} = 2\pi DRc_L,$$

$$\alpha^* = \frac{\alpha}{c_L} = 2\pi DR. \tag{4.17}$$

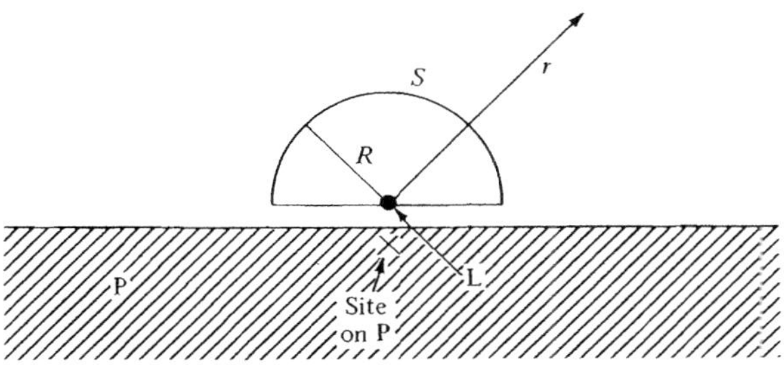

Fig. 1-6. Hemispherical capture "window" for an atomic ligand L around its binding site on P. See text for details.

Many of the basic equations in the preceding subsection still apply here, though the meaning of U is more complicated (as already discussed). The equations referred to are (4.1)–(4.4), (4.8), (4.9), and (4.13). As before, we can find β from α^* and K:

$$\beta = \frac{\alpha^*}{K} = \frac{2\pi DRe^{U/kT}}{q_{\mathrm{CMv}}\Lambda^3}. \tag{4.18}$$

Eyring's rate theory is not valid for diffusion processes. However, a modification does apply: the frequency factor must be changed from kT/h to $D/R\Lambda$. In the present example, the reader can easily verify that α^* and β in Eqs. (4.17) and (4.18) satisfy

$$\alpha^* = \frac{D}{R\Lambda} \cdot \frac{q_{\mathrm{L}}^{\ddagger}}{(q_{\mathrm{L}}/V)}, \tag{4.19}$$

$$\beta = \frac{D}{R\Lambda} \cdot \frac{q_{\mathrm{L}}^{\ddagger}}{q_{\mathrm{L}}(\mathrm{P})}. \tag{4.20}$$

Comparison should be made with Eqs. (4.11) and (4.12). The frequency factor kT/h is "universal": it is the same for all problems encompassed by the original Eyring theory. This is obviously not true of $D/R\Lambda$ because D and R depend on the particular case.

This modification of the Eyring theory was introduced in Refs. 3 and 5, where further details are available.

Arbitrary Ligand in Solution

We return now to the more general binding model considered in Eqs. (3.15)–(3.29). In estimating the binding rate constant, we make the approximation that the translational and rotational diffusion of L can be treated separately. As far as translation is concerned, the problem remains essentially as in Fig. 1-6. The only modification needed is shown in Fig. 1-7: when the binding site on L is within the lower "capture" hemisphere, the center of mass of L is within the upper hemisphere, about the origin of coordinates. Both hemispheres have radius R and surface area $S = 2\pi R^2$. However, it is no longer sufficient for binding that the center of mass of L should have crossed the surface $r = R$. In addition, the rotational orientation of L must be close enough to the optimal orientation or capture will not occur. We represent the allowed range in orientation (when $r = R$) for capture to occur by $4\pi g_{\theta\phi}^{\ddagger}$ for θ and ϕ and by $2\pi g_{\psi}^{\ddagger}$ for ψ [see the paragraph prior to Eq. (3.25)]. The fractions $g_{\theta\phi}^{\ddagger}$ and $g_{\psi}^{\ddagger}$ are analogous to $g_{\theta\phi}$ and g_{ψ} for the rocking motion when L is actually established in the binding site. We would usually expect $g_{\theta\phi}^{\ddagger} > g_{\theta\phi}$ and $g_{\psi}^{\ddagger} > g_{\psi}$ because of the much closer intermolecular contacts and restraints when L is actually bound compared to being incipiently bound (at $r = R$).

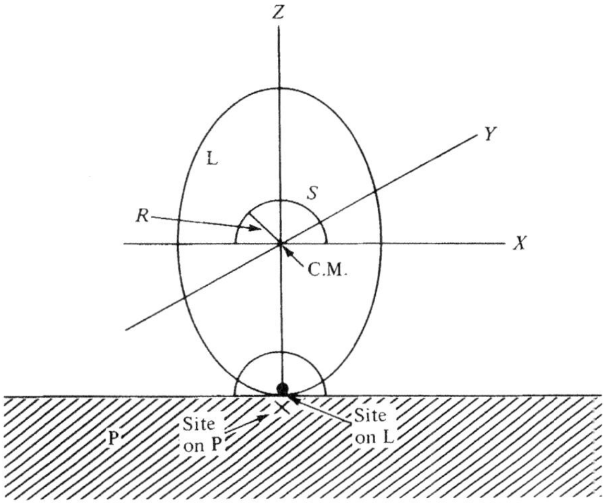

Fig. 1-7. Lower capture hemisphere for ligand site around the site on P. The upper hemisphere is the corresponding figure for the center of mass (C.M.). See text for details.

The rate constant α^* in Eq. (4.17) becomes here

$$\alpha^* = 2\pi DRg^{\ddagger}_{\theta\phi}g^{\ddagger}_{\psi}, \tag{4.21}$$

where $g^{\ddagger}_{\theta\phi}g^{\ddagger}_{\psi}$ is, in effect, the probability that a ligand whose center of mass crosses $r = R$ has a rotational orientation suitable for actual capture of L by the site on P. For a bulky ligand, the factor $g^{\ddagger}_{\theta\phi}g^{\ddagger}_{\psi}$ could reduce α^* considerably. From [Eq. (3.25)]

$$K = q_{\text{CMv}}\Lambda^3 g_{\theta\phi}g_{\psi}e^{-U/kT}, \tag{4.22}$$

we then have

$$\beta = \frac{\alpha^*}{K} = \frac{2\pi DR}{q_{\text{CMv}}\Lambda^3}\left(\frac{g^{\ddagger}_{\theta\phi}g^{\ddagger}_{\psi}}{g_{\theta\phi}g_{\psi}}\right)e^{U/kT}. \tag{4.23}$$

The g ratio here is new compared to Eq. (4.18) for a monatomic ligand. We expect this ratio to be > 1.

Equations (4.19) and (4.20) still apply, with

$$q_{\text{L}} = (V/\Lambda^3)q_{\text{rot}}q_{\text{vib}}, \tag{4.24}$$

$$q_{\text{L}}(\text{P}) = q_{\text{CMv}}(q_{\text{rot}}g_{\theta\phi}g_{\psi})q_{\text{vib}}e^{-U/kT}, \tag{4.25}$$

$$q^{\ddagger}_{\text{L}} = (S/\Lambda^2)(q_{\text{rot}}g^{\ddagger}_{\theta\phi}g^{\ddagger}_{\psi})q_{\text{vib}}. \tag{4.26}$$

It is easy to verify that Eqs. (4.19)–(4.26) are all self-consistent.

As a numerical example, let us continue with the spherical protein ligand considered in Eqs. (3.30)–(3.36), where $T = 300$ °K, $K = 5.95 \times 10^4$ M^{-1}, and $g_{\theta\phi}g_{\psi} = \frac{1}{1200}$. Here we suppose that the molecular weight of L is 50,000, $R = 5$ Å, $D = 6 \times 10^{-7}$ cm^2 s^{-1}, and $g^{\ddagger}_{\theta\phi}g^{\ddagger}_{\psi} = \frac{1}{60}$, i.e., 20 $g_{\theta\phi}g_{\psi}$. Then $2\pi DR = 1.14 \times 10^8$ M^{-1} s^{-1} and $\alpha^* = 2\pi DR/60 = 1.89 \times 10^6$ M^{-1} s^{-1}. From K and α^*, $\beta = 31.8$ s^{-1}. The value of Λ is 4.51×10^{-3} Å and the frequency factor is $D/R\Lambda = 2.66 \times 10^{11}$ s^{-1}. For comparison, $kT/h = 6.25 \times 10^{12}$ s^{-1}, which is larger by a factor of 23.5.

5. Perturbation of Rate Constants

In a number of examples in later chapters we have to deal with the effect of some perturbation on the forward and backward rate constants between two states of a molecule, usually a macromolecule. The perturbation will generally shift the free energy change and equilibrium constant between the states, but we shall need to know in addition how the perturbation of the free energy change is distributed between forward and backward transitions, i.e., how it affects the rate constants. A simple formalism is introduced in this section to take care of this problem.

We consider two states 1 and 2 of, say, a macromolecule P with unperturbed (or normal) first-order rate constants $\alpha_o(1 \rightarrow 2)$ and $\beta_o(2 \rightarrow 1)$. The equilibrium constant is $K_o = \alpha_o/\beta_o$, which is dimensionless. When perturbed, the rate constants are α and β, and $K = \alpha/\beta$. The following are examples of the kinds of perturbations contemplated: one or more other macromolecules are close enough to P to interact with it; one or more ligand molecules are bound on P; P is charged and, in the transitions $1 \rightleftarrows 2$, moves in an electric field (e.g., by rotation in a membrane); and P is a subunit in a long linear polymer under tension, and states 1 and 2 have different axial lengths (e.g., helix and coil states).

Figure 1-8 shows schematically the Gibbs free energy per molecule of P as a function of the reaction coordinate for the transitions $1 \rightleftarrows 2$. The full free energy curve relates to an unperturbed P: G_1 and G_2 are initial and final free energies, and $G^{o\ddagger}$ is the free energy at the free energy barrier (transition state). If the two states are in equilibrium with each other,

$$\frac{p_2^{oe}}{p_1^{oe}} = \frac{e^{-G_2/kT}}{e^{-G_1/kT}}, \tag{5.1}$$

where p_1^{oe} and p_2^{oe} are the fractions of P in states 1 and 2. Also, at equilibrium, there is detailed balance: $\alpha_o p_1^{oe} = \beta_o p_2^{oe}$. Thus

$$K_o = \frac{p_2^{oe}}{p_1^{oe}} = e^{(G_1 - G_2)/kT} = \frac{\alpha_o}{\beta_o}. \tag{5.2}$$

The perturbation, of whatever nature, alters the free energy of state 1 from G_1 to $G_1 + w_1$ and alters G_2 for state 2 to $G_2 + w_2$. In Fig. 1-8, w_1 and w_2 are both negative, and $w_1 > w_2$ (algebraically), but these conditions are arbitrary. In the presence of the perturbation, we find, instead of Eq. (5.2),

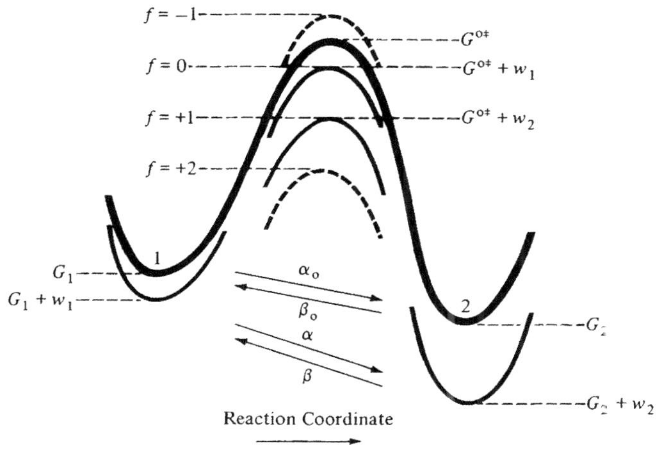

Fig. 1-8. Effects of a perturbation (w_1, w_2) on the Gibbs free eneger change $G_2 - G_1$ for a state transition pair $1 \leftrightarrows 2$ of a macromolecule. The effect of the perturbation on the transition state free energy $G^{o\ddagger}$ is given by Eq. (5.4), which defines the scale for the parameter f.

$$K = \frac{p_2^e}{p_1^e} = e^{(G_1-G_2)/kT}e^{(w_1-w_2)/kT} = \frac{\alpha}{\beta}$$
$$= K_o e^{(w_1-w_2)/kT} = \frac{\alpha_o}{\beta_o} e^{(w_1-w_2)/kT}. \tag{5.3}$$

Equation (5.3) shows how the perturbation free energy $w_1 - w_2$ alters the ratio α_o/β_o to give α/β. However, the effect of the perturbation on α_o and β_o *separately* depends on how $G^{o\ddagger}$ is influenced by the perturbation. There is a continuum of possibilities, all of which can be encompassed by the introduction of a single dimensionless parameter f, which is characteristic of P, of the particular states 1 and 2 of P, and of the nature of the perturbation. That is, f would have different values in different examples. We define f by

$$G^{\ddagger} = G^{o\ddagger} + (1 - f)w_1 + fw_2, \tag{5.4}$$

where $G^{\ddagger}$ is the free energy at the *perturbed* free energy barrier. For example, if $f = \frac{1}{2}$, the change in $G^{o\ddagger}$ is the average of the respective changes w_1 and w_2 in G_1 and G_2. This value of f might be expected in a symmetrical transition such as a conformational change, where the structure of the transition state is intermediate between initial and final states (i.e., states 1 and 2).

The relative values of α_o and α depend on the heights of the two "forward" free energy barriers:

$$\alpha_o \propto e^{-(G^{o\ddagger}-G_1)/kT},$$
$$\alpha \propto e^{-(G^{\ddagger}-G_1-w_1)/kT}. \tag{5.5}$$

Upon using Eq. (5.4), we find

$$\alpha = \alpha_o e^{f(w_1 - w_2)/kT}. \tag{5.6}$$

Similarly,

$$\begin{aligned} \beta_o &\propto e^{-(G^{o\ddagger} - G_2)/kT}, \\ \beta &\propto e^{-(G^{\ddagger} - G_2 - w_2)/kT} \end{aligned} \tag{5.7}$$

and then

$$\beta = \beta_o e^{(f-1)(w_1 - w_2)/kT}. \tag{5.8}$$

As a check, we note that α/β from Eqs. (5.6) and (5.8) agrees with Eq. (5.3). Instead of defining f by Eq. (5.4), we could have defined it at the outset by Eq. (5.6).

Figure 1-8 shows the position of the top of the free energy barrier for a few values of f. Although values of f between 0 and 1 are probably most common, there is no requirement that f be in this range. Note that if $f = 0$, the potential barrier for the forward transition is the same with and without the perturbation; hence $\alpha = \alpha_o$. Conversely, if $f = 1$, the backward transition has an unaltered free energy barrier, and $\beta = \beta_o$. In either of these cases, the opposite rate constant takes up the full free energy effect $w_1 - w_2$.

Thermodynamic Perturbations

The above approach to the effect of perturbations on rate constants, which is the one we shall use in this book almost exclusively, is based on elementary states and transitions, and on detailed balance at the corresponding microscopic level. However, there are cases in which such a detailed analysis is not practical. Then rate constant perturbations must be handled at a macroscopic or thermodynamic level. This is equivalent to a statistical mechanical averaging over the discrete states and transitions of the more fundamental microscopic approach. An explicit simple example illustrating the macroscopic–microscopic relationship will be found in Section 26, and a more complicated example appears in Section 34.

We give two examples of the thermodynamic approach here, both concerned with the binding of a ligand at concentration c_L in solution onto a one-or two-dimensional lattice of equivalent binding sites on a surface. In the first example, we assume that the solution is very dilute in ligand but that there are intermolecular pair interactions between bound ligand molecules on nearest-neighbor sites of the lattice. Molecules bound on a lattice of sites are sometimes called a lattice gas; this then is an "imperfect" lattice gas. That is, the lattice gas is not ideal because of the presence of intermolecular interactions. These intermolecular interactions are the perturbation in this problem. In the second example, the sitution is reversed: the lattice gas is ideal (no intermolecular interactions among bound ligand molecules), but the ligand solution is imperfect: c_L is large enough so that intermolecular forces between ligand molecules in solution require the use of the activity a_L in place of c_L, as in Eq. (3.10). Again, interligand forces are the perturbation.

In the first example, because of the discrete nature of the states of the lattice gas, it is not necessary to use the thermodynamic approach to the perturbation of rate constants, but we pursue it anyhow as a convenient illustration. In the microscopic approach to this example, in order to use the method of the preceding subsection, each state and possible transition of the entire lattice gas must be taken into account. Figure 1-9 shows an illustrative possible state for a small lattice. The transitions of binding to or removing a ligand from site X, for example, are perturbed by the presence of two nearest neighbors that will interact with a ligand at X. Such details, for all the states, can be managed analytically for small groups of sites or in certain one-dimensional lattice problems, or by Monte Carlo computational methods otherwise. Detailed examples will be found in Parts II and III.

Handling the perturbation (intermolecular forces) at the thermodynamic level is formally quite simple but not particularly useful except in important special cases mentioned below. Binding on a lattice of independent sites (i.e., the ideal lattice gas case) is represented, as in Section 3, by

$$K = \frac{\alpha^*}{\beta} = \frac{\theta}{c_L(1-\theta)}, \tag{5.9}$$

$$\alpha^* c_L(1-\theta) = \beta\theta, \tag{5.10}$$

where the latter is the detailed balance relation that equates the average on and off rates at a particular site. The corresponding chemical potential of the bound ligand molecules is given in Eq. (3.14). With intermolecular forces between bound ligand molecules included, Eq. (3.14) is modified to

$$\mu_L^b = -kT \ln q_L(P) + kT \ln\left(\frac{\theta\gamma_b}{1-\theta}\right), \tag{5.11}$$

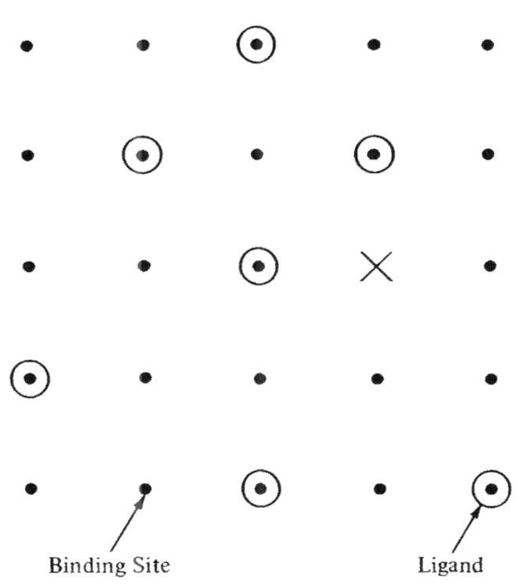

Fig. 1-9. Binding of a ligand to or removal of a ligand from site X (for example) would be perturbed by the presence of two nearest neighbors.

where γ_b is an activity coefficient, a function of θ, which we do not attempt to specify here. It will, in fact, depend on the particular lattice and on the magnitude of the nearest-neighbor ligand interaction. In any case, though, $\gamma_b \to 1$ as $\theta \to 0$. Correspondingly, from Eq. (3.13), Eq. (5.9) becomes

$$K = \frac{\alpha^*}{\beta} = \frac{\theta\gamma_b}{c_L(1 - \theta)}. \tag{5.12}$$

The K in Eq. (5.12) is a true thermodynamic equilibrium constant, related to standard chemical potentials, and is independent of θ; K is a function of T, as are α^* and β. On the other hand, the analogue of the K in Eq. (5.3) is

$$\frac{\theta}{1 - \theta} = \frac{\alpha^* c_L}{\beta} \cdot \frac{1}{\gamma_b}, \tag{5.13}$$

where $\alpha^* c_L$ and β are the unperturbed first-order rate constants and γ_b^{-1} corresponds to the perturbation factor $e^{(w_1 - w_2)/kT}$ in Eq. (5.3). If we split γ_b^{-1} between the on and off rate constants, just as in Eqs. (5.6)–(5.8), the detailed balance relation, Eq. (5.10), is modified by the perturbation to read

$$(\alpha^* \gamma_b^{-\bar{f}_b}) c_L (1 - \theta) = (\beta \gamma_b^{1-\bar{f}_b})\theta, \tag{5.14}$$

where $\alpha^* \gamma_b^{-\bar{f}_b}$ and $\beta \gamma_b^{1-\bar{f}_b}$ are the perturbed rate constants corresponding to the unperturbed constants α^* and β. The parameter $\bar{f}_b$ is the analogue of f in Eqs. (5.4), (5.6), and (5.8), but the bar on $\bar{f}_b$ emphasizes that this is an *averaged* version of f because the "detailed" balance in Eq. (5.14) is actually not so detailed. That is, Eq. (5.14) equates the *mean* on and off rates per site of the lattice, not the on and off rates for a particular site with some definite perturbation (as in Fig. 1-9, for example). If we use a square lattice (Fig. 1-9) as an illustration, the number of nearest-neighbor bound ligands of a particular site (e.g., X in Fig. 1-9) may be 0, 1, 2, 3, or 4. The "detailed" balance in Eq. (5.14) is an average over these five possibilities, each with its own microscopic detailed balance, with proper (but not simple) statistical–mechanical weights for each case. Thus $\bar{f}_b$ is a macroscopic, not a microscopic, parameter. In general, both γ_b and $\bar{f}_b$ are functions of θ. In the important special case that the on, or binding, rate constant does not depend on θ, e.g., if the binding process is strictly diffusion controlled, then $\bar{f}_b = 0$ and Eq. (5.14) becomes

$$\alpha^* c_L (1 - \theta) = (\beta\gamma_b)\theta, \tag{5.15}$$

where $\beta\gamma_b$ is the perturbed off rate constant. The full effect of the activity coefficient is on β. Explicit examples relating microscopic and macroscopic approaches will be found in Sections 26 and 34.

In the second example, the lattice gas is ideal but the ligand solution is imperfect, as in Eq. (3.10). We define the activity coefficient γ of L in solution by $a_L = c_L\gamma$, as usual, where a_L is the activity and γ can be expressed as a power series in c_L with leading term unity. That is, $\gamma \to 1$ as $c_L \to 0$. In place of Eqs. (5.12) and (5.13), we now have

$$K = \frac{\alpha^*}{\beta} = \frac{\theta}{c_L \gamma (1 - \theta)}, \tag{5.16}$$

$$\frac{\theta}{1 - \theta} = \frac{\alpha^* c_L}{\beta} \cdot \gamma. \tag{5.17}$$

The perturbation factor here is γ so that the "detailed" balance relation is

$$(\alpha^* \gamma^{\bar{f}}) c_L (1 - \theta) = (\beta \gamma^{\bar{f}-1}) \theta, \tag{5.18}$$

where the perturbed rate constants are $\alpha^* \gamma^{\bar{f}}$ and $\beta \gamma^{\bar{f}-1}$. In general, both γ and $\bar{f}$ are functions of c_L. Because the imperfection is in the solution and not in the lattice in this example, the most important special case would be $\bar{f} = 1$, so that the off rate constant is independent of c_L. Then Eq. (5.18) becomes

$$(\alpha^* \gamma) c_L (1 - \theta) = \beta \theta, \tag{5.19}$$

with the full effect of the solution imperfection (perturbation) placed on α^*. The general case, Eq. (5.18), is not useful, except conceptually, because the c_L dependence of $\bar{f}$ is not available. In contrast, the θ dependence of $\bar{f}_b$ in Eq. (5.14) can be calculated from the underlying microscopic details, as already described.

References

1. Hill, T. L. (1956) *Statistical Mechanics* (McGraw-Hill, New York).
2. Hill, T. L. (1960) *Introduction to Statistical Thermodynamics* (Addison-Wesley, Reading, MA).
3. Hill, T. L. (1975) Proc. Natl. Acad. Sci. USA **72**, 4918.
4. Hill, T. L. and Eisenberg, E. (1980) Biophys. Chem. **11**, 271.
5. Hill, T. L. (1976) Proc. Natl. Acad. Sci. USA **73**, 679.

2

Simple Equilibrium Binding

In this chapter we introduce the grand partition function for use in equilibrium binding problems. This method is not needed for the simple problems we consider in this chapter; in fact, the main results are already contained in Eqs. (3.4)–(3.14) and were obtained by thermodynamic methods. However, with the background acquired here, we shall see that the grand partition function approach converts the more complicated and interesting problems in Chapters 4 and 6 into straightforward exercises.

6. Binding of a Ligand on Independent Equivalent Sites

As in Section 3, we consider the binding of a ligand L from solution onto independent and equivalent binding sites on other molecules P, which are usually larger. There is one site for L per P molecule. The P molecules may be moving freely in a solution (dilute with respect to P), or they may be immobile, for example, in a one- or two-dimensional lattice or structure. An example would be the actin subunits of an actin filament. Another possibility is that the P molecules have mobility in a membrane. However, in all cases (in this chapter) the P molecules are independent of each other: inter-P interactions are negligible.

In Section 3 we examined the theoretical basis of the binding constant K for L on P. Here we take K as given and study the statistics of the equilibrium binding process.

The partition functions q_L, q_P, and q_{LP} for L, P, and LP in solution have been discussed in Section 2. If P and LP are immobilized, as in a lattice, q_P and q_{LP} would contain vibrational and rocking factors (Chapter 1). If P and LP are in a membrane, the solvent (Section 2) would be lipid rather than water. In any case, each P, with its binding site for L, is an independent binding system that is in the presence of a reservoir of L molecules at chemical potential μ. [Here we begin to simplify the notation for the ligand that was needed in Chapter 1: we shall now use μ for μ_L; c for c_L; and q for $q_L(P) \equiv q_{LP}/q_P$.] As such, this system, at equilibrium, can be represented by a grand partition function, as in Eq. (1.8). While it is true that Eq. (1.8) would conventionally be applied to a macroscopic system, it actually applies to small systems as well, such as the present one (i.e., to P, with its binding site either empty or occupied). In the present problem, the only possible values of N in Eq. (1.8) are $N = 0$ (site empty) or $N = 1$ (site occupied by L). That is, this small system is open with respect to the ligand L. When $N = 0$, the canonical partition function is q_P; when $N = 1$, it is q_{LP}. Thus, in this case, Eq. (1.8) becomes

$$\xi = q_P + q_{LP}\lambda \qquad (\lambda \equiv e^{\mu/kT}). \tag{6.1}$$

We have changed Ξ to ξ as a reminder that this is a small, not a macroscopic, system. We shall make this distinction in notation throughout the book.

From Eq. (1.13),

$$P_0 = \frac{q_P}{\xi}, \qquad P_1 = \frac{q_{LP}\lambda}{\xi}. \tag{6.2}$$

These are the respective probabilities that the site is empty or occupied, at equilibrium. Equations (6.2) can be simplified to

$$P_0 = \frac{1}{1 + q\lambda}, \qquad P_1 = \frac{q\lambda}{1 + q\lambda}, \tag{6.3}$$

if we define (as already mentioned)

$$q \equiv q_L(P) \equiv q_{LP}/q_P, \tag{6.4}$$

as in Eq. (3.11).

Equations (6.3) suggest that if our only interest is in the extent of binding, ξ in Eq. (6.1) can be rewritten as $q_P(1 + q\lambda)$, and then the factor q_P might as well be omitted, as a convenience, because it appears in both terms, q_P and $q_P q\lambda$, and has no effect on the P_N distribution. This is true of any grand partition function: if all of its terms are multiplied by a constant, mean values calculated from these terms are not altered.

Thus we can start over and, in place of Eq. (6.1), define ξ as

$$\xi = 1 + q\lambda, \tag{6.5}$$

with

$$P_0 = \frac{1}{\xi}, \qquad P_1 = \frac{q\lambda}{\xi}. \tag{6.6}$$

The two terms in Eq. (6.5) give the relative weights of the two possible states of occupation of the system:

$$\text{site empty} \propto 1, \qquad \text{site occupied} \propto q\lambda. \tag{6.7}$$

This is a special case of the discussion following Eq. (1.13). This equilibrium probability property of the separate terms of a grand partition function (the analogue for an open system of the Boltzmann distribution) allows one to write immediately expressions for mean values of various quantities of interest.

In this simple example P_1 itself is just what we called θ in Section 3 (the probability that one site is occupied, or the fraction of all sites that are occupied). Note that P_1, in this example, is also equal to $\overline{N}$ in Eq. (1.11). Thus, from Eq. (6.6),

$$\theta = \frac{q\lambda}{1 + q\lambda} \tag{6.8}$$

or

$$q\lambda = \frac{\theta}{1 - \theta}. \tag{6.9}$$

We shall return to these important relations below.

The chemical potential μ, or $\lambda = e^{\mu/kT}$, refers to both the binding system and the solution (reservoir) of L molecules, because of the equilibrium. In the binding system, from Eq. (6.9), μ is related to θ by

$$\mu = -kT \ln q + kT \ln\left(\frac{\theta}{1 - \theta}\right). \tag{6.10}$$

In the solution, from Eq. (2.18), μ is related to the ligand concentration c by

$$\mu = -kT \ln(q_{\mathrm{L}}/V) + kT \ln c. \tag{6.11}$$

If the solution is not sufficiently dilute, c must be replaced by the concentration activity a. From Eq. (6.11),

$$\lambda = \frac{c}{(q_{\mathrm{L}}/V)}, \qquad q\lambda = \frac{qc}{(q_{\mathrm{L}}/V)} = Kc, \tag{6.12}$$

having used

$$K = \frac{q}{(q_{\mathrm{L}}/V)} = \frac{q_{\mathrm{LP}}/q_{\mathrm{P}}}{(q_{\mathrm{L}}/V)} \tag{6.13}$$

from Eq. (3.11). Thus, we can replace $q\lambda$ by Kc (or by Ka) in Eqs. (6.8) and (6.9). (The reason λ is called an absolute "activity" is because $\lambda \propto c$ or, more generally, $\lambda \propto a$.)

Returning to Eqs. (6.8) and (6.9):

$$\theta = \frac{Kc}{1 + Kc}, \qquad 1 - \theta = \frac{1}{1 + Kc}, \tag{6.14}$$

$$Kc = \frac{\theta}{1 - \theta}, \tag{6.15}$$

in agreement with Eq. (3.7). Half-saturation, $\theta = \frac{1}{2}$, occurs when $Kc = 1$, or $c = 1/K$. Also, we can put Eqs. (6.5) and (6.6) into the form

$$\xi = 1 + Kc, \tag{6.16}$$

$$P_0 = \frac{1}{\xi}, \qquad P_1 = \frac{Kc}{\xi}. \tag{6.17}$$

This is the practical way (i.e., Kc in place of $q\lambda$) in which we shall generally write grand partition functions for small binding systems: K and c are familiar operational quantities, while q and λ are not.

As a self-consistency check, we can use Eqs. (6.5) and (6.16) in

$$\theta = \lambda\left(\frac{\partial \ln \xi}{\partial \lambda}\right)_T = c\left(\frac{\partial \ln \xi}{\partial c}\right)_T, \tag{6.18}$$

which is a special case of Eq. (1.11), to recover Eqs. (6.8) and (6.14). A second differentiation gives, from Eq. (1.14),

$$\sigma_1^2 = \lambda\left(\frac{\partial \theta}{\partial \lambda}\right)_T = c\left(\frac{\partial \theta}{\partial c}\right)_T \tag{6.19}$$

$$= \frac{Kc}{(1 + Kc)^2} = \theta(1 - \theta), \tag{6.20}$$

where σ_1^2 is the variance for a single-site system (the subscript refers to the number of sites). We note, for future reference, that the variance is a maximum at half-saturation and has the value there of

$$\sigma_1^2 = \left(\frac{\partial \theta}{\partial \ln c}\right)_{\theta=1/2} = \frac{1}{4}. \tag{6.21}$$

It should be noted that we have not specifically used the condition that P is a much larger molecule than L, though this is the main case of interest. If L, P, and LP are all present in solution, the definition $q = q_{LP}/q_P$ applies for any relative sizes of L and P. However, the simple interpretation of q as relating essentially to the motion of L in a site (including perturbations of L and of P), as in Eq. (2.35), is possible only if P is much larger than L. Indeed, as suggested by Eqs. (3.7)–(3.9), if L and P are comparable in size, we could just as well reverse the roles of L and P (e.g., use q_{LP}/q_L). However, to avoid confusion, we shall not pursue this point any further.

Macroscopic Systems

For the sceptical reader who feels somewhat uncomfortable with the application of the grand partition function to such a small system as one P molecule with a single binding site for L, we give a brief treatment in this subsection of the two macroscopic systems that are equivalent to the single-site system already considered. The same results are obtained. This confirms that it suffices to work with the much simpler small system ξ whenever the macroscopic system is a collection of *independent* small systems.

First consider a large lattice or structure comprised of M equivalent and independent immobilized P molecules, each with one binding site for L. In the absence of any L molecules in this system, the canonical partition function would be simply q_{P}^M, one factor q_{P} for each P molecule. However, if there are N bound ligand molecules distributed among the M sites, this partition function would be generalized to

$$Q_N(M, T) = \frac{M!\, q_{\mathrm{LP}}^N q_{\mathrm{P}}^{M-N}}{N!\,(M-N)!}. \tag{6.22}$$

That is, this system is now a mixture (a solid solution) of N molecules of type LP and $M - N$ molecules of type P, each molecule contributing its own q_{LP} or q_{P} to Q_N. The factorial expression is the number of ways of distributing the N ligand molecules among the M distinguishable sites.

If we now substitute the above Q_N into Eq. (1.8), we obtain for this macroscopic system, which is open with respect to the number N of bound ligand molecules,

$$\begin{aligned}\Xi &= \sum_{N=0}^{M} \frac{M!\, q_{\mathrm{P}}^M}{N!\,(M-N)!}\left(\frac{q_{\mathrm{LP}}\lambda}{q_{\mathrm{P}}}\right)^N \\ &= q_{\mathrm{P}}^M(1 + q\lambda)^M = \xi^M,\end{aligned} \tag{6.23}$$

where the ξ here is the original ξ introduced in Eq. (6.1). Thus there is one factor ξ in Ξ for each P; this is a consequence of the independence of the separate Ps. The extensive thermodynamic properties obtained here will therefore simply be M times larger than for a single-site system, but the intensive properties will be the same and determined by ξ. To verify this for a few properties, we need the appropriate analogues of Eqs. (1.9) and (1.10):

$$\Phi M = kT \ln \Xi(\mu, M, T), \tag{6.24}$$

$$d(\Phi M) = S dT + \Phi dM + \bar{N} d\mu. \tag{6.25}$$

Here Φ (called the spreading pressure) corresponds to p, though Φ has dimensions of energy, and M is the equivalent of V. From Eqs. (6.18), (6.23)–(6.25), and (1.11), we have

$$\bar{N} = \lambda\left(\frac{\partial \ln \Xi}{\partial \lambda}\right)_{T,M} = M\lambda\left(\frac{\partial \ln \xi}{\partial \lambda}\right)_T = M\theta, \tag{6.26}$$

where θ is the same as in Eq. (6.8).

The probability distribution in N for this open system is, from Eq. (1.13),

$$P_N = \frac{M!\,(q\lambda)^N}{N!\,(M-N)!\,(1+q\lambda)^M}. \tag{6.27}$$

For very large M, and a given λ, this is a very sharp Gaussian distribution with maximum at

$$N_{\max} = \overline{N} = M\theta = \frac{Mq\lambda}{1 + q\lambda}. \tag{6.28}$$

The variance in N, denoted by σ_M^2, follows from Eqs. (1.14) and (6.20):

$$\sigma_M^2 = \overline{N^2} - \overline{N}^2 = M\lambda\left(\frac{\partial\theta}{\partial\lambda}\right)_T$$
$$= M\theta(1 - \theta) = M\sigma_1^2. \tag{6.29}$$

The variance of the M-site system is M times the variance of the one-site system because the M sites are independent of each other. The quantity σ_M is the standard deviation of the Gaussian P_N. The magnitude of σ_M relative to the mean $\overline{N}$ is

$$\frac{\sigma_M}{\overline{N}} = \left[\frac{M\theta(1 - \theta)}{M^2\theta^2}\right]^{1/2} = \left(\frac{1 - \theta}{M\theta}\right)^{1/2}. \tag{6.30}$$

This is of order $M^{-1/2}$ and approaches zero as $M \to \infty$.

In the second macroscopic system we consider, there are M mobile and independent P molecules, each with one binding site for L. The P molecules are in a solution of volume V (or may be in a membrane). In the absence of L molecules, the canonical partition function for the collection of P molecules is $q_P^M/M!$ The division by $M!$ is included because these M molecules are indistinguishable by virtue of their mobility in V (unlike the lattice system above). In the presence of ligand molecules, if N ligands are bound on the M sites, the corresponding canonical partition function is

$$Q(N, M, V, T) = \frac{q_{LP}^N}{N!} \cdot \frac{q_P^{M-N}}{(M - N)!}. \tag{6.31}$$

This is the partition function for an ideal binary mixture, in solution, of P and LP molecules. This differs from Eq. (6.22) by a factor $M!$ We are interested here in a system that is open with respect to the N ligand molecules but closed with respect to the M host molecules P. The appropriate semigrand partition function is

$$\Psi' = \sum_{N=0}^{M} Q(N, M, V, T)\lambda^N, \tag{6.32}$$

which is somewhat related to Ψ used in Eqs. (2.11)–(2.14). On substituting Eq. (6.31) in Eq. (6.32), we obtain

$$\Psi' = \xi^M/M!,$$
$$\xi = q_P(1 + q\lambda). \tag{6.33}$$

This is the same ξ as in Eqs. (6.1) and (6.23). The new factor $M!$ in Eq.

(6.33), compared to Eq. (6.23), has no effect on P_N, $\overline{N}$, θ, or σ_M, so we need not pursue this case any further.

The essential point is that, in either of the above macroscopic systems, each P molecule accounts for one ξ factor, the same ξ found for a one-site system. Consequently, one may as well, at the outset, obtain the same results by treating the conceptually much simpler small system that consists of only one P molecule, with its binding site, and that has the grand partition function $\xi = 1 + q\lambda$ or $\xi = 1 + Kc$. This approach is also the method of choice for more complicated small systems (Chapters 4 and 6), provided that the small systems are independent of each other. If the small systems interact, the full Ξ or Ψ' must be used and not ξ (Chapter 8).

7. Two Variations on Simple Binding

Two Ligands Compete for the Same Site

The simplest generalization of the above binding problem is the case in which two ligands L_1 and L_2 compete for the same binding site on P: only one can be bound at a time to form either L_1P or L_2P. The partition function notation we use is

$$q_1 \equiv q_{L_1P}/q_P, \qquad q_2 \equiv q_{L_2P}/q_P. \tag{7.1}$$

Other parameters, with obvious notation, are λ_1, λ_2; c_1, c_2; and K_1, K_2.

The basic grand partition function we start with, for a single P molecule, possibly with L_1 or L_2 bound, is Eq. (2.9). The only possible values of N_1, N_2 in this equation are 0, 0; 1, 0; and 0, 1. Thus, instead of Eq. (6.1), we have

$$\begin{aligned}\xi &= q_P + q_{L_1P}\lambda_1 + q_{L_2P}\lambda_2 \\ &= q_P(1 + q_1\lambda_1 + q_2\lambda_2).\end{aligned} \tag{7.2}$$

Again, q_P is a common "background" factor in all terms. Hence, it is of no practical consequence and can be omitted insofar as binding studies are concerned. Consequently we begin again, omit q_P, and at the same time replace $q_1\lambda_1$ by K_1c_1 and $q_2\lambda_2$ by K_2c_2. Thus

$$\xi = 1 + K_1c_1 + K_2c_2. \tag{7.3}$$

State:	○	①	②
Weight:	1	K_1c_1	K_2c_2

In Eq. (7.3) we emphasize that the site on P, in this model, can exist in three possible states, empty (○), occupied by L_1 (①), or occupied by L_2 (②); that each state is represented by a term in ξ; and that the relative weights or populations of the three states, at equilibrium, are determined by the values of the corresponding terms in ξ [as in Eq. (1.13)]. Thus, we can write immediately, from Eq. (7.3),

$$\theta_1 = K_1c_1/\xi, \qquad \theta_2 = K_2c_2/\xi, \tag{7.4}$$
$$1 - \theta_1 - \theta_2 = 1/\xi,$$

where θ_i is the fraction of sites occupied by L_i ($i = 1, 2$), and $1 - \theta_1 - \theta_2$ is the fraction of empty sites.

Equations (7.3) and (7.4) illustrate the general procedure we shall use in more complicated cases in Chapters 4 and 6: first, adopt a model for the small system; second, enumerate all possible states consistent with the model; third, construct ξ as a sum with one term per state; and fourth, obtain the desired average properties of the model using the terms in ξ as relative weights.

Equations (7.4) can be confirmed from Eqs. (2.10) and (7.3):

$$\theta_1 = \lambda_1\left(\frac{\partial \ln \xi}{\partial \lambda_1}\right)_{\lambda_2, T} = c_1\left(\frac{\partial \ln \xi}{\partial c_1}\right)_{c_2, T}, \tag{7.5}$$

$$\theta_2 = \lambda_2\left(\frac{\partial \ln \xi}{\partial \lambda_2}\right)_{\lambda_1, T} = c_2\left(\frac{\partial \ln \xi}{\partial c_2}\right)_{c_1, T}. \tag{7.6}$$

Binding within a Structured System

We first digress on a more general topic. If a macromolecule can exist in two different states, conformations, or structures, A and B, with partition functions q_A and q_B, then the complete canonical partition function is

$$Q = q_A + q_B. \tag{7.7}$$

This is a special case of Eq. (1.1) in which the microscopic substates j can be divided into two classes, A and B. From the Boltzmann distribution, then, the relative populations of states A and B, at equilibrium, are proportional to q_A and q_B, respectively. Thus,

$$Q = q_A(1 + K), \tag{7.8}$$
$$K = \frac{q_B}{q_A} = \frac{p_B^e}{p_A^e},$$

where K is the equilibrium constant (dimensionless) for A $\rightleftarrows$ B and the ps are equilibrium probabilities. The sum $1 + K$ is formally similar to $\xi = 1 + Kc$ above, but the present system is closed, not open. The analogy goes a little further [see Eq. (6.14)]:

$$p_B^e = \frac{K}{1 + K}, \qquad p_A^e = \frac{1}{1 + K}. \tag{7.9}$$

Also, as in Eq. (6.18),

$$K\frac{\partial \ln Q}{\partial K} = \frac{K}{1 + K} = p_B^e. \tag{7.10}$$

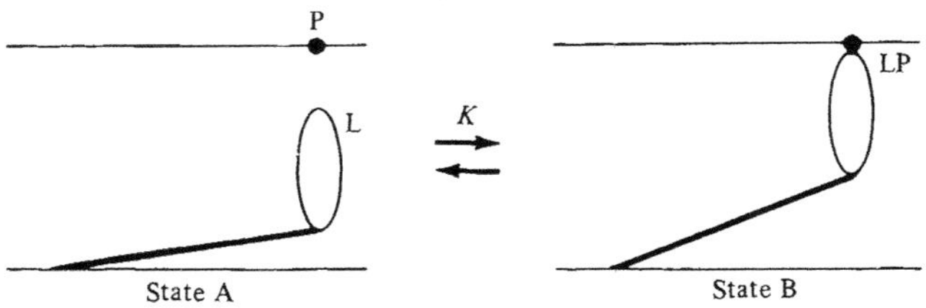

Fig. 2-1. "Internal" binding of L on P within a structured system.

This kind of formal operation is sometimes useful in more complicated problems.

We now consider a special case of the above. In some macromolecular structures, there is "internal" binding. Figure 2-1 shows an example: in the myofilament structure, a myosin molecule (L) may or may not be bound (attached) to a particular nearby site on actin (P). If attached, we call the state LP. Then, q_A and q_B in the above notation are

$$q_A = q_L q_P, \qquad q_B = q_{LP}, \tag{7.11}$$

and

$$K = \frac{q_{LP}}{q_L q_P} = \frac{(q_{LP}/q_P)}{q_L} = \frac{q}{q_L}. \tag{7.12}$$

The partition function q_L here does not include completely free translational motion, as usual. Hence the factor V is absent. The binding fraction θ in this example is

$$\theta = p_B^e = \frac{K}{1+K}. \tag{7.13}$$

In internal binding, K (dimensionless) replaces Kc (also dimensionless) in Section 6. There is no "concentration" of L in the internal binding problem; each L simply changes its state within the same structure (A $\rightleftarrows$ B in Fig. 2-1). Further details on this particular example will be found in Ref. 1.

Reference

1. Hill, T. L. and Eisenberg, E. (1980) Biophys. Chem. **11,** 271.

General Reference

Hill, T. L. (1960) *Introduction to Statistical Thermodynamics* (Addison-Wesley, Reading, MA).

3

Introduction to Steady-State Systems

In the first two chapters we have discussed certain topics that are fundamental for equilibrium systems, especially binding. In this chapter we introduce simple steady-state systems. Chapters 4 and 5 will continue with somewhat more complicated equilibrium and steady-state systems, but without explicit introduction of interaction free energies. Then, in Parts II and III, we turn to the main theme of the book: systems of both classes comprised of subsystems that interact with each other, with the interactions treated explicitly.

8. Illustrative Steady-State Systems

The purpose of this section is to introduce the reader to a few simple and explicit models of steady-state systems, and to relate these models to the corresponding equilibrium systems as limiting cases. In each example, the rate constants are specified and related to the thermodynamic force. After this introduction, in Sections 9 and 10, we shall derive the fundamental properties of any independent steady-state system that can be represented by two-state and three-state cycles. These results, especially on two-state cycles, will be needed repeatedly in the remainder of the book.

One-Site Membrane Transport

The first and simplest model is introduced in Fig. 3-1, and the corresponding two-state kinetic cycle is shown in Fig. 3-2(a). A protein molecule E is present in a membrane (mem in Fig. 3-1) that separates two solutions, A and B, both of which contain a ligand L at concentration c_A and c_B, respectively. The ligand can bind onto E to form EL, either from side A or from side B. Thus E can exist in the two states, E and EL. The possible on and off

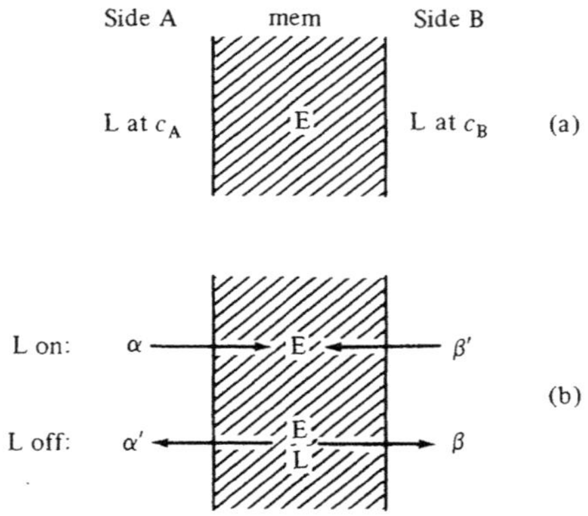

Fig. 3-1. (a) Binding of ligand L on E in membrane, from either side. (b) Physical significance of rate constants.

transitions for L, with first-order rate constants, are indicated in Fig. 3-1(b). The cycle in Fig. 3-2(a) connects the two states of E and contains the same rate constants.

The usual dominant direction of cyclic activity in Fig. 3-2(a) is counterclockwise. This is arbitrarily called the positive direction (for force and flux, defined below). This corresponds to net transport of L in the direction $A \rightarrow B$. The rate constants in this direction are α and β; the inverse transitions have primed rate constants, α' and β'. The binding, or on, rate constants are α and β'; the off rate constants are α' and β. All of these are first-order rate constants. That is, α and β' (binding) are pseudo first order: $\alpha = \alpha^* c_A$ and $\beta' = \beta^* c_B$, where α^* and β^* are second-order binding rate constants.

When $c_A > c_B$, there will be net transport of L from side A to side B, via binding of L on E, that is, using net cyclic activity in Fig. 3-2(a) in the positive direction. If $c_A < c_B$, the net transport will be in the opposite direction. If $c_A = c_B$, there is no net transport: the system will be at equilibrium.

We have been referring to the activity of a single molecule E in the membrane. Ordinarily, of course, there will be a large ensemble of independent E molecules in the membrane. Our equations below usually refer to

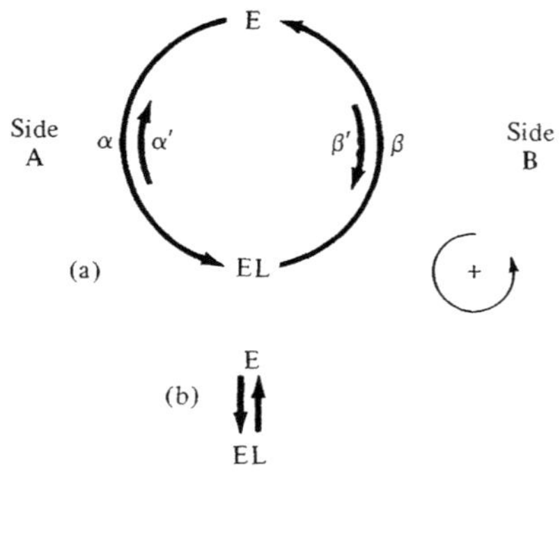

Fig. 3-2. (a) Two-state kinetic cycle corresponding to model in Fig. 3-1. (b) Contrast with equilibrium binding of L on E.

ensemble averages, though the same equations would also apply to a single E molecule if it were observed over a long time period (i.e., the ensemble and long-time averages would be the same for equilibrium or steady-state systems).

The thermodynamic force X driving L in the direction A $\rightarrow$ B is the chemical potential difference of L on the two sides:

$$\mu_{\mathrm{A}} = \mu^{\circ} + kT \ln c_{\mathrm{A}}, \tag{8.1}$$

$$\mu_{\mathrm{B}} = \mu^{\circ} + kT \ln c_{\mathrm{B}}, \tag{8.2}$$

$$X = \mu_{\mathrm{A}} - \mu_{\mathrm{B}} = kT \ln (c_{\mathrm{A}}/c_{\mathrm{B}}). \tag{8.3}$$

When $c_{\mathrm{A}} > c_{\mathrm{B}}$, with net transport A $\rightarrow$ B, we have $\mu_{\mathrm{A}} > \mu_{\mathrm{B}}$ and $X > 0$. We want to see how the thermodynamic quantity X is related to the kinetic parameters of the model that transports L, that is, to the rate constants.

We begin at equilibrium, and denote the probabilities of the two states $p_{\mathrm{E}}^{\mathrm{e}}$ and $p_{\mathrm{EL}}^{\mathrm{e}} = 1 - p_{\mathrm{E}}^{\mathrm{e}}$. In Chapter 2, we used $1 - \theta$ and θ for these quantities, but we change notation here to allow later generalization to kinetic cycles or diagrams (more than one cycle) with more than two states. Also, at equilibrium, $c_{\mathrm{A}} = c_{\mathrm{B}}$, both of which we denote by c. At equilibrium, there will be detailed balance in both transition pairs:

$$\alpha^* c p_{\mathrm{E}}^{\mathrm{e}} = \alpha' p_{\mathrm{EL}}^{\mathrm{e}}, \tag{8.4}$$

$$\beta p_{\mathrm{EL}}^{\mathrm{e}} = \beta^* c p_{\mathrm{E}}^{\mathrm{e}}. \tag{8.5}$$

If we multiply these equations together, c and the ps cancel and we find

$$\alpha^* \beta = \alpha' \beta^* \qquad \text{or} \qquad \frac{\alpha^*}{\alpha'} = \frac{\beta^*}{\beta} = K. \tag{8.6}$$

That is, although the separate rate constants will in general be different for binding from side A or B (the membrane is generally unsymmetrical), $\alpha^* \neq \beta^*$ and $\alpha' \neq \beta$, the binding constant K is the same from the two sides. The reason for this is that the thermodynamic process is the same from either side: E + L $\leftrightarrows$ EL. Closely analogous relationships also obtain for the on–off transitions of subunits at the two ends of a linear equilibrium aggregate with intrinsic directionality (and, therefore, two different ends).

The relation between rate constants in Eq. (8.6) is an intrinsic property of the model, independent of the values of c_{A} and c_{B}, even though this relation was derived from the special case $c_{\mathrm{A}} = c_{\mathrm{B}}$.

The distinction between Figs. 3-2(a) and 3-2(b) should be emphasized. In Fig. 3-2(b), as in Chapter 2, we have a two-state system with one pair of inverse transitions between the two states. The only steady-state available to this system is the equilibrium state (i.e., equilibrium binding of L on E). In contrast, in Fig. 3-2(a), there are two distinct pairs of inverse transitions between the two states. This system can have nonequilibrium steady states in which there is a nonzero thermodynamic force X that produces a nonzero net flux J (see below) of ligand molecules across the membrane. Correspondingly, there is a flux around the cycle in Fig. 3-2(a) at steady state when $X \neq 0$. That is, there is a net rate J of completing cycles in Fig. 3-2(a). Each + cycle completion transports one L from A to B. In Fig. 3-2(b), there is no cycle and no possible net mean flux at equilibrium.

We return now to an arbitrary steady state, $c_A \neq c_B$. We define (for any cycle) Π_+ and Π_- as the products of first-order rate constants around the cycle in the two directions. For the cycle in Fig. 3-2(a),

$$\Pi_+ \equiv \alpha\beta, \qquad \Pi_- \equiv \alpha'\beta', \tag{8.7}$$

$$\frac{\Pi_+}{\Pi_-} = \frac{\alpha\beta}{\alpha'\beta'} = \frac{\alpha^* c_A \beta}{\alpha'\beta^* c_B} = \frac{c_A}{c_B}, \tag{8.8}$$

having used Eq. (8.6). On comparison with Eq. (8.3), we see that

$$\frac{\Pi_+}{\Pi_-} = \frac{\alpha\beta}{\alpha'\beta'} = e^{X/kT}. \tag{8.9}$$

This shows how the thermodynamic force X, which is a property only of the ligand in the two baths, is related to the rate constants of the transport system. Thermodynamics, through X, imposes a restraint on the rate constants of the model.

At equilibrium, $c_A = c_B = c, X = 0$, and $\Pi_+ = \Pi_-$. Also, at equilibrium, $\xi = 1 + Kc$ as in Section 6. The weights 1 and Kc are consistent with p_E^e, p_{EL}^e, and K in Eqs. (8.4)–(8.6):

$$\frac{p_{EL}^e}{p_E^e} = \frac{\beta^* c}{\beta} = \frac{\alpha^* c}{\alpha'} = Kc.$$

As will be shown in Section 9, the flux J (the rate at which L crosses the membrane) is proportional to[1] $\Pi_+ - \Pi_-$. Hence, $J = 0$ when $X = 0$ (equilibrium), as expected.

One-Site Transport with Membrane Potential

This is a convenient example of the perturbation of rate constants (Section 5). Suppose that L is an ion with charge $z\epsilon$ (ϵ = charge on proton), and that bath A is at electrostatic potential ψ, whereas bath B is at potential zero (i.e., B is chosen as reference). The binding site for L (on E) in the membrane is at potential ψ_m. Figure 3-3 shows a hypothetical free energy profile, across the membrane, with and without the membrane potential ψ. The electrochemical potentials of L on the two sides, and the thermodynamic force, are now

$$\mu_A = \mu^o + kT \ln c_A + z\epsilon\psi, \tag{8.10}$$

$$\mu_B = \mu^o + kT \ln c_B, \tag{8.11}$$

$$X = \mu_A - \mu_B = kT \ln\left(\frac{c_A e^{z\epsilon\psi/kT}}{c_B}\right). \tag{8.12}$$

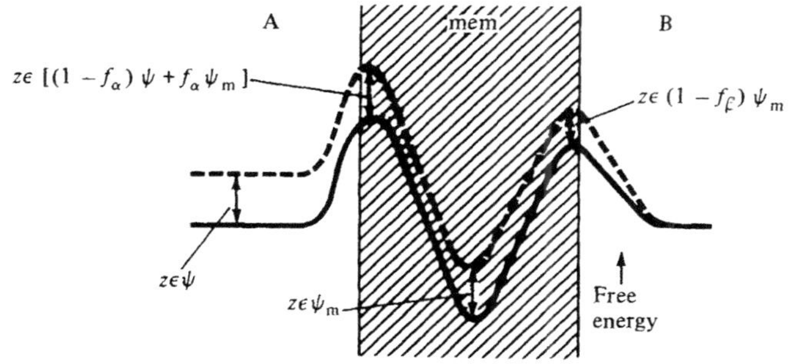

Fig. 3-3. Free energy profile, without (solid curve) and with membrane potential ψ, for binding ligand, with charge $z\epsilon$, on E in membrane. See text for details.

At equilibrium, $\mu_A = \mu_B$ and

$$c_B = c_A e^{z\epsilon\psi/kT}, \qquad X = 0. \tag{8.13}$$

When $\psi = 0$, all of the rate constants and the equilibrium constant in the preceding subsection apply, and are here designated with a subscript o. Thus [Eq. (8.6)]

$$\frac{\alpha_o^*}{\alpha_o'} = \frac{\beta_o^*}{\beta_o} = K_o. \tag{8.14}$$

In the binding of L from side A in the presence of ψ, the free energy perturbation $w_1 - w_2$ in Eq. (5.3) is here $z\epsilon(\psi - \psi_m)$. Hence, the binding constant from side A is now

$$K_A = K_o e^{z\epsilon(\psi-\psi_m)/kT} = \frac{\alpha^*}{\alpha'} = \frac{\alpha_o^*}{\alpha_o'} e^{z\epsilon(\psi-\psi_m)/kT}. \tag{8.15}$$

Similarly, the binding constant from side B is

$$K_B = K_o e^{-z\epsilon\psi_m/kT} = \frac{\beta^*}{\beta} = \frac{\beta_o^*}{\beta_o} e^{-z\epsilon\psi_m/kT}. \tag{8.16}$$

Note that $K_A \neq K_B$ unless $\psi = 0$.

The rate constant products Π_+ and Π_- now lead to

$$\frac{\Pi_+}{\Pi_-} = \frac{\alpha\beta}{\alpha'\beta'} = \frac{\alpha^* c_A \beta}{\alpha'\beta^* c_B} = \frac{K_A c_A}{K_B c_B} = \frac{c_A e^{z\epsilon\psi/kT}}{c_B} = e^{X/kT}. \tag{8.17}$$

This same relation between Π_+/Π_- and X is, in fact, found for any cycle (see Ref. 1, Chapter 3). Also, as for any cycle,[1] the flux J is proportional to $\Pi_+ - \Pi_-$: $J = 0$ when $X = 0$.

At equilibrium, $X = 0$ and $K_A c_A = K_B c_B$. The grand partition function for the binding site is thus

$$\xi = 1 + K_A c_A = 1 + K_B c_B. \tag{8.18}$$

Equations (8.15) and (8.16) show how the membrane potential ψ perturbs rate constant ratios. To express the effect of ψ on the individual rate constants, we introduce parameters f_α and f_β, as is done in Eqs. (5.6) and (5.8). We choose (this is a convention we adhere to throughout the book) the "forward" direction in each transition pair to correspond to the + direction in the cycle [Fig. 3.2(a)]. That is, α and β are "forward," whereas α' and β' are not. Recall that α is a binding rate constant but that β is a release rate constant. Then, as in Eqs. (5.6) and (5.8),

$$\alpha^* = \alpha_o^* e^{f_\alpha z\epsilon(\psi-\psi_m)/kT}, \tag{8.19}$$

$$\alpha' = \alpha_o' e^{(f_\alpha-1)z\epsilon(\psi-\psi_m)/kT}, \tag{8.20}$$

$$\beta = \beta_o e^{f_\beta z\epsilon\psi_m/kT}, \tag{8.21}$$

$$\beta^* = \beta_o^* e^{(f_\beta-1)z\epsilon\psi_m/kT}. \tag{8.22}$$

Note that one f is introduced for each transition pair. Because α^* and β^* are both binding rate constants, we might expect that $f_\alpha = 1 - f_\beta$, or $f_\alpha + f_\beta = 1$, though this is not necessary. This kind of relation in a two-state cycle leads to important simplifications in some problems, as we shall see. For this model, as in most, it would be reasonable to take f_α and f_β as constants, but in principle they might be functions of ψ. The effects of ψ on the free energy barriers for binding are expressed in terms of f_α and f_β in Fig. 3-3.

Two-Site Membrane Transport

Here we consider transport of L across a membrane using a two-site channel such that at most only one L can be in a channel at a time. We omit the membrane potential. The model and cycle are shown in Fig. 3-4. The two sites are labelled A and B, and are generally different. Site A is accessible for binding of L from side A only, and site B is accessible only from side B. The binding constants for L on the two sites are K_A and K_B (not related to the K_A and K_B in the previous subsection). The three possible states of occupation of the channel are called 1, 2, 3 and are indicated in Fig. 3-4(a). The corresponding three-state cycle for the channel, with first-order rate constant notation, is shown in Fig. 3-4(b). The constants κ and κ' refer to jumps of L from one site to another within the channel.

Equations (8.1)–(8.3) for the chemical potentials of L and the force X are still applicable here. Also, $\alpha = \alpha^* c_A$ and $\beta' = \beta^* c_B$, as in Figs. 3-1 and 3-2.

From Fig. 3-4(b), the three detailed balance relations, at equilibrium ($c_A = c_B = c$), are

$$\alpha^* c p_1^e = \alpha' p_2^e, \qquad \kappa p_2^e = \kappa' p_3^e, \qquad \beta p_3^e = \beta^* c p_1^e. \tag{8.23}$$

If we multiply these three equations together and then cancel c and the ps, we have

$$\alpha^* \kappa \beta = \alpha' \kappa' \beta^* \tag{8.24}$$

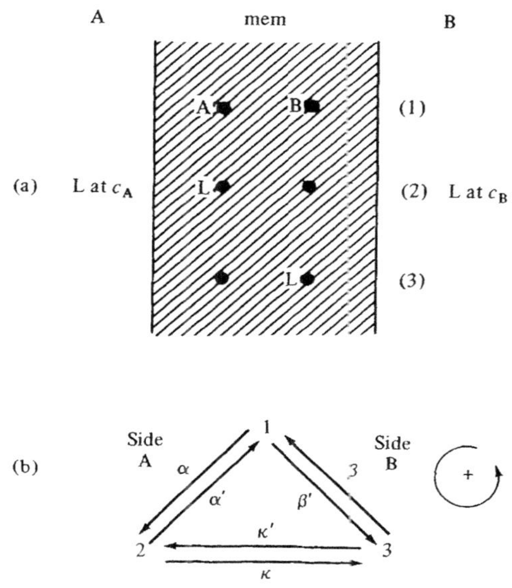

Fig. 3-4. (a) Model for binding ligand on two sites labelled A and B. At most one site can be occupied at a time. The three possible states are (1), (2), and (3). (b) Corresponding kinetic cycle with first-order rate constants.

as a required relation among the rate constants of the model. If we define the dimensionless equilibrium constant for jumping ($2 \rightleftarrows 3$) as $K = \kappa/\kappa'$, and use the binding constants $K_A = \alpha^*/\alpha'$ and $K_B = \beta^*/\beta$, Eq. (8.24) is equivalent to

$$K_A K = K_B. \tag{8.25}$$

Thus if $K_A \neq K_B$, then $K \neq 1$.

At a steady state, with arbitrary c_A and c_B, the ratio of Π_+ to Π_- for this model is then

$$\frac{\Pi_+}{\Pi_-} = \frac{\alpha\kappa\beta}{\alpha'\kappa'\beta'} = \frac{\alpha^* c_A \kappa\beta}{\alpha'\kappa'\beta^* c_B} = \frac{c_A}{c_B} = e^{X/kT}. \tag{8.26}$$

Again X is related in the usual way[1] to Π_+/Π_-, and thus to the rate constants of the model.

The equilibrium grand partition function for the system is

$$\xi = 1 + K_A c + K_B c, \tag{8.27}$$

where the successive terms give the relative weights of states 1, 2, and 3. These relative weights can be confirmed by rearranging Eqs. (8.23):

$$\frac{p_2^e}{p_1^e} = \frac{\alpha^* c}{\alpha'} = K_A c, \qquad \frac{p_3^e}{p_1^e} = \frac{\beta^* c}{\beta} = K_B c. \tag{8.28}$$

Also, p_3^e/p_2^e is equal to κ/κ' or K in Eq. (8.23) and, in addition, is equal to K_B/K_A, according to Eqs. (8.28). These two results are consistent with each other, in view of Eq. (8.25). Whereas Eq. (7.3) refers to two ligands competing for one site, Eq. (8.27) refers to two sites competing for one ligand.

Two-Conformation Membrane Transport

Suppose a protein E in a membrane exists in two conformations E and E* such that its binding site for L is exposed or accessible to side A when in conformation E and to side B when in conformation E*. The cycle of states for the protein, in this case, is a four-state cycle, shown in Fig. 3-5(a). In this diagram, a line is shorthand for the usual pair of opposing arrows with rate constants. If the transitions between the states E and E* are relatively fast, the four-state cycle in Fig. 3-5(a) degenerates into the three-state cycle in Fig. 3-5(b), where (E, E*) represents a fast equilibrium between these two species. This three-state cycle is then formally indistinguishable from the cycle in Fig. 3-4(b); in fact, the rate constant notation in Fig. 3-4(b) can be taken over in Fig. 3-5(b). Of course the actual values of the rate constants would be different because the molecular details are different, but the formal analysis of the two cases is identical.

In Fig. 3-4(a), access to different baths is a consequence of two spatially different sites in the channel; in Fig. 3-5(b), there is only one binding site on the protein [as in Fig. 3-1(a)], but a change in conformation alters bath access.

Two-Site Double-Occupancy Transport

In Fig. 3-6(a) we have the same model as in Fig. 3-4(a), except that now both sites can be occupied at the same time. The channel has four states; the kinetic diagram is given in Fig. 3-6(b). This diagram has three cycles, shown in Fig. 3-6(c), and labelled a, b, c. If we take the counterclockwise direction as positive in all three cycles, the cycles a, b, c have forces X, $-X$, and 0, respectively, where X is defined as usual by Eq. (8-3) (X is the force $\mu_A - \mu_B$ in the direction A → B). That is, in the counterclockwise direction in each case, one completed cycle of type a moves one L from A to B, one cycle of type b moves one L from B to A, and one cycle of type c accomplishes no net transport across the membrane.

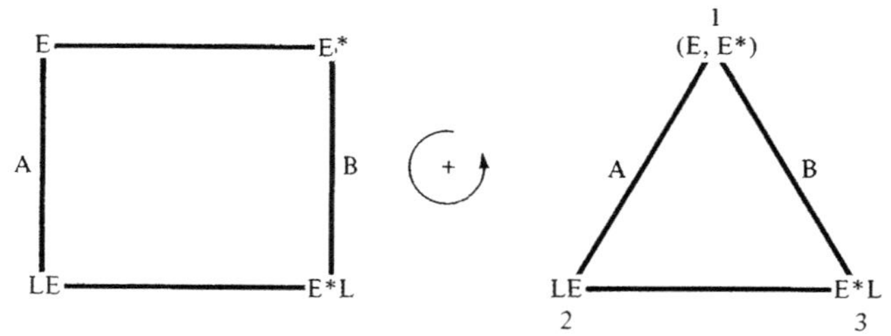

(a) (b)

Fig. 3-5. (a) Kinetic cycle for an enzyme like the one in Figs. 3-1 and 3-2(a), except that here E exists in two conformations. See text for details. (b) Reduced cycle if E ⇄ E* transitions are relatively fast.

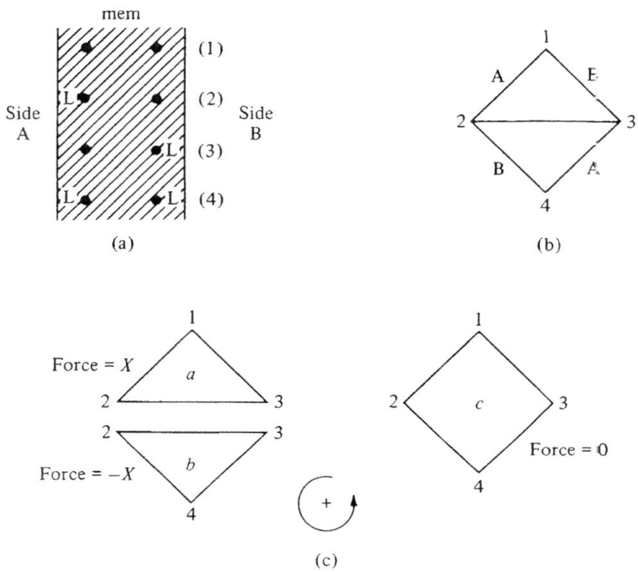

Fig. 3-6. (a) Four possible states for a transport model with two sites, both of which can be occupied by ligand at the same time. (b) Kinetic diagram for the model. (c) Three cycles of the model and the force in each cycle.

Systems of this type (with many sites) are discussed in some detail in Chapter 5, so we do not pursue this model any further here (see, also, Section 27).

Multicycle diagrams are common in systems with more than one thermodynamic force.[1]

Enzyme Molecule in Solution

As a final illustration of a steady-state system, we consider independent enzyme molecules E in solution, each of which catalyzes the reaction S (substrate) $\rightleftarrows$ P (product). The reaction is negligibly slow without the intervention of the enzyme. The three-state cycle is shown in Fig. 3-7(a). We use the same rate constant notation as in Fig. 3-4(b), but the system is of course very different.

There are now two ligands, S and P, with chemical potentials in solution

$$\mu_S = \mu_S^o + kT \ln c_S,$$

$$\mu_P = \mu_P^o + kT \ln c_P.$$

The thermodynamic force X driving the reaction $S \to P$ and the enzymatic cycle in Fig. 3-7(a) is

$$X = \mu_S - \mu_P = \mu_S^o - \mu_P^o + kT \ln(c_S/c_P). \tag{8.29}$$

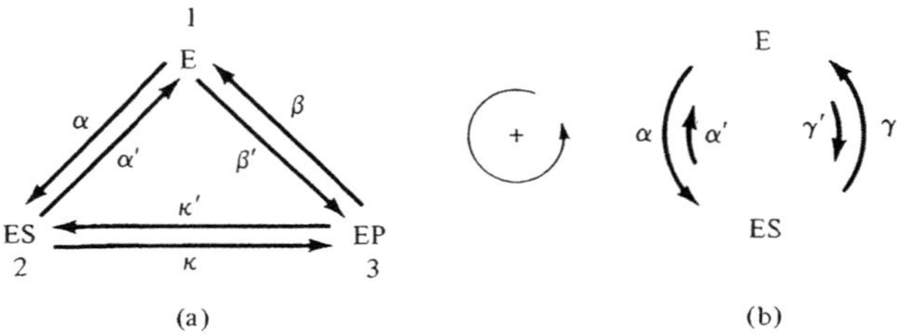

Fig. 3-7. (a) Three-state cycle for an enzyme E, in solution, that catalyzes S → P. (b) Reduced cycle when state EP is a transient intermediate.

At equilibrium, $\mu_S = \mu_P$, $X = 0$, and

$$\frac{c_P^e}{c_S^e} = e^{-(\mu_P^o - \mu_S^o)/kT} \equiv K_{SP}. \tag{8.30}$$

Here K_{SP} and $\mu_P^o - \mu_S^o$ are the conventional equilibrium constant and standard free energy change for the reaction S $\leftrightarrows$ P. If we use Eq. (8.30) in Eq. (8.29), other expressions for X are

$$X = kT \ln\left(\frac{c_S}{c_P} \cdot \frac{c_P^e}{c_S^e}\right) = kT \ln\left(\frac{K_{SP} c_S}{c_P}\right). \tag{8.31}$$

The binding rate constants are related to concentrations by

$$\alpha = \alpha^* c_S, \qquad \beta' = \beta^* c_P. \tag{8.32}$$

The three detailed balance relations, at equilibrium, are then

$$\alpha^* c_S^e p_1^e = \alpha' p_2^e, \qquad \kappa p_2^e = \kappa' p_3^e, \qquad \beta p_3^e = \beta^* c_P^e p_1^e. \tag{8.33}$$

On multiplying these three relations together, we find that the rate constants must satisfy

$$\frac{c_P^e}{c_S^e} = K_{SP} = \frac{\alpha^* \kappa \beta}{\alpha' \kappa' \beta^*}. \tag{8.34}$$

The binding constants for S and P on E are $K_S = \alpha^*/\alpha'$ and $K_P = \beta^*/\beta$. If we define $K = \kappa/\kappa'$ (i.e., for the equilibrium $2 \rightleftarrows 3$), then Eq. (8.34) can also be written

$$K_{SP} = K_S K / K_P. \tag{8.35}$$

The usual quotient of rate constant products at an arbitrary steady state is here

$$\frac{\Pi_+}{\Pi_-} = \frac{\alpha\kappa\beta}{\alpha'\kappa'\beta'} = \frac{\alpha^* c_S \kappa\beta}{\alpha'\kappa'\beta^* c_P}$$

$$= \frac{c_S}{c_P} \cdot \frac{c_P^e}{c_S^e} = \frac{K_{SP} c_S}{c_P} = e^{X/kT}. \tag{8.36}$$

This is again the standard result.

The equilibrium grand partition function for this system is

$$\xi = 1 + K_S c_S^e + K_P c_P^e. \tag{8.37}$$

The three terms give the relative weights of states 1, 2, and 3, respectively. These weights are consistent with Eqs. (8.33). Equation (8.37) relates to competitive binding of S and P for one site on L, with S and P at concentrations c_S^e and c_P^e (in equilibrium with each other).

A common special case in enzymology occurs when the state EP in Fig. 3-7(a) is a transient intermediate, that is, when EP has a negligible population at steady state relative to E and ES. This will occur when $\kappa' + \beta$ is much larger than both κ and β'. In this case, the three-state cycle in Fig. 3-7(a) simplifies to the effective two-state cycle in Fig. 3-7(b) [compare, also, Fig. 3-2(a)]. The new rate constants γ and γ' are composites of the old ones (see Ref. 1, Appendix 1):

$$\gamma = \frac{\kappa\beta}{\kappa' + \beta}, \qquad \gamma' = \frac{\kappa'\beta'}{\kappa' + \beta}. \tag{8.38}$$

Note that Π_+/Π_- is unchanged:

$$\frac{\Pi_+}{\Pi_-} = \frac{\alpha\gamma}{\alpha'\gamma'} = \frac{\alpha\kappa\beta}{\alpha'\kappa'\beta'} = e^{X/kT}. \tag{8.39}$$

This is as expected because the force X is a thermodynamic property of S and P in solution and must be invariant to kinetic details.

A "Michaelis–Menten" enzyme, often discussed in textbooks, is the special case of Fig. 3-7(b) in which γ' is relatively small and can be neglected for most purposes. However, one cannot set $\gamma' = 0$ in Eq. (8.39), because X is finite [(Eq. (8.29)].

9. Two-State Steady-State System

In this section we discuss general properties of any steady-state system that can be represented by the two-state cycle in Fig. 3-8(a). The rate constants are all first order, as usual. The two states are 1 and 2; deliberately we do not identify states 1 and 2 with any particular model. Because of this, we define the thermodynamic force here by

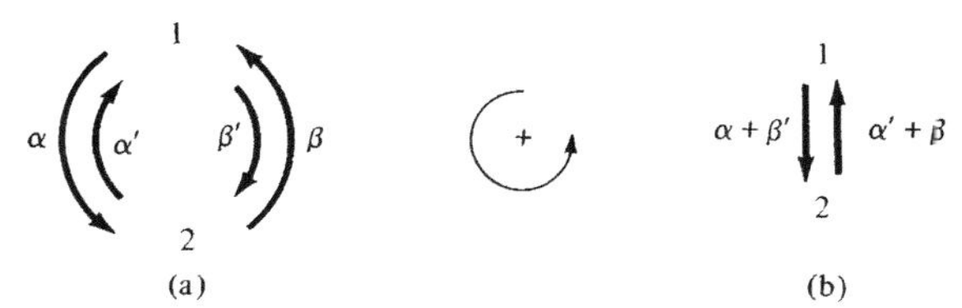

Fig. 3-8. (a) Two-state cycle for an arbitrary model, with first-order rate constants. (b) Quasiequilibrium between states 1 and 2. See text for details.

$$\Pi_+/\Pi_- = \alpha\beta/\alpha'\beta' = e^{X/kT}, \tag{9.1}$$

instead of employing the usual thermodynamic definition of X in terms of the chemical potentials of the small molecules that interact with the cycling enzyme, macromolecule, channel, etc.

In an ensemble of these two-state systems, p_1 and p_2 are the fractions in states 1 and 2 (with $p_1 + p_2 = 1$). From Fig. 3-8(a),

$$\frac{dp_1}{dt} = (\alpha' + \beta)p_2 - (a + \beta')p_1 = -\frac{dp_2}{dt}. \tag{9.2}$$

At steady state, $dp_1/dt = 0$. Hence

$$p_1 = (\alpha' + \beta)/\Sigma, \qquad p_2 = (\alpha + \beta')/\Sigma, \qquad \Sigma \equiv \alpha + \alpha' + \beta + \beta'. \tag{9.3}$$

The steady-state flux J (mean net number of cycles completed in the positive direction) is then

$$J = \alpha p_1 - \alpha' p_2 = \beta p_2 - \beta' p_1. \tag{9.4}$$

At steady-state, the net flux on the two sides of the cycle (α and β) must be the same. From either of these expressions, and Eq. (9.3), we find

$$J = \frac{\alpha\beta - \alpha'\beta'}{\Sigma} = \frac{\alpha'\beta'(e^{X/kT} - 1)}{\Sigma}. \tag{9.5}$$

At equilibrium, $X = 0$, $\alpha\beta = \alpha'\beta'$, and $J = 0$. Also, J and X always have the same sign when $X \neq 0$.

X can be varied by varying ligand concentrations, membrane potential, etc., which in turn affect one or more of the four rate constants.

At any X, even at equilibrium, cycles can be completed in either direction. If we denote the mean rates of cycle completions in the two directions, at steady state, by J_+ and J_- (always positive quantities), then $J = J_+ - J_-$. J can be positive or negative. In view of the form of Eq. (9.5), it is plausible to assume that

$$J_+ = \frac{\alpha\beta}{\Sigma}, \qquad J_- = \frac{\alpha'\beta'}{\Sigma}. \tag{9.6}$$

A general theorem, applicable to arbitrary kinetic diagrams, shows that this assumption is indeed correct.[1] For this simple cycle, it is also easy to prove Eq. (9.6) by a stochastic argument, as follows. Suppose we start with a system in state 1 and calculate J_+. The mean lifetime in state 1, before a transition to state 2, is $1/(\alpha + \beta')$. The probability that this transition is via α (i.e., in the + cycle direction) is $\alpha/(\alpha + \beta')$. Having reached state 2, the mean lifetime of the system in this state, before a transition back to state 1, is $1/(\alpha' + \beta)$. The probability that this second transition is via β (again in the + direction) is $\beta/(\alpha' + \beta)$. Thus, in the total mean time

$$\tau = \frac{1}{\alpha + \beta'} + \frac{1}{\alpha' + \beta}, \tag{9.7}$$

the probability P_+ that a + cycle has been completed is

$$P_+ = \frac{\alpha}{\alpha + \beta'} \cdot \frac{\beta}{\alpha' + \beta}. \tag{9.8}$$

Hence the rate of + cycle completions is

$$J_+ = P_+/\tau = \alpha\beta/\Sigma. \tag{9.9}$$

The same type of argument can be used for J_-.

At equilibrium,

$$\alpha\beta = \alpha'\beta', \qquad J_+^e = J_-^e, \qquad J^e = J_+^e - J_-^e = 0. \tag{9.10}$$

That is, cycles are completed at equilibrium at the rates given by Eqs. (9.6); but in this case the set of rate constants is such that $J_+^e = J_-^e$.

Equation (9.5) can now be written

$$J = J_-(e^{X/kT} - 1) = J_+(1 - e^{-X/kT}). \tag{9.11}$$

The flux-force relation very near equilibrium (i.e., $|X|/kT \ll 1$) is found by expanding $e^{X/kT}$:

$$J = J_\pm^e \cdot \frac{X}{kT} \qquad (X \to 0), \tag{9.12}$$

where $J_\pm^e$ refers to either J_+^e or J_-^e. Thus, near equilibrium, there is a linear flux-force relation. The conventional notation is $J = LX$, where here $L = J_\pm^e/kT$. L is a kinetic property of the system *at equilibrium*. For a two-state cycle, the explicit expressions for $J_\pm^e$ are given in Eq. (9.6). Actually, because for *any* cycle[1]

$$\frac{\Pi_+}{\Pi_-} = e^{X/kT} = \frac{J_+}{J_-}, \qquad J = J_+ - J_-, \tag{9.13}$$

Eqs. (9.11) and (9.12) hold for any cycle, even for a cycle embedded in a multicycle diagram. However, the explicit expressions for $J_\pm^e$ in terms of rate constants depend on the particular case.[1]

The coefficients L_{ij} in the linear flux-force relations for an arbitrary diagram are composites[2,3] of the $J_\pm^e/kT$ for the various cycles of the diagram. Onsager reciprocal relations, $L_{ij} = L_{ji}$, are an automatic consequence. This provides a physical interpretation of the L_{ij} for any biochemical diagram.

Quasiequilibrium at Steady State

It is useful to notice that the state distribution in a two-state cycle has a quasiequilibrium form. That is, Eqs. (9.2) and (9.3) apply, without change, to the simple two-state equilibrium in Fig. 3-8(b), with effective rate constants $\alpha + \beta'$ and $\alpha' + \beta$. This formal equivalence exists for *two*-state cycles only.

10. Three-State Steady-State System

Here we summarize the basic properties for an arbitrary three-state system, with rate constant notation given in Fig. 3-9. As in Section 9, we define X by

$$\frac{\Pi_+}{\Pi_-} = \frac{\alpha_{12}\alpha_{23}\alpha_{31}}{\alpha_{21}\alpha_{32}\alpha_{13}} = e^{X/kT}. \tag{10.1}$$

We cannot relate the α_{ij} to ligand concentrations, etc., without a specific model.

The state probabilities are p_1, p_2, and p_3. Any two kinetic equations for the p_i are independent, for example,

$$\begin{aligned} \frac{dp_1}{dt} &= \alpha_{21}p_2 + \alpha_{31}p_3 - (\alpha_{12} + \alpha_{13})p_1, \\ \frac{dp_2}{dt} &= \alpha_{12}p_1 + \alpha_{32}p_3 - (\alpha_{21} + \alpha_{23})p_2. \end{aligned} \tag{10.2}$$

At steady state, the derivatives are zero, and we find, using $p_1 + p_2 + p_3 = 1$,

$$\begin{aligned} p_1 &= \frac{\alpha_{32}\alpha_{21} + \alpha_{23}\alpha_{31} + \alpha_{21}\alpha_{31}}{\Sigma}, \\ p_2 &= \frac{\alpha_{12}\alpha_{32} + \alpha_{13}\alpha_{32} + \alpha_{31}\alpha_{12}}{\Sigma}, \\ p_3 &= \frac{\alpha_{12}\alpha_{23} + \alpha_{13}\alpha_{23} + \alpha_{21}\alpha_{13}}{\Sigma}, \end{aligned} \tag{10.3}$$

where Σ is the sum of the three numerators (i.e., Σ has nine terms). Equations (10.3) can be deduced by straightforward algebra (i.e., we have three linear algebraic equations in three unknowns) or, more simply, by using the King–Altman method.[1,4]

The steady-state flux in the cycle (mean net rate of cycle completions in

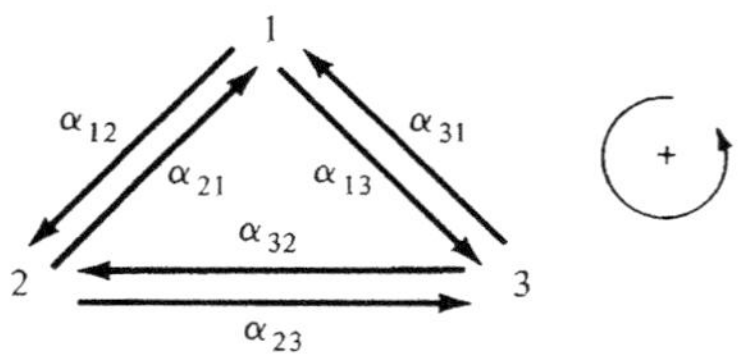

Fig. 3-9. Cycle and rate constants for an arbitrary three-state enzyme.

the + direction) is

$$
\begin{aligned}
J &= \alpha_{12} p_1 - \alpha_{21} p_2 \\
&= \alpha_{23} p_2 - \alpha_{32} p_3 \\
&= \alpha_{31} p_3 - \alpha_{13} p_1.
\end{aligned}
\tag{10.4}
$$

Any of these leads to

$$
\begin{aligned}
J &= \frac{\alpha_{12}\alpha_{23}\alpha_{31} - \alpha_{21}\alpha_{32}\alpha_{13}}{\Sigma} \\
&= \frac{\Pi_+ - \Pi_-}{\Sigma} = J_+ - J_-,
\end{aligned}
\tag{10.5}
$$

where

$$
J_+ = \frac{\Pi_+}{\Sigma}, \qquad J_- = \frac{\Pi_-}{\Sigma}. \tag{10.6}
$$

Alternatively, Eq. (10.5) follows immediately from the diagram method for cycles[1] that is analogous to the King–Altman method for states.[1,4] Equations (9.11)–(9.13) all hold here.

References

1. Hill, T. L. (1977) *Free Energy Transduction in Biology* (Academic, New York).
2. Hill, T. L. (1982) Nature **299**, 84.
3. Hill, T. L. (1983) Proc. Natl. Acad. Sci USA **80**, 2589.
4. King, E. L. and Altman, C. (1956) J. Phys. Chem. **60**, 1375.

4

Equilibrium Systems without Explicit Interactions

Chapters 1 and 2 introduced the statistical mechanical fundamentals of equilibrium binding. Here the emphasis is more practical: we employ the grand partition function ξ (Chapter 2) for numerous simple binding models in order to indoctrinate the reader in the use of this very simple but powerful device. In this chapter we treat models without explicit interactions; Chapter 6 is a continuation of this discussion, where intermolecular interactions are introduced explicitly. However, we do include in this chapter a discussion of two limiting cases in which interactions are implicit and extreme: the A. V. Hill binding equation, and the Monod–Wyman–Changeux "concerted" conformation binding model.

The naturalness of Gibbs' grand partition function for binding problems in biology is evidenced by the rediscovery of what is essentially the grand partition function for this particular type of problem by various physical biochemists, including E. Q. Adams, G. S. Adair, H. S. Simms, K. Linderstrom-Lang, and, especially, J. Wyman. These treatments, however, were empirical or thermodynamic in content, that is, expressed from the outset in terms of thermodynamic equilibrium constants. The advantage of the explicit use of the actual grand partition function is that it is more general: it includes everything in the empirical or thermodynamic approach, plus providing, when needed, the background molecular theory (as statistical mechanics always does). This latter point is especially important in Chapter 6, where intermolecular interactions appear explicitly in the true grand partition function but are embedded and lost sight of in equilibrium constants in the empirical formulation. In this connection, it should be mentioned that throughout this book, the emphasis is on explicit molecular models (equilibrium or steady state) and on the corresponding molecular theory. An exception is Section 15 at the end of this chapter.

11. Single Conformation Binding Models

The three sections that follow this one relate to macromolecular subunits with two or more conformations. As a preliminary, the models treated in this section are all one-conformation models.

Binding on a Single Site (Review)

As we have already seen [Eq. (6.16)], if a single ligand L at concentration c can bind on a single site on the host molecule P, with binding constant K, then the grand partition function is

$$\underset{\bigcirc}{\xi = 1} + \underset{\text{Ⓛ}}{Kc}. \tag{11.1}$$

The individual terms in ξ always give the relative weights of the states at equilibrium. Thus the fractional occupation of the site by L is

$$\theta = Kc/(1 + Kc). \tag{11.2}$$

If two ligands L_1 and L_2 at concentrations c_1 and c_2 compete for one site [Eq. (7.3)], the grand partition function is

$$\xi = \underset{\bigcirc}{1} + \underset{\text{①}}{K_1c_1} + \underset{\text{②}}{K_2c_2}. \tag{11.3}$$

There is one term in ξ for each possible state of the site. Then the fractional binding of L_i on the site is

$$\theta_i = K_ic_i/\xi \qquad (i = 1, 2). \tag{11.4}$$

Binding on Two Sites

If P has two independent and different sites, one for L_1 and one for L_2, then

$$\xi = \underset{\boxed{}}{1} + \underset{\boxed{1}}{K_1c_1} + \underset{\boxed{2}}{K_2c_2} + \underset{\boxed{1\ 2}}{K_1c_1 \cdot K_2c_2} \tag{11.5}$$

$$= (1 + K_1c_1)(1 + K_2c_2). \tag{11.6}$$

In this case the pair of sites has four states, and there are four terms in ξ. We can verify this, in more detail, starting with Eq. (2.9) with $N_1, N_2 = 0,0$; $1,0$; $0,1$; $1,1$ [compare Eq. (7.2)]:

$$\xi = q_P + q_{L_1P}\lambda_1 + q_{PL_2}\lambda_2 + q_{L_1PL_2}\lambda_1\lambda_2. \tag{11.7}$$

Using

$$q_1 \equiv \frac{q_{L_1P}}{q_P}, \qquad q_2 \equiv \frac{q_{PL_2}}{q_P}, \qquad q_1q_2 = \frac{q_{L_1PL_2}}{q_P}, \tag{11.8}$$

Eq. (11.7) becomes

$$\xi = q_P(1 + q_1\lambda_1 + q_2\lambda_2 + q_1q_2\lambda_1\lambda_2). \tag{11.9}$$

Equation (11.9) is the same as Eq. (11.5) if we omit q_P as usual [see Eq. (7.2)]. The last relation in Eq. (11.8) deserves further comment: using the definitions of q_1 and q_2 in the same equation, this relation is equivalent to

$$q_{L_1P}q_{PL_2} = q_{L_1PL_2}q_P, \tag{11.10}$$

which is a way of stating that the binding of L_1 or L_2 on P is independent of the binding of the other. The nonindependent case is treated in Chapter 6.

The fractional binding of, say, L_1 on its site follows from Eq. (11.5):

$$\theta_1 = \frac{K_1c_1 + K_1c_1 \cdot K_2c_2}{\xi} = \frac{K_1c_1}{1 + K_1c_1}. \tag{11.11}$$

This result is, of course, to be expected because of the assumed independent binding. Equation (11.11) also follows from Eq. (7.5).

Because the binding in this example is independent, it would, of course, have been simpler to treat each site separately, as in Eq. (11.1), from the outset.

Binding on the Ligand

In this example we have "piggy-back" binding. That is, L binds on a site on P and R binds on a site on L. The concentrations of free L and R in solution are c_L and c_R. Because of the binding equilibrium of R on L, the chemical potentials of R, L, and LR in solution are related by

$$\mu_R + \mu_L = \mu_{LR}, \qquad \lambda_R\lambda_L = \lambda_{LR}. \tag{11.12}$$

We consider the grand partition function for P and its binding site. The site may be empty, may be occupied by L, or may be occupied by LR. That is, L and LR compete for the site, but L and LR are in equilibrium [Eq. (8.37) is somewhat analogous]. Thus

$$\begin{aligned}\xi &= q_P + q_{PL}\lambda_L + q_{PLR}\lambda_{LR} \\ &= q_P\left(1 + \frac{q_{PL}}{q_P}\lambda_L + \frac{q_{PLR}}{q_P}\lambda_L\lambda_R\right).\end{aligned} \tag{11.13}$$

As in Eqs. (6.12) and (6.13),

$$\frac{q_{PL}}{q_P}\lambda_L = \frac{q_{PL}c_L}{q_P(q_L/V)} = K_Lc_L, \tag{11.14}$$

where K_L is the binding constant of L on P. Similarly,

$$\frac{q_{PLR}}{q_{PL}}\lambda_R = K_R^{(PL)}c_R, \tag{11.15}$$

where $K_R^{(PL)}$ is the binding constant of R on PL. If we substitute Eqs. (11.14) and (11.15) in Eq. (11.13), and drop the factor q_P as usual, we have

$$\underset{\text{P}}{\xi = 1} + \underset{\text{PL}}{K_L c_L}(1 + \underset{\text{PLR}}{K_R^{(PL)} c_R}). \tag{11.16}$$

Actually, Eq. (11.16) could have been written immediately if we consider that the three states of P, namely, P, PL, and PLR, are generated, first, by binding L on P to form PL and then by binding R on PL to form PLR. At each stage we introduce the by now familiar combination $1 + Kc$. Thus, if there were three stages of this type:

$$\xi = 1 + K_1c_1[1 + K_2c_2(1 + K_3c_3)]. \tag{11.17}$$

Here, 1 binds on the P site, 2 binds on 1, and 3 binds on 2. The concentrations c_1, c_2, and c_3 refer to the free (uncombined) species 1, 2, 3 in solution.

If $K_R^{(L)}$ is the binding constant of R on L and K_{LR} is the binding constant of LR on P, then the four binding constants that have been introduced are interrelated by $K_L K_R^{(PL)} = K_R^{(L)} K_{LR}$. The two sides of this equation represent two different sequences in putting L and R on P.

The three terms in Eq. (11.16) give the relative equilibrium weights, as always. Thus the fraction of P sites on which L is bound is

$$\theta_L = K_L c_L(1 + K_R^{(PL)} c_R)/\xi, \tag{11.18}$$

and the fraction on which R is bound is

$$\theta_R = K_L c_L K_R^{(PL)} c_R/\xi. \tag{11.19}$$

The fraction of empty sites is $1/\xi$.

Also, as in Eq. (11.2), the fractional binding of R on L to form LR is

$$\theta = K_R^{(L)} c_R/(1 + K_R^{(L)} c_R). \tag{11.20}$$

This does not follow from Eq. (11.16), however. Equation (11.16) is the grand partition function for one P and its site; P is not involved in Eq. (11.20).

One Ligand on Multiple Sites

Suppose P contains m independent sites for binding L, which is at concentration c in solution. The sites are labelled $i = 1, 2, \ldots, m$; the binding constant for the ith site is K_i. In general, this is "heterogeneous" binding, because the Ks can be different. The grand partition function for P is then

$$\xi = \prod_{i=1}^{m} (1 + K_i c). \tag{11.21}$$

That is, there is a factor $1 + K_i c$ for each site. Simple multiplication of these factors reflects the independence of the sites, just as in Eq. (11.6). This can be confirmed by using Eq. (1.11):

$$\bar{s} = \lambda \frac{\partial \ln \xi}{\partial \lambda} = c \frac{\partial \ln \xi}{\partial c}$$

$$= c \frac{\partial}{\partial c} \sum_i \ln(1 + K_i c)$$

$$= \sum_i \frac{K_i c}{1 + K_i c} = \sum_i \theta_i, \tag{11.22}$$

where $\bar{s}$ is the mean number of occupied sites per P molecule and θ_i is the fractional occupation of sites of type i. When $c \to \infty$, $\bar{s} \to m$, the number of sites per P. Equation (11.22) is simply the sum of terms of the type given in Eq. (11.2). This is just what one would expect for a collection of independent binding sites. From Eq. (1.14), the variance in s, the number of L molecules on P, is

$$\sigma_m^2 = \overline{s^2} - \bar{s}^2 = c \frac{\partial \bar{s}}{\partial c} = \sum_i \theta_i (1 - \theta_i). \tag{11.23}$$

This is a generalization of Eq. (6.20).

Two Ligands on a Pair of Sites

In this case, each P has a pair of equivalent sites that can be occupied by two competing ligands, 1 and 2, which are at concentrations c_1 and c_2. Ligand 2 is larger and can occupy either one site (K_2) or both sites (K_2'). Ligand 1 occupies only one site. Binding on the two sites is independent (except for K_2'). The 10 possible states of P are shown in Fig. 4-1, together with the corresponding terms in ξ. Thus

$$\xi = (1 + K_1 c_1 + K_2 c_2)^2 + K_2' c_2. \tag{11.24}$$

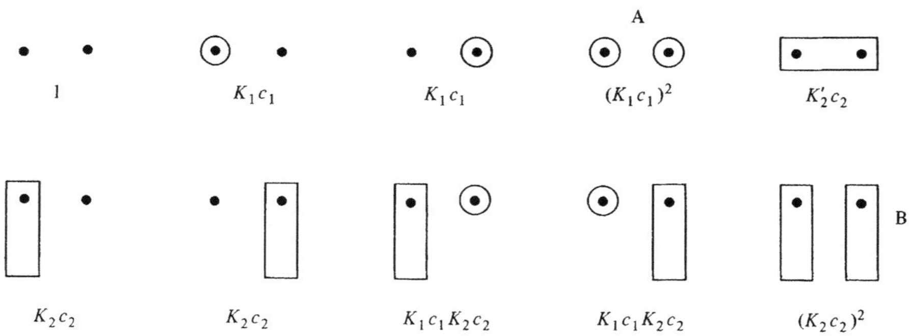

Fig. 4-1. Ten possible states of a protein molecule P with two binding sites for two competing ligands of different sizes. States A and B are mentioned in the text.

This is the grand partition function for two independent and equivalent sites, each site with competitive binding and three states, plus an additional state for double binding (K_2'). This model is suggested by the competitive binding of two myosin fragments on actin (HMM with two "heads" for attachment to actin and S1 with one head). See Fig. 8-9 and Section 35 for a more detailed treatment (with different notation).

If $\bar{s}_i$ ($i = 1, 2$) is the mean number of sites occupied by species i, we can find $\bar{s}_i$ using the weights in Fig. 4-1, together with an extra factor of two for states A and B, because two ligand molecules of the same type are bound in these states. Alternatively, one can obtain the same results from

$$\bar{s}_1 = c_1 \frac{\partial \ln \xi}{\partial c_1} = \frac{2(1 + K_1c_1 + K_2c_2)K_1c_1}{\xi}, \tag{11.25}$$

$$\bar{s}_2 = c_2 \frac{\partial \ln \xi}{\partial c_2} = \frac{2(1 + K_1c_1 + K_2c_2)K_2c_2 + K_2'c_2}{\xi}. \tag{11.26}$$

If c_2 is finite and $c_1 \to \infty$, the c_1^2 terms dominate and $\bar{s}_1 \to 2$, $\bar{s}_2 \to 0$. Conversely, if c_1 is finite and $c_2 \to \infty$, the c_2^2 terms dominate and $\bar{s}_2 \to 2$, $\bar{s}_1 \to 0$. That is, state A or state B, in Fig. 4-1, dominates in the respective cases.

Equivalent Sites or Equivalent Subunits

If P has m equivalent and independent sites for binding L (concentration c, binding constant K), Eq. (11.21) simplifies to

$$\xi = (1 + Kc)^m. \tag{11.27}$$

Alternatively, if P is comprised of m equivalent and independent subunits, each with one binding site for L, then Eq. (11.27) is again the grand partition function for the binding of L on P. In either case, the binding is on m sites and the extent of binding at c is

$$\bar{s} = c\frac{\partial \ln \xi}{\partial c} = \frac{mKc}{1 + Kc} = m\theta. \tag{11.28}$$

Although the context is different, Eq. (11.28) is essentially the same as Eq. (6.26). The maximum value of $\bar{s}$ is m (when $c \to \infty$). The variance in s is

$$\sigma_m^2 = \overline{s^2} - \bar{s}^2 = c\frac{\partial \bar{s}}{\partial c} = m\theta(1 - \theta). \tag{11.29}$$

The grand partition function, Eq. (11.27), for $m = 4$ subunits, each with one site for binding L, is illustrated in Fig. 4-2. The integers, 1, 4, 6, 4, 1 are degeneracies (e.g., in the term $4Kc$, the single L can be located on any one of four subunits).

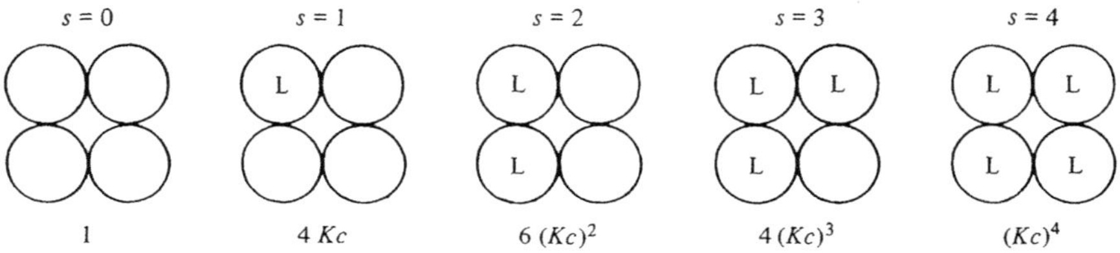

Fig. 4-2. Grand partition function terms for four subunits, each with one binding site for L.

The A. V. Hill Binding Equation

In the preceding subsection, the binding isotherm is the Langmuir equation, $\theta = Kc/(1 + Kc)$ [Eq. (11.28)] for any number of sites m, because the sites bind independently. If we plot θ against $\ln Kc$ (Fig. 4-3), the curve is symmetrical about $\theta = \frac{1}{2}$ and $\ln Kc = 0$. To be more precise, $\theta - \frac{1}{2}$ is an odd function of $\ln Kc$. The odd function property can be confirmed after putting $Kc = e^{\ln Kc}$ in the expression for $\theta - \frac{1}{2}$.

For a number of binding models with positive or negative cooperativity (Chapter 6), the curve θ vs $\ln c$ has the same kind of symmetry but is more or less steep than $\theta = Kc/(1 + Kc)$ at the symmetry point ($\theta = \frac{1}{2}$). Symmetry *per se* is discussed in Section 15. In the limit of extreme positive cooperativity for such models, the only terms that remain in ξ [Eq. (11.27)] are 1 and $K^m c^m$. That is, either all sites are empty or all sites are occupied, with no intermediate configurations (see Fig. 4-2). This resembles a phase transition. This extreme model of positive cooperativity was first introduced

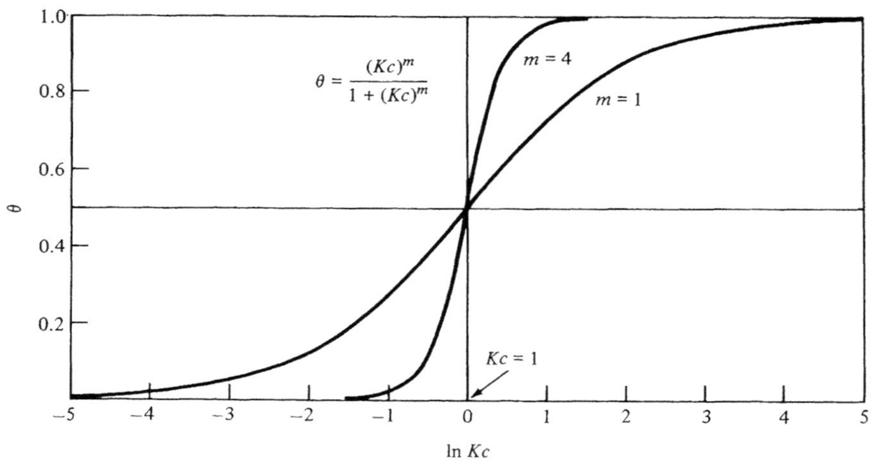

Fig. 4-3. The Langmuir binding isotherm ($m = 1$) and the A. V. Hill binding isotherm with $m = 4$.

by A. V. Hill in 1913. Unfortunately, it is still used by some scientists as a *general* empirical model of cooperativity in binding systems, without realizing that it is anything but general: it is valid only in a very limiting case. Even worse, this very limited *equilibrium* model is sometimes applied to steady-state enzyme kinetics.

The grand partition function and binding isotherm for the A. V. Hill model are

$$\xi = 1 + K^m c^m, \qquad \theta = \frac{K^m c^m}{1 + K^m c^m}. \tag{11.30}$$

An explicit model that could lead to these equations is discussed in connection with Eq. (14.10). The symmetry of $\theta(\ln c)$ here is the same as for Eq. (11.28), as illustrated in Fig. 4-3 for $m = 4$. Using $\bar{s} = m\theta$, we find for the variance in s,

$$\sigma_m^2 = \overline{s^2} - \bar{s}^2 = c\frac{\partial \bar{s}}{\partial c} = m^2\theta(1 - \theta). \tag{11.31}$$

This is larger than in Eq. (11.29) by a factor of m: the fluctuations in s are extreme, between $s = 0$ (probability $1 - \theta$) and $s = m$ (probability θ). Because $\bar{s} = m\theta$ in both cases, we note from Eqs. (11.29) and (11.31) that

$$\text{Langmuir:} \quad \frac{\partial \theta}{\partial \ln c} = \theta(1 - \theta) \qquad (\text{any } m), \tag{11.32}$$

$$\text{A. V. Hill:} \quad \frac{\partial \theta}{\partial \ln c} = m\theta(1 - \theta). \tag{11.33}$$

Both of these slopes (and hence also the variances in s) have a maximum at $\theta = \frac{1}{2}$, as is evident from Fig. 4-3:

$$\text{Langmuir:} \quad \left(\frac{\partial \theta}{\partial \ln c}\right)_{\theta=1/2} = \frac{1}{4} \qquad (\text{any } m) \tag{11.34}$$

$$\text{A. V. Hill:} \quad \left(\frac{\partial \theta}{\partial \ln c}\right)_{\theta=1/2} = \frac{m}{4}. \tag{11.35}$$

The Hill plot,

$$\ln[\theta/(1 - \theta)] \text{ vs } \ln c,$$

is much used, especially for experimental data with the symmetry of Fig. 4-3 (see Section 15 in this connection). This plot was originally suggested by Eq. (11.30), because it can be put in the form

$$K^m c^m = \theta/(1 - \theta). \tag{11.36}$$

Although this particular equation is not generally useful, even misleading if its limitations are not understood, the slope of the Hill plot,

$$\frac{\partial \ln[\theta/(1-\theta)]}{\partial \ln c} \tag{11.37}$$

is definitely worth examining. It should be emphasized, however, that this slope, obtained say from experimental data, has its own intrinsic interest and actually has nothing to do with the A. V. Hill equation, Eq. (11.36). This slope is in fact closely related to the simpler derivative $\partial\theta/\partial \ln c$, which in turn is related to fluctuations in the extent of binding [Eq. (1.14)]. That is,

$$\begin{aligned}\frac{\partial \ln[\theta/(1-\theta)]}{\partial \ln c} &= \frac{\partial \ln \theta}{\partial \ln c} - \frac{\partial \ln(1-\theta)}{\partial \ln c} \\ &= \frac{1}{\theta(1-\theta)}\frac{\partial\theta}{\partial \ln c}.\end{aligned} \tag{11.38}$$

At the symmetry point ($\theta = \frac{1}{2}$) of a symmetrical $\theta(\ln c)$ curve,

$$n \equiv \left(\frac{\partial \ln[\theta/(1-\theta)]}{\partial \ln c}\right)_{\theta=1/2} = 4\left(\frac{\partial\theta}{\partial \ln c}\right)_{\theta=1/2}, \tag{11.39}$$

where n is generally called the Hill coefficient.

What the above discussion suggests is that, in the treatment of *symmetrical* equilibrium binding data (see Section 15) with possible cooperativity, the Hill equation and Hill plot should be abandoned, and the simpler and more fundamental plot of θ vs $\ln c$ should be used in their place. The Hill coefficient is then defined by

$$n \equiv 4\left(\frac{\partial\theta}{\partial \ln c}\right)_{\theta=1/2} \tag{11.40}$$

and is still a very useful index of the degree of cooperativity in simple binding systems (but see Section 24). Furthermore, n has the physical significance of

$$n = 4\sigma_m^2/m \qquad (\theta = 1/2), \tag{11.41}$$

where σ_m^2 is the variance in s (number of ligands bound) at $\theta = \frac{1}{2}$ in a system with m binding sites. If n is evaluated experimentally [Eq. (11.40)] and m is known, then $\sigma_m^2 = nm/4$ at $\theta = \frac{1}{2}$.

The definition of n for *unsymmetrical* systems is discussed in connection with Eq. (12.44).

In the independent (Langmuir) binding case, Eq. (11.34), $n = 1$ (no cooperativity). In the case of extreme positive cooperativity (A. V. Hill), Eq. (11.35), $n = m$. That is, m is the maximum possible value of n or, conversely, m is at least as large as n. For real systems with positive cooperativity, $1 < n < m$; for systems with negative cooperativity, or in some cases with heterogeneous binding [Eq. (11.21)], $0 < n < 1$. We shall return to this topic in Chapter 6.

12. Two Conformations Modulated by Length-Tension and Ligand Binding

We consider here and in Section 13 a collection of M independent and distinguishable macromolecular units, each of which can be in two different conformations or states, 1 and 2, with different properties. The partition function for one unit in conformation 1 is j_1 and for one unit in conformation 2 is j_2. These quantities are functions of T. The equilibrium constant for $1 \rightleftarrows 2$ is $K = j_2/j_1$ (dimensionless). This simple model was introduced in Eqs. (7.7)–(7.10) with different notation. In this section, we study a more complicated version of this model in which the M units occur in a linear chain, and each unit has a length l_1 or l_2, depending on the conformation or state. We take $l_2 > l_1$. If the chain of units is placed under a positive tension τ, tending to increase its length, the equilibrium $1 \rightleftarrows 2$ will be shifted toward the longer state 2. If, in addition, each unit has m equivalent and independent sites for binding a ligand L from solution, and if the binding constants K_1 and K_2 are different in the two conformations, then the $1 \rightleftarrows 2$ equilibrium can also be modulated by ligand binding. Binding of ligand will shift the equilibrium toward the state with the higher binding constant.

The above-described model is of interest for biopolymers like collagen.

A similar problem is considered in Section 13: the M units are in an electric field, for example, in a membrane; and the two states of a unit have different polarizabilities so that the $1 \rightleftarrows 2$ equilibrium can be shifted either by the electric field or by ligand binding.

In either of the two systems above (units under tension or in an electric field), if the $1 \rightleftarrows 2$ modulation is limited to binding effects only, we might anticipate from the results in Section 11 that the partition function $j_1 + j_2$ [Eq. (7.7)] in the absence of ligand would become

$$j_1(1 + K_1c)^m + j_2(1 + K_2c)^m \tag{12.1}$$

in the presence of ligand at concentration c. This is indeed the case, but we provide a proof here. The proof is hardly more complicated if we include the length-tension (or electric field) aspect already mentioned, so we return now to the length-tension problem.

To begin with, we omit binding of ligand. If the number of units in state 2 is M_2, then the total length of the chain is

$$l = l_1(M - M_2) + l_2M_2. \tag{12.2}$$

The basic thermodynamic equation for the chain of M units is

$$dE = TdS + \tau dl + \mu dM. \tag{12.3}$$

With M constant, in Eq. (12.2), l can vary only by changing M_2 (l_1 and l_2 are considered constants). For our statistical mechanical purposes, it is convenient to change independent variables from l to M_2. On using Eq. (12.2) to eliminate dl from Eq. (12.3), we find

$$dE = TdS + \tau(l_2 - l_1)dM_2 + (\mu + \tau l_1)dM, \tag{12.4}$$

$$dA = -SdT + \tau(l_2 - l_1)dM_2 + (\mu + \tau l_1)dM, \tag{12.5}$$

where $A = E - TS$. The free energy A is a function of T, M_2, and M. The corresponding partition function for the chain of M units is

$$Q(M_2, M, T) = \frac{M!\, j_1^{M-M_2} j_2^{M_2}}{M_2!\,(M - M_2)!}, \tag{12.6}$$

$$A = -kT \ln Q. \tag{12.7}$$

A more useful partition function is one that is "open" with respect to M_2, i.e., one which allows fluctuations in M_2. For this we need to exchange the roles of τ and M_2 in Eq. (12.5):

$$G \equiv A - \tau(l_2 - l_1)\overline{M}_2, \tag{12.8}$$

$$dG = -SdT - \overline{M}_2(l_2 - l_1)d\tau + (\mu + \tau l_1)\, dM, \tag{12.9}$$

$$G = -kT \ln Y, \tag{12.10}$$

$$Y(\tau, M, T) = \sum_{M_2=0}^{M} Q(M_2, M, T)\eta^{M_2}, \tag{12.11}$$

where

$$\eta \equiv e^{\tau(l_2 - l_1)/kT}. \tag{12.12}$$

The summand used in Eq. (12.11) follows from Eq. (12.8) (to see this, put $Q = e^{-A/kT}$). The factor η^{M_2} is the analogue of λ^N in a grand partition function. If we now substitute Eq. (12.6) in Eq. (12.11), the sum is easy to carry out:

$$Y = j_1^M \sum_{M_2=0}^{M} \frac{M!\,(j_2\eta/j_1)^{M_2}}{M_2!\,(M - M_2)!} \tag{12.13}$$

$$= j_1^M\left(1 + \frac{j_2\eta}{j_1}\right)^M \tag{12.14}$$

$$= (j_1 + j_2\eta)^M. \tag{12.15}$$

Equation (12.14) can also be written in the form

$$Y = j_1^M y^M, \tag{12.16}$$

$$y = 1 + K\eta, \qquad K = j_2/j_1, \tag{12.17}$$

where y resembles a grand partition function for one unit. The first term in y (i.e., 1) is the weight for state 1, and the second term ($K\eta$) is the weight for state 2. This assignment can be confirmed formally as follows.

From Eqs. (12.9) and (12.10),

$$\overline{M}_2(l_2 - l_1) = kT\left(\frac{\partial \ln Y}{\partial \tau}\right)_{T,M}. \tag{12.18}$$

Also, Eq. (12.12) gives

$$\frac{d\eta}{\eta} = \frac{(l_2 - l_1)\,d\tau}{kT} \qquad (T \text{ constant}). \tag{12.19}$$

Then, from Eqs. (12.16), (12.18), and (12.19),

$$\overline{M}_2 = \eta\left(\frac{\partial \ln Y}{\partial \eta}\right)_{T,M} = M\eta\left(\frac{\partial \ln y}{\partial \eta}\right)_T, \tag{12.20}$$

$$\delta \equiv \frac{\overline{M}_2}{M} = \eta\left(\frac{\partial \ln y}{\partial \eta}\right)_T = \frac{K\eta}{1 + K\eta}, \tag{12.21}$$

where δ is the mean fraction of units in state 2. This expression for δ confirms the assertion above that the terms 1 and $K\eta$ in y are the respective equilibrium weights for states 1 and 2. The formal resemblance of Eq. (12.21) to Eqs. (6.14) and (6.18) is striking. Of course the K here is a dimensionless equilibrium constant, not a binding constant.

If $\tau = 0$, then $\eta = 1$ and $\delta = K/(1 + K)$, as in Eq. (7.9). When $\tau > 0$, $\eta > 1$, and δ is increased: the equilibrium is shifted toward the longer state 2, as expected, when the chain is under tension.

Introduction of Ligand Binding

We now assume that each unit has m independent and equivalent sites for binding a ligand L, with binding constants K_1 and K_2 in states 1 and 2. The concentration of free L is c, its chemical potential is μ_L, and $\lambda_L \equiv e^{\mu_L/kT}$. The partition functions for a bound L are, as usual [Eq. (6.4)], denoted by q_1 and q_2. Then $q_1\lambda_L = K_1 c$ and $q_2\lambda_L = K_2 c$ [Eq. (6.12)].

The chain of M units now contains another component (L). The generalization of Eq. (12.5) for this system is

$$dA = -S\,dT + \tau(l_2 - l_1)dM_2 + (\mu + \tau l_1)dM + \mu_L dN, \tag{12.22}$$

where N is the number of L molecules bound on the total of mM sites ($N \leq mM$). These N molecules will be distributed between units in states 1 and 2, with $N_2 \leq mM_2$ on state 2 units and $N - N_2 \leq m(M - M_2)$ on state 1 units. For given M_2 and N in Eq. (12.22), N_2 fluctuates. This is reflected in the Q that corresponds to Eq. (12.22), by summing over N_2:

$$A = -kT \ln Q, \tag{12.23}$$

$$Q(M_2, M, N, T) = \frac{M!\, j_1^{M-M_2} j_2^{M_2}}{M_2!\,(M - M_2)!} \times \sum_{N_2=0}^{mM_2} \frac{[m(M - M_2)]!\, q_1^{N-N_2}}{(N - N_2)!\,[m(M - M_2) - (N - N_2)]!} \cdot \frac{(mM_2)!\, q_2^{N_2}}{N_2!\,(mM_2 - N_2)!}. \tag{12.24}$$

It is much easier to deal with the grand partition function for this system than with Q: the system is considered open with respect to both M_2 and N. This is in fact the actual experimental situation when τ and μ_L (i.e., c) are fixed: M_2 (number of units in state 2) and N (number of bound ligand molecules) are free to fluctuate. The basic equations are then

$$G' \equiv A - \tau(l_2 - l_1)\overline{M}_2 - \overline{N}\mu_L, \tag{12.25}$$

$$dG' = -SdT - \overline{M}_2(l_2 - l_1)d\tau + (\mu + \tau l_1)dM - \overline{N}d\mu_L, \tag{12.26}$$

$$G' = -kT \ln \Xi, \tag{12.27}$$

$$\Xi(\tau, M, \mu_L, T) = \sum_{N=0}^{mM} \sum_{M_2=0}^{M} Q(M_2, M, N, T)\eta^{M_2}\lambda_L^N. \tag{12.28}$$

It is simpler to sum first over N in Eq. (12.28) for a fixed value of M_2. Because Q already includes a sum over N_2, the two sums over N_2 and N are equivalent to two sums over $N_1 = N - N_2$ and N_2, where $N_1 \leq m(M - M_2)$ and $N_2 \leq mM_2$. If we write $\lambda_L^N = \lambda_L^{N_1+N_2}$, we obtain, after the N_1 and N_2 sums,

$$\Xi = \sum_{M_2} \frac{M!\, j_1^{M-M_2} j_2^{M_2} \eta^{M_2}}{M_2!\,(M - M_2)!}(1 + q_1\lambda_L)^{m(M-M_2)}(1 + q_2\lambda_L)^{mM_2}. \tag{12.29}$$

The sum over M_2 is now immediate:

$$\Xi = [j_1(1 + q_1\lambda_L)^m]^M \left[1 + \frac{j_2\eta(1 + q_2\lambda_L)^m}{j_1(1 + q_1\lambda_L)^m}\right]^M \tag{12.30}$$

$$= [j_1(1 + K_1c)^m + j_2\eta(1 + K_2c)^m]^M \tag{12.31}$$

$$= j_1^M \xi^M, \tag{12.32}$$

where

$$\xi = (1 + K_1c)^m + K\eta(1 + K_2c)^m. \tag{12.33}$$

This confirms Eq. (12.1) and is the generalization of Eq. (12.17) that is to be expected. The fraction of units in state 2 is then

$$\delta = \frac{\overline{M}_2}{M} = \frac{K\eta(1 + K_2c)^m}{\xi}. \tag{12.34}$$

We see that δ is increased by τ and also by c if $K_2 > K_1$.

Of course the most common case has nothing to do with an elastic chain of units. If we have any collection of independent macromolecules, each of which can exist in two conformations and has m independent and equivalent binding sites for a ligand, then Eq. (12.34) applies with $\eta = 1$:

$$\delta = \frac{K(1 + K_2c)^m}{(1 + K_1c)^m + K(1 + K_2c)^m}. \tag{12.35}$$

The ratio of state 2 to state 1 units is

$$\frac{\delta}{1-\delta} = \frac{K(1 + K_2 c)^m}{(1 + K_1 c)^m}. \tag{12.36}$$

The right-hand side is an effective equilibrium constant for $1 \rightleftarrows 2$: K is corrected for the effect of ligand at c. The units are equally divided between states 1 and 2 ($\delta = \frac{1}{2}$) at $c = c'$, where

$$c' = \frac{1 - K^{1/m}}{K_2 K^{1/m} - K_1}. \tag{12.37}$$

The terms in Eq. (12.33) are illustrated in Fig. 4-4 for $m = 2$, $\eta = 1$. It will be seen from this figure that ξ could have been written immediately without reliance on our rather complicated (but reassuring) derivation.

Equation (12.34) follows in a more formal way from

$$\overline{M}_2(l_2 - l_1) = kT\left(\frac{\partial \ln \Xi}{\partial \tau}\right)_{T,M,\mu_L}, \tag{12.38}$$

or

$$\delta = \eta\left(\frac{\partial \ln \xi}{\partial \eta}\right)_{T,c}. \tag{12.39}$$

Similarly, the amount of ligand binding is given by

$$\overline{N} = kT\left(\frac{\partial \ln \Xi}{\partial \mu_L}\right)_{T,M,\tau}, \tag{12.40}$$

$$\bar{s} = \frac{\overline{N}}{M} = m\theta = c\left(\frac{\partial \ln \xi}{\partial c}\right)_{T,\eta}. \tag{12.41}$$

From Eq. (12.33), we then find

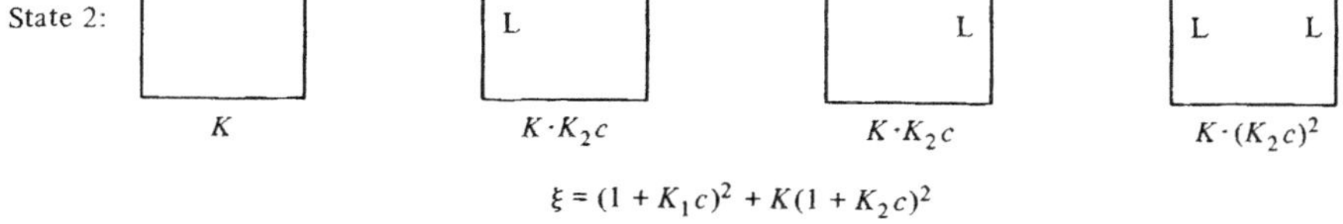

Fig. 4-4. States and grand partition function of a macromolecule with two different conformations and with two binding sites for a ligand L. This is the case $m = 2$, $\eta = 1$ (see text).

$$\theta = \frac{(1 + K_1c)^{m-1}K_1c + K\eta(1 + K_2c)^{m-1}K_2c}{\xi}$$

$$= (1 - \delta)\theta_1 + \delta\theta_2, \tag{12.42}$$

where

$$\theta_1 = \frac{K_1c}{1 + K_1c}, \qquad \theta_2 = \frac{K_2c}{1 + K_2c}. \tag{12.43}$$

Equation (12.42) is what one would expect intuitively.

It is worth noting that even if one is interested only in the $\eta = 1$ case, it is convenient to have $\eta \neq 1$ properties available in order to use Eq. (12.39) to find δ [compare Eq. (7.10)].

The above results were first obtained[1] in 1952 and rederived in 1960 (Ref. 2, pp. 140–143). The same formal results (in the $\eta = 1$ special case) were also found[3] in 1965 by Monod, Wyman, and Changeux, but for a conceptually quite different model. In the present case we have a collection of two-conformation units, each with m binding sites for a ligand. There is no interactive cooperativity at all in this sytem. In the MWC model, a unit is comprised of m identical two-conformation subunits, each with one binding site for a ligand. This system has the same formal properties as above if and only if we invoke extreme positive cooperativity among the m subunits (as MWC did) so that the subunits are either all in conformation 1 or all in conformation 2, with no mixed forms allowed. We shall return to this topic in Section 14 and in Chapter 6.

Numerical Example

To illustrate Eqs. (12.35) and (12.42), we use a model suggested by the binding of the myosin fragment S1 on actin sites of tropomyosin-troponin-actin ("regulated" actin). In effect, regulated actin is a linear chain of two-conformation units, each unit of which contains $m = 7$ sites for the binding of S1. There is no elasticity or length change involved ($\eta = 1$). In the absence of Ca^{2+}, rather realistic parameters are $K = 0.02$, $K_1 = 0.004\ \mu M^{-1}$, and $K_2 = 1.0\ \mu M^{-1}$, with c in units of μM. The real system has strong interactive cooperativity (Section 36) between units (not within a unit), but we omit this feature here. In Fig. 4-5, we show δ, θ, and θ_2 as functions of $\ln K_2c$. The $\theta_2(\ln K_2c)$ curve [Eq. (12.43)] is symmetrical about $\theta = \frac{1}{2}$, $K_2c = 0$, with a maximum slope of 0.25 at the symmetry point [Eq. (11.34)]. The θ and δ curves are unsymmetrical and steeper, with maximum slopes 0.53 and 0.76, respectively. There is no interactive cooperativity in this model, but there is an effective "group" cooperativity contributing to these larger slopes: $m = 7$ binding sites change state simultaneously when $1 \rightleftarrows 2$.

At low c, $\delta \rightarrow 0.02/1.02$. That is, state 1 is much more stable and highly favored in the absence of ligand; but $K_2/K_1 = 250$, so an increase in c shifts the equilibrium toward state 2. The value of c at which δ reaches the value $\delta = \frac{1}{2}$ is $c' = 0.754\ \mu M$.

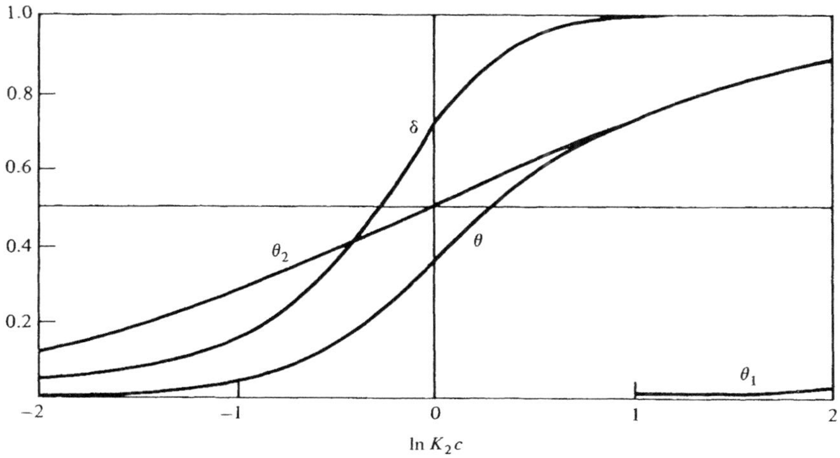

Fig. 4-5. Numerical illustration of Eqs. (12.35) and (12.42) with parameters $\eta = 1$, $m = 7$, $K = 0.02$, $K_1 = 0.004\ \mu\text{M}^{-1}$, $K_2 = 1.0\ \mu\text{M}^{-1}$, and c in units of μM. See the text for further details.

Extension of the Hill Coefficient Definition

As a useful comparative index of cooperativity, the definition of n in Eq. (11.40) can be extended to include unsymmetrical binding systems (this section is an example) by writing

$$n \equiv 4\left(\frac{\partial \theta}{\partial \ln c}\right)_{\substack{\max \\ \min}}. \tag{12.44}$$

That is, the inflection point in the $\theta(\ln c)$ plot is used, even though it does not occur at $\theta = \frac{1}{2}$. In some cases, there may be several inflection points. In the above numerical example (Fig. 4-5), $n = 2.13$ at $\theta = 0.358$ and $\delta = 0.713$. Calculating n, for an unsymmetrical system, from the slope at $\theta = \frac{1}{2}$ is inappropriate because this slope is generally not an extremum slope and hence is not a suitable cooperativity index because of its arbitrariness: there is nothing special about $\theta = \frac{1}{2}$ in such a case.

13. Two Conformations Modulated by an Electric Field and Ligand Binding

The beginning of Section 12 provides an introduction to this section. We consider a collection of M independent units (e.g., channels in a membrane), each of which can exist in two states, 1 and 2, with different polarizabilities α_1 and α_2. There may be several kinds of contribution to the polarizabilities,[4] but we do not pursue this topic here. There are m binding sites for L on each unit, as in Section 12. However, we omit binding to begin with.

We shall treat the M units as immobile and distinguishable, but if they were mobile and indistinguishable (e.g., free to move about in a membrane), we would merely have to correct with an inconsequential partition function factor $M!$, just as in Eqs. (6.31) and (6.33) compared with Eqs. (6.22) and (6.23).

There is an electric field of strength $\mathscr{E}$ in a particular direction relative to the units (e.g., normal to the plane of the membrane). The sign of $\mathscr{E}$ can be positive or negative. A field $\mathscr{E}$ induces an electric moment $\alpha_i\mathscr{E}$ in a unit of type i ($i = 1, 2$). The work done on the unit to create the moment is $-\alpha_i\mathscr{E}^2/2$. This contributes $-\alpha_i\mathscr{E}^2/2$ to the internal energy of the unit (i.e., the energy is lowered). Hence the partition function j_i, in the absence of a field, becomes $j_i e^{\alpha_i\mathscr{E}^2/2kT}$ in the presence of a field (i.e., the new effective j_i is increased).

For simplicity, we assume that the two states of a unit have the same net charge and same permanent dipole moment. Hence there are no differential effects from these sources, and we omit such contributions from the partition functions (but see Section 24).

The fundamental thermodynamic equations for the system of M units, in the absence of ligand binding, are

$$dE = TdS - \bar{P}d\mathscr{E} + \mu dM, \tag{13.1}$$

$$dA = -SdT - \bar{P}d\mathscr{E} + \mu dM, \tag{13.2}$$

where $\bar{P}$ is the mean total polarization. Because of the equilibrium $1 \rightleftarrows 2$ within the M units, M_2 and P fluctuate. The relation between $\bar{P}$ and $\bar{M}_2$ is clearly

$$\begin{aligned}\bar{P} &= (M - \bar{M}_2)\alpha_1\mathscr{E} + \bar{M}_2\alpha_2\mathscr{E} \\ &= M[(1 - \delta)\alpha_1 + \delta\alpha_2]\mathscr{E}.\end{aligned} \tag{13.3}$$

The canonical partition function for this system is

$$A = -kT \ln Q, \tag{13.4}$$

$$Q = \sum_{M_2=0}^{M} \frac{M!\,(j_1 e^{\alpha_1\mathscr{E}^2/2kT})^{M-M_2}(j_2 e^{\alpha_2\mathscr{E}^2/2kT})^{M_2}}{M_2!\,(M - M_2)!} \tag{13.5}$$

$$= (j_1 e^{\alpha_1\mathscr{E}^2/2kT} + j_2 e^{\alpha_2\mathscr{E}^2/2kT})^M \equiv z^M. \tag{13.6}$$

This is the analogue of Eq. (12.15); in fact, Eq. (12.15) could have been put in the closer form

$$(j_1 e^{\tau l_1/kT} + j_2 e^{\tau l_2/kT})^M. \tag{13.7}$$

The total polarization, from Eqs. (13.2), (13.4), and (13.6), is

$$\bar{P} = kT\left(\frac{\partial \ln Q}{\partial \mathscr{E}}\right)_{T,M} = MkT\left(\frac{\partial \ln z}{\partial \mathscr{E}}\right)_T \tag{13.8}$$

$$= M\left(\frac{j_1 e^{\alpha_1\mathscr{E}^2/2kT}}{z}\cdot \alpha_1\mathscr{E} + \frac{j_2 e^{\alpha_2\mathscr{E}^2/2kT}}{z}\cdot \alpha_2\mathscr{E}\right). \tag{13.9}$$

Comparison of this result with Eq. (13.3) shows that

$$\delta = \frac{\overline{M}_2}{M} = \frac{j_2 e^{\alpha_2 \mathscr{E}^2/2kT}}{z}$$

$$= \frac{K e^{(\alpha_2 - \alpha_1)\mathscr{E}^2/2kT}}{1 + K e^{(\alpha_2 - \alpha_1)\mathscr{E}^2/2kT}}, \tag{13.10}$$

where $K = j_2/j_1$. This is the analogue of Eq. (12.21) and could have been written immediately from the terms in z in Eq. (13.6), without resorting to Eq. (13.8). Indeed, z could have been written without using Q. In this chapter, we have been starting with macroscopic partition functions in a variety of cases to convince the reader that it is equally legitimate, and of course much simpler, to employ appropriate partition functions for *single* units and to assign equilibrium occupation weights to the individual terms in these single-unit partition functions in order to deduce average properties of interest.

According to Eq. (13.10), the equilibrium between states 1 and 2 will be shifted toward state 2 by an electric field (of either sign) if state 2 is more polarizable ($\alpha_2 > \alpha_1$) than state 1.

Introduction of Ligand Binding

The argument here is very similar to that in Section 12, so we can be brief. The $N = N_1 + N_2$ bound ligand molecules are again distributed between $M - M_2$ units in state 1 and M_2 units in state 2. The relevant thermodynamic equations are

$$dA = -SdT - \overline{P}d\mathscr{E} + \mu dM + \mu_L dN, \tag{13.11}$$

$$d(A - \overline{N}\mu_2) = -SdT - \overline{P}d\mathscr{E} + \mu dM - \overline{N}d\mu_L. \tag{13.12}$$

The partition function Q, which is related to Eq. (13.11) by $A = -kT \ln Q$, is the product of the sum in Eq. (12.24) by the sum in Eq. (13.5). That is, this Q already contains summations over N_2 and M_2. The additional sum over N appears in Ξ:

$$A - \overline{N}\mu_L = -kT \ln \Xi, \tag{13.13}$$

$$\Xi(\mathscr{E}, M, \mu_L, T) = \sum_{N=0}^{mM} Q(\mathscr{E}, M, N, T)\lambda_L^N. \tag{13.14}$$

This is in fact the same triple sum as in Eq. (12.28), except that $\eta = 1$ and $j_i \rightarrow j_i e^{\alpha_i \mathscr{E}^2/2kT}$. Hence we have immediately

$$\Xi = \xi^M, \tag{13.15}$$

$$\xi = j_1 e^{\alpha_1 \mathscr{E}^2/2kT}(1 + K_1 c)^m + j_2 e^{\alpha_2 \mathscr{E}^2/2kT}(1 + K_2 c)^m. \tag{13.16}$$

This is the generalization of Eq. (13.6), to include binding, that one would expect.

The mean properties of this system are obviously [compare Eqs. (12.36) and (12.42)]

$$\frac{\delta}{1-\delta} = \frac{Ke^{(\alpha_2-\alpha_1)\mathscr{E}^2/2kT}(1 + K_2c)^m}{(1 + K_1c)^m}, \tag{13.17}$$

$$\theta = (1-\delta)\theta_1 + \delta\theta_2. \tag{13.18}$$

These relations can be confirmed by using Eqs. (13.12), (13.13), and (13.16). The effective equilibrium constant for $1 \rightleftarrows 2$ [right-hand side of Eq. (13.17)] depends on $\mathscr{E}$ and c.

14. Multisubunits with Two Conformations and Ligand Binding

In Sections 12 and 13, we examined the properties of two-conformation units, each with m binding sites for a ligand L. Regulated actin, with $m = 7$, is an example. Here we treat the more common case of units comprised of m identical two-conformation subunits, each with one binding site for L (or for the competitive binding of two ligands). Hemoglobin is an example. These systems are more interesting and realistic when intersubunit interactions are included explicitly, but we postpone this aspect of the problem until Chapter 6. Thus this section is, in effect, an introduction to parts of Chapter 6. However, we do include here a brief discussion of the Monod–Wyman–Changeux (MWC) model[3], which like the A. V. Hill equation, is a limiting special case that corresponds to extreme positive cooperativity.

The basic model we consider is the following: each unit of a collection of M independent units is comprised of m equivalent and independent subunits; each subunit has two conformations or states, 1 and 2; each subunit has one binding site for L (concentration c), with binding constants K_1 and K_2; and the partition functions for an empty (no bound ligand) subunit are j_1 and j_2, with $K = j_2/j_1$ (the equilibrium constant for $1 \rightleftarrows 2$).

The special case $m = 2$ is shown in Fig. 4-6. Because the m subunits are independent of each other, the grand partition function ξ for one unit is simply the subunit grand partition function ϵ to the power m [compare Eq. (11.27)]:

$$\begin{aligned} \text{Unit: } \xi &= [(1 + K_1c) + K(1 + K_2c)]^m \\ &= \epsilon^m, \end{aligned} \tag{14.1}$$

$$\text{Subunit: } \epsilon = (1 + K_1c) + K(1 + K_2c). \tag{14.2}$$

○ Ⓛ □ [L]

A further step in the hierarchy is $\Xi = \xi^M$, for M independent and distinguishable units. When interactions between subunits are included in the model (Chapter 6), ξ for one unit cannot be put in the simple form of Eq. (14.1).

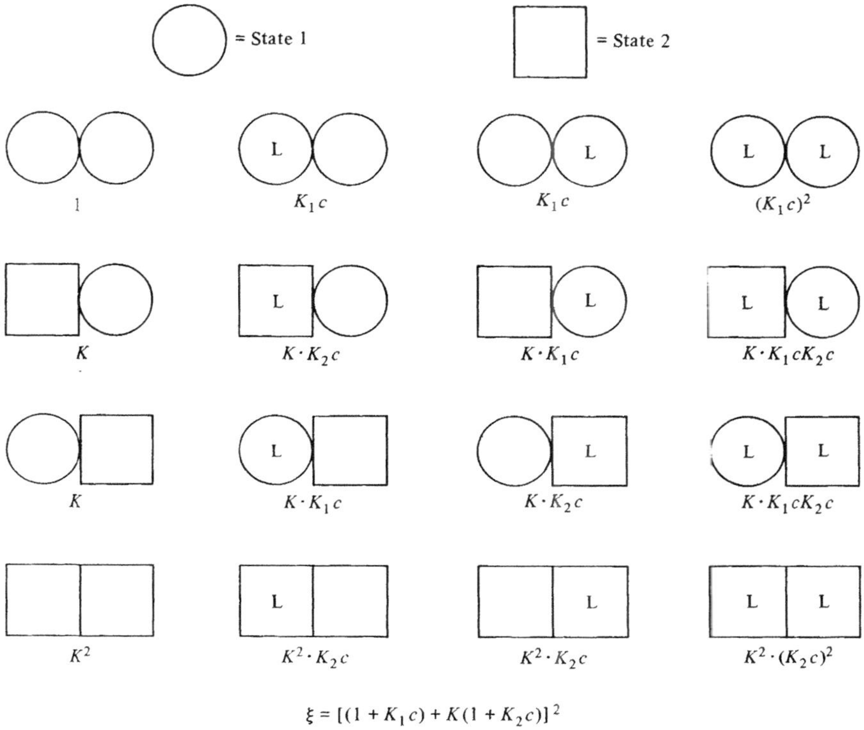

Fig. 4-6. States and grand partition function for a unit with $m = 2$ subunits. Each subunit has two possible conformations or states, each with one binding site for a ligand L. See the text for further details.

Because the subunits are independent, average properties follow directly from the terms in Eq. (14.2):

$$\delta = K(1 + K_2c)/\epsilon, \tag{14.3}$$

$$\theta = (K_1c + KK_2c)/\epsilon,$$

$$= (1 - \delta)\theta_1 + \delta\theta_2, \tag{14.4}$$

where

$$\theta_i = K_ic/(1 + K_ic) \qquad (i = 1, 2). \tag{14.5}$$

These are the same properties found in Eqs. (12.33), (12.34), and (12.42) in the special case $\eta = 1, m = 1$. However, Eqs. (14.3) and (14.4) hold for any m: they are the properties of individual independent subunits; the number of subunits in a unit is immaterial.

The mean value of and variance in s, the number of ligands bound per unit, are:

$$\bar{s} = c\frac{\partial \ln \xi}{\partial c} = mc\frac{\partial \ln \epsilon}{\partial c} = m\theta, \tag{14.6}$$

$$\sigma_m^2 = \overline{s^2} - \bar{s}^2 = c\frac{\partial \bar{s}}{\partial c} = \frac{m(1 + K)(K_1 c + KK_2 c)}{\epsilon^2}$$
$$= m\theta(1 - \theta). \tag{14.7}$$

This has the same form as Eq. (11.29) (m subunits, only one conformation) even though θ is different. The maxima in σ_m^2 and in $\partial\theta/\partial \ln c$ are at $\theta = \frac{1}{2}$. In fact, $\partial\theta/\partial \ln c = \frac{1}{4}$ at $\theta = \frac{1}{2}$. This suggests that $\theta(\ln c)$ here has the same symmetry as in the simple case $\theta = Kc/(1 + Kc)$. This is confirmed if we note that Eq. (14.4) can also be written as

$$\theta = \bar{K}c/(1 + \bar{K}c), \tag{14.8}$$

where

$$\bar{K} = \frac{K_1 + KK_2}{1 + K}. \tag{14.9}$$

Note that $\bar{K}$ is a weighted average of K_1 and K_2.

Section 15 deals with symmetry in a more general way.

Induced Conformational Change

A special case of considerable importance (developed especially by Koshland) is one in which, say, conformation 1 completely dominates when the binding site on a subunit is empty ($K \ll 1$) and conformation 2 completely dominates when the binding site is occupied ($KK_2 \gg K_1$). Thus K_2 must greatly exceed K_1. In effect, then, the binding of L on the site induces the conformational change 1 → 2. On dropping the small terms, Eqs. (14.1) and (14.2) become

$$\xi = (1 + KK_2 c)^m = \epsilon^m, \tag{14.10}$$

$$\epsilon = 1 + KK_2 c. \tag{14.11}$$

This is essentially the same as Eq. (11.27) with the effective binding constant KK_2. If there is extreme positive cooperativity among the m subunits, which is *not* part of Koshland's proposals, the important terms in ξ are 1 and $(KK_2)^m c^m$. This then brings us back to the A. V. Hill equation, Eq. (11.30). Here there is an explicit though unlikely physical model that leads to the equation: the intermolecular interactions among the subunits are such that the complex is very unstable unless all subunits are in conformation 1 or all are in conformation 2. We shall return to this model in Chapter 6, where finite interactions among the subunits will be introduced.

It is possible that two different ligands, A and B, might compete for the one binding site on a subunit, and each induce a different change in conformation.

This appears to be the case in the binding of H^+ and Ca^{2+} on the Ca-ATPase of sarcoplasmic reticulum, a strongly cooperative system (Section 25). Here, as usual, we omit interactions. To begin with, suppose each subunit can exist in three conformations, 0, 1, and 2, and ligands A and B compete for the one binding site per subunit. Thus the generalization of Eq. (14.1), in rather obvious notation, is

$$\xi = [1 + K_{A0}c_A + K_{B0}c_B) + K_{01}(1 + K_{A1}c_A + K_{B1}c_B) + K_{02}(1 + K_{A2}c_A + K_{B2}c_B)]^m. \tag{14.12}$$

For example, K_{02} is the equilibrium constant for $0 \rightleftarrows 2$ and K_{A2} is the binding constant of A (concentration c_A) on conformation 2. At this point we assume that conformation 0 dominates when no ligand is bound, that the binding of A induces the conformational change $0 \rightarrow 1$, and that the binding of B induces the change $0 \rightarrow 2$. Then, in this special case, the only important terms in Eq. (14.12) are

$$\xi = (1 + K_{01}K_{A1}c_A + K_{02}K_{B2}c_B)^m = \epsilon^m, \tag{14.13}$$

$$\epsilon = 1 + K_{01}K_{A1}c_A + K_{02}K_{B2}c_B. \tag{14.14}$$

Equation (14.14), for one subunit, is essentially the same as Eq. (11.3) (simple competitive binding). In Chapter 6, we shall return to this model and include interactions among the three kinds of subunits (0, 1, 2).

The Monod–Wyman–Changeux (MWC) Model[3]

We return now to Eq. (14.1), which applies to a unit with m subunits, each with one binding site for L. If we assume that there are extremely strong interactions among the subunits such that either all subunits will be in state 1 or all in state 2 (mixtures are of negligible importance), then Eq. (14.1) simplifies to

$$\xi = (1 + K_1c)^m + K^m(1 + K_2c)^m. \tag{14.15}$$

This is the "concerted" MWC model: that is, all subunits change state ($1 \rightleftarrows 2$) together. Figure 4-7 shows the individual states and weights in the special case $m = 2$. Note that these are just the states in the top and bottom rows of Fig. 4-6. The kinetic or transition diagram (Chapter 3) for this case is presented in Fig. 4-8. Each line in the diagram represents a possible forward or backward transition between two states of a unit. The double conformational change ($11 \rightleftarrows 22$, horizontal lines) can occur in any one of four binding modes.

As already pointed out in Section 12, Eq. (14.15) is the same, mathematically, as Eq. (12.33) with $\eta = 1$. The relation between the two models is made clear by a comparison of Figs. 4-4 and 4-7. In Eq. (12.33) and Fig. 4-4, K refers to one unit; in Eq. (14.15) and Fig. 4-7, the corresponding

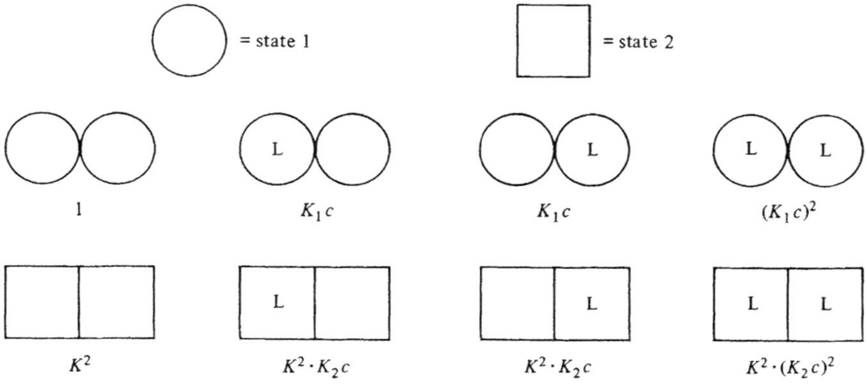

$$\xi = (1 + K_1c)^2 + K^2(1 + K_2c)^2$$

Fig. 4-7. The MWC "concerted" special case of Fig. 4-6 in which both subunits ($m = 2$) must be in the same state.

quantity for one unit (m subunits) is K^m. There is no interactive cooperativity in Fig. 4-4, but such cooperativity is implicit and so strong in Fig. 4-7 that the two rows of mixed states in Fig. 4-6 are reduced to negligible populations in the equilibrium distribution.

Because of the mathematical identity referred to above, the properties of the MWC model are to be found in Section 12. However, for convenience, we give a summary here:

$$\text{State 2:}\ \delta = \frac{K^m(1 + K_2c)^m}{\xi}, \tag{14.16}$$

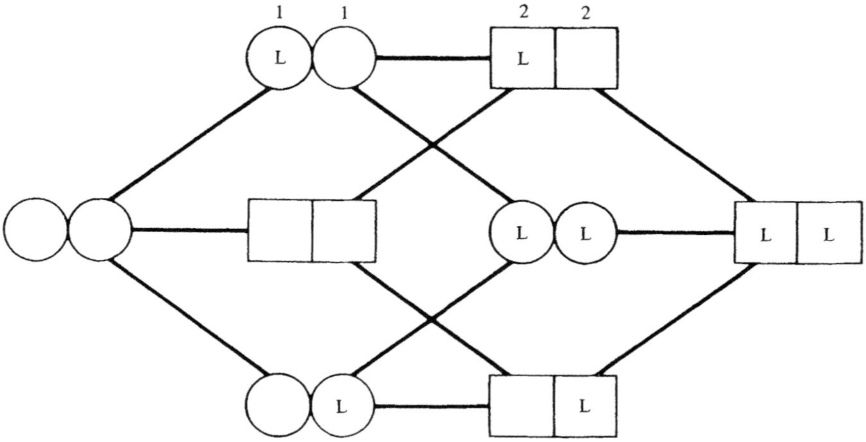

Fig. 4-8. The kinetic diagram showing possible transitions between states for the MWC special case in Fig. 4-7.

$$\text{Binding: } \theta = (1 - \delta)\theta_1 + \delta\theta_2, \tag{14.17}$$

$$\delta = \frac{1}{2}: \quad c' = \frac{1 - K}{K_2 K - K_1}, \tag{14.18}$$

where ξ is given in Eq. (14.15), θ_i is given in Eq. (14.5), and c' is the ligand concentration at which $\delta = \frac{1}{2}$.

Two Conformations and Two Ligands

An important generalization of the model represented by Eq. (14.1) is one in which each subunit has two independent binding sites, one for a ligand A and one for a ligand B. If, for example, A binds much more strongly on a state 1 subunit and B binds much more strongly on a state 2 subunit, then the $1 \rightleftarrows 2$ equilibrium can be shifted one way or the other by the appropriate ligand (called activator and inhibitor; the assignment depends on which state one wants to "activate"). From Eqs. (11.21) and (14.1), the new grand partition function is

$$\xi = \epsilon^m = [(1 + K_{A1}c_A)(1 + K_{B1}c_B) + K(1 + K_{A2}c_A)(1 + K_{B2}c_B)] \tag{14.19}$$

Equation (14.12) is closely related, but applies when there is competitive binding of A and B on only one site. Also, three conformations are involved in Eq. (14.12).

The fraction of state 2 subunits is

$$\delta = \frac{K(1 + K_{A2}c_A)(1 + K_{B2}c_B)}{\epsilon}. \tag{14.20}$$

The amount of binding of A is

$$\theta_A = \frac{\bar{s}_A}{m} = \frac{1}{m} \cdot c_A \frac{\partial \ln \xi}{\partial c_A} = c_A \frac{\partial \ln \epsilon}{\partial c_A}$$
$$= (1 - \delta)\theta_{A1} + \delta\theta_{A2}, \tag{14.21}$$

where

$$\theta_{A1} = \frac{K_{A1}c_A}{1 + K_{A1}c_A}, \text{ etc.} \tag{14.22}$$

Similarly, for B,

$$\theta_B = (1 - \delta)\theta_{B1} + \delta\theta_{B2}. \tag{14.23}$$

In the special case mentioned above, where $K_{A2} \cong 0$ and $K_{B1} \cong 0$ (A binds on 1 only; B binds on 2 only),

$$\epsilon = (1 + K_{A1}c_A) + K(1 + K_{B2}c_B), \tag{14.24}$$

$$\delta = \frac{K(1 + K_{B2}c_B)}{\epsilon}, \tag{14.25}$$

$$\theta_A = (1 - \delta)\theta_{A1}, \qquad \theta_B = \delta\theta_{B2}. \tag{14.26}$$

In this case, A at c_A shifts the $1 \rightleftarrows 2$ equilibrium toward state 1, and B at c_B shifts it toward state 2. The effective modified equilibrium constant for $1 \rightleftarrows 2$ is

$$\frac{\delta}{1-\delta} = \frac{K(1 + K_{B2}c_B)}{(1 + K_{A1}c_A)}. \tag{14.27}$$

Finally, we consider the MWC concerted version of the Eq. (14.19) model. In this case, there are extremely strong interactions among the subunits so that either all subunits are in state 1 or all are in state 2. Then

$$\xi = (1 + K_{A1}c_A)^m(1 + K_{B1}c_B)^m + K^m(1 + K_{A2}c_A)^m(1 + K_{B2}c_B)^m.$$

We then have

$$\delta = \frac{K^m(1 + K_{A2}c_A)^m(1 + K_{B2}c_B)^m}{\xi}, \tag{14.28}$$

$$\theta_A = (1 - \delta)\theta_{A1} + \delta\theta_{A2}, \tag{14.29}$$

$$\theta_B = (1 - \delta)\theta_{B1} + \delta\theta_{B2}. \tag{14.30}$$

The important special case $K_{A2} \cong 0$, $K_{B1} \cong 0$ (see above) follows easily.

Most of the results in this chapter are not new, and we have not discussed particular applications. Instead, our object has been to show that, after a little practice, the grand partition function for models of this general type can be written easily and the equilibrium properties obtained with little effort. The essential point is the realization that the individual states of a small open system are associated with particular terms in a grand partition function and, furthermore, that these terms are proportional to the equilibrium populations of the corresponding states. Average system properties of interest then follow from these state populations. It is not necessary to contend explicitly with successive multiple binding equilibria, etc., as in the empirical thermodynamic approach.

15. Systems with Symmetrical Binding Isotherms

Consider *any* system of independent units with a total of m binding sites per unit for a ligand at concentration c. The units may contain subunits. In the formal discussion here, we do not exclude interactions within a unit. That is, in this section, the treatment is "empirical." For any such system, the grand partition function for one unit can be written in the form

$$\xi = \alpha_0 + \alpha_1 c + \alpha_2 c^2 + \cdots + \alpha_m c^m, \tag{15.1}$$

where the α_i depend on the particular model. We have seen a number of examples in this chapter; there are many more in Chapter 6 (with cooperativity). For example, for the model in Fig. 4-4,

$$\xi = (1 + K_1 c)^2 + K(1 + K_2 c)^2, \tag{15.2}$$

so that

$$\alpha_0 = 1 + K, \qquad \alpha_1 = 2(K_1 + KK_2), \qquad \alpha_2 = K_1^2 + KK_2^2. \tag{15.3}$$

The question we consider here is: how must the α_i in Eq. (15.1) be interrelated so that $\theta(\ln c)$, derived from Eq. (15.1), is symmetrical about $\theta = \frac{1}{2}$? It is convenient to have available such a test of the α_i so that the symmetry or lack of symmetry of a given model can be established easily. Here a useful way to express this symmetry is the following. Let c_o be the value of c at which $\theta = \frac{1}{2}$; that is, $\theta(c_o) = \frac{1}{2}$. Then, for any constant $a \geq 1$,

$$\theta(ac_o) + \theta(c_o/a) = 1. \tag{15.4}$$

This relation is clear from Fig. 4-9 for a symmetrical curve.

The definition of the Hill coefficient n in Eq. (11.40) can be used for all systems with the above symmetry.

The successive terms in Eq. (15.1) are the relative weights in the equilibrium distribution among states for the different degrees of binding in a unit, $s = 0, 1, \ldots, m$. That is, $\alpha_r c^r$ is the weight for $s = r$ at concentration c. Consider the pairs of terms: α_0, $\alpha_m c^m$; $\alpha_1 c$, $\alpha_{m-1}c^{m-1}$; etc. The first pair relates to $s = 0, m$; the second pair to $s = 1, m - 1$; etc. If, at $c = c_o$, *the two members of every pair have equal values,* then the equilibrium numbers of units with $s = 0$ and m will be equal; the numbers with $s = 1$ and $m - 1$ will be equal; etc. Hence θ will necessarily have the value $\frac{1}{2}$ (because the weighted s value for each pair is $m/2$). Furthermore, $\theta(c)$ will have the required symmetry. This can be seen as follows.

Consider the arbitrary pair $\alpha_r c^r$ and $\alpha_{m-r}c^{m-r}$. At $c = c_o$, we have, by assumption,

$$\alpha_r c_o^r = \alpha_{m-r} c_o^{m-r}. \tag{15.5}$$

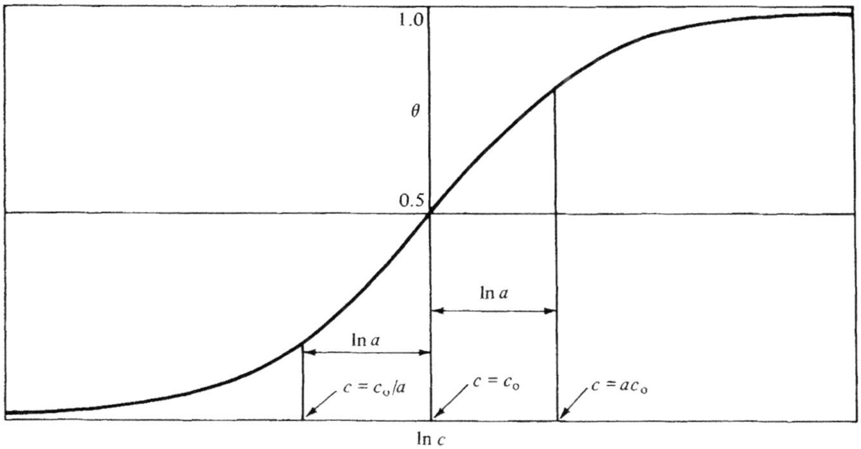

Fig. 4-9. Illustration of Eq. (15.4) for a symmetrical binding curve $\theta(\ln c)$.

When $c = ac_o$, these two terms (no longer equal) become

$$\alpha_r c_o^r a^r \qquad \text{and} \qquad \alpha_{m-r} c_o^{m-r} a^{m-r}. \tag{15.6}$$

Their ratio is a^r to a^{m-r}, in view of Eq. (15.5). When $c = c_o/a$, the two terms are

$$\alpha_r c_o^r a^{-r} \qquad \text{and} \qquad \alpha_{m-r} c_o^{m-r} a^{r-m}. \tag{15.7}$$

If we multiply both quantities in Eq. (15.7) by a^m and use Eq. (15.5), we see that the ratio of these two terms is a^{m-r} to a^r. Thus, when ac_o is changed to c_o/a, the relative weights of the two terms are just reversed. Because we are assuming Eq. (15.5) for every r, the relative weights in every pair will be reversed when $ac_o \to c_o/a$. Hence, if the fractional binding is θ for $c = ac_o$, it will become $1 - \theta$ for $c = c_o/a$, as in Eq. (15.4).

Thus we have shown that if Eq. (15.5) holds for every r, $m - r$ pair in Eq. (15.1), then $\theta(\ln c)$ is symmetrical about $\theta = \frac{1}{2}$. Equation (15.5) is necessary as well as sufficient, for if Eq. (15.5) does not hold for any r, the required symmetry will be spoiled.

If m is even, the central term in ξ ($s = m/2$) will not have a partner in the r, $m - r$ pairing. Consequently, its coefficient, $\alpha_{m/2}$, may have any value without destroying the symmetry.

We now consider a few examples. For $m = 1$,

$$\xi = \alpha_0 + \alpha_1 c. \tag{15.8}$$

Equation (14.2) (for ϵ rather than ξ) is an example. In this case, the equality $\alpha_0 = \alpha_1 c_o$ gives $c_o = \alpha_0/\alpha_1$ as the value of c at which $\theta = \frac{1}{2}$. A special case is the familiar $\xi = 1 + Kc$, with $c_o = 1/K$. There is no other α_i condition to be met, so symmetry is automatic (Fig. 4-3).

For $m = 2$,

$$\xi = \alpha_0 + \alpha_1 c + \alpha_2 c^2. \tag{15.9}$$

Here the equality $\alpha_0 = \alpha_2 c_o^2$ gives $c_o = (\alpha_0/\alpha_2)^{1/2}$ as the value of c at which $\theta = \frac{1}{2}$. Again there is no other α_i condition to be met; symmetry is automatic for any α_1.

Because every system with $m = 2$ is symmetrical, we can derive an expression for the Hill coefficient n from Eq. (11.41) that will apply to any $m = 2$ system; but first we digress to note that, for *any* m at $c = c_o$ and $\theta = \frac{1}{2}$,

$$n = \frac{4\sigma_m^2}{m} = \frac{4(\overline{s^2} - \bar{s}^2)}{m} = \frac{4[\overline{s^2} - (m/2)^2]}{m}, \tag{15.10}$$

where

$$\overline{s^2} = \sum_{s=1}^{m} s^2 \alpha_s c_o^s / \xi(c_o). \tag{15.11}$$

To find a compact expression for n, we have to express $\overline{s^2}$ in terms of the α_i

by eliminating c_o and by using any additional relations among the α_i required by symmetry (see below).

In the $m = 2$ case, we merely put $c_o = (\alpha_0/\alpha_2)^{1/2}$ in

$$\overline{s^2} = \frac{\alpha_1 c_o + 4\alpha_2 c_o^2}{\alpha_0 + \alpha_1 c_o + \alpha_2 c_o^2} \tag{15.12}$$

and find

$$n = 2(\overline{s^2} - 1) = \frac{2}{1 + [\alpha_1/2(\alpha_0\alpha_2)^{1/2}]} \qquad (m = 2). \tag{15.13}$$

The maximum value of n is $m = 2$, as expected [Eq. (11.41)], which occurs when $\alpha_1 = 0$ (i.e., very strong positive cooperativity; only the $s = 0$ and $s = 2$ terms remain in ξ). If α_1 is large enough, i.e., $\alpha_1 > 2(\alpha_0\alpha_2)^{1/2}$, we can have $n < 1$ (negative cooperativity or heterogeneous binding). Note that in the special case

$$\alpha_0 = 1, \qquad \alpha_1 = 2K, \qquad \alpha_2 = K^2, \tag{15.14}$$
$$\xi = (1 + Kc)^2,$$

we find $n = 1$ as expected (Section 11).

In the case of $m = 2$ heterogeneous binding,

$$\xi = (1 + K_1 c)(1 + K_2 c) \tag{15.15}$$

and

$$n = \frac{2}{1 + [(K_1 + K_2)/2(K_1K_2)^{1/2}]}. \tag{15.16}$$

If $K_1 \neq K_2$, then

$$K_1 + K_2 > 2(K_1K_2)^{1/2} \qquad \text{and} \qquad n < 1.$$

As is well known, heterogeneous binding is difficult to distinguish from negative cooperativity, which would also give $n < 1$ (Chapter 6).

For $m = 3$,

$$\xi = \alpha_0 + \alpha_1 c + \alpha_2 c^2 + \alpha_3 c^3. \tag{15.17}$$

For this system to be symmetrical, we need (at $\theta = \frac{1}{2}$)

$$\alpha_0 = \alpha_3 c_o^3 \qquad \text{and} \qquad \alpha_1 c_o = \alpha_2 c_o^2 \tag{15.18}$$

Thus

$$c_o = (\alpha_0/\alpha_3)^{1/3} \qquad \text{and} \qquad \alpha_1^3\alpha_3 = \alpha_2^3\alpha_0. \tag{15.19}$$

The latter relation is the required connection among the four coefficients in order to have symmetry about $c = c_o$, $\theta = \frac{1}{2}$. If we use Eqs. (15.18) and (15.19) in Eq. (15.10) for n, when there is symmetry, we find, after a little algebra,

$$n = \frac{4}{3}\left(\overline{s^2} - \frac{9}{4}\right) = \frac{3}{1 + [8\phi/(9\alpha_0 + \phi)]} \qquad (m = 3), \quad (15.20)$$

where

$$\phi \equiv \alpha_1\left(\frac{\alpha_0}{\alpha_3}\right)^{1/3} = \alpha_1 c_o. \quad (15.21)$$

When α_1 (and α_2) = 0 (extreme cooperativity), we have $\phi = 0$ and $n = 3 = m$. In the special case

$$\xi = (1 + Kc)^3,$$

$$\alpha_0 = 1, \qquad \alpha_1 = 3K, \qquad \alpha_2 = 3K^2, \qquad \alpha_3 = K^3, \quad (15.22)$$

we find $\phi = 3$ and $n = 1$.

Next we consider $m = 4$:

$$\xi = \alpha_0 + \alpha_1 c + \alpha_2 c^2 + \alpha_3 c^3 + \alpha_4 c^4. \quad (15.23)$$

The symmetry conditions are

$$\alpha_0 = \alpha_4 c_o^4 \qquad \text{and} \qquad \alpha_1 c_o = \alpha_3 c_o^3. \quad (15.24)$$

Thus

$$c_o = (\alpha_0/\alpha_4)^{1/4} \qquad \text{and} \qquad \alpha_1^2\alpha_4 = \alpha_3^2\alpha_0. \quad (15.25)$$

For symmetry, the coefficients must satisfy the latter relation; the value of α_2 is immaterial insofar as symmetry is concerned. Equations (15.10), (15.24), and (15.25) then give, for the Hill coefficient, when there is symmetry,

$$n = \overline{s^2} - 4 = \frac{4}{1 + [(3\psi + 2\omega)/(4\alpha_0 + \psi)]} \qquad (m = 4), \quad (15.26)$$

where

$$\psi \equiv \alpha_1\left(\frac{\alpha_0}{\alpha_4}\right)^{1/4} = \alpha_1 c_o, \quad (15.27)$$

$$\omega \equiv \alpha_2\left(\frac{\alpha_0}{\alpha_4}\right)^{1/2} = \alpha_2 c_o^2. \quad (15.28)$$

Again the usual special cases are verified: if $\alpha_1 = \alpha_2 = \alpha_3 = 0$, then $\psi = \omega = 0$ and $n = 4 = m$; if $\xi = (1 + Kc)^4$, then $\alpha_0 = 1$, $\psi = 4$, $\omega = 6$, and $n = 1$.

Finally, for $m = 6$, we give results only: the symmetry conditions are

$$c_o = (\alpha_0/\alpha_6)^{1/6}, \qquad \alpha_6^2\alpha_1^3 = \alpha_0^2\alpha_5^3, \qquad \alpha_6\alpha_2^3 = \alpha_0\alpha_4^3, \quad (15.29)$$

and, at $\theta = \frac{1}{2}$, with symmetry

$$n = \frac{6}{1 + v}, \quad (15.30)$$

where

$$v \equiv \frac{10\alpha_1 c_o + 16\alpha_2 c_o^2 + 9\alpha_3 c_o^3}{2(9\alpha_0 + 4\alpha_1 c_o + \alpha_2 c_o^2)}. \quad (15.31)$$

Here c_o may be replaced by $(\alpha_0/\alpha_6)^{1/6}$, if desired.

In summary, it is easy to test the coefficients of ξ for a small ligand-binding system to see if $\theta(\ln c)$ is symmetrical about $\theta = \frac{1}{2}$; if there is symmetry, the Hill coefficient n (defined at $\theta = \frac{1}{2}$) can be expressed, with little effort, in terms of the coefficients of ξ, using the relation between $\partial\theta/\partial \ln c$ and σ_m^2 [Eqs. (11.40) and (11.41)]; if symmetry does not exist, the relation

$$\sigma_m^2 = m \frac{\partial\theta}{\partial \ln c} \quad (15.32)$$

still holds at any θ, but it should not be used at $\theta = \frac{1}{2}$ to define n in such cases, as already discussed in connection with Eq. (12.44).

References

1. Hill, T. L. (1952) J. Chem. Phys. **20,** 1259.
2. Hill, T. L. (1960) *Introduction to Statistical Thermodynamics* (Addison-Wesley, Reading, MA).
3. Monod, J., Wyman, J., and Changeux, J. P. (1965) J. Molec. Biol. **12,** 88.
4. Hill, T. L. and Chen, Y. (1972) Proc. Natl. Acad. Sci. USA **69,** 1723.

General References

Levitzki, A. (1978) *Quantitative Aspects of Allosteric Mechanisms* (Springer-Verlag, New York).

Wyman, J. (1968) Q. Revs. Biophys. **1,** 1.

Koshland, D. E., Nemethy, G., and Filmer, D. (1966) Biochemistry **5,** 365.

Weber, G. (1972) Biochemistry **11,** 864.

Poland, D. (1978) *Cooperative Equilibria in Physical Biochemistry* (Clarendon, Oxford).

Hess, V. L. and Szabo, A. (1979) J. Chem. Educ. **56,** 289.

5

Steady-State Systems without Explicit Interactions

Beginning with the next chapter, the remainder of the book will be concerned with the explicit treatment of intermolecular interaction effects on equilibrium and steady-state systems. In this chapter we deal with several kinds of perturbation of steady-state enzyme activity, but we do not introduce intermolecular interaction free energies explicitly. In the models of Sections 17, 18, and 19, such interactions are present, but they are considered to be so strong that they can be handled by the complete omission of certain states or transitions from the kinetic diagram. These models are thus somewhat analogous to the A. V. Hill and Monod–Wyman–Changeux equilibrium models treated in Chapter 4. In contrast, other kinds of enzyme interaction or perturbation are considered in Sections 16, 20, and 21.

Section 16 is concerned primarily with the perturbation of the activity of a single enzyme molecule by the binding of a ligand on the enzyme or by the binding of the enzyme molecule itself onto a site. In Sections 17–20, we consider interactive effects in several kinds of models for enzyme *complexes,* that is, units comprised of more than one molecule in which the molecules interfere with each other in some way. Section 21 treats a very different kind of interenzyme interaction: if two (or more) types of enzyme are present in a closed vesicular membrane, and if these different enzymes share one or more ligands, substrates, etc., from the interior of the vesicle, then there is an effective coupling among the different enzymes by virtue of this sharing. The various systems that are present in the mitochondrial inner membrane provide an important example.

16. Perturbations of a Single Enzyme Molecule

Before turning to details in the subsections below, we give a qualitative review of the perturbations under consideration. We do this with reference to the unperturbed three-state enzyme E in Fig. 5-1(a). This case is of course

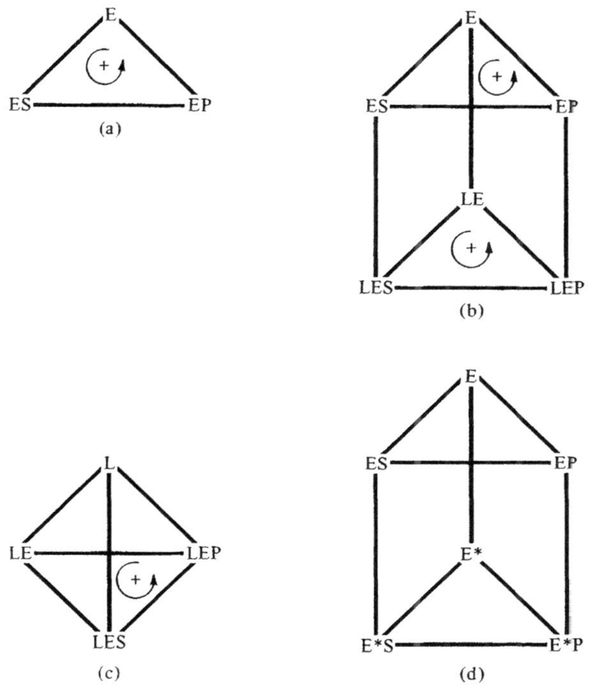

Fig. 5-1. (a) Diagram for unperturbed three-state enzyme. (b) Modified diagram if E binds a ligand L. (c) Diagram for a site L on a surface, etc., that can bind an enzyme E. The top state is the empty site. (d) Generalization of (a) if E has two conformations, E and E*.

simply an illustration. The unperturbed cycle or diagram (with more than one cycle) could be much more complicated.

The cycle in Fig. 5-1(a) is the same as that in Fig. 3-7(a). The system has steady-state properties summarized in Eqs. (8.29)–(8.37) and (10.5). Here the unperturbed enzyme molecules are considered to be moving freely in solution. If a ligand L that binds on E is now introduced into the solution, the kinetic diagram becomes that shown in Fig. 5-1(b). The "perturbation" here is the ligand L bound on E: bound L may affect the rate constants of the $S \to P$ cycle. The catalysis of $S \to P$ by the enzyme might now be accomplished by E, by LE, or by a combination of E and LE (see below). The introduction of L might increase (activate), decrease (inhibit), or have no effect on the flux or rate J (of $S \to P$) per enzyme molecule.

Another possible kind of perturbation of the activity of free E above is to introduce in, or in contact with, the solution a one-, two-, or three-dimensional lattice or assembly of independent sites that can bind E molecules. Examples are, respectively, a long polymer molecule with a linear array of sites for E (e.g., F-actin with sites for myosin S1), a surface or membrane, or a crystalline matrix that is penetrable by E. As will be shown below, and is perhaps intuitively obvious, this case is in fact formally the same as the one above: Fig. 5-1(b) applies, with L understood to mean a binding site for E on the lattice. Figure 5-1(b) is the kinetic diagram for an enzyme molecule; correspondingly, Fig. 5-1(c) is the diagram for a binding site (i.e., each site can exist in four possible states, connected by transitions).

If the enzyme E of Fig. 5-1(a) has two possible conformations, E and E*, the diagram is generalized to Fig. 5-1(d). Figure 5-1(d) is topologically equivalent to Fig. 5-1(b) and would have the same formal kinetics, expressed in terms of the first-order rate constants of the diagram. However, Fig. 5-1(d) can hardly be considered an example of a "perturbation" because the E, E* interconversion is intrinsic to E and is uncontrollable—unlike the effect of L in Fig. 5-1(b). Indeed, an enzyme E might have a second conformation E* that has not been detected yet; operationally, E would then be treated as a single-conformation enzyme. In any case, E in Fig. 5-1(a) is a complicated protein molecule, and each of its three states in the cycle has many substates. The observed rate constants are averages over more microscopic rate constants for substates. The question is whether these substates have been observed to be grouped into two major sets of states, E, ES, EP and E*, E*S, E*P. If so, Fig. 5-1(d) is appropriate and useful—but there is no "perturbation" of E in the sense we are using the term.

In discussing Fig. 5-1(b), it was mentioned that a combination of E and LE might be used in the catalysis of S → P. Figure 5-2(a) shows an example (solid lines in the diagram). In fact, this cycle has been used to represent the mechanism of ATPase activity of the myosin S1 fragment in the presence of F-actin binding sites (S = ATP; P = ADP, P_i; L = actin site for S1). A rather similar phenomenon occurs in the aggregation of actin monomers to form F-actin or in the aggregation of tubulin to form microtubules. Either case is shown very schematically (as a single polymer strand) in Fig. 5-2(b). Possible biochemical details, on gain or loss of a monomer from either end of the polymer, are indicated in Fig. 5-2(c), where Λ = monomer, T = ATP (actin) or GTP (tubulin), and D = ADP (actin) or GDP (tubulin). The principal point is that the NTPase activity in either the top or bottom cycle, alone, of Fig. 5-2(c) is very feeble (note the dotted lines), but the large combination cycle (solid lines) has very effective NTPase activity. The top cycle (for the free monomer) in Fig. 5-2(c) is the unperturbed cycle; in the bottom cycle, the activity of the end monomer of the polymer is perturbed by the presence of its neighboring monomer (or monomers) in the polymer. The two boxed states in Fig. 5-2(c) are thought to be the most important states in the large cycle. We shall not pursue this particular example any further; it has been discussed in detail by Hill and Kirschner.[1]

Three-State Enzyme Perturbed by Ligand

The analysis of this case [Fig. 5-1(b)] is a bit complicated and is in fact already available in Ref. 2, pp. 91–101, with several numerical examples.

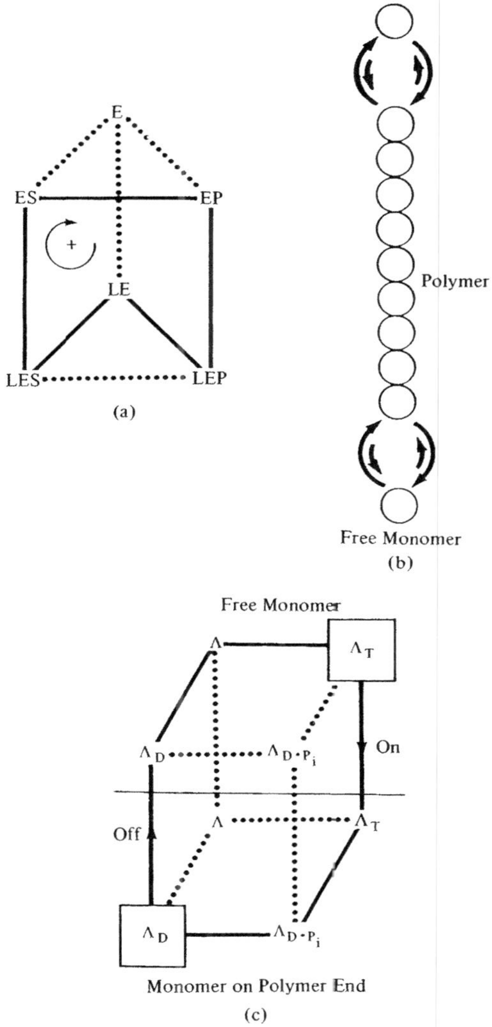

Fig. 5-2. (a) Example of a combination cycle. (b) Schematic representation of aggregation of actin or tubulin, with NTPase activity. (c) Combination cycle (solid lines) used in NTPase activity in (b).

Hence we give only a brief qualitative discussion here and reserve a detailed treatment (below) for the corresponding two-state enzyme problem.

If J_o is the S → P flux in the absence of L [Fig. 5-1(a)] and J is the same flux in the presence of L (ligand free in solution, or binding sites for E—as explained above), it is easy to see that J may be larger than J_o (L activates E) or smaller than J_o (L inhibits E). For this purpose, suppose that L is present at a sufficient concentration to practically saturate E with L. Then the top cycle in Fig. 5-1(b) (in the absence of L; flux J_o) is essentially replaced by the bottom cycle in the same figure (in the presence of saturating L; flux J). It is then obvious that, depending on the nature of E and L, the LE cycle (bottom) might have larger or smaller rate constants than the E cycle (top); hence J can be larger or smaller than J_o. Offhand it might be thought that making additional pathways available for S → P when L is introduced [Fig. 5-1(a) → Fig. 5-1(b)], could only increase J_o (i.e., $J > J_o$). This is not correct, however, because the total number of E molecules is conserved. If the concentration of L is large enough, all E is in the form LE, and only the LE (bottom) cycle is available.

At intermediate concentrations of L, a more general approach is needed. The diagram is shown again in Fig. 5-3, which also contains the 14 cycles that belong to this diagram. Cycles a through h all have the same thermodynamic force, $X = \mu_S - \mu_P$ [Eq. (8.29)]. Cycles a and b are the "top" and "bottom" cycles we have been discussing above; cycles c through h are combination cycles, using parts of cycles a and b. The solid-line cycle in Fig. 5-2(a) is cycle d here. Cycles i through n have force $X = 0$ (the reaction S → P does not occur); hence the corresponding mean cycle fluxes at steady state are all zero (see Section 8, or Chapter 3 of Ref. 2).

The total steady-state flux per enzyme molecule can be written as the sum

$$J = J_a + \cdots + J_h, \tag{16.1}$$

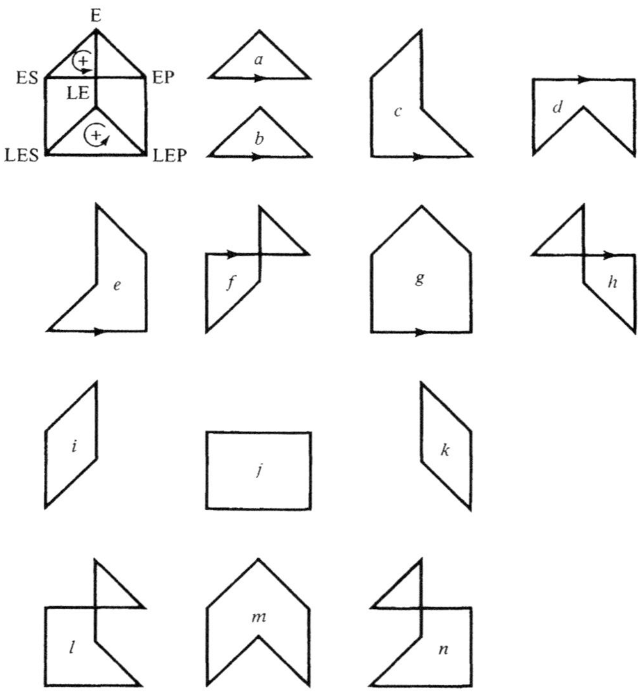

Fig. 5-3. Fourteen cycles belonging to the diagram at upper left. Arrows in cycles a through h show + direction (S → P). These cycles have force $X = \mu_S - \mu_P$. The remaining cycles have $X = 0$.

where J_a, J_b, . . . are individual cycle fluxes. J_i through J_n are omitted because they all have the value zero. The cycle fluxes, in turn, can be expressed explicitly in terms of the rate constants of the diagram [see Chapter 2 of Ref. 2, or Eqs. (16.39)–(16.42) below for a simpler example]. The pseudo-first-order rate constants for adding L (vertical-down transitions in Fig. 5-3) are all proportional to c_L, the concentration of free L (i.e., of free L in solution or of empty sites for binding E, expressed as a volume concentration). Hence all of the cycle fluxes J_a, ... , J_h and J are functions of c_L. As already mentioned, when $c_L \rightarrow 0$,

$$J = J_a \text{ (at } c_L = 0) = J_o, \tag{16.2}$$

and when $c_L \rightarrow \infty$,

$$J = J_b \text{ (at } c_L \rightarrow \infty). \tag{16.3}$$

The other cycle fluxes drop out of Eq. (16.1) in these special cases. At intermediate values of c_L, one or more of cycles c through h may be much larger than either J_a or J_b, though this need not be the case.

Even without more explicit details it should be clear that the availability of a free ligand L or of binding sites for E can alter the unperturbed flux J_o, and that the perturbed flux J at arbitrary c_L can be comprehended in terms of additive cycle contributions.

In the next five subsections we present a rather detailed treatment of a closely related but simpler model: we assume that, in Fig. 5-1(b), the states EP and LEP are transient intermediates that can be omitted from the diagram. That is, the two three-state S → P cycles (for E or LE) are reduced to two-state cycles, as already discussed at the end of Section 8. The new diagram is then that shown in Fig. 5-4, with four states. The same reaction S → P is catalyzed, and the presence of a ligand L (or sites for binding E) perturbs the activity of E as before. We begin with the equilibrium properties of this model, after first digressing to emphasize the formal equivalence of free L binding on E and of E binding on sites L.

Equilibrium Binding of L to E or E to L

We consider the equilibrium shown in Fig. 5-5(a), with equilibrium constant K_L. The conventional case is Fig. 5-5(b), where the three dilute species E, L,

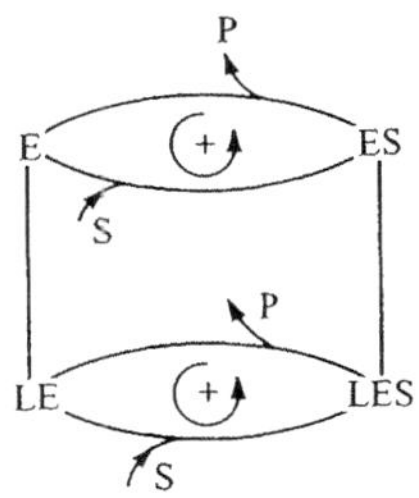

Fig. 5-4. Reduction of Fig. 5-1(b) using two-state S → P cycles.

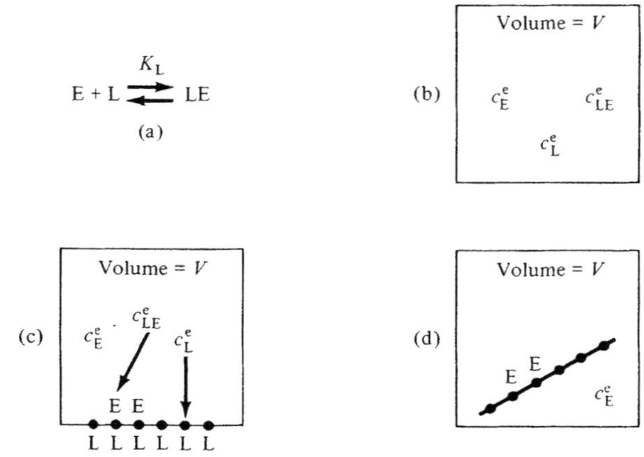

Fig. 5-5. (a) Binding equilibrium. (b) (c) (d) Three different but formally equivalent modes of binding L on E or E on L. See text for details.

and LE all move freely in a volume V and have concentrations c_E^e, c_L^e, and c_{LE}^e at equilibrium. This case is described as "L binding on E," though the formal symmetry with respect to L and E has already been mentioned in Eqs. (3.7)–(3.9).

In Fig. 5-5(c), there are independent sites on a surface for the binding of E. These sites are called L. At equilibrium, the concentration of free E in V is c_E^e. Also, if the number of empty sites is N_L^e, then we define the effective concentration of L as $c_L^e \equiv N_L^e/V$. That is, this would be the concentration of empty sites if they were dispersed in V. Similarly, if the number of sites occupied by E is N_{LE}^e, then the effective concentration of LE is $c_{LE}^e \equiv N_{LE}^e/V$. If the independent binding sites for E are actually moving around in the solution, for example, on polymer molecules [Fig. 5-5(d)], the same definitions of c_L^e and c_{LE}^e can be used. Thus the three physically different cases [Figs. 5-5(b), 5-5(c), and 5-5(d)] can be treated together using notation that is most natural for Fig. 5-5(b).

We define θ, as usual, as the fraction of E on which L is bound. Also, θ' is defined as the fraction of L on which E is bound. Then

$$K_L = \frac{c_{LE}^e}{c_E^e c_L^e} = \frac{\theta}{(1-\theta)c_L^e} = \frac{\theta'}{(1-\theta')c_E^e}. \tag{16.4}$$

For completeness, we include an easily derived symmetrical third expression:

$$K_L = \frac{\theta\theta'}{(1-\theta)(1-\theta')c_{LE}^e}. \tag{16.5}$$

Equilibrium Properties of the Enzyme-Ligand Diagram

We return now to Fig. 5-4, which is as simple a model as possible for the effect of a ligand on enzyme activity. However, in this subsection, we confine ourselves to the properties of this system at equilibrium. That is, the concentrations of free S and P in solution satisfy [Eq. (8.30)]

$$c_P^e/c_S^e = K_{SP}, \tag{16.6}$$

where K_{SP} is the equilibrium constant for S → P. The thermodynamic force X [Eqs. (8.29) and (8.31)] driving S → P is zero in this special case. In the usual steady state, c_S and c_P have values that correspond to $X > 0$ [Eq. (8.31)]; that is, $c_S/c_P > c_S^e/c_P^e$.

Figure 5-6(a) shows the equilibrium constant notation we shall use, and Fig. 5-6(b) gives the first-order rate constant notation. Figure 5-6(b) applies at an arbitrary steady state; equilibrium is a special case. Six of the 12 rate constants are pseudo first order because of binding of S, P, or L. The associated second-order rate constants are denoted with an asterisk:

$$\begin{aligned} \alpha &= \alpha^* c_S, & \alpha_L &= \alpha_L^* c_S, & \beta' &= \beta^* c_P, \\ \beta_L' &= \beta_L^* c_P, & \kappa &= \kappa^* c_L, & \kappa_S &= \kappa_S^* c_L. \end{aligned} \tag{16.7}$$

These relations apply at arbitrary concentrations c_S, c_P, and c_L.

At equilibrium, we have the fundamental relations

$$\text{E} + \text{S} \rightleftarrows \text{ES}; \qquad K_S = \frac{c_{ES}^e}{c_E^e c_S^e} = \frac{\alpha^*}{\alpha'}, \tag{16.8}$$

$$\text{ES} \rightleftarrows \text{E} + \text{P}; \qquad K = \frac{c_E^e c_P^e}{c_{ES}^e} = \frac{\beta}{\beta^*}, \tag{16.9}$$

$$\text{LE} + \text{S} \rightleftarrows \text{LES}; \qquad K_S' = \frac{c_{LES}^e}{c_{LE}^e c_S^e} = \frac{\alpha_L^*}{\alpha_L'}, \tag{16.10}$$

$$\text{LES} \rightleftarrows \text{LE} + \text{P}; \qquad K' = \frac{c_{LE}^e c_P^e}{c_{LES}^e} = \frac{\beta_L}{\beta_L^*}, \tag{16.11}$$

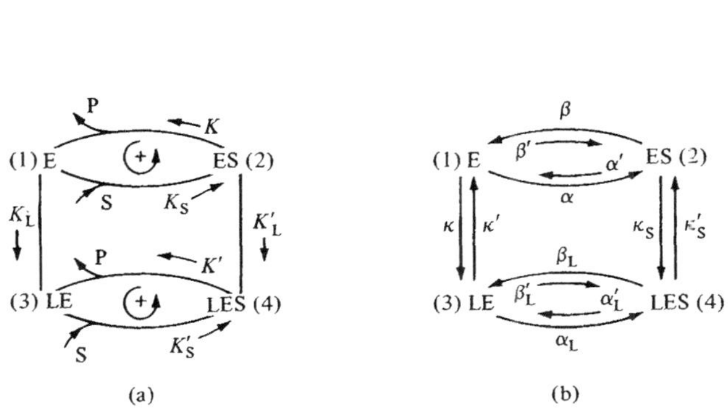

Fig. 5-6 (a) Equilibrium constant notation used for Fig. 5-4. The arrow on each K shows the forward direction chosen in the definition of K. (b) First-order rate constant notation for the same system, at arbitrary c_S, c_P, and c_L.

$$\mathrm{E} + \mathrm{L} \rightleftarrows \mathrm{LE}; \qquad K_{\mathrm{L}} = \frac{c_{\mathrm{LE}}^{\mathrm{e}}}{c_{\mathrm{E}}^{\mathrm{e}} c_{\mathrm{L}}^{\mathrm{e}}} = \frac{\kappa^*}{\kappa'}, \tag{16.12}$$

$$\mathrm{ES} + \mathrm{L} \rightleftarrows \mathrm{LES}; \qquad K_{\mathrm{L}}' = \frac{c_{\mathrm{LES}}^{\mathrm{e}}}{c_{\mathrm{ES}}^{\mathrm{e}} c_{\mathrm{L}}^{\mathrm{e}}} = \frac{\kappa_{\mathrm{S}}^*}{\kappa_{\mathrm{S}}'}. \tag{16.13}$$

The connections with rate constants follow from detailed balance at equilibrium for each of the six transition pairs in the diagram [Fig. 5-6(b)]. For example, in Eq. (16.8),

$$(\alpha^* c_{\mathrm{S}}^{\mathrm{e}}) c_{\mathrm{E}}^{\mathrm{e}} = \alpha' c_{\mathrm{ES}}^{\mathrm{e}}. \tag{16.14}$$

This equates the rate E $\longrightarrow$ ES via α to the rate ES $\longrightarrow$ E via α', at equilibrium.

There are two interconnections among the six equilibrium constants and among the 12 rate constants in Eqs. (16.8)–(16.13). We can derive these as follows. There are two paths between E and LES in Fig. 5-6(a), in the arrowed directions. From these two paths we find

$$K_{\mathrm{S}} K_{\mathrm{L}}' = K_{\mathrm{S}}' K_{\mathrm{L}} = \frac{\alpha^* \kappa_{\mathrm{S}}^*}{\alpha' \kappa_{\mathrm{S}}'} = \frac{\alpha_{\mathrm{L}}^* \kappa^*}{\alpha_{\mathrm{L}}' \kappa'}. \tag{16.15}$$

Similarly, there are two arrowed paths between ES and LE:

$$K K_{\mathrm{L}} = K' K_{\mathrm{L}}' = \frac{\beta \kappa^*}{\beta^* \kappa'} = \frac{\beta_{\mathrm{L}} \kappa_{\mathrm{S}}^*}{\beta_{\mathrm{L}}^* \kappa_{\mathrm{S}}'}. \tag{16.16}$$

Equations (16.15) and (16.16) are the desired results. These relations must be satisfied when assigning parameters to the model. On combining Eqs. (16.15) and (16.16), we obtain the alternative but nonindependent relations [compare Eq. (8.34)]

$$\frac{c_{\mathrm{P}}^{\mathrm{e}}}{c_{\mathrm{S}}^{\mathrm{e}}} = K_{\mathrm{SP}} = K_{\mathrm{S}} K = K_{\mathrm{S}}' K' = \frac{\alpha^* \beta}{\alpha' \beta^*} = \frac{\alpha_{\mathrm{L}}^* \beta_{\mathrm{L}}}{\alpha_{\mathrm{L}}' \beta_{\mathrm{L}}^*}. \tag{16.17}$$

This equation will prove useful in the next subsection on steady states.

The relations among the concentrations of the four states E, ES, LE, and LES in Fig. 5-6(a), at equilibrium, follow from Eqs. (16.8)–(16.13). Alternatively and more simply, one can use the grand partition function for E to obtain the same results. Thus,

$$\underset{\mathrm{E}}{\xi = 1} + \underset{\mathrm{ES}}{K_{\mathrm{S}} c_{\mathrm{S}}^{\mathrm{e}}} + \underset{\mathrm{LE}}{K_{\mathrm{L}} c_{\mathrm{L}}^{\mathrm{e}}} + \underset{\mathrm{LES}}{\begin{cases} K_{\mathrm{S}} c_{\mathrm{S}}^{\mathrm{e}} \cdot K_{\mathrm{L}}' c_{\mathrm{L}}^{\mathrm{e}} \\ K_{\mathrm{S}}' c_{\mathrm{S}}^{\mathrm{e}} \cdot K_{\mathrm{L}} c_{\mathrm{L}}^{\mathrm{e}} \end{cases}}. \tag{16.18}$$

As usual, the separate terms in ξ are proportional to relative concentrations or probabilities. For example,

$$c_{\mathrm{ES}}^{\mathrm{e}} = c_{\mathrm{E}}^{\mathrm{e}} \cdot K_{\mathrm{S}} c_{\mathrm{S}}^{\mathrm{e}}. \tag{16.19}$$

Alternatively, because S can be added to E or LE via P as well as via S, $K_{\mathrm{S}} c_{\mathrm{S}}^{\mathrm{e}}$

in Eq. (16.18) (two places) may be replaced by $K^{-1}c_P^e$, and $K_S'c_S^e$ may be replaced by $K'^{-1}c_P^e$. These substitutions are consistent with Eq. (16.17).

In the special case that the binding of L and S are independent of each other, we would have $K_S = K_S'$, $K_L = K_L'$, $K = K'$, and

$$\xi = (1 + K_S c_S^e)(1 + K_L c_L^e).$$

Finally, in this subsection, we note that Eqs. (16.4) and (16.5) can be generalized to the present more complicated system. L can now be bound to either E or ES. We define θ as the fraction of E and ES on which L is bound. Similarly, θ' is the fraction of L occupied by E or ES. Then we define a mean binding constant $\overline{K}_L$ by

$$\overline{K}_L \equiv \frac{(c_{LE}^e + c_{LES}^e)}{(c_E^e + c_{ES}^e)c_L^e} = \frac{\theta}{(1-\theta)c_L^e} = \frac{\theta'}{(1-\theta')(c_E^e + c_{ES}^e)}. \tag{16.20}$$

From these relations, we also find

$$\overline{K}_L = \frac{\theta\theta'}{(1-\theta)(1-\theta')(c_{LE}^e + c_{LES}^e)}. \tag{16.21}$$

The physical significance of $\overline{K}_L$ is clarified if we use, from Eq. (16.18),

$$c_{LE}^e = c_E^e \cdot K_L c_L \qquad \text{and} \qquad c_{LES}^e = c_{ES}^e \cdot K_L' c_L^e$$

in Eq. (16.20). Then we obtain

$$\overline{K}_L = \frac{c_E^e K_L + c_{ES}^e K_L'}{c_E^e + c_{ES}^e}. \tag{16.22}$$

Hence $\overline{K}_L$ is a weighted average of K_L and K_L', as one would expect. The weights are c_E^e and c_{ES}^e, as might have been anticipated from Fig. 5-6(a).

State Probabilities at Steady State

We continue the discussion of the model in Fig. 5-6 but turn now to steady-state properties. State probabilities are considered in this subsection and cycle fluxes in the following subsection. At steady state, the concentrations c_S and c_P are arbitrary. The thermodynamic force X for the top and bottom cycles of Fig. 5-6(b) is

$$\begin{aligned} e^{X/kT} &= \frac{\alpha\beta}{\alpha'\beta'} = \frac{\alpha_L\beta_L}{\alpha_L'\beta_L'} \\ &= \frac{\alpha^*\beta c_S}{\alpha'\beta^* c_P} = \frac{\alpha_L^*\beta_L c_S}{\alpha_L'\beta_L^* c_P} \\ &= \frac{c_S}{c_P}\cdot\frac{c_P^e}{c_S^e} = \frac{K_{SP}c_S}{c_P}, \end{aligned} \tag{16.23}$$

just as in Eq. (8.36). In the last line we have made use of Eq. (16.17).

The enzyme E has four possible states in the diagram in Fig. 5-6. As usual, we denote the steady-state probability of state i by p_i. It is convenient here to number the states (see Fig. 5-6). Of course p_i is proportional to the concentration c_i used above (c_E, c_{ES}, etc.):

$$p_i = \frac{c_i}{c_1 + c_2 + c_3 + c_4} \qquad (i = 1, 2, 3, 4), \tag{16.24}$$

where the denominator is the total concentration of E in all forms.

For given concentrations c_S, c_P, and c_L, the rate constants of the diagram in Fig. 5-6(b) are determined by Eqs. (16.7). The state probabilities at steady state then follow from the linear algebraic equations

$$\frac{dp_1}{dt} = 0 = (\alpha' + \beta)p_2 + \kappa' p_3 - (\alpha + \beta' + \kappa)p_1, \tag{16.25}$$

$$\frac{dp_2}{dt} = 0 = (\alpha + \beta')p_1 + \kappa_S' p_4 - (\alpha' + \beta + \kappa_S)p_2, \tag{16.26}$$

$$\frac{dp_3}{dt} = 0 = (\alpha_L' + \beta_L)p_4 + \kappa p_1 - (\alpha_L + \beta_L' + \kappa')p_3, \tag{16.27}$$

$$\frac{dp_4}{dt} = 0 = (\alpha_L + \beta_L')p_3 + \kappa_S p_2 - (\alpha_L' + \beta_L + \kappa_S')p_4, \tag{16.28}$$

$$p_1 + p_2 + p_3 + p_4 = 1. \tag{16.29}$$

Actually, one of Eqs. (16.25)–(16.28) is redundant. Equation (16.29) and three of the other four equations are used to find the p_i.

At the end of Section 9, it was pointed out that the state distribution in a two-state cycle has a quasiequilibrium form at steady state. The question arises whether the diagram in Fig. 5-6(b) also has a quasiequilibrium state distribution. In other words, do Eqs. (16.25)–(16.29) have a detailed balance type of solution at an arbitrary steady state? This seems plausible because this diagram is a composite of two two-state cycles.

If there is a detailed balance solution, the equations

$$\begin{aligned} \kappa p_1 &= \kappa' p_3, & (\alpha + \beta')p_1 &= (\alpha' + \beta)p_2, \\ \kappa_S p_2 &= \kappa_S' p_4, & (\alpha_L + \beta_L')p_3 &= (\alpha_L' + \beta_L)p_4, \end{aligned} \tag{16.30}$$

provide that solution. If we express p_2, p_3, and p_4 (in this order) in terms of p_1, using Eqs. (16.30), we see that there are two possible expressions for p_4. These two expressions are consistent if

$$\frac{\kappa(\alpha_L + \beta_L')}{\kappa'(\alpha_L' + \beta_L)} = \frac{\kappa_S(\alpha + \beta')}{\kappa_S'(\alpha' + \beta)},$$

or if

$$\frac{\kappa^*(\alpha_L^* c_S + \beta_L^* c_P)}{\kappa'(\alpha_L' + \beta_L)} = \frac{\kappa_S^*(\alpha^* c_S + \beta^* c_P)}{\kappa_S'(\alpha' + \beta)}. \tag{16.31}$$

Upon using Eq. (16.17), it is easy to show that this equation reduces to Eq. (16.15) *at equilibrium,* that is, if we replace c_S and c_P by c_S^e and c_P^e. However, for arbitrary c_S and c_P (i.e., at an arbitrary steady state), Eq. (16.31) does *not* hold. Hence Eqs. (16.30) do not provide a self-consistent solution of Eqs. (16.25)–(16.28): these latter equations do not have, in general, a quasi-equilibrium solution.

However, there *is* a quasiequilibrium solution (as the reader can verify) in either of the special cases $\kappa^* = 0$, $\kappa' = 0$ or $\kappa_S^* = 0$, $\kappa_S' = 0$. In these cases, the diagram is opened up, as shown in Fig. 5-7, and it now has a linear (rather than a closed-ring) topology.

Also, there is a true equilibrium (not quasiequilibrium) solution in the special case that all four βs in Fig. 5-6(b) are very small compared to the other eight rate constants. In this case, there is a relatively fast binding equilibrium of S and L on E, not appreciably perturbed by the four β transitions. Equation (16.31) is valid at arbitrary c_S if the four β terms are omitted; this follows from Eq. (16.15). This type of assumption (the βs are small) is in fact made implicitly and routinely by most biochemists, presumably to achieve simplicity. One of our objectives in this book is to try to deal with actual steady-state problems without making such assumptions.

We are forced then to return to general methods to solve Eqs. (16.25)–(16.28). The diagram method (see Chapter 1 of Ref. 2) is the simplest way to deduce an algebraic solution. (Of course a numerical solution is easy to obtain by computer.) We shall merely sketch the method in order to avoid lengthy algebra. The full diagram has 12 partial diagrams, as shown in Fig. 5-8. The probability p_i is proportional to the sum of the corresponding 12 directional diagrams for state i [arrows are introduced on all lines of each partial diagram, "flowing" toward state i; each arrow represents a rate constant factor, as provided by Fig. 5-6(b)]. There are a total of 48 directional diagrams (4 × 12); their sum is denoted Σ. That is, Σ is comprised of 48 terms. Each normalized p_i is then equal to the sum of 12 directional diagrams for state i, divided by Σ.

There is an important special case for which the algebra is much simpler: the top and bottom cycles in Fig. 5-6(b) are one-way cycles. That is α', β', α_L', and β_L' are all negligible. In this case, two-thirds of the terms drop out, and we find, from the directional diagrams,

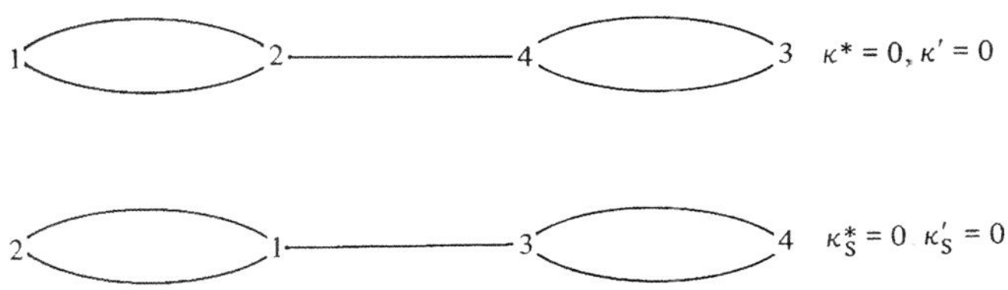

Fig. 5-7. Linear topology of the diagram in Fig. 5-6 in the two special cases indicated.

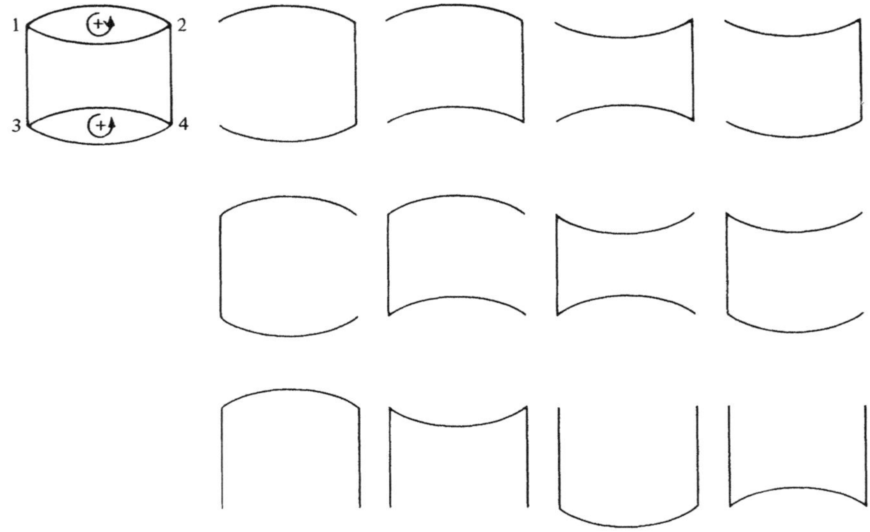

Fig. 5-8. The 12 partial diagrams belonging to the full diagram shown [see, also, Fig. 5-6(b)].

$$\text{E: } p_1 = [\beta(\alpha_L \kappa_S' + \beta_L \kappa' + \kappa' \kappa_S') + \beta_L \kappa' \kappa_S]/\Sigma, \tag{16.32}$$

$$\text{ES: } p_2 = [\alpha(\alpha_L \kappa_S' + \beta_L \kappa' + \kappa' \kappa_S') + \alpha_L \kappa \kappa_S']/\Sigma, \tag{16.33}$$

$$\text{LE: } p_3 = [\beta_L(\alpha \kappa_S + \beta \kappa + \kappa \kappa_S) + \beta \kappa \kappa_S']/\Sigma, \tag{16.34}$$

$$\text{LES: } p_4 = [\alpha_L(\alpha \kappa_S + \beta \kappa + \kappa \kappa_S) + \alpha \kappa' \kappa_S]/\Sigma, \tag{16.35}$$

where Σ is the sum of the four numerators (16 terms).

The dependence of the above p_i on c_L is of particular interest. It should be recalled [Eq. (16.7)] that κ and κ_S are proportional to c_L. Thus the numerators of p_1 and p_2 have constant and linear terms in c_L, whereas the numerators of p_3 and p_4 have linear and quadratic terms in c_L. In the limit $c_L \to 0$, the constant terms lead to

$$\begin{aligned} p_1 &= \beta/(\alpha + \beta), \\ p_2 &= \alpha/(\alpha + \beta). \end{aligned} \tag{16.36}$$

That is, only E(1) and ES(2) are present. In the limit $c_L \to \infty$, the quadratic terms give

$$\begin{aligned} p_3 &= \beta_L/(\alpha_L + \beta_L), \\ p_4 &= \alpha_L/(\alpha_L + \beta_L). \end{aligned} \tag{16.37}$$

In this case, only LE(3) and LES(4) are present. At intermediate values of c_L, all four states would generally have significant probabilities.

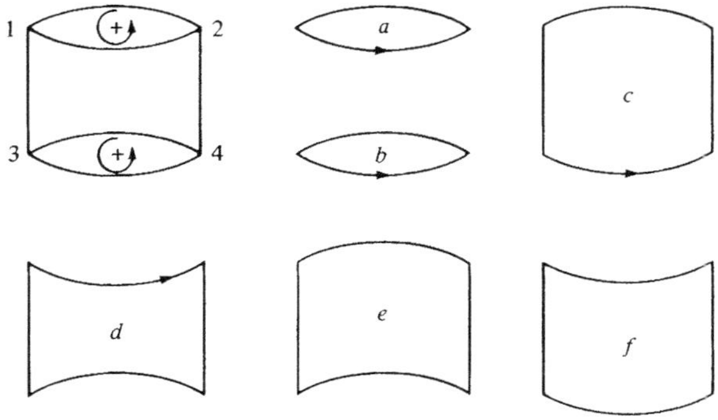

Fig. 5-9. The six cycles for the diagram shown. Cycles a through d have $X = \mu_S - \mu_P$ in the + direction, indicated by an arrow. Cycles e and f have $X = 0$.

Cycle Fluxes at Steady State

The cycles for the diagram we are discussing are shown in Fig. 5-9. We have already derived [Eq. (16.23)] an expression for the force X in cycles a and b. It is easy to show, using Eqs. (16.15) and (16.16), that Π_{c+}/Π_{c-} and Π_{d+}/Π_{d-}, for cycles c and d, also give the same $e^{X/kT}$ as in Eq. (16.23). On the other hand, cycles e and f have $X = 0$; hence $J_e = J_f = 0$. Thus the total flux is

$$J = J_a + J_b + J_c + J_d. \tag{16.38}$$

The separate cycle fluxes can be obtained from the flux diagrams (see Chapter 2 of Ref. 2). There are 12 of these, shown in Fig. 5-10. Thus (writing terms in the order given in Fig. 5-10),

$$J_a = (\alpha\beta - \alpha'\beta')(\beta_L\kappa' + \alpha_L'\kappa' + \beta_L'\kappa_S' + \alpha_L\kappa_S' + \kappa'\kappa_S')/\Sigma, \tag{16.39}$$

$$J_b = (\alpha_L\beta_L - \alpha_L'\beta_L')(\alpha'\kappa + \beta\kappa + \alpha\kappa_S + \beta'\kappa_S + \kappa\kappa_S)/\Sigma, \tag{16.40}$$

$$J_c = (\alpha_L\kappa_S'\beta\kappa - \alpha_L'\kappa_S\beta'\kappa')/\Sigma, \tag{16.41}$$

$$J_d = (\alpha\kappa_S\beta_L\kappa' - \alpha'\kappa_S'\beta_L'\kappa)/\Sigma, \tag{16.42}$$

where Σ is the 48-term Σ referred to in the discussion of Fig. 5-8. The *relative* magnitudes of the cycle fluxes are determined by the numerators above; it is not necessary to know Σ. Recalling that κ and κ_S are proportional to c_L, we see that the numerator of J_a is independent of c_L, the numerator of J_b has linear and quadratic terms in c_L, and the numerators of J_c and J_d both have only linear terms in c_L. Thus, in the limit $c_L \to 0$,

$$J = J_a = J_o = \frac{\alpha\beta - \alpha'\beta'}{\alpha + \alpha' + \beta + \beta'}, \tag{16.43}$$

as in Eq. (9.5). In the limit $c_L \to \infty$, because of the quadratic term in J_b,

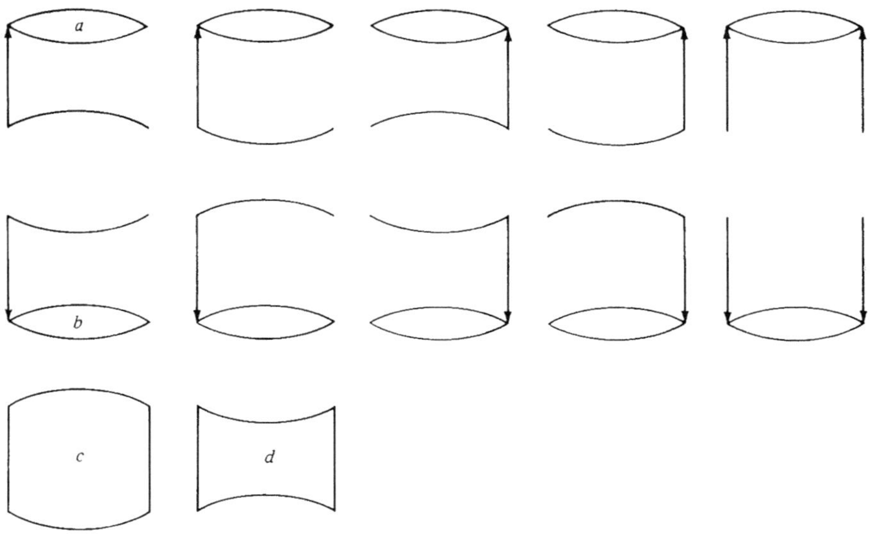

Fig. 5-10. Flux diagrams belonging to cycles a through d in Fig. 5-9.

$$J = J_b = \frac{\alpha_L \beta_L - \alpha'_L \beta'_L}{\alpha_L + \alpha'_L + \beta_L + \beta'_L}. \tag{16.44}$$

Very qualitatively, as c_L increases from $c_L = 0$, J starts with $J_a = J_o$, then cycles b, c, and d may all contribute (linear terms in c_L) to J, and finally J_b dominates (quadratic term in c_L). An inspection of the vertical lines in Fig. 5-8 reveals that the full 48-term Σ has constant, linear, and quadratic terms in c_L. Hence the complete J, from Eqs. (16.38)–(16.42), has the form

$$J = \frac{a_0 + a_1 c_L + a_2 c_L^2}{b_0 + b_1 c_L + b_2 c_L^2}, \tag{16.45}$$

where a_0/b_0 is the flux in Eq. (16.43) and a_2/b_2 is the flux in Eq. (16.44). Some possible shapes of $J(c_L)$ curves are shown, schematically, in Fig. 5-11.

Simple Inhibition and Activation by L

Suppose that the S $\rightleftarrows$ P transitions are inhibited when L is bound on E. The diagram and cycles that remain to catalyze S $\rightarrow$ P are shown in Fig. 5-12. In the absence of L, cycle a operates. As c_L increases, cycle c contributes. However, when c_L is large, although $J_c \gg J_a$, J_c itself becomes very small because the enzyme is almost always confined to states LE and LES. Algebraically, when c_L is large,

$$J_a = O(\text{const}/c_L^2) \rightarrow 0, \qquad J_c = O(\text{const} \cdot c_L/c_L^2) \rightarrow 0. \tag{16.46}$$

The quadratic terms in Σ are responsible for this behavior.

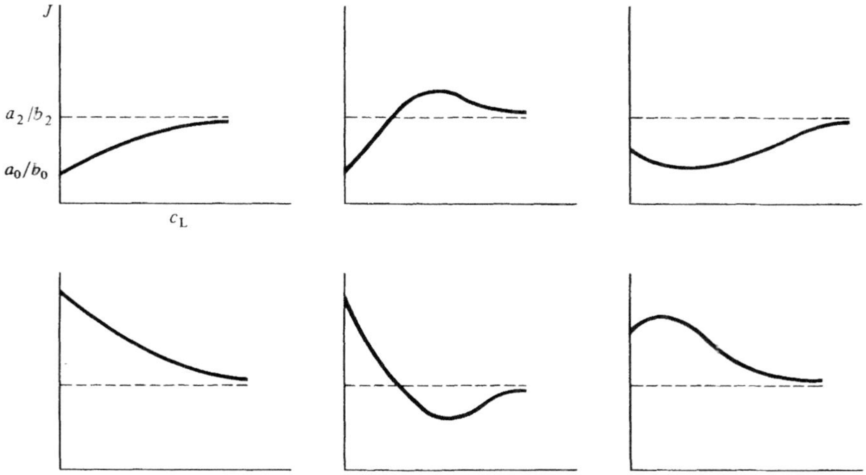

Fig. 5-11. Some possible shapes of flux curves $J(c_L)$ from Eq. (16.45).

The reverse case, in which L activates E, is shown in Fig. 5-13. In the absence of L, there is no operable S → P cycle. As c_L increases, first cycle *d* operates and then it is joined by cycle *b*, which eventually completely dominates when $c_L \to \infty$.

Two-Conformation Enzyme Perturbed by L

In the remainder of this section, we discuss a more complicated case. This is a generalization of Figs. 5-4 and 5-6 in which the enzyme E has two conformations, E and E*, possibly with very different properties (rate constants, equilibrium constants). The kinetic diagram is shown in Fig. 5-14(a). Note that, in general, all of E, LE, E*, and LE* can catalyze S → P; in each case, X is given by Eq. (16.23). Whereas any effect of the bound ligand L on the enzymatic activity of E in Figs. 5-4 and 5-6 is direct, here the effect of L can be indirect ("allosteric") by shifting the E, E* distribution one way or the other. An extreme example of this is the induced conformational change in Fig. 5-14(b): the binding of L to E or to ES immediately induces the E → E*

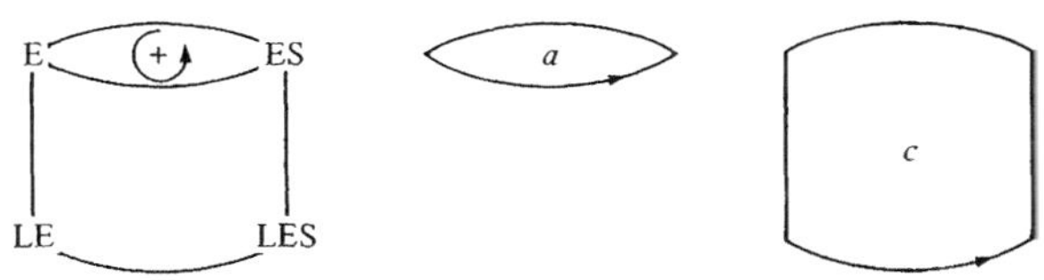

Fig. 5-12. Simplification of Fig. 5-9 if S ⇄ P transitions are inhibited when L is bound to E.

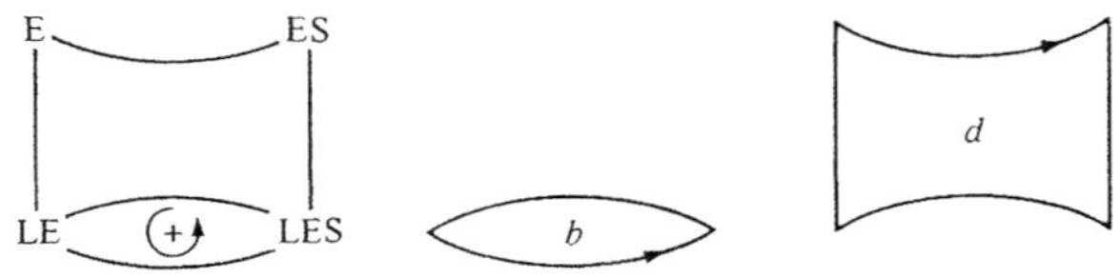

Fig. 5-13. Reverse case to Fig. 5-12: S ⇄ P transitions are inhibited unless L is bound to E.

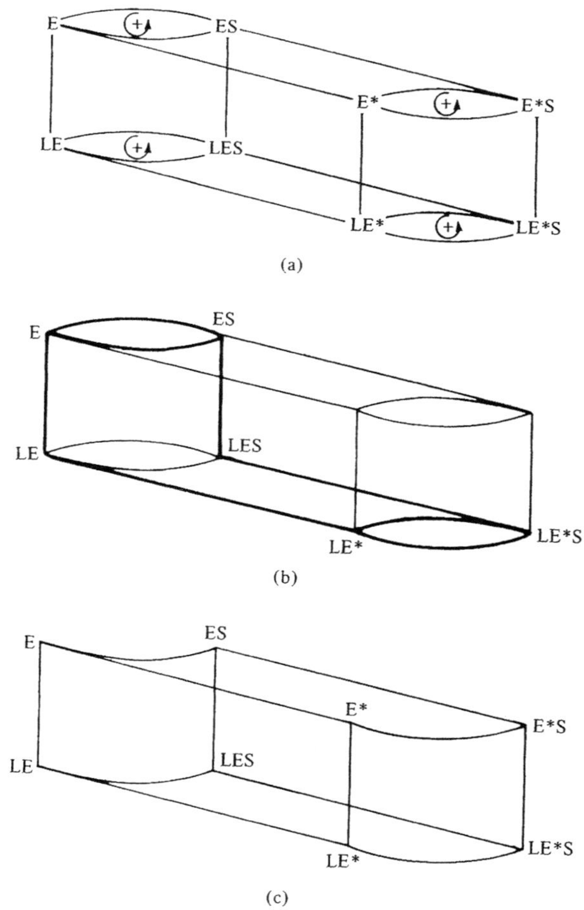

Fig. 5-14. (a) Generalization of Figs. 5-4 and 5-6 in which E has two conformations, E and E*. All four small cycles can, in general, catalyze S → P with $X = \mu_S - \mu_P$. (b) Special case in which binding of L on E or ES induces an immediate conformational change. Without L bound, E* and E*S are unstable. (c) Diagram that excludes S ⇄ P. This would be an equilibrium system with binding of S and L, and the conformational change E ⇄ E*.

conformational change. The heavy lines in Fig. 5-14(b) show the effective kinetic diagram (LE and LES are transient intermediates). This diagram is topologically the same as that in Figs. 5-4 or 5-6 (hence the formal kinetics would be the same). However, here L might be bound on E some distance from the site for S yet influence the properties of the S site by inducing a large-scale change in structure of E, E $\rightarrow$ E*.

In general, the effect of bound L need not be so drastic. If the catalytic activity is different in the four small cycles of Fig. 5-14(a), bound L (as c_L increases) can increase or decrease the overall S $\rightarrow$ P flux J by shifting the steady-state distribution of the eight states to emphasize the lower-left S $\rightarrow$ P cycle or the lower-right cycle. These lower cycles may be more or less effective than the upper cycles in catalyzing S $\rightarrow$ P.

Complexes of several (e.g., a dimer, $m = 2$, or tetramer, $m = 4$) of these enzymes provide a prototype for steady-state cooperative systems. In general, because of intermolecular interactions, the rate constants and equilibrium constants in the diagram for a given enzyme molecule in the complex at a particular time depend on which of the eight states the neighbors of the given molecule are in at that time. This is the starting point for our general treatment of steady-state cooperative systems, beginning in Chapter 7. In the much simplified Monod–Wyman–Changeux version of this problem, which we consider in Section 19, two assumptions are made: all members of the complex change conformation (E $\rightleftarrows$ E*) together; and the binding of L and S on E, E*, at any c_L and c_S, can be treated as an equilibrium problem rather than as a steady-state problem. For a single enzyme molecule ($m = 1$), this amounts to calculating the state probabilities from Fig. 5-14(c) rather than from the actual diagram, Fig. 5-14(a). The equilibrium assumption is valid only if the eight rate constants belonging to the four missing lines in Fig. 5-14(c) (these relate to ES $\rightleftarrows$ E + P) are all very small compared to the 16 rate constants belonging to the eight lines that are included in Fig. 5-14(c), or if $c_S/c_P = c_S^e/c_P^e$ (i.e., the steady state is an equilibrium state).

The above discussion of complexes ($m > 1$) provides additional motivation for the study of the basic and more elementary $m = 1$ case in this and the following subsections.

Equilibrium Constant Relations

Here we use the simplified notation for equilibrium constants presented in Fig. 5-15. Just as in Eqs. (16.15) and (16.16), we deduce the required relations

$$
\begin{aligned}
K_1K_6 &= K_3K_5, & K_{11}K_{16} &= K_{13}K_{15},\\
K_2K_5 &= K_4K_6, & K_{12}K_{15} &= K_{14}K_{16},\\
K_7K_{15} &= K_5K_9, & K_8K_{16} &= K_6K_{10},\\
K_1K_8 &= K_7K_{11}, & K_1K_2 &= K_{11}K_{12}.
\end{aligned}
\tag{16.47}
$$

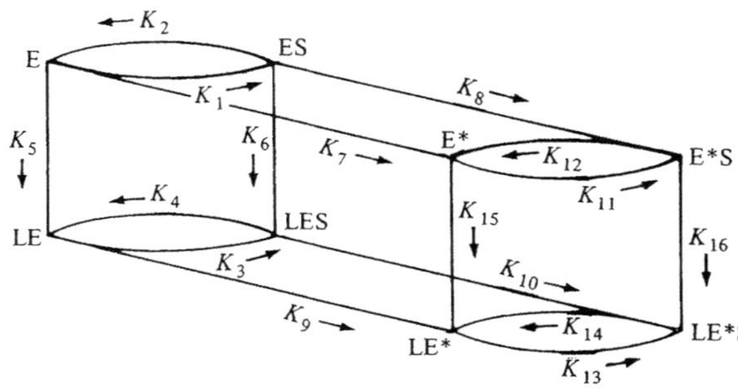

Fig. 5-15. Definitions of equilibrium constants for processes in the directions shown by arrows.

The last equation ensures that the force is the same in all four small cycles. There are, of course, many combinations of equilibrium constants other than those used in Eqs. (16.47). In place of Eq. (16.17), we have

$$c_P^e/c_S^e = K_{SP} = K_1K_2 = K_3K_4 = K_{11}K_{12} = K_{13}K_{14}. \tag{16.48}$$

Corresponding to Eq. (16.18), the grand partition function for E at equilibrium is

$$\begin{aligned}\xi = {} & \underset{\text{E}}{1} + \underset{\text{ES}}{K_1c_S^e} + \underset{\text{LE}}{K_5c_L^e} + \underset{\text{LES}}{K_1c_S^e \cdot K_6c_L^e} \\ & + K_7(\underset{\text{E*}}{1} + \underset{\text{E*S}}{K_{11}c_S^e} + \underset{\text{LE*}}{K_{15}c_L^e} + \underset{\text{LE*S}}{K_{11}c_S^e \cdot K_{16}c_L^e}).\end{aligned} \tag{16.49}$$

In view of Eqs. (16.47), alternate choices of some equilibrium constants could have been made here. The relative magnitudes of the terms in Eq. (16.49) determine the equilibrium probabilities p_i^e of the eight states. Equation (14.19) is rather similar to Eq. (16.49). In the MWC model for this system at steady state, Eq. (16.49) is assumed to give the state probabilities at *arbitrary* c_S, c_P, and c_L [there is no problem about arbitrary c_L; the notation c_L^e is being used only for symmetry, starting with Eq. (16.4)]. That is, as already mentioned, in effect Fig. 5-14(c) is used to determine state probabilities even at steady state.

Steady-State Probabilities of States

We shall not introduce a complete rate constant notation. If such a notation were available, Eqs. (16.47) would provide a number of rate-constant interconnections, just as in Eqs. (16.15) and (16.16). Also, nine equations (with one redundancy), analogous to Eqs. (16.25)–(16.29), that determine the steady-state p_i could be written by inspection of the diagram (Fig. 5-15). This diagram is sufficiently complicated, however, that the diagram method (di-

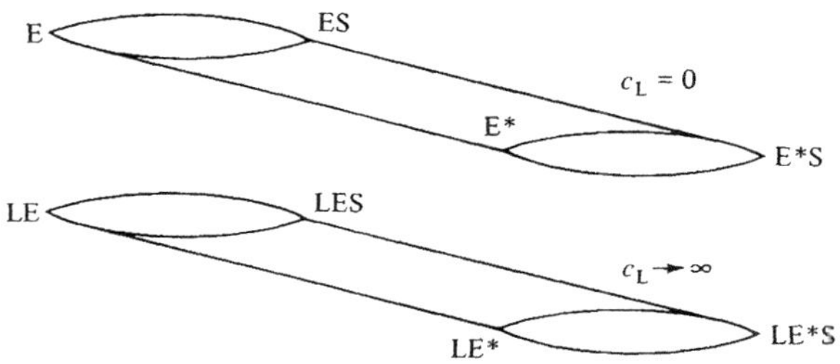

Fig. 5-16. Special cases of Fig. 5-14(a) when ligand is absent or present in excess. See text for further details.

rectional diagrams, etc.) is quite impractical as a method for obtaining the p_i. Instead, the p_i would have to be found numerically by solving eight linear algebraic equations in eight unknowns, using matrix inversion on a computer.

In the special case $c_L = 0$, the p_i are determined from the upper diagram in Fig. 5-16. Similarly, when $c_L \to \infty$, the p_i are to be found from the lower diagram in Fig. 5-16. Both cases are topologically equivalent to Fig. 5-6(b), so equations like Eqs. (16.25)–(16.29) would have to be solved. This is barely practical by the diagram method (Σ has 48 terms); on the other hand, a numerical solution by computer is easy.

S → P Flux and Cycles

Besides the four small cycles in Fig. 5-14 or 5-15, there are 80 additional cycles with force $X = \mu_S - \mu_P$. These additional cycles are all "combination" cycles, with the E + S → ES step occurring in one small cycle and the ES → E + P step occurring in one of the other three small cycles. Although it is an interesting exercise to enumerate all of these cycles, there are too many to provide a useful expression for the total S → P flux J. Instead, the practical way to write J is in terms of the steady-state p_i, found numerically (see above). The labelling of states and rate constants (in the small cycles) is given in Fig. 5-17. This follows the notational system in Fig. 5-6(b), with the

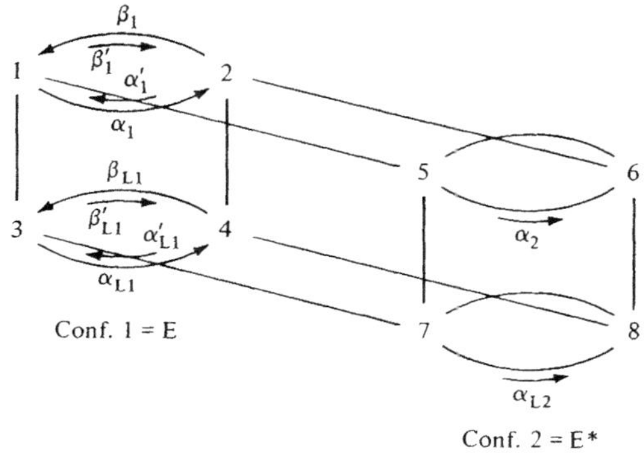

Fig. 5-17. First-order rate constant notation and numbering of states. E* is conformation 2. The rate constants for E* (right-hand side) are denoted as on the left except for use of subscript 2.

addition of a subscript on the rate constants that depend on the conformation: E = 1 or E* = 2. The flux is then

$$J = (\alpha_1 p_1 - \alpha_1' p_2) + (\alpha_{L1} p_3 - \alpha_{L1}' p_4) + (\alpha_2 p_5 - \alpha_2' p_6) + (\alpha_{L2} p_7 - \alpha_{L2}' p_8). \tag{16.50}$$

This is the total net mean rate, per enzyme molecule, at which the reaction E + S → ES occurs. At steady state, this must equal the net rate of ES → E + P. Hence an alternative expression for J is

$$J = (\beta_1 p_2 - \beta_1' p_1) + (\beta_{L1} p_4 - \beta_{L1}' p_3) + (\beta_2 p_6 - \beta_2' p_5) + (\beta_{L2} p_8 - \beta_{L2}' p_7). \tag{16.51}$$

As already mentioned, in the limits $c_L = 0$ and $c_L \to \infty$, the diagram simplifies to the four-state diagrams in Fig. 5-16. Each of these has four cycles with X, analogous to cycles a, b, c, d in Fig. 5-9. If E is much more stable than E*, or vice versa, one of the upper small cycles would dominate at $c_L = 0$. If LE is much more stable than LE*, or vice versa, one of the lower small cycles would dominate at $c_L \to \infty$. Thus the binding of L might shift the dominant cycle from E to LE* or from E* to LE, if L binds much more strongly on the less stable conformation (E* or E, respectively). These cases can correspond to inhibition or activation of the enzyme by L, depending on the relative activity of the initial ($c_L = 0$) and final ($c_L \to \infty$) small cycles. These are classical examples, but the behavior of J, on adding L, would generally be more complicated and less dramatic.

Special Cases

Two extreme but important special cases are shown in Fig. 5-18. In Fig. 5-18(a), L activates the enzyme E. The catalytic step (ES → E + P) occurs only if both L is bound and E is in conformation E*. Also, E is much more stable than E*, but L binds very much more strongly to E* than to E. Thus, at $c_L = 0$, E and ES are the dominant states ($J \cong 0$), whereas when $c_L \to \infty$, the dominant states are LE* and LE*S, with a full enzymatic cycle ($J > 0$) connecting these latter states.

In Fig. 5-18(b), L inhibits E. In this example, the only catalytic step in the diagram is E*S → E* + P. Also, E* is much more stable than E, but L binds very much more strongly on E than on E*. Consequently the E*, E*S cyclic activity ($J > 0$) at $c_L = 0$ is abolished when the enzyme is saturated with L, because states LE and LES dominate. Of course the assignment of roles to E and E* in these two examples is arbitrary; the assignments could have been reversed.

Incidentally, in either Fig. 5-18(a) or 5-18(b), there are 17 possible cycles with the usual force $X = \mu_S - \mu_P$: one small cycle and 16 combination cycles.

Even simpler versions of the two cases in Fig. 5-18 are shown in Fig. 5-19.

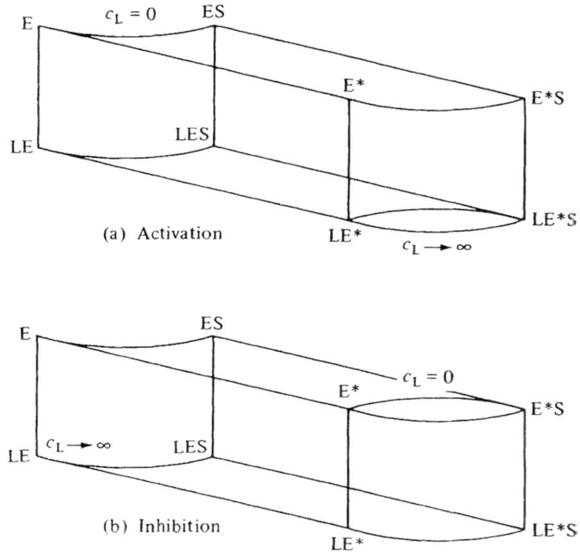

Fig. 5-18. (a) L activates E. The catalytic step ES → E + P occurs only if both L is bound and E is in conformation E*. (b) L inhibits E. The only catalytic step is E*S → E* + P. See text for further details.

The additional features in Fig. 5-19(a) [compared to Fig. 5-18(a)] are that L binds only on E* and that S binds only on LE*. Similarly, in Fig. 5-19(b), the new features [compared to Fig. 5-18(b)] are: L binds only on E and S binds only on E*. The algebra is simple for these two cases and is presented in the final two subsections.

Simple Example of Activation by L

Here we study the model in Fig. 5-19(a). The two equilibrium constants (K_*, K_{L2}) are the same as K_7 and K_{15} in Fig. 5-15. As described above, $K_* \ll 1$.

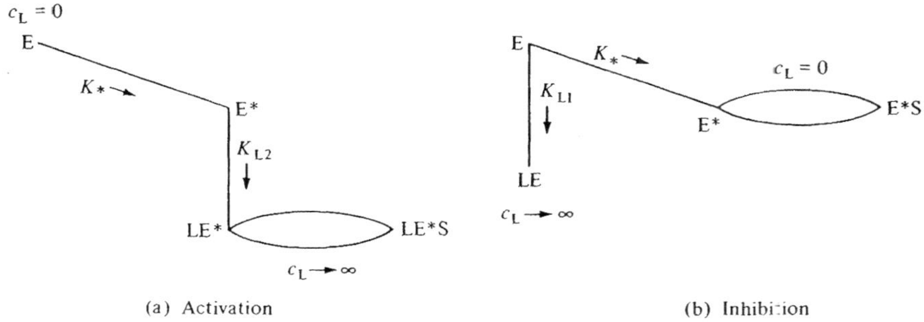

Fig. 5-19. Simpler versions of Fig. 5-18. See text for details.

Hence state E dominates at $c_L = 0$. The rate constants in the cycle (see Fig. 5-17) are designated α_{L2}, α'_{L2}, β_{L2}, β'_{L2}. Because the topology of the diagram is linear and the cycle is two state, there is a quasiequilibrium state distribution at steady state (as a consequence of this, we do not need the separate rate constants associated with K_* and K_{L2}).

There is a quasidetailed balance between LE* and LE*S at steady state:

$$\frac{p_{LE^*S}}{p_{LE^*}} = \frac{\alpha_{L2} + \beta'_{L2}}{\alpha'_{L2} + \beta_{L2}} = \frac{\alpha^*_{L2} c_S + \beta^*_{L2} c_P}{\alpha'_{L2} + \beta_{L2}}. \tag{16.52}$$

The right-hand expression here is the analogue of Kc_S in the simple equilibrium binding of S [as in Eqs. (3.7) and (6.15)], if K is the binding constant. To simplify notation, we write

$$\frac{\alpha^*_{L2} c_S + \beta^*_{L2} c_P}{\alpha'_{L2} + \beta_{L2}} \equiv a_L c_S + b_L c_P. \tag{16.53}$$

In this case, then, there is an effective steady-state "grand partition function" ξ_{ss} that establishes the relative state probabilities of the states (the subscript ss on ξ_{ss} refers to the steady state):

$$\xi_{ss} = 1 + K_*[1 + K_{L2} c_L (1 + a_L c_S + b_L c_P)]. \tag{16.54}$$

E E* LE* LE*S

For example, $p_E = 1/\xi_{ss}$. Equation (16.54) is very similar to Eq. (11.17). The reader should use the diagram to write out the four steady-state linear equations in the p_i, analogous to Eqs. (16.25)–(16.28), and confirm that the p_i from Eq. (16.54) are indeed the solution of these equations.

The steady-state S → P flux can be obtained from either of

$$\begin{aligned} J &= \alpha^*_{L2} c_S p_{LE^*} - \alpha'_{L2} p_{LE^*S} \\ &= \beta_{L2} p_{LE^*S} - \beta^*_{L2} c_P p_{LE^*}. \end{aligned} \tag{16.55}$$

In either case, we find

$$J = \frac{K_* K_{L2} c_L (\alpha^*_{L2} \beta_{L2} c_S - \alpha'_{L2} \beta^*_{L2} c_P)}{(\alpha'_{L2} + \beta_{L2}) \xi_{ss}}. \tag{16.56}$$

The simple dependence of J on c_L and c_S can be seen from Eqs. (16.56) and (16.54) (the c_P terms are usually negligible). J has the form

$$J = \frac{A c_L c_S}{B + C c_L + D c_L c_S}. \tag{16.57}$$

When either c_L or c_S is held constant, the dependence of J on the other concentration has the conventional form

$$J = \frac{A' c}{B' + C' c}. \tag{16.58}$$

When both c_L and c_S are large,

$$J \to A/D = \beta_{L2}, \tag{16.59}$$

as expected.

Simple Example of Inhibition by L

The model for this case is Fig. 5-19(b), with $K_* \gg 1$. K_{L1} is the same as K_5 in Fig. 5-15. The rate constants in the cycle are α_2, α_2', β_2, β_2'. This is also a quasiequililbrium system. Corresponding to Eqs. (16.52) and (16.53), we have, at steady state,

$$\frac{p_{E^*S}}{p_{E^*}} = \frac{\alpha_2^* c_S + \beta_2^* c_P}{\alpha_2' + \beta_2} \equiv a c_S + b c_P. \tag{16.60}$$

The effective steady-state "grand partition function" is

$$\begin{aligned} \xi_{ss} = \underset{\text{E}}{1} + \underset{\text{LE}}{K_{L1} c_L} + \underset{\text{E*}}{K_*}&\underset{\text{E*S}}{(1 + a c_S + b c_P)}. \end{aligned} \tag{16.61}$$

The separate terms determine the four state probabilities.

The steady-state flux is

$$\begin{aligned} J &= \alpha_2^* c_S p_{E^*} - \alpha_2' p_{E^*S} \\ &= \beta_2 p_{E^*S} - \beta_2^* c_P p_{E^*} \\ &= \frac{K_* (\alpha_2^* \beta_2 c_S - \alpha_2' \beta_2^* c_P)}{(\alpha_2' + \beta_2)\xi_{ss}}. \end{aligned} \tag{16.62}$$

If we again ignore the usually small terms in c_P, the dependence of J on c_L and c_S is

$$J = \frac{A c_S}{B + C c_L + D c_S}. \tag{16.63}$$

When $c_L = 0$ or $c_L =$ constant, the dependence of J on c_S is conventional [Eq. (16.58)]. At constant c_S, J is maximized at $c_L = 0$ and decreases as c_L increases. When $c_L \to \infty$, $J \to 0$. In this limit, the enzyme is all in state LE: ligand binding removes E from the active cycle E*, E*S.

17. One-Conformation Complexes with Strong Interactions: Missing States

In this section and the next we study examples of steady-state systems that have three features in common: the independent unit is an enzyme complex comprised of two or more enzyme molecules ("subunits") that interact with each other; these intermolecular interactions are so strong that certain states or transitions of the kinetic diagram of the complex are completely missing (i.e., excluded by the interactions); and each of the enzyme molecules of the complex exists in only one conformation. Two-conformation examples are more complicated and are reserved for Section 19.

The examples in this section illustrate diagrams with missing states. In the next section we examine a few cases in which the diagram has missing transitions. To keep the analysis simple, in all of the examples in both sections, we use an all-or-none kind of assumption: intermolecular interactions completely remove some states or transitions from the diagram, but they have no effect on the remaining rate constants. That is, the rate constants for transitions still left in the diagram are unperturbed rate constants. The more general quantitative approach to these problems, in which neither extreme assumption (all-or-none states or transitions) need be made, is presented in Chapter 7.

Two-State Dimer ($\rho = 2$, $m = 2$)

We use m for the number of enzyme molecules or subunits in the complex and ρ for the number of states in each enzymatic cycle. We first consider a complex (dimer) of two enzymes E_1 and E_2, shown schematically in Fig. 5-20(a). In the most important special case, E_1 and E_2 are the same enzyme, E, but it is easy to be more general, so we start in this way. E_1 catalyzes $S_1 \rightarrow P_1$, whereas E_2 catalyzes $S_2 \rightarrow P_2$. The cycles for the separate enzymes are shown in Fig. 5-20(b) with first-order rate constant notation. In both cases, E is called state 1 and ES is state 2. The complex pictured in Fig. 5-20(a) then has four states, labelled 11, 12, 21, 22, as in Fig. 5-20(c). The left-hand index gives the state of E_1 and the right-hand index refers to E_2.

The rate constants in Fig. 5-20(c) (this is a kinetic diagram for the dimer) are the same as in Fig. 5-20(b) (primed rate constants are associated with the inverse arrows). Because of this, Fig. 5-20(c) represents a dimer in which one enzyme has no kinetic influence on the other, despite the proximity of the two molecules. In Chapter 7 we shall consider the general case in which the rate constants of E_1 depend in a quantitative way, through intermolecular forces, on the state of E_2, and vice versa. Here we confine ourselves to a particular special case: we assume that state 22 (i.e., $E_1S_1 \cdot E_2S_2$) is relatively unstable and can be ignored (omitted from the diagram), but that otherwise the two enzymes do not perturb each other. For example, the two bound substrates in state 22 might interfere with each other. Thus the diagram we deal with is that shown in Fig. 5-20(d) (again primed rate constants are associated with inverse arrows).

Figure 5-20(d) is an elementary example of so-called half-of-the-sites reactivity. In this case, both subunits of the dimer can cycle with unperturbed rate constants, but only one subunit can cycle at a time because of the instability of state 22. This will reduce the fluxes J_1 and J_2 below their unperturbed

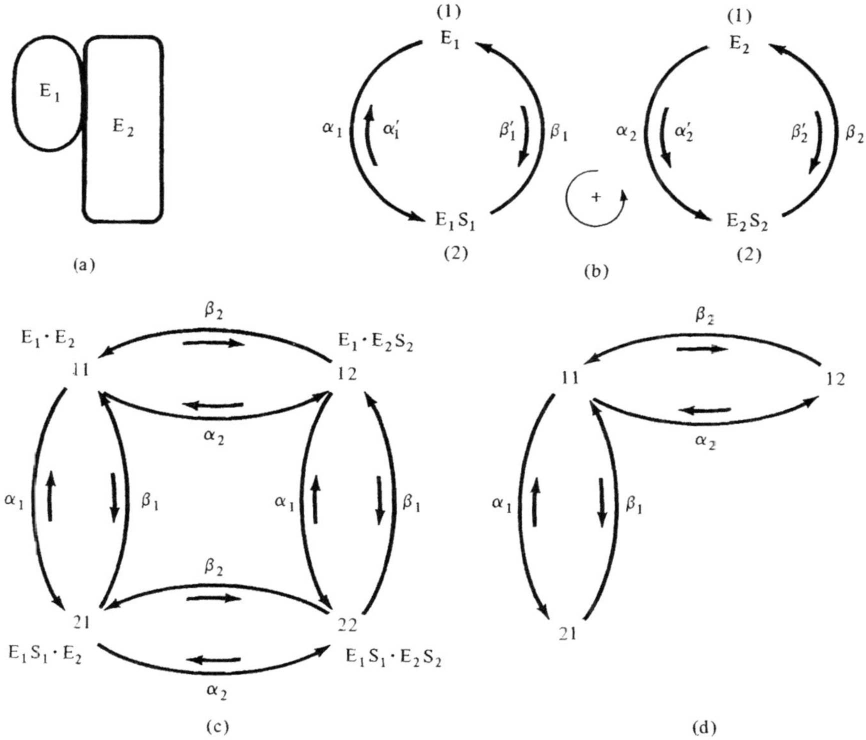

Fig. 5-20. (a) Two enzyme molecules in a complex (dimer). (b) Cycles and first-order rate constant notation for the separate enzymes. (c) Diagram and rate constants for dimer if E_1 and E_2 do not perturb each other. Primed inverse rate constants have been omitted to simplify the figure. (d) Special case in which state 22 is relatively unstable and can be omitted.

values [Fig. 5-20(b)] J_{o1} and J_{o2}. We now derive the simple relations between these fluxes.

The unperturbed fluxes are, from Fig. 5-20(b) [see Eq. (9.5)],

$$J_{o1} = \frac{\alpha_1\beta_1 - \alpha_1'\beta_1'}{\alpha_1 + \beta_1' + \alpha_1' + \beta_1} = \frac{\alpha_1\beta_1 - \alpha_1'\beta_1'}{(1 + x_1)(\alpha_1' + \beta_1)}, \tag{17.1}$$

$$J_{o2} = \frac{\alpha_2\beta_2 - \alpha_2'\beta_2'}{\alpha_2 + \beta_2' + \alpha_2' + \beta_2} = \frac{\alpha_2\beta_2 - \alpha_2'\beta_2'}{(1 + x_2)(\alpha_2' + \beta_2)}, \tag{17.2}$$

where

$$x_1 \equiv \frac{\alpha_1 + \beta_1'}{\alpha_1' + \beta_1}, \qquad x_2 \equiv \frac{\alpha_2 + \beta_2'}{\alpha_2' + \beta_2}, \tag{17.3}$$

and [see Eq. (16.7)]

$$\alpha_1 = \alpha_1^* c_{S_1}, \qquad \beta_1' = \beta_1^* c_{P_1}, \qquad \alpha_2 = \alpha_2^* c_{S_2}, \qquad \beta_2' = \beta_2^* c_{P_2}. \tag{17.4}$$

The second form used for both J_{o1} and J_{o2} is for comparison below. As in Eqs. (16.52) and (16.60),

$$x_1 = \frac{\alpha_1^* c_{S_1} + \beta_1^* c_{P_1}}{\alpha_1' + \beta_1}, \qquad x_2 = \frac{\alpha_2^* c_{S_2} + \beta_2^* c_{P_2}}{\alpha_2' + \beta_2}. \tag{17.5}$$

As mentioned following Eq. (16.52), both x_1 and x_2 are steady-state analogues of Kc_S in the equilibrium binding of S with binding constant K. If β_1' and β_2' are negligible (the Michaelis–Menten case), then $x_1 \propto c_{S_1}$ and $x_2 \propto c_{S_2}$.

The above unperturbed fluxes, J_{o1} and J_{o2}, also follow from the diagram for the dimer in Fig. 5-20(c). The Fig. 5-20(c) diagram "factors," so to speak, into the two diagrams in Fig. 5-20(b). If the two molecules in the dimer operate independently of each other, there is of course no practical reason why the full dimer diagram need be considered. It was introduced here simply for pedagogical purposes.

Turning now to the case of real interest, Fig. 5-20(d), we note that this is another diagram that is topologically linear. Consequently, the solution of the steady-state linear equations in the p_{ij} has a "detailed balance" form:

$$p_{21} = x_1 p_{11}, \qquad p_{12} = x_2 p_{11}. \tag{17.6}$$

Hence

$$\begin{aligned} p_{11} &= \frac{1}{1 + x_1 + x_2}, \\ p_{21} &= \frac{x_1}{1 + x_1 + x_2}, \\ p_{12} &= \frac{x_2}{1 + x_1 + x_2}. \end{aligned} \tag{17.7}$$

The denominator has the appearance of a grand partition function for a site with competitive equilibrium binding [Eq. (7.3)]. The steady-state competition is more complicated [Eq. (17.5)] than simple equilibrium binding, but the overall result (binding of S_1 or S_2) is similar.

The two steady-state fluxes are then

$$\begin{aligned} J_1 &= \alpha_1 p_{11} - \alpha_1' p_{21} \\ &= \frac{\alpha_1 \beta_1 - \alpha_1' \beta_1'}{(1 + x_1 + x_2)(\alpha_1' + \beta_1)}, \end{aligned} \tag{17.8}$$

$$J_2 = \frac{\alpha_2 \beta_2 - \alpha_2' \beta_2'}{(1 + x_1 + x_2)(\alpha_2' + \beta_2)}. \tag{17.9}$$

These differ from Eqs. (17.1) and (17.2) only in the "grand partition functions" in the denominators:

$$\frac{J_1}{J_{o1}} = \frac{1 + x_1}{1 + x_1 + x_2}, \tag{17.10}$$

$$\frac{J_2}{J_{o2}} = \frac{1 + x_2}{1 + x_1 + x_2}. \tag{17.11}$$

As expected, $J_1 \leq J_{o1}$ and $J_2 \leq J_{o2}$, because of the steady-state competition between states 21 and 12 in Fig. 5-20(d). The result in Eqs. (17.10) and (17.11) is due simply to the spreading of the steady-state probability distribution over more states.

If the two subunits of the dimer are identical,

$$\frac{J}{J_o} = \frac{1 + x}{1 + 2x}, \tag{17.12}$$

$$x = \frac{\alpha + \beta'}{\alpha' + \beta} = \frac{\alpha^* c_S + \beta^* c_P}{\alpha' + \beta}. \tag{17.13}$$

If β' is negligible in Eq. (17.13) (Michaelis–Menten), $x \propto c_S$. When x is small, $J/J_o \rightarrow 1$ (when $x \rightarrow 0$, state 22 is hardly used even if not excluded by intermolecular forces). When x is very large, $J/J_o \rightarrow \frac{1}{2}$. In the dimer, at saturation with S, each E is in state ES only half the time.

If both states 11 and 22 are unstable compared to 21 and 12, and are eliminated from the diagram [see Fig. 5-20(c)], cycling could occur only by concerted transitions 12 → 21 and 21 → 12. These transitions are not included in Fig. 5-20(c). Such strictly concerted transitions are rather implausible. Instead, a more likely mechanism would be from 12 to 21 or 21 to 12 via 11 and 22 as transient intermediates. This problem will be dealt with in Chapter 7.

Three-State Dimer ($\rho = 3$, $m = 2$)

We again consider the hypothetical dimer in Fig. 5-20(a), but now we assume that both E_1 and E_2 require a three-state cycle to represent the kinetics adequately. Figure 5-21(a) shows the unperturbed cycles, the numbering of the states, and the first-order rate-constant notation. We are again using the reactions $S_1 \rightarrow P_1$ and $S_2 \rightarrow P_2$ as an illustration, but other three-state cycles would be treated in essentially the same way. The pseudo-first-order rate constants are

$$\alpha_1 = \alpha_1^* c_{S_1}, \qquad \gamma_1' = \gamma_1^* c_{P_1}, \qquad \alpha_2 = \alpha_2^* c_{S_2}, \qquad \gamma_2' = \gamma_2^* c_{P_2}. \tag{17.14}$$

The complete diagram for the dimer is shown in Fig. 5-21(b), where the left-hand index refers to E_1, the right-hand index to E_2, and the arrows show the positive cycle direction. This full diagram is needed to handle arbitrary interaction effects between E_1 and E_2 (Chapter 7). In this subsection, we shall examine three special cases in which strong interactions between the subunits

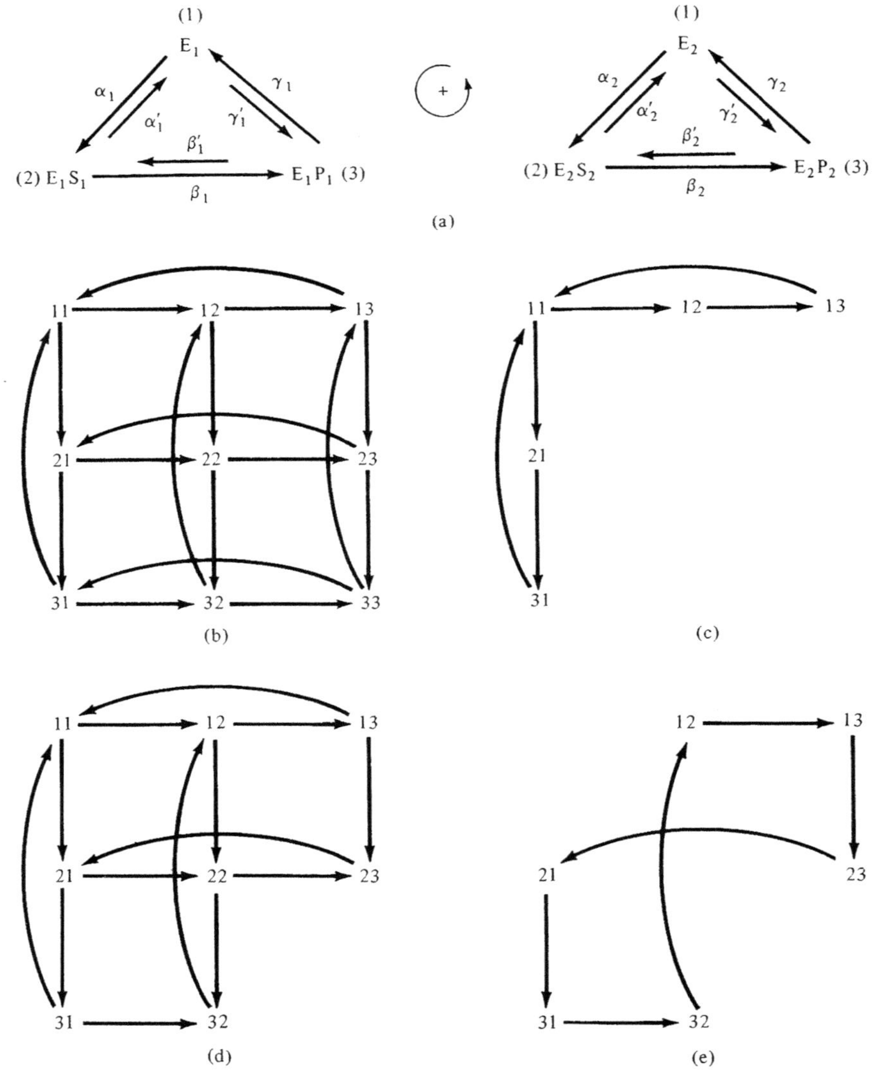

Fig. 5-21. (a) The enzymes in Fig. 5-20(a) require three-state cycles here. Unperturbed first-order rate constants and numbering of states. (b) Complete diagram for dimer. State 12 is $E_1 \cdot E_2S_2$, etc. The arrows show the positive cycle directions (in general, inverse transitions are included). (c) Special case in which states 22, 23, 32, and 33 are unstable. (d) Special case in which state 33 is unstable. (e) Special case in which states 11, 22, and 33 are unstable. See text for further details.

of the dimer have the effect of deleting one or more states from the full diagram, but otherwise leave the rate constants [Fig. 5-21(a)] unchanged. The three cases are shown in Figs. 5-21(c), 5-21(d), and 5-21(e). In Fig. 5-21(c), there is again half-of-the-sites reactivity, but now with three-state cycles. The missing states in the diagram are 22, 32, 23, and 33. This could arise, for example, if bound substrates and products interfere with each other. Figure 5-21(d) refers to a less extreme case in which only state 33 is excluded (e.g., the two bound products interfere with each other, or the bound products induce new conformations $E_1^*P_1$ and $E_2^*P_2$ that interfere with each other). In Fig. 5-21(e), all like states of the dimer (11, 22, 33) are unstable and are omitted from the diagram. It was mentioned at the end of the preceding subsection that, with two-state cycles, if like states (11, 22) are excluded, cycling can no longer occur without the introduction of new and implausible concerted transitions. However, as can be seen in Fig. 5-21(e), this feature does not arise with three-state cycles: there remains in Fig. 5-21(e) a single six-state cycle in which both E_1 and E_2 can cycle by alternating transitions. Note that if all of states 13, 23, 31, and 32 were transient intermediates, we would be left with a quasiconcerted two-state cycle, $12 \rightarrow 21 \rightarrow 12$.

We return now to Fig. 5-21(c) to work out the details of this first case. The diagram is rewritten in Fig. 5-22; the inverse arrows have primed rate constants. The five linear equations in the p_{ij} are most easily solved in this case by the diagram method (see Ref. 2). The result is

$$
\begin{aligned}
p_{21} &= t_{21}r_2/\Sigma, \qquad p_{31} = t_{31}r_2/\Sigma, \\
p_{12} &= t_{12}r_1/\Sigma, \qquad p_{13} = t_{13}r_1/\Sigma, \\
p_{11} &= r_1r_2/\Sigma, \\
\Sigma &= r_1r_2 + (t_{21} + t_{31})r_2 + (t_{12} + t_{13})r_1,
\end{aligned}
\tag{17.15}
$$

where

$$
\begin{aligned}
r_1 &= \beta_1\gamma_1 + \alpha_1'\beta_1' + \alpha_1'\gamma_1, \\
r_2 &= \beta_2\gamma_2 + \alpha_2'\beta_2' + \alpha_2'\gamma_2,
\end{aligned}
\tag{17.16}
$$

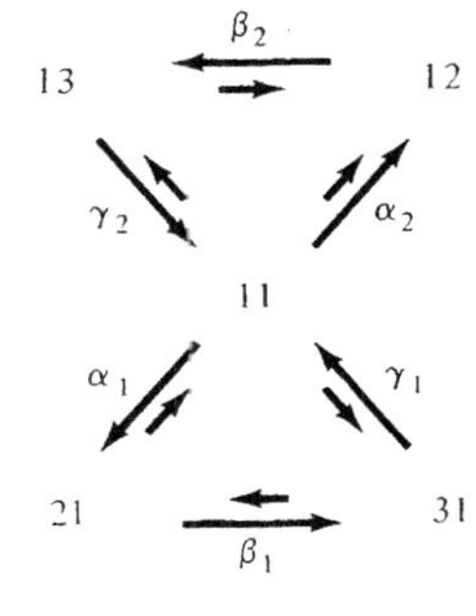

Fig. 5-22. This is the same as Fig. 5-21(c). Primed inverse rate constants are associated with the interior arrows.

$$\begin{aligned}
t_{21} &= \alpha_1\beta_1' + \beta_1'\gamma_1' + \alpha_1\gamma_1, \\
t_{31} &= \alpha_1\beta_1 + \beta_1\gamma_1' + \alpha_1'\gamma_1', \\
t_{12} &= \alpha_2\beta_2' + \beta_2'\gamma_2' + \alpha_2\gamma_2, \\
t_{13} &= \alpha_2\beta_2 + \beta_2\gamma_2' + \alpha_2'\gamma_2'.
\end{aligned} \tag{17.16}$$

The physical significance of these six quantities is illustrated in Fig. 5-23 for r_1 and t_{21}.

The steady-state flux for E_1 follows from the diagram method or from

$$\begin{aligned}
J_1 &= \alpha_1 p_{11} - \alpha_1' p_{21} \\
&= \frac{\alpha_1\beta_1\gamma_1 - \alpha_1'\beta_1'\gamma_1'}{r_1(1 + t_{21}r_1^{-1} + t_{31}r_1^{-1} + t_{12}r_2^{-1} + t_{13}r_2^{-1})}.
\end{aligned} \tag{17.17}$$

This is the analogue of Eq. (17.8) for two-state cycles. The sum in the denominator is Σ/r_1r_2. Similarly,

$$J_2 = \frac{\alpha_2\beta_2\gamma_2 - \alpha_2'\beta_2'\gamma_2'}{r_2(1 + t_{21}r_1^{-1} + t_{31}r_1^{-1} + t_{12}r_2^{-1} + t_{13}r_2^{-1})}. \tag{17.18}$$

For comparison, the unperturbed fluxes may be found by putting $\alpha_2 = 0$, $\gamma_2' = 0$ in J_1 and $\alpha_1 = 0$, $\gamma_1' = 0$ in J_2:

$$t_{12} = 0, \qquad t_{13} = 0,$$

$$J_{o1} = \frac{\alpha_1\beta_1\gamma_1 - \alpha_1'\beta_1'\gamma_1'}{r_1(1 + t_{21}r_1^{-1} + t_{31}r_1^{-1})}, \tag{17.19}$$

and

$$t_{21} = 0, \qquad t_{31} = 0,$$

$$J_{o2} = \frac{\alpha_2\beta_2\gamma_2 - \alpha_2'\beta_2'\gamma_2'}{r_2(1 + t_{12}r_2^{-1} + t_{13}r_2^{-1})}. \tag{17.20}$$

Equations (17.19) and (17.20) are of course examples of Eq. (10.5). Upon examining J_1 and J_{o1}, we see that the expressions are the same except for missing terms in the denominator of J_{o1}. Thus,

$$\frac{J_1}{J_{o1}} = \frac{1 + t_{21}r_1^{-1} + t_{31}r_1^{-1}}{1 + t_{21}r_1^{-1} + t_{31}r_1^{-1} + t_{12}r_2^{-1} + t_{13}r_2^{-1}}, \tag{17.21}$$

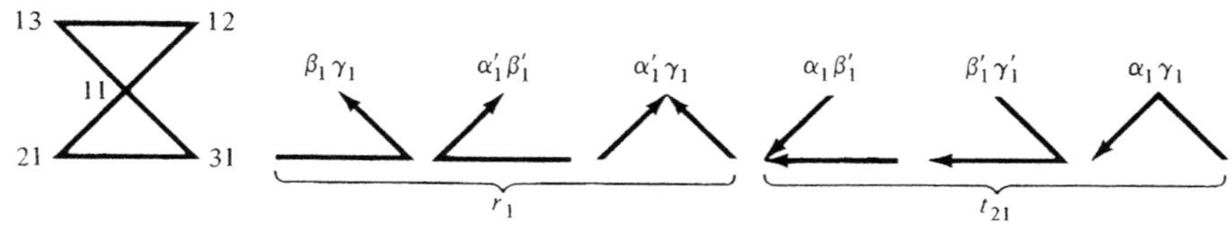

Fig. 5-23. Graphical interpretation of the terms r_1 and t_{21} in Eq. (17.16).

with a similar equation for J_2/J_{o2}. Hence $J_1 \leq J_{o1}$ and $J_2 \leq J_{o2}$. The sums here also resemble grand partition functions in that the steady-state probabilities of states are proportional to the separate terms in the sums. The leading term in both sums relates to state 11; the other terms can be associated with states from the subscripts on t_{ij}. Again, as in Eqs. (17.10) and (17.11), the probability distribution is spread over more states in the case of J_1 and J_2 (Fig. 5-22) than in the case of J_{o1} and J_{o2} [Fig. 5-21(a)].

If the two enzyme molecules in the dimer are identical, we can drop the subscripts on all of the rate constants. Then

$$r_1 = r_2 \equiv r, \qquad t_{21} = t_{12}, \qquad t_{31} = t_{13}, \tag{17.22}$$

$$J = \frac{\alpha\beta\gamma - \alpha'\beta'\gamma'}{r(1 + 2t_{21}r^{-1} + 2t_{31}r^{-1})}, \tag{17.23}$$

$$J_o = \frac{\alpha\beta\gamma - \alpha'\beta'\gamma'}{r(1 + t_{21}r^{-1} + t_{31}r^{-1})}. \tag{17.24}$$

These are fluxes per subunit of the dimer. The dependences on c_S and c_P are $\alpha = \alpha^* c_S$ and $\gamma' = \gamma^* c_P$. Both t_{21} and t_{31} are linear in c_S and c_P [Eqs. (17.16)], but r is independent of c_S and c_P. In the special case where γ' is negligibly small,

$$t_{21} = \alpha(\beta' + \gamma), \qquad t_{31} = \alpha\beta, \tag{17.25}$$

$$\frac{J}{J_o} = \frac{1 + \alpha^* c_S(\beta' + \gamma)r^{-1} + \alpha^* c_S \beta r^{-1}}{1 + 2\alpha^* c_S(\beta' + \gamma)r^{-1} + 2\alpha^* c_S \beta r^{-1}}. \tag{17.26}$$

Consequently,

$$\begin{aligned} J/J_o &\to 1 \quad \text{when} \quad c_S \to 0, \\ J/J_o &\to \tfrac{1}{2} \quad \text{when} \quad c_S \to \infty. \end{aligned} \tag{17.27}$$

When c_S is very large, the state probability distribution is

$$\begin{aligned} p_{21} = p_{12} &= (\beta' + \gamma)/2(\beta' + \gamma + \beta), \\ p_{31} = p_{32} &= \beta/2(\beta' + \gamma + \beta). \end{aligned} \tag{17.28}$$

The next example is Fig. 5-21(d). However, we consider only a very limited special case of this model: the two subunits of the dimer are identical; all inverse (primed) rate constants are negligible (the cycles are one way); and $\alpha = \beta = \gamma$. This special case (which we shall use below for more complicated models) is adequate to give a qualitative idea of the effect on the flux per molecule of eliminating states from the full diagram. This case is also shown in Fig. 5-24(c), where every arrow has a rate constant α and there are no inverse transitions. For comparison, Figs. 5-24(b) and 5-24(d) also have a rate constant α on each arrow, but no states are missing in Fig. 5-24(b) and four states are missing in Fig. 5-24(d). Thus Figs. 5-24(b), 5-24(c), and 5-24(d) represent an ordered sequence. Actually, Fig. 5-24(b) is equivalent

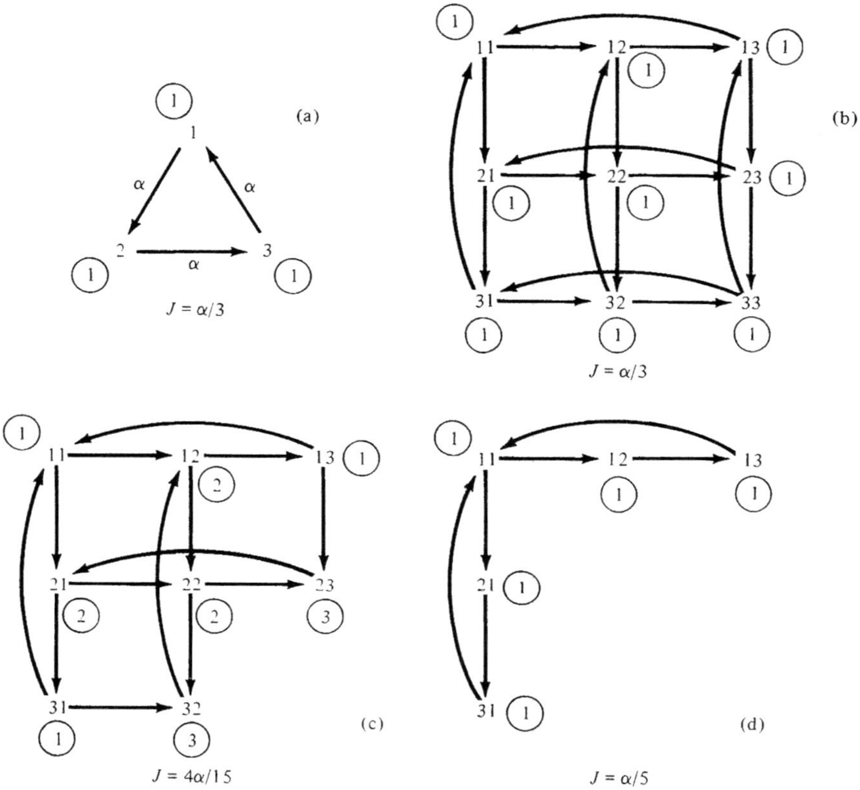

Fig. 5-24. (a) Monomer with one-way cycle and all rate constants α. The numbers in circles are the relative probabilities of states at steady state. J is the steady-state flux. (b) (c) (d) Three special cases for the dimer, the latter two with missing states. J is the flux per subunit. See text for further details.

to Fig. 5-24(a) [i.e., the two molecules in the dimer in Fig. 5-24(b) are independent of each other] and Fig. 5-24(d) is a special case of Fig. 5-22, just considered.

In all of the diagrams of Fig. 5-24, the numbers in circles are the *relative* probabilities of the states at steady state if all arrows have the same rate constant α. The only nontrivial case is Fig. 5-24(c). The circled numbers have to be assigned in such a way that the flow of circled numbers (state probabilities) into any state is equal to the flow out of the state. This simple rule applies only because all rate constants are α. For example, for state 23 the flow in is $1 + 2$ and the flow out is 3, and for state 22 the flow in is $2 + 2$ and the flow out is $2 + 2$.

The sum of the circled numbers in Fig. 5-24(c) is 15. Hence $p_{11} = \frac{1}{15}$, $p_{12} = \frac{2}{15}$, etc. The total rate of production of product, P, by the dimer is

$$2J = \alpha(p_{13} + p_{31} + p_{23} + p_{32}) = 8\alpha/15.$$

Hence the flux per subunit is $J = 4\alpha/15$, as indicated in the figure. In the sequence of Figs. 5-24(b), 5-24(c), and 5-24(d), the fluxes are $5\alpha/15$, $4\alpha/15$, and $3\alpha/15$, decreasing as we should expect as more states are excluded from the diagram. The flux for Fig. 5-24(d) ($J = \alpha/5$) also follows from Eq. (17.23) on putting $r = t_{21} = t_{31} = \alpha^2$, and the flux for Fig. 5-24(b) is also given by Eq. (17.24).

The third example of a three-state dimer with missing states owing to strong interactions is Fig. 5-21(e); states 11, 22, and 33 are excluded. We consider first the special case: the two subunits are identical; and the inverse (primed) rate constants are negligible. Figure 5-25(a) shows the unperturbed diagram for one molecule and Figs. 5-25(b) and 5-25(c) give two equivalent representations of the diagram for the dimer. Using the diagram method (see Appendix 4 of Ref. 2), we find that the rate at which the dimer completes the six-state cycle in Fig. 5-25(c) is

$$\frac{\alpha\beta\gamma}{2(\alpha\beta + \alpha\gamma + \beta\gamma)}.$$

However, in each of these six-state cycles, two Ps are produced. Hence the dimer P flux is twice the above expression. Therefore the P flux *per subunit* is

$$J = \frac{\alpha\beta\gamma}{2(\alpha\beta + \alpha\gamma + \beta\gamma)}. \tag{17.29}$$

On the other hand, the unperturbed flux per subunit [Fig. 5-25(a)] is

$$J_o = \frac{\alpha\beta\gamma}{\alpha\beta + \alpha\gamma + \beta\gamma}. \tag{17.30}$$

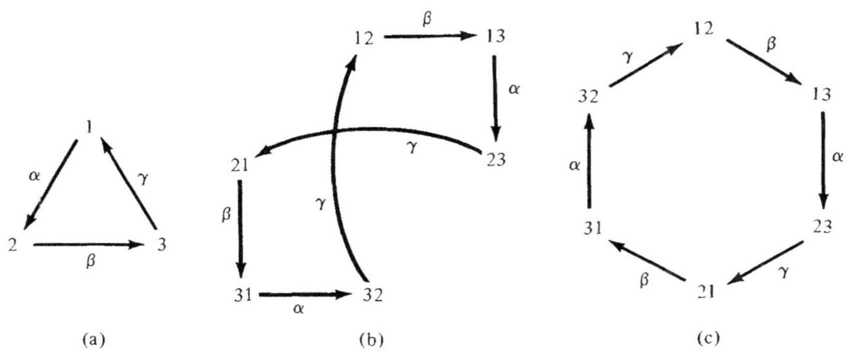

Fig. 5-25. (a) Monomer with one-way cycle. (b) (c) Two equivalent representations of the diagram for a dimer, with missing states. See text for further details.

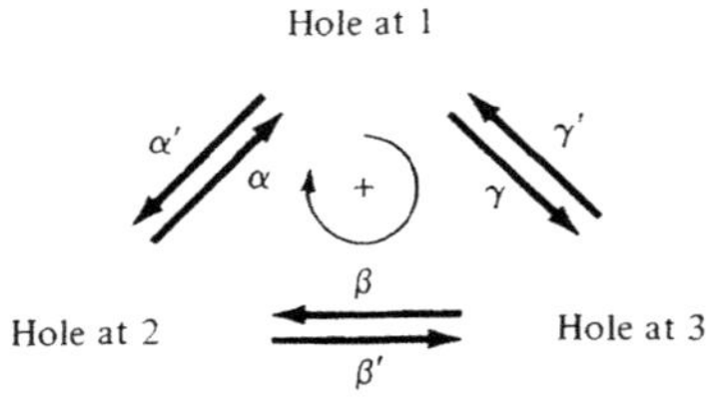

Fig. 5-26. Inclusion of inverse rate constants in Figs. 5-25(b) and 5-25(c); treatment of kinetics by following missing enzyme state (a "hole"). See text for details.

Thus $J = J_o/2$: the missing states cut the flux in half. Because of the missing states, the two subunits are obliged to make alternate transitions in Figs. 5-25(b) and 5-25(c); hence the factor of two is intuitively plausible.

For comparison with the fluxes in Fig. 5-24, we note that if $\alpha = \beta = \gamma$, Eq. (17.29) gives $J = \alpha/6$, a smaller flux than in Fig. 5-24(d).

The model in Figs. 5-25(b) and 5-25(c) can also be handled easily if inverse transitions are included. This is best done by following the vacant enzyme state (a "hole") in Figs. 5-25(b) and 5-25(c). Thus, for dimer state 12, the hole is at 3; for dimer state 13, the hole is at 2; etc. The hole has a cycle with rate constants shown in Fig. 5-26. Note that the sequence of rate constants in the + direction is α, γ, β compared to α, β, γ in Fig. 5-21(a). One P is produced *by the dimer* per hole cycle (clockwise). The hole cycle flux is given by Eq. (10.5). The subunit P flux J then follows upon division by two. Thus

$$J = \frac{\alpha\beta\gamma - \alpha'\beta'\gamma'}{2(\alpha\beta + \beta'\gamma' + \alpha\gamma' + \alpha'\beta + \alpha'\gamma' + \beta\gamma + \alpha\gamma + \alpha'\beta' + \beta'\gamma)}. \tag{17.31}$$

Incidentally, the nine terms in the denominator are not all the same as in the denominator of J_o from Fig. 5-25(a) if primed rate constants are included there. Hence J is not exactly $J_o/2$ when inverse transitions are included. Equation (17.31) reduces to Eq. (17.29) if the primed rate constants are negligible.

Four-State Dimer *($\rho = 4$, $m = 2$)*

In this subsection we extend the above discussion, rather briefly, to a complex of two four-state enzyme molecules. For simplicity, we consider one-way cycles only. Figure 5-27(a) shows a plausible illustrative mechanism for $S \rightarrow P$ for each subunit of the dimer. The kinetic diagram for the dimer appears in Fig. 5-27(b); this is an obvious extension of Fig. 5-21(b). The five special cases treated below are, in the order considered: the unperturbed enzyme (E_1 or E_2) with the Fig. 5-27(a) cycle; cases a, b, and c in Fig. 5-28, where 9, 4, or 1 states are excluded from the diagram because of strong intermolecular interactions; and Fig. 5-29, where the two enzyme molecules are unstable if both are in the same state (dimer states 11, 22, 33, and 44 are excluded).

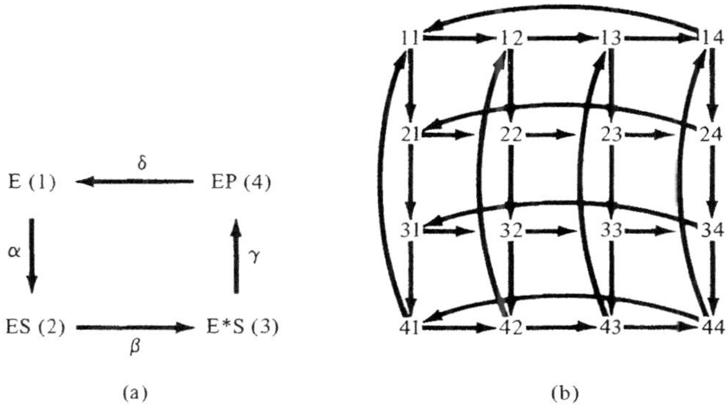

Fig. 5-27. (a) One-way four-state cycle for a monomer. (b) Corresponding diagram for a dimer.

The unperturbed flux for E_1 or E_2 follows easily from Fig. 5-27(a) and the diagram method:

$$J_{oi} = \frac{\alpha_i \beta_i \gamma_i \delta_i}{\beta_i \gamma_i \delta_i + \alpha_i \gamma_i \delta_i + \alpha_i \beta_i \delta_i + \alpha_i \beta_i \gamma_i} \qquad (i = 1, 2)$$

$$= \frac{\alpha_i}{1 + y_i} \qquad (i = 1, 2), \tag{17.32}$$

where

$$y_i = \frac{\alpha_i}{\beta_i} + \frac{\alpha_i}{\gamma_i} + \frac{\alpha_i}{\delta_i} \qquad (i = 1, 2). \tag{17.33}$$

Again, $1 + y_i$ resembles a grand partition function: the successive terms in $1 + y_1$, for example, are proportional to the steady-state probabilities of states 1, 2, 3, and 4 for E_1 in Fig. 5-27(a). In the special reference case in

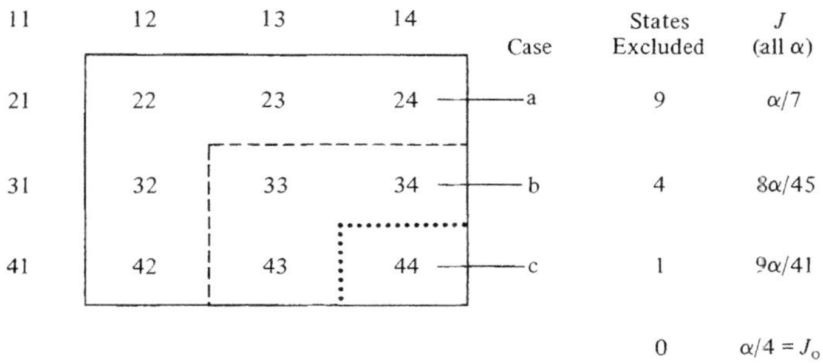

Fig. 5-28. Flux per subunit J for different sets of excluded states, with all rate constants α in Fig. 5-27.

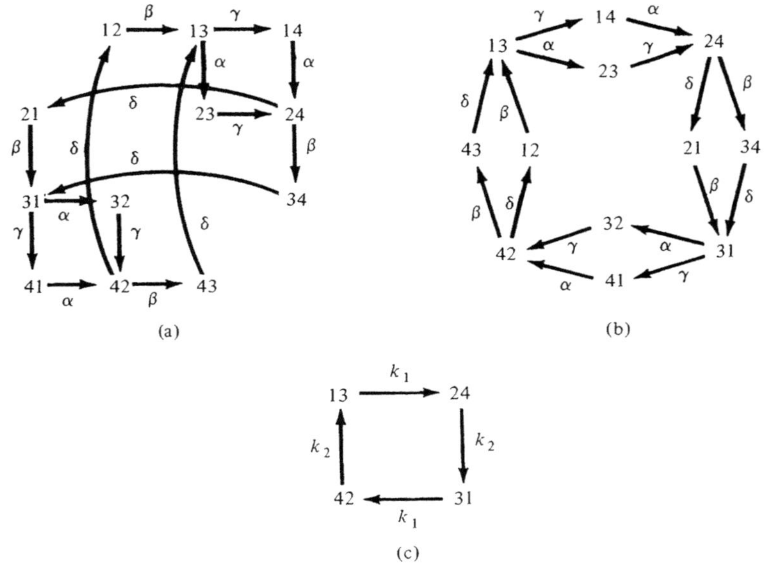

Fig. 5-29. (a) Diagram for two identical subunits, one-way transitions, and missing states 11, 22, 33, and 44. (b) Topologically equivalent diagram. (c) Condensed version of (b) with effective rate constants k_1 and k_2.

which all rate constants are equal to α, $J_o = \alpha/4$ (this is included in Fig. 5-28).

The diagram for case a (Fig. 5-28) is shown more explicitly in Fig. 5-30. Only one subunit of the pair can cycle at a time. The steady-state probabilities are easy to find. These are proportional to terms in an effective grand partition function:

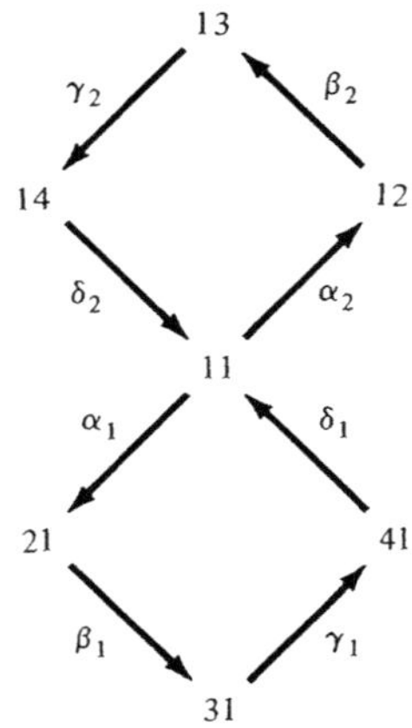

Fig. 5-30. Diagram for case a in Fig. 5-28 (i.e., nine dimer states are excluded), with two different subunits.

$$1 + y_1 + y_2 = 1 + \frac{\alpha_1}{\beta_1} + \frac{\alpha_1}{\gamma_1} + \frac{\alpha_1}{\delta_1} + \frac{\alpha_2}{\beta_2} + \frac{\alpha_2}{\gamma_2} + \frac{\alpha_2}{\delta_2}. \tag{17.34}$$

11 21 31 41 12 13 14

The flux for E_i ($i = 1, 2$) in Fig. 5-30 is then

$$J_i = \frac{\alpha_i}{1 + y_1 + y_2} \qquad (i = 1, 2) \tag{17.35}$$

and

$$\frac{J_i}{J_{oi}} = \frac{1 + y_i}{1 + y_1 + y_2} \qquad (i = 1, 2). \tag{17.36}$$

This is the same type of result as in Eqs. (17.10) and (17.11); necessarily $J_i \leq J_{oi}$. In the special case of Eq. (17.35), where all rate constants are α, $J = \alpha/7$ (included in Fig. 5-28).

We now turn to cases b and c in Fig. 5-28, but limit ourselves to the special case in which the two subunits are identical and all rate constants are α. Thus we are considering generalizations of Fig. 5-24(c). The discussion of this figure should be reviewed. The relative steady-state probabilities of states, expressed as integers, are found to be:

Case b, Fig. 5-28:				Case c, Fig. 5-28:				
1	4	2	1	7	8	14	7	
4	4	6	7	8	8	11	9	(17.37)
2	6			14	11	11	20	
1	7			7	9	20		

These numbers are arranged in the same order as in Fig. 5-28, and their correctness can easily be verified. For example, in the right-hand case, the flow into state 31 is $8 + 20 = 28$ and the flow out of state 31 is $14 + 14 = 28$. The steady-state fluxes per subunit in the above two cases are then found to be $8\alpha/45$ and $36\alpha/164 = 9\alpha/41$ (obtained, for example, by dividing the sum of the right-hand column of the array by the sum of the whole array). These fluxes are included in Fig. 5-28. It will be seen in this figure that, as expected, the flux per subunit in these all-α cases decreases as the number of excluded states increases.

The final case in this subsection is presented in Fig. 5-29(a): the two subunits are identical; there are one-way transitions only; and the two subunits are never in the same state. Figure 5-29(b) is merely a convenient rearrangement of Fig. 5-29(a). Figure 5-29(c) makes the point that there is, in effect, a single rate constant k_1 for $13 \rightarrow 24$ and $31 \rightarrow 42$ and another constant k_2 for $24 \rightarrow 31$ and $42 \rightarrow 13$. If we take k_1 as the reciprocal of the mean time to pass from 13 to 24, then we find

$$k_1 = \frac{(\alpha + \gamma)\alpha\gamma}{\alpha^2 + \alpha\gamma + \gamma^2} \tag{17.38}$$

and, similarly,

$$k_2 = \frac{(\beta + \delta)\beta\delta}{\beta^2 + \beta\delta + \delta^2}. \tag{17.39}$$

Let J_{cy} be the cycle flux in Fig. 5-29(c). Upon going around this cycle once, two S → P conversions are accomplished. Hence the S → P flux is $2J_{cy}$, for a dimer. For a single subunit, then, the S → P flux is $J = J_{cy}$. From Fig. 5-29(c) and the diagram method,

$$\begin{aligned} J &= \frac{k_1^2 k_2^2}{2k_1 k_2^2 + 2k_1^2 k_2} \\ &= \frac{k_1 k_2}{2(k_1 + k_2)}. \end{aligned} \tag{17.40}$$

In the all-α special case,

$$\begin{aligned} k_1 &= k_2 = 2\alpha/3, \\ J &= \alpha/6. \end{aligned} \tag{17.41}$$

This should be compared with the all-α values of J in Fig. 5-28.

Five-State Dimer ($p = 5$, $m = 2$)

In this case we merely give the one-way all-α results. The relative probability arrays corresponding to Eq. (17.37) are

1	8	4	2	1	3	4	12	6	3	20	23	25	40	20
8	8	12	14	15	4	4	8	7	5	23	23	24	32	26
4	12				12	8	8	15	20	25	24	24	28	27
2	14				6	7	15			40	32	28	28	55
1	15				3	5	20			20	26	27	55	

(17.42)

The fluxes per subunit for the various cases are:

$$\begin{aligned} &\text{16 states excluded,} \quad J = \alpha/9 = 0.1111\alpha; \\ &\text{9 states excluded,} \quad J = 16\alpha/121 = 0.1322\alpha; \\ &\text{4 states excluded,} \quad J = 4\alpha/25 = 0.1600\alpha; \\ &\text{1 state excluded,} \quad J = 128\alpha/695 = 0.1842\alpha; \\ &\text{0 states excluded,} \quad J_o = \alpha/5 = 0.2000\alpha. \end{aligned} \tag{17.43}$$

For comparison, the flux when all like dimer states are excluded (11, . . . , 55) is $J = 3\alpha/20 = 0.1500\alpha$.

Two-State Trimer ($\rho = 2$, $m = 3$)

This case is a straightforward generalization of Fig. 5-20 from a dimer to a trimer, as illustrated in Fig. 5-31. The three molecules in the complex are generally different. The complete diagram is shown in Fig. 5-31(a), where each line represents a pair of inverse transitions. The geometry of the trimer is not specified at this stage. If the subunits do not alter each other's rate constants, then Fig. 5-31(a) "factors" into three unperturbed cycles, as in Fig. 5-20(b) (one for each of E_1, E_2, E_3).

Figures 5-31(b), 5-31(c), and 5-31(d) represent a hypothetical special case in which only one of the three molecules can be in state 2 (E_iS_i) at a time. This could conceivably arise from strong neighbor interactions in a triangular shaped complex. Instead of half-of-the-sites reactivity, this is third-of-the-sites reactivity. Figure 5-31(c) shows the part of the full diagram [Fig. 5-31(a)] that is used, and Fig. 5-31(d) gives the rate constants [assumed unchanged from Fig. 5-20(b)].

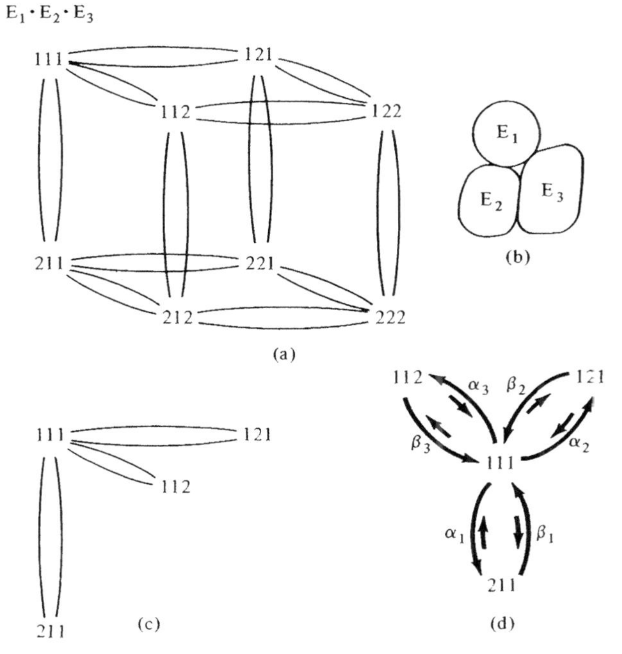

Fig. 5-31. (a) Generalization of Fig. 5-20 to a trimer of two-state enzyme molecules. Each line represents a pair of inverse transitions. (b) (c) (d) Because of a tight fit, only one enzyme can be in state 2 (E_iS_i) at a time. Primed rate constants belong to inverse arrows in (d).

It is easy to see that, at steady state,

$$p_{211} = x_1 p_{111}, \qquad p_{121} = x_2 p_{111}, \qquad p_{112} = x_3 p_{111}, \tag{17.44}$$

just as in Eqs. (17.3) and (17.6). Thus

$$p_{111} = \frac{1}{1 + x_1 + x_2 + x_3}, \qquad p_{211} = \frac{x_1}{1 + x_1 + x_2 + x_3}, \tag{17.45}$$

etc., and

$$J_i = \frac{\alpha_i \beta_i - \alpha_i' \beta_i'}{(1 + x_1 + x_2 + x_3)(\alpha_i' + \beta_i)} \qquad (i = 1, 2, 3), \tag{17.46}$$

$$\frac{J_i}{J_{oi}} = \frac{1 + x_i}{1 + x_1 + x_2 + x_3} \qquad (i = 1, 2, 3), \tag{17.47}$$

where J_{oi} is the same as in Eqs. (17.1) and (17.2). The flux J_i is reduced even more than in Eqs. (17.10) and (17.11) because each enzyme molecule has to share cycling time with two other molecules: only one out of three can cycle at a time.

Three-State Trimer *(ρ = 3, m = 3)*

Here we discuss a triangular trimer of three identical enzyme molecules E [Fig. 5-32(a)], each of which has, effectively, a one-way three-state enzymatic cycle [Fig. 5-32(b)]. The complete diagram for the complex, with

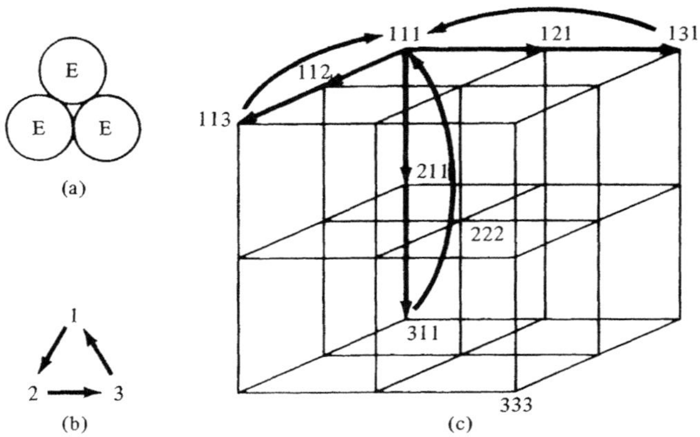

Fig. 5-32. (a) Three identical subunits forming trimer. (b) One-way enzymatic cycle for each independent subunit. (c) Schematic complete diagram for trimer with only partial labelling.

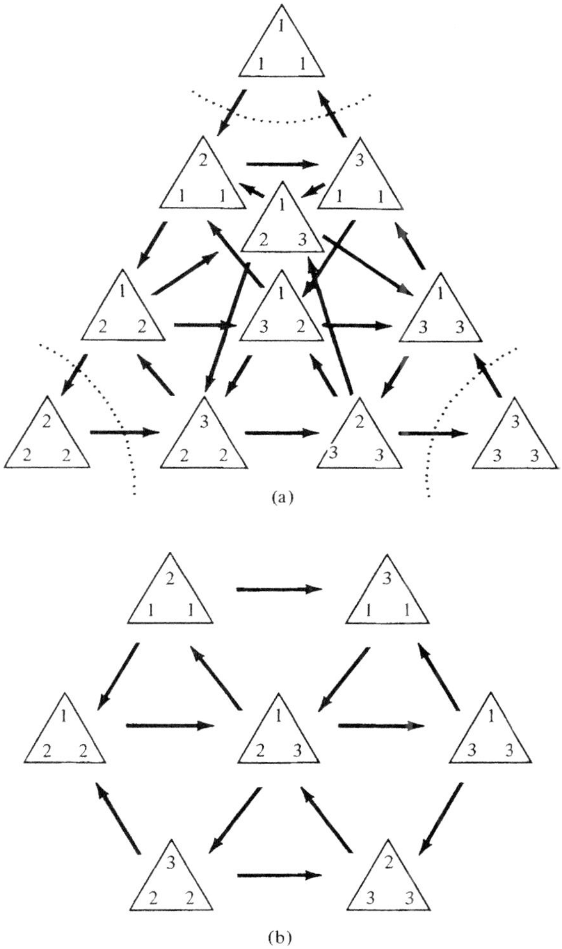

Fig. 5-33. (a) Diagram equivalent to Fig. 5-32(c). See text for details. (b) Special case of (a), discussed in text.

$3^3 = 27$ states, is shown rather schematically in Fig. 5-32(c). A different but equivalent representation of the complete diagram is given in Fig. 5-33(a). Each of the eight nonvertex states in Fig. 5-33(a) has a rotational degeneracy of 3 (e.g., $1^2 1$, $2^1 1$, $1^1 2$), whereas the three vertex states are nondegenerate. Hence the total number of states is $8 \times 3 + 3 = 27$, as in Fig. 5-32(c). Rate constants assigned to the arrows in Fig. 5-33(a) must, of course, take degeneracies into account. The two interior states ($2^1 3$, $3^1 2$) differ from each other because the upper side and lower side (relative to the plane of the paper) of the trimer in Fig. 5-32(a) are generally different. All trimers in Fig. 5-33(a) have the same side facing upward. Thus, turning state 2^13 over, like a pancake, does not produce state 3^12.

If the complex is moving freely in solution, or has rotational mobility in a membrane, then the three members of each of the eight degenerate triplets in Fig. 5-33(a) (e.g., 2^13, 1^32, 3^21) are indistinguishable. If, on the other hand, the complex has a fixed orientation in a membrane, then the three members of each triplet would be distinguishable even though they would have the same kinetic properties [we are assuming that the three Es in Fig. 5-32(a) are identical; this implies that the membrane environment of the complex does not alter the *kinetic* identity of the three Es].

We reserve an analytical discussion of a special case of this model for Chapter 7 because explicit treatment of neighbor interactions cannot be avoided. This subsection is confined to introductory and qualitative remarks.

Gresser, Myers, and Boyer[3] have proposed a mechanism for the H^+-ATPase of the mitochondrial inner membrane that involves (in the present nomenclature) a trimer with three identical subunits, trimer states with no two subunits in the same subunit state, and simultaneous (concerted) subunit state changes [Fig. 5-32(b)] in all three subunits to maintain this restriction. This complex has a fixed orientation in the membrane. An example of simultaneous subunit state changes is

$$2^13 \rightarrow 3^21 \rightarrow 1^32 \rightarrow 2^13. \tag{17.48}$$

This mechanism implies that like pairs in the trimer (11, 22, 33) are very unstable compared to unlike pairs (12, 13, 23).

The occurrence of three simultaneous events in each (or any) of the three steps in Eq. (17.48) is implausible physically. Instead, "simultaneous" can be interpreted to mean a sequence [e.g., for the first step in Eq. (17.48)] such as

$$2^13 \rightarrow \left(2^11\right) \rightarrow \left(2^21\right) \rightarrow 3^21, \tag{17.49}$$

in which the two central trimer states are unstable transient intermediates owing to the presence of an unstable 11 or 22 pair.

In the Boyer example, the three vertex states in Fig. 5-33(a) would be extremely unstable (note the dotted lines) because of the three like neighbor pairs in each of these three trimer states. If we omit these states and, furthermore, for simplicity (though this is not unrealistic), assume that, say, 2^13 is much more stable than 3^12, then Fig. 5-33(a) reduces to Fig. 5-33(b) as the effective kinetic diagram. In this diagram, 21 of the original 27 states remain (7×3), and the degenerate triplet in the center dominates the steady-state probability distribution. These are the three states in Eq. (17.48). From Fig. 5-33(b) it is clear that there are six possible ways [Eq. (17.49) shows one way] to accomplish each of the three steps in Eq. (17.48), i.e., by making use of any one of the six arrowed triangles in Fig. 5-33(b). Hence the effective rate constant for a step in Eq. (17.48) would be a stochastic composite of all of the rate constants in Fig. 5-33(b) (Chapter 7).

Four-State Trimer ($\rho = 4$, $m = 3$)

We mention only one special case in this subsection. We have a triangular trimer of three identical enzyme molecules [Fig. 5-32(a)], each with the one-way four-state cycle shown in Figs. 5-27(a) and 5-34(a). However, the three subunits interact in such a way that no two subunits can be in the same state at the same time; otherwise the rate constants are unaffected by interactions. Unlike the preceding case, we have here $\rho > m$. Hence it is possible to pass directly from one allowed trimer state to another without resorting to unstable transient intermediates, as in Eq. (17.49). The same point was made previously in the subsection on three-state dimers ($\rho = 3$, $m = 2$).

If we start, for example, in trimer state $2^1 3$, the only possible transition (to avoid duplicate subunit states) is $3 \rightarrow 4$ with rate constant γ. The first few transitions of the allowed sequence, with rate constants, are given in Fig. 5-34(b). After a total of 12 transitions, the complex returns to state $2^1 3$. The missing state, or hole, is included in Fig. 5-34(b), and the hole cycle (compare Fig. 5-26) is shown, with rate constants, in Fig. 5-34(c). The flux in the hole cycle is easily seen to be the same (J_o) as in the unperturbed cycle, Fig. 5-34(a). The expression for J_o appears in Eq. (17.32) (omit the subscript i). Because the hole cycle applies to the trimer, the flux per subunit is then $J = J_o/3$. Hence the restriction of the complex to only those states with no duplicate subunit states reduces the steady-state enzymatic activity per sub-

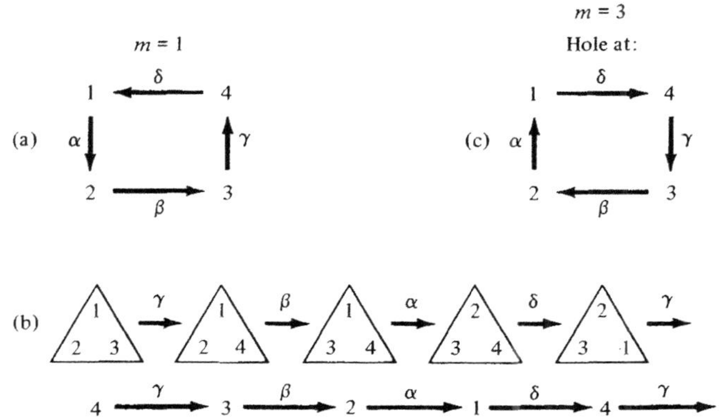

Fig. 5-34. (a) One-way four-state cycle for an independent subunit of a trimer with three identical subunits. (b) Transition sequence of trimer if no two subunits can be in the same state at the same time. (c) Cycle for hole (missing state), corresponding to (b).

unit by a factor of 3. Correspondingly, it will be noticed in Fig. 5-34(b) that the three molecules of the complex take turns in undergoing transitions.

Closed Chain of Two-State Enzyme Molecules

This is a considerably generalized two-state enzyme interaction model that is closely related to the equilibrium Ising problem and the matrix method to be discussed in Chapter 8. We have a closed linear (circular) array of m identical two-state enzyme molecules, each with the cycle and rate constants shown in Fig. 5-35(a). Each subunit in the circle or ring has two nearest neighbors. There are strong repulsive interactions between nearest neighbors when both are in state 2; consequently, nearest-neighbor pairs of type 22 do not occur at all. Otherwise the unperturbed rate constants in Fig. 5-35(a) apply. As an example, Fig. 5-35(b) shows four possible states (out of a total of 18) for a ring with $m = 6$.

We have already encountered the special cases $m = 1$, 2, and 3. The $m = 1$ case [see Eqs. (17.1)–(17.3)] is the unperturbed enzyme [Fig. 5-35(a)] with, at steady state,

$$(\alpha + \beta')(1 - \theta) = (\alpha' + \beta)\theta, \tag{17.50}$$

$$J_o = \frac{\alpha\beta - \alpha'\beta'}{(1 + x)(\alpha' + \beta)}, \tag{17.51}$$

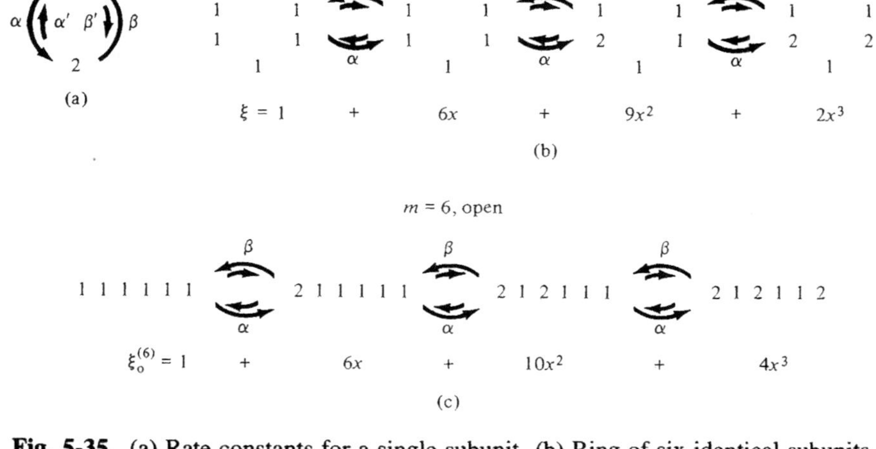

Fig. 5-35. (a) Rate constants for a single subunit. (b) Ring of six identical subunits with the same rate constants, but nearest-neighbor pairs of type 22 are excluded. The effective steady-state grand partition function is given. See text for details. (c) Same as (b) except for an open chain of $m = 6$ subunits.

$$x \equiv \frac{\alpha + \beta'}{\alpha' + \beta}. \tag{17.52}$$

In Eq. (17.50), $\theta \equiv p_2$, the probability that the enzyme is in state 2, and $1 - \theta \equiv p_1$. Equation (17.50) exhibits the quasiequilibrium "detailed balance" that exists between states 1 and 2 at steady state. This equation leads to

$$\theta = \frac{x}{1 + x}, \qquad J_o = \frac{\alpha\beta - \alpha'\beta'}{\alpha' + \beta} \cdot \frac{\theta}{x}. \tag{17.53}$$

In the $m = 2$ case [Eqs. (17.8) and (17.9)],

$$J = \frac{\alpha\beta - \alpha'\beta'}{(1 + 2x)(\alpha' + \beta)} \tag{17.54}$$

and

$$\theta = \frac{1}{2}(p_{21} + p_{12}) = \frac{x}{1 + 2x}. \tag{17.55}$$

When $m > 1$, θ is defined as the probability that any particular subunit in the ring of m subunits (they are all equivalent) is in state 2. More precisely, because this is a stochastic system, θ is the fraction of time, at steady state, that any particular subunit spends in state 2. Note that J in Eq. (17.54) can be put in the same form as in Eq. (17.53) (though θ is different). This is in fact the case for arbitrary m, as will be shown below.

In the $m = 3$ case [Eqs. (17.45) and (17.46)],

$$J = \frac{\alpha\beta - \alpha'\beta'}{(1 + 3x)(\alpha' + \beta)} = \frac{\alpha\beta - \alpha'\beta'}{\alpha' + \beta} \cdot \frac{\theta}{x}, \tag{17.56}$$

$$\theta = \frac{1}{3}(p_{211} + p_{121} + p_{112}) = \frac{x}{1 + 3x}. \tag{17.57}$$

With the above introduction, we now consider a ring of subunits with arbitrary m. With a ring of m subunits, there are also m nearest-neighbor pairs of subunits and m nearest-neighbor triplets of subunits. We define P_{11} as the probability that any particular nearest-neighbor pair of subunits is in state 11 at steady state. Similarly, P_{111} is the probability that any nearest-neighbor triplet is in state 111. A particular subunit in state 1 can make a transition to state 2 only if it is in the center of a triplet of type 111, because of the restriction on 22 pairs. On the other hand, any subunit in state 2 can make a $2 \to 1$ transition. Hence the net steady-state flux J per subunit is

$$J = \alpha P_{111} - \alpha'\theta = \beta\theta - \beta' P_{111}. \tag{17.58}$$

This equation expresses the fact that, at steady state, the net α flux $1 \to 2$ must equal the net β flux $2 \to 1$. From Eq. (17.58) we then find

$$P_{111} = \frac{\theta}{x}, \qquad J = \frac{\alpha\beta - \alpha'\beta'}{\alpha' + \beta} \cdot \frac{\theta}{x}. \tag{17.59}$$

These relations hold for arbitrary m, but of course we have yet to find $\theta(x)$ for arbitrary m.

In considering the $1 \rightleftarrows 2$ transitions between states of the ring with different numbers of subunits in state 2, as in Fig. 5-35(b), we recognize that calculation of steady-state probabilities of ring states (with rate constants $\alpha + \beta'$ and $\alpha' + \beta$) does not differ formally from the corresponding equilibrium calculation (say with inverse rate constants α and α', and $x \equiv \alpha/\alpha'$). That is, this steady-state problem has a quasiequilibrium solution. This solution is indicated, for $m = 6$, in Fig. 5-35(b) by the terms of the effective grand partition function. Thus, for $m = 6$, the all-1 ring state has relative probability 1; there are six ring states with one subunit in state 2, each with relative probability x, etc. Simpler examples ($m = 1, 2, 3$) occur in Eqs. (17.53), (17.55), and (17.57), where ξ is $1 + x$, $1 + 2x$, and $1 + 3x$, respectively.

With the quasiequilibrium ξ available, θ follows in the usual way [Eq. (11.28)] from

$$\theta = \frac{x}{m\xi}\frac{\partial \xi}{\partial x}. \tag{17.60}$$

In the $m = 6$ example [Fig. 5-35(b)],

$$\theta = \frac{x(1 + 3x + x^2)}{1 + 6x + 9x^2 + 2x^3}, \tag{17.61}$$

$$P_{111} = \frac{\theta}{x} = \frac{1 + 3x + x^2}{1 + 6x + 9x^2 + 2x^3}. \tag{17.62}$$

By actual enumeration of states, we find, up to $m = 8$,

$$\begin{array}{ll} m = 1 & \xi = 1 + x \\ 2 & 1 + 2x \\ 3 & 1 + 3x \\ 4 & 1 + 4x + 2x^2 \\ 5 & 1 + 5x + 5x^2 \\ 6 & 1 + 6x + 9x^2 + 2x^3 \\ 7 & 1 + 7x + 14x^2 + 7x^3 \\ 8 & 1 + 8x + 20x^2 + 16x^3 + 2x^4. \end{array} \tag{17.63}$$

The corresponding grand partition function $\xi_o^{(m)}$ for an *open* chain of m subunits (not closed in a ring), also with no 22 nearest-neighbor pairs, is closely related to ξ (see below). The special case $\xi_o^{(6)}$ is illustrated in Fig.

5-35(c). Note that the last configuration shown in this figure is allowed for the open chain but would not be allowed if the chain were closed (because of the 22 pair that would be formed by the two end subunits). As to notation, the quantities ξ, θ, P_{11}, P_{111}, and J all refer to a closed ring of size m, but the m dependence is implicit [e.g., $\xi(x; m)$]; however, for an open chain, it is convenient to indicate the value of m explicitly in $\xi_o^{(m)}$.

By actual enumeration, we find, for an open chain,

$$\begin{aligned} \xi_o^{(1)} &= 1 + x, \\ \xi_o^{(2)} &= 1 + 2x, \\ \xi_o^{(3)} &= 1 + 3x + x^2, \\ \xi_o^{(4)} &= 1 + 4x + 3x^2, \\ \xi_o^{(5)} &= 1 + 5x + 6x^2 + x^3, \\ \xi_o^{(6)} &= 1 + 6x + 10x^2 + 4x^3, \\ \xi_o^{(7)} &= 1 + 7x + 15x^2 + 10x^3 + x^4, \\ \xi_o^{(8)} &= 1 + 8x + 21x^2 + 20x^3 + 5x^4. \end{aligned} \qquad (17.64)$$

It will be noticed in both Eqs. (17.63) and (17.64) that the numerical coefficients in one column provide the *differences* between coefficients in the next column. This observation leads to the general formulas

$$\xi = 1 + mx + \frac{m(m-3)}{2!}x^2 + \frac{m(m-4)(m-5)}{3!}x^3 + \frac{m(m-5)(m-6)(m-7)}{4!}x^4 + \cdots, \qquad (17.65)$$

$$\xi_o^{(m)} = 1 + mx + \frac{(m-1)(m-2)}{2!}x^2 + \frac{(m-2)(m-3)(m-4)}{3!}x^3 + \cdots. \qquad (17.66)$$

These are both polynomials that break off at the first zero term. We shall return to this problem in Section 34.

If a particular subunit of a *closed* chain is fixed in state 1, then the grand partition function for all possible states of the remaining $m - 1$ subunits is $\xi_o^{(m-1)}$, because it is possible for both ends of the remaining $m - 1$ chain to be in state 2 (as in an open chain). Because grand partition function terms are proportional to probabilities of states, $\xi_o^{(m-1)}$ is the weight to be assigned to all possible configurations of the closed chain of m subunits when one particular subunit is fixed in state 1. Without this restraint, the corresponding weight is the complete grand partition function ξ. Therefore, the probability $1-\theta$ that the particular subunit (or *any* subunit) is in state 1 is

$$1 - \theta = \frac{\xi_{\text{o}}^{(m-1)}}{\xi}. \tag{17.67}$$

Then

$$\theta = \frac{\xi - \xi_{\text{o}}^{(m-1)}}{\xi}. \tag{17.68}$$

Exactly the same argument can be applied to a pair of subunits of the closed chain fixed in state 11, or to a triplet fixed in state 111, leading to

$$P_{11} = \frac{\xi_{\text{o}}^{(m-2)}}{\xi}, \qquad P_{111} = \frac{\theta}{x} = \frac{\xi_{\text{o}}^{(m-3)}}{\xi}. \tag{17.69}$$

These equations relate θ for a closed chain of m subunits to $\xi_{\text{o}}^{(m-1)}$ and $\xi_{\text{o}}^{(m-3)}$. There is another such relation for $\xi_{\text{o}}^{(m-2)}$, which is easy to derive. Let P_{12} be the probability that a nearest-neighbor pair of the closed chain is in state 12, and similarly for P_{21}. By symmetry, $P_{12} = P_{21}$. Then

$$\begin{aligned} P_{11} + 2P_{12} &= 1, \\ \theta = \tfrac{1}{2}(P_{12} + P_{21}) &= P_{12}, \end{aligned} \tag{17.70}$$

and

$$P_{11} = 1 - 2\theta = \xi_{\text{o}}^{(m-2)}/\xi. \tag{17.71}$$

Because θ is also related to ξ alone through Eq. (17.60), elimination of θ between Eq. (17.60) and one of Eqs. (17.68), the second of Eqs. (17.69), and Eq. (17.71) produces a differential equation relating ξ, $\partial\xi/\partial x$, and one of $\xi_{\text{o}}^{(m-1)}$, $\xi_{\text{o}}^{(m-2)}$, and $\xi_{\text{o}}^{(m-3)}$. The simplest case is

$$\theta = \frac{x}{m\xi}\frac{\partial \xi}{\partial x} = \frac{x\xi_{\text{o}}^{(m-3)}}{\xi}, \qquad \frac{1}{m}\frac{\partial \xi}{\partial x} = \xi_{\text{o}}^{(m-3)},$$

which, after integration, leads to

$$\xi = 1 + m \int_0^x \xi_{\text{o}}^{(m-3)}(x')dx'. \tag{17.72}$$

Thus $\xi(x; m)$ is easily found from $\xi_{\text{o}}^{(m-3)}(x)$.

Several recursion relations can be derived by eliminating θ among Eqs. (17.67)–(17.71):

$$\begin{aligned} \xi &= \xi_{\text{o}}^{(m-1)} + x\xi_{\text{o}}^{(m-3)}, \\ \xi &= \xi_{\text{o}}^{(m-2)} + 2x\xi_{\text{o}}^{(m-3)}, \\ \xi &= 2\xi_{\text{o}}^{(m-1)} - \xi_{\text{o}}^{(m-2)}, \\ \xi_{\text{o}}^{(m)} &= \xi_{\text{o}}^{(m-1)} + x\xi_{\text{o}}^{(m-2)}, \\ \xi(m) &= \xi(m-1) + x\xi(m-2). \end{aligned} \tag{17.73}$$

In summary, with Eqs. (17.63)–(17.66) available, the simplest expressions for θ and J for a closed chain are:

$$\theta = \frac{x\xi_o^{(m-3)}}{\xi}, \tag{17.74}$$

$$J = \frac{\alpha\beta - \alpha'\beta'}{\alpha' + \beta} \cdot \frac{\xi_o^{(m-3)}}{\xi}, \tag{17.75}$$

$$\frac{J}{J_o} = \frac{\theta(1 + x)}{x} = \frac{(1 + x)\xi_o^{(m-3)}}{\xi}, \tag{17.76}$$

where J_o is given in Eq. (17.51). When x is large, $J/J_o \to \theta$. When m is even, $\theta \to \frac{1}{2}$ when $x \to \infty$. That is, alternate subunits are in state 2. When m is odd, $\theta \to (m-1)/2m$ in the same limit. This also approaches $\theta = \frac{1}{2}$ when m is large.

As an illustration of Eq. (17.76), Fig. 5-36 shows J/J_o as a function of x for $m = 8$. Note that J/J_o undershoots $\theta = \frac{1}{2}$ a little (to $\theta = 0.475$) and then returns asymptotically to $\theta = \frac{1}{2}$. J is of course less than J_o because of the many excluded states in the complex of eight subunits.

Finally, although we have been discussing a steady-state system, it should be noted that all of Eqs. (17.60)–(17.74) are actually equilibrium relations (with, say, $x \equiv \alpha/\alpha'$) that pertain to a particular finite one-dimensional Ising problem (Chapter 8). In this connection, note that the relation $P_{111} = \theta/x$ follows, in the equilibrium case, from [Eq. (17.58)]

$$J = 0 = \alpha P_{111} - \alpha'\theta. \tag{17.77}$$

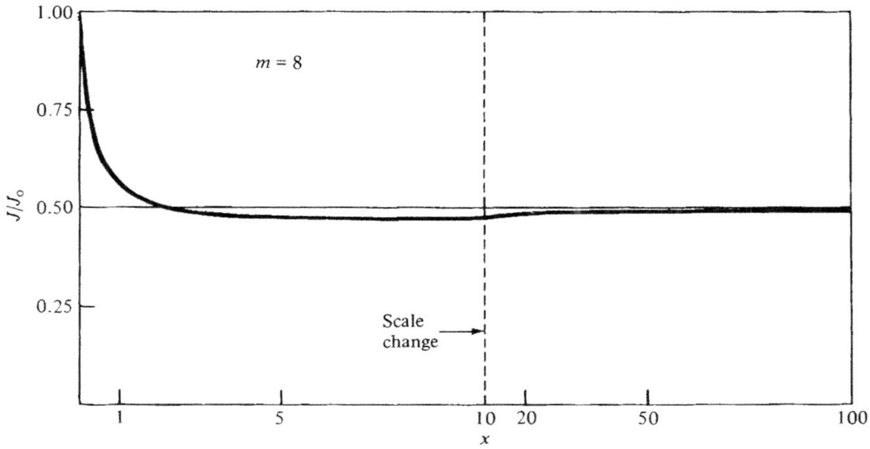

Fig. 5-36. J/J_o as a function of x for a closed chain of $m = 8$ subunits [Eq. (17.76)].

18. One-Conformation Complexes with Strong Interactions: Missing Transitions

The introduction to Section 17 applies here as well. In this section we provide a few examples in which strong interactions between neighbors in a complex lead to missing transitions in the kinetic diagram rather than to missing states. States are missing in a diagram (Section 17) because of strong repulsive interactions between neighboring subunits that raise the free energy of these states considerably. Transitions are missing, rather than states, when the intersubunit interactions raise the free energy of the activated complex (or transition state) between pairs of states. That is, the two states of such a pair are both present in the diagram, but a given transition (and its inverse) between the two states is not accessible because of some significant intersubunit free energy barrier that arises in the course of the transition. For example, a normal binding site for substrate may be available on a particular subunit, but the substrate's access to this site is blocked by a neighboring subunit of the complex.

Two-State Dimer with Missing Transition

As an example, consider the system in Figs. 5-20(a) and 5-20(c). We suppose that the left-hand molecule (E_1) in state E_1 has a shape or structure that blocks access of the product P_2 to or from the substrate binding site on the right-hand molecule E_2. However, this inhibition is not present when molecule E_1 is in state E_1S_1, because the binding of S_1 induces a structural change in E_1 that allows access of P_2 to or from the substrate site on E_2. Aside from the above effect, the rate constants are assumed to be unperturbed. The resulting diagram is shown in Fig. 5-37(a), with one-way transitions only (to simplify the algebra).

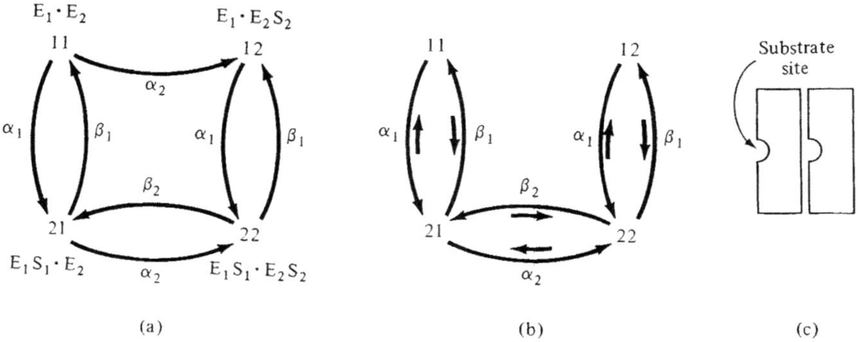

Fig. 5-37. (a) System as in Figs. 5-20(a) and 5-20(c), except one-way cycles and one missing transition. See text for details. (b) Same example except two-way cycles and a pair of missing transitions. Primed rate constants go with inverse arrows. (c) Heterologous dimer in which two identical subunits are not equivalent.

The steady-state probabilities of the four states in Fig. 5-37(a) are easily found from, for example,

$$\begin{aligned}(\alpha_1 + \alpha_2)p_{11} &= \beta_1 p_{21},\\ (\beta_1 + \beta_2)p_{22} &= \alpha_1 p_{12} + \alpha_2 p_{21},\\ \alpha_1 p_{12} &= \alpha_2 p_{11} + \beta_1 p_{22},\\ p_{11} + p_{12} + p_{21} + p_{22} &= 1.\end{aligned} \tag{18.1}$$

The result is

$$\begin{aligned}p_{11} &= \frac{\alpha_1\beta_1\beta_2}{\Sigma},\\ p_{21} &= \frac{\alpha_1\beta_2(\alpha_1 + \alpha_2)}{\Sigma},\\ p_{12} &= \frac{\alpha_2\beta_1(\alpha_1 + \alpha_2 + \beta_1 + \beta_2)}{\Sigma},\\ p_{22} &= \frac{\alpha_1\alpha_2(\alpha_1 + \alpha_2 + \beta_1)}{\Sigma},\\ \Sigma &= (\alpha_1 + \beta_1)[\alpha_2(\alpha_1 + \alpha_2 + \beta_1) + \beta_2(\alpha_1 + \alpha_2)].\end{aligned} \tag{18.2}$$

The $S_1 \rightarrow P_1$ flux for molecule E_1 is

$$\begin{aligned}J_1 &= \alpha_1(p_{11} + p_{12}) = \beta_1(p_{21} + p_{22})\\ &= \alpha_1\beta_1/(\alpha_1 + \beta_1) = J_{o1},\end{aligned} \tag{18.3}$$

as expected [Eq. (17.1)], because the rate constants in the E_1 cycle are the same whatever the state of E_2. The flux for E_2 is

$$J_2 = \beta_2 p_{22} = \frac{\alpha_1\alpha_2\beta_2(\alpha_1 + \alpha_2 + \beta_1)}{\Sigma}. \tag{18.4}$$

It is easy to show that J_2 is less than

$$J_{o2} = \frac{\alpha_2\beta_2}{\alpha_2 + \beta_2}, \tag{18.5}$$

as expected, because of the missing transition in Fig. 5-37(a).

If the two subunits of the dimer are identical, Eq. (18.4) (for the right-hand subunit) becomes

$$J_2 = \frac{\alpha\beta(2\alpha + \beta)}{(\alpha + \beta)(2\alpha + 3\beta)}. \tag{18.6}$$

The left-hand subunit has a different (larger) flux:

$$J_1 = \frac{\alpha\beta}{\alpha + \beta} = J_o. \tag{18.7}$$

This is a so-called heterologous dimer in which the two identical subunits are not equivalent in the dimer because of the symmetry. Fig. 5-37(c) shows an example.

Two-State Dimer with Two Missing Transitions

Suppose, in the above example, that [see Figs. 5-20(a) and 5-20(c)] E_1 blocks access of both S_2 and P_2 to or from the substrate site on E_2, but that E_1S_1 does not block either access. Thus the complete E_2 cycle is inhibited by state E_1. In this case the diagram is that shown in Fig. 5-37(b). All four states are still present, but one complete cycle is missing. Because this diagram has a quasiequilibrium steady-state solution, it is easy to include two-way transitions. Equations (17.1)–(17.9) are closely related. We find

$$\begin{aligned}
p_{11} &= \frac{1}{(1 + x_1)(1 + x_2)}, \\
p_{21} &= \frac{x_1}{(1 + x_1)(1 + x_2)}, \\
p_{12} &= \frac{x_2}{(1 + x_1)(1 + x_2)}, \\
p_{22} &= \frac{x_1 x_2}{(1 + x_1)(1 + x_2)},
\end{aligned} \tag{18.8}$$

where x_1 and x_2 are defined in Eq. (17.3). The two fluxes are then

$$J_1 = \alpha_1(p_{11} + p_{12}) - \alpha_1'(p_{21} + p_{22}) = J_{o1}, \tag{18.9}$$

$$J_2 = \alpha_2 p_{21} - \alpha_2' p_{22} = \frac{x_1 J_{o2}}{1 + x_1}, \tag{18.10}$$

where J_{o1} and J_{o2} are given in Eqs. (17.1) and (17.2). The E_1 flux is unperturbed by the presence of E_2 [see the assumed rate constants in Fig. 5-37(b)]. However, J_2 is less than J_{o2} by the factor $x_1/(1 + x_1)$, which is the probability that E_1 is in state 2 (i.e., in state E_1S_1) irrespective of the state of E_2: E_2 cycles only when E_1 is in state 2.

If the two subunits are identical,

$$J_1 = J_o, \qquad J_2 = xJ_o/(1 + x), \tag{18.11}$$

$$J_o = (\alpha\beta - \alpha'\beta')/(\alpha + \beta' + \alpha' + \beta),$$

$$x = (\alpha + \beta')/(\alpha' + \beta).$$

The probability that either subunit is in state 2 is $x/(1 + x)$ [see Eqs. (18.8)], but the two fluxes are different.

A Simple Model of Adenylate Cyclase Activity

To illustrate the above subsections, we consider an oversimplified model of the adenylate cyclase complex, based on the description in Ref. 4. This complex is a trimer (RNC) with subunits R (receptor for a hormone, H), N (nucleotide regulatory subunit, which has GDP or GTP bound), and C (catalytic subunit, which can bind ATP and convert it into cyclic AMP and PP_i). Each of the subunits may exist in either of two states (we assume): R and RH; ND and NT; and C and CT. The receptor R can either bind or release H. ND and NT are shorthand for N-GDP and N-GTP. The transition ND → NT occurs by exchanging GTP for GDP; the N cycle is completed by the hydrolysis of GTP: NT → ND, with release of P_i. The T in CT refers to ATP. ATP is bound to C in the first step of the C cycle: C → CT. The C cycle is completed by the formation and release of cyclic AMP and PP_i: CT → C.

When the complex is written in the order RNC, the exchange reaction on N, ND → NT, occurs only when N's left-hand neighbor, R, is in state RH (hormone bound). Further, the catalytic subunit C is active only when its left-hand neighbor, N, is in state NT (GTP bound). Presumably, these observed discriminations are due to conformational changes.

The above-described model (with $m = 3$, $\rho = 2$) is represented in Fig. 5-38. The front square of four states has the receptor in state RH; in the rear square the state is R. In the rear square, the two exchange arrows (ND → NT) are omitted because exchange does not occur in the absence of bound H. The left-hand square of four states has N in state ND; in the right-hand square, N is in state NT. Thus, in the right-hand square, the two C, CT cycles are intact (bind ATP; produce cyclic AMP, labelled cA in the figure). In the left-hand square, because GTP is not bound to N, the two C, CT cycles are not active. Two "inactive" possibilities are represented: if the two dashed arrows are included, ATP is bound on C even when N is in state ND, but the formation of cyclic AMP and PP_i is inhibited; if the two dashed arrows are omitted, both steps of the C, CT cycle are suppressed by the absence of GTP on N.

The four straight diagonal lines in Fig. 5-38 relate to the binding and release of H to and from R. These transitions are not cyclical, as are those associated with N and C. The front square of four states dominates when the concentration of H, c_H, is large enough to nearly saturate R with H. The rear square of states dominates when $c_H \rightarrow 0$. In this case, because of the two single GTP hydrolysis arrows, the steady-state population is all in states R·ND·C and R·ND·CT (or only in the latter state if the dashed arrows are included). Thus, when $c_H \rightarrow 0$, there is no significant production of cyclic AMP.

When c_H is large and the front square of states dominates, the operative

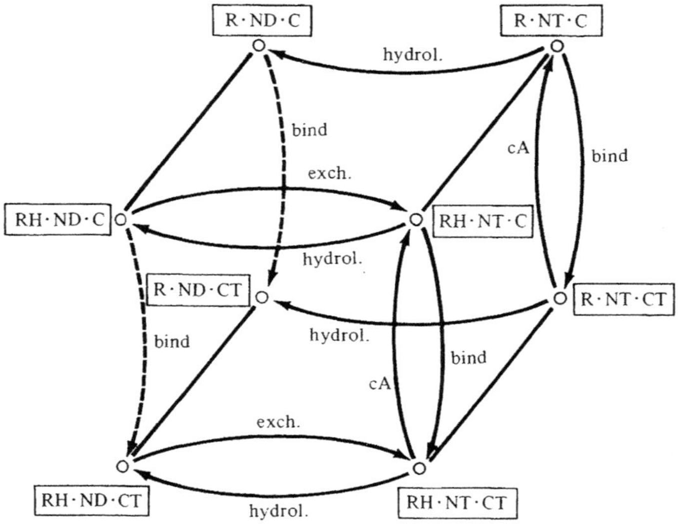

Fig. 5-38. Diagram for a simple model of adenylate cyclase activity. Details are given in the text.

diagram has the same form as in Fig. 5-37(a) (if dashed arrows are included) or as in Fig. 5-37(b) (if dashed arrows are omitted). Thus the flux of cyclic AMP is given by Eq. (18.4) or Eq. (18.10), respectively.

At intermediate values of c_H, the full eight-state diagram must be used to calculate the cyclic AMP flux. This calculation is straightforward but would require a full set of rate constants and solution of eight linear equations in eight unknowns by computer.

Open Chain of Two-State Molecules with Missing Transitions

This model is somewhat related to the adenylate cyclase complex, where $m = 3$. Here we consider an open chain of m two-state enzyme molecules, $E_1, E_2, \ldots, E_m$, complexed to each other in this order, with interactions such that the transitions in the cycle of enzyme i ($i = 2, \ldots, m$) are inhibited unless the molecule on the left of i (i.e., E_{i-1}) is in state 2. Aside from this inhibition, the rate constants [Fig. 5-39(a)] are assumed to be unperturbed by interactions. Figure 5-37(b) and Eqs. (18.8)–(18.11) pertain to the $m = 2$ special case.

The kinetic diagrams for $m = 2$, 3, and 4 are shown in Figs. 5-39(b), 5-39(c), and 5-39(d), where a line between states is shorthand here for a cycle between states [compare Figs. 5-37(b) and 5-39(b) for $m = 2$]. Only allowed transitions are included in these diagrams. For any m, the steady state is a quasiequilibrium with all states accessible to each other (directly or indi-

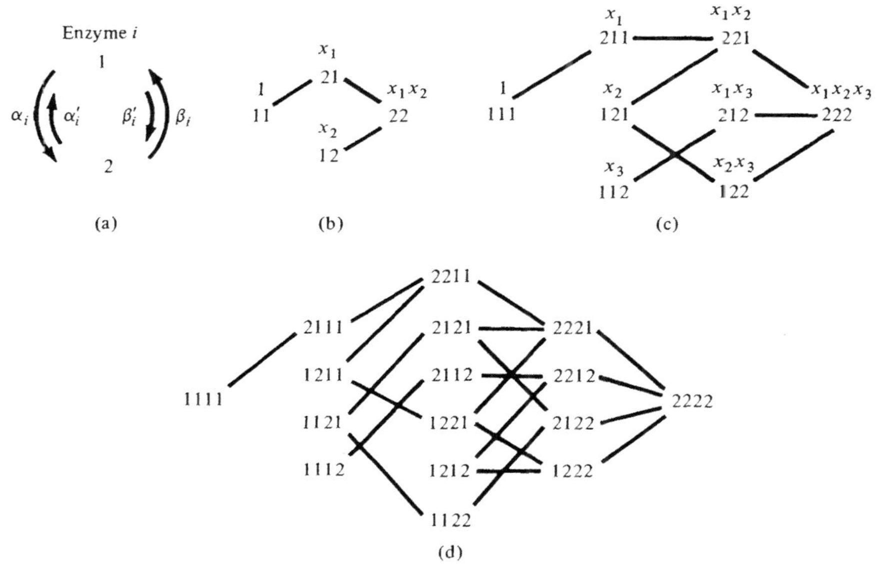

Fig. 5-39. (a) First-order rate constants for enzyme i in an open chain of m enzymes. (b) Diagram for $m = 2$ if enzyme $i = 2$ operates only when $i = 1$ is in state 2. A single line in this figure represents a complete cycle, as in (a). The quantities 1, x_1, x_2, and x_1x_2 are relative probabilities of states at steady state. See text for details. (c) Same for $m = 3$. Molecules $i = 2, 3$ operate only if molecules $i = 1, 2$, respectively, are in state 2. (d) Same for $m = 4$.

rectly), despite the missing transitions. The steady-state system is equivalent to an equilibrium system with a forward rate constant for (E_i in state 1) $\rightarrow$ (E_i in state 2) of $\alpha_i + \beta_i'$ and an inverse rate constant of $\alpha_i' + \beta_i$. The quotient $(\alpha_i + \beta_i')/(\alpha_i' + \beta_i)$ is defined as x_i, as usual. The relative steady-state probabilities of the states are included in Figs. 5-39(b) and 5-39(c). For any m, the sum of these relative probabilities is the grand partition function

$$\xi = (1 + x_1)(1 + x_2) \ldots (1 + x_m). \tag{18.12}$$

Because ξ is a product of separate factors [see Eq. (11.6)], the $1 \leftrightarrows 2$ quasi-equilibrium distribution for any E_i is independent of the other members of the chain. Hence the probability that any E_i is in state 2 is $x_i/(1 + x_i)$.

The enzymatic operation of E_1 is unperturbed, irrespective of the states of the other molecules. Hence $J_1 = J_{o1}$, as in Eqs. (17.1) and (18.9). For any other member of the chain,

$$J_i = \frac{x_{i-1}J_{oi}}{1 + x_{i-1}} \qquad (i = 2, \cdots, m) \tag{18.13}$$

because E_i operates only if E_{i-1} (the left-hand neighbor) is in state 2. Thus, because of missing transitions, $J_i \leq J_{oi}$. The explicit expression for J_{oi} [Fig. 5-39(a)] is

$$J_{oi} = (\alpha_i\beta_i - \alpha_i'\beta_i')/(a_i + \beta_i' + \alpha_i' + \beta_i). \tag{18.14}$$

The adenylate cyclase model in Fig. 5-38 is similar to the $m = 3$ case here, but is definitely not the same. In order to convert Fig. 5-38 into Fig. 5-39(c), the two dashed ATP binding transitions and also the two single GTP hydrolysis transitions (rear square) in Fig. 5-38 have to be omitted. This leads to Fig. 5-40, which is topologically equivalent to Fig. 5-39(c). Note, however, that the x_1 lines in Fig. 5-40 refer to binding and release of H to and from R, which is *not* a cyclic process. Hence x_1 in Fig. 5-40 is defined as $x_1 \equiv Kc_H$, where K is the binding constant of H on R [see Eqs. (3.7) and (6.15)]. The cyclic AMP flux in this modified adenylate cyclase model is then [Eq. (18.13)]

$$J_3 = x_2J_{o3}/(1 + x_2). \tag{18.15}$$

This is an unsatisfactory result because J_3 does not depend on x_1 (i.e., on c_H), as it should. This lack of dependence on c_H is a consequence of the omission of the two single GTP hydrolysis transitions that are in Fig. 5-38. These transitions are needed in the model to ensure that there is no cyclic AMP production ($J_3 \rightarrow 0$) when $c_H \rightarrow 0$.

Three-State Dimer with Missing Transitions

We limit ourselves here to a few very specialized cases that are analogous to those in Fig. 5-24 (missing states). The enzymatic cycles are shown in Fig.

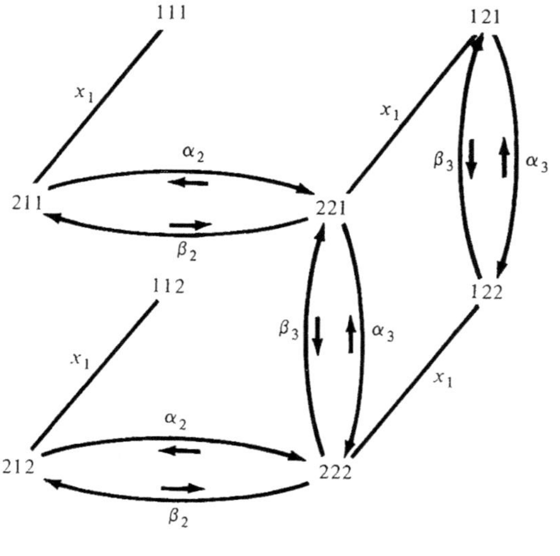

Fig. 5-40. Model related to Figs. 5-38 and 5-39(c). See text for explanation. Inverse arrows belong to primed rate constants.

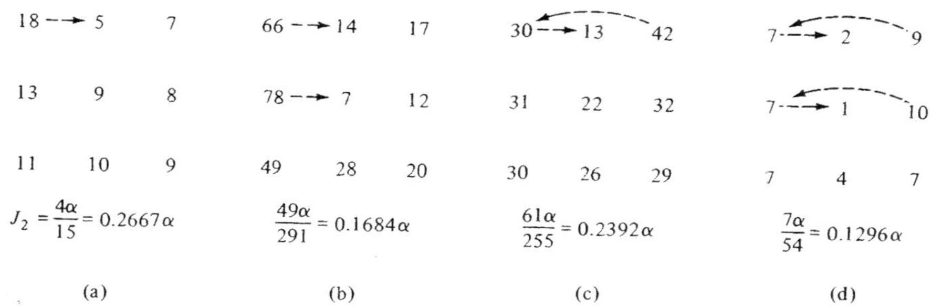

Fig. 5-41. These are special cases of Fig. 5-24(b) in which some transitions are missing (dashed arrows) for enzyme 2 in the dimer. All other transitions are one way and have rate constant α. The flux for enzyme 2 (J_2) is given in each case. The integers are proportional to the steady-state probabilities of states. (a) One transition (state 11 → state 12) missing. (b) (c) Two transitions missing. (d) Four transitions missing.

5-21(a), but we consider one-way transitions only, with α for all allowed rate constants. Thus the reference case (no missing transitions) is that in Figs. 5-24(a) and 5-24(b), with

$$J_1 = J_2 = J_{o1} = J_{o2} = \alpha/3. \tag{18.16}$$

The first example is summarized in Fig. 5-41(a). All transitions in Fig. 5-24(b) are present *except* the one indicated by the dashed arrow in Fig. 5-41(a). This is the transition 11 → 12 [Fig. 5-24(b)]. That is, the substrate S_2 cannot bind on E_2 when E_1 is in state 1 (i.e., in state E_1); S_2 *can* bind when E_1 is in state 2 (E_1S_1) or 3 (E_1P_1). If all allowed transitions have the rate constant α, the steady-state probabilities of the nine states are proportional to the integers given in Fig. 5-41(a) [the same system was used in Eqs. (17.37) and (17.42)]. From these integers we find $J_1 = J_{o1} = \alpha/3$ and $J_2 = 4\alpha/15$. Thus J_1 is unperturbed by the missing transition, but J_2 is reduced from $J_{o2} = \alpha/3$.

Figures 5-41(b), 5-41(c), and 5-41(d) present three other examples of the same type; the dashed arrows show the missing transitions and the sets of integers are proportional to steady-state probabilities of states. In Fig. 5-41(d), for example, the binding of S_2 to E_2 and the release of P_2 from E_2P_2 are both inhibited unless E_1 is in state 3 (E_1P_1). In all cases, $J_1 = J_{o1} = \alpha/3$, but J_2 is reduced below $\alpha/3$, the more so the more transitions are missing.

19. Two-Conformation Complexes with Strong Interactions

In the previous two sections we studied one-conformation complexes of two or more subunits in which strong interactions had the effect of removing states or transitions from the kinetic diagram. If the subunits of the complex can exist in two different conformations, the possible examples of the above type proliferate considerably and generally become sufficiently complicated that the steady-state properties must be found numerically by computer. Because of the quite restrictive all-or-none type of assumption explained at the beginning of Section 17, and used in Sections 17, 18, and in this section, it does not seem worthwhile to give an extensive survey of the two-conformation class of models. The more general approach to these problems, with a quantitative treatment of intersubunit interactions, is included in Chapter 7. We limit the discussion in this brief section to qualitative remarks and to one explicit example that can be treated analytically.

Dimer with Two-Conformation Subunits

We consider a dimer of two identical subunits, each with a two-state enzymatic cycle and two conformations. For a single subunit, if E catalyzes the usual S → P, the diagram is shown in Fig. 5-42(a), where E = ○ and E* = □ are the two conformations. This diagram is a two-state simplification of Fig. 5-1(d). The corresponding diagram for the dimer is given in Fig. 5-42(b). The four-state upper left section of Fig. 5-42(b) is the same as Fig. 5-20(c) (with $E_1 = E_2$): by itself, this is the diagram for a one-conformation dimer with two-state subunits. The conformational changes then multiply the possibilities by a factor of 4. The full diagram thus has 16 states. Incidentally, these 16 states are the analogues of those in Fig. 4-6 (S replaces L). The transitions in the diagram are either part of an enzymatic cycle (curved lines) or are conformational changes (straight lines).

More generally, if there are m subunits in the complex and each subunit has a ρ-state cycle (or diagram) and two conformations, then the total number of

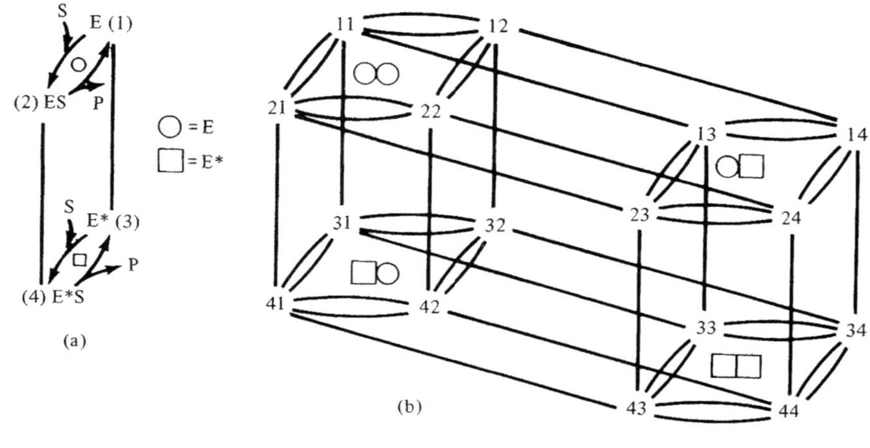

Fig. 5-42. (a) Diagram for a single two-state subunit with two conformations. (b) Corresponding diagram for a dimer. See text for further details.

states in the full diagram of the complex is $(2\rho)^m$. In Fig. 5-42(b), $\rho = 2$ and $m = 2$. If each subunit of the dimer has a three-state cycle (e.g., E, ES, EP), then the full diagram ($\rho = 3$, $m = 2$) has 36 states [in Fig. 5-42(b), each of the four squares becomes the 3×3 diagram in Fig. 5-21(b)]. As another example, if the two-state enzyme is modified by a ligand, as in Fig. 5-4, then the full diagram for a dimer ($\rho = 4$, $m = 2$) has 64 states [16 states in each of the four corners of Fig. 5-42(b)].

We return now to the specific example in Fig. 5-42(b). If the two subunits do not perturb each other at all, the full diagram "factors" into two single-subunit diagrams [Fig. 5-42(a)]; the full diagram need not be considered in this case. If there are intermolecular interactions of arbitrary strength between the subunits, the rate constants of the full diagram will reflect this (Chapter 7). If there are very strong intersubunit interactions (as in Sections 17 and 18), there are a number of possibilities: a particular dimer state, e.g., 22, might be missing in all four conformation pairs (○○, etc), or it might be missing in only certain pairs of conformations, say, ○○ and □□; particular transitions, e.g., 11 $\rightleftarrows$ 12, might be missing in all four conformation pairs, or in only certain of them; entire quartets of states might be missing from the diagram (except as transient intermediates) owing to the instability of certain conformation pairs (e.g., ○○ and □□, or ○□ and □○); or there might be combinations of the above.

Example with Missing Conformation Pairs

Here we examine a special case of Fig. 5-42(b) in which the conformation pairs ○○ and □□ are unstable and serve only as transient intermediates in "concerted" conformational changes between the two remaining pairs, ○□ and □○, which are assumed to be equivalent (i.e., this is a so-called isologous dimer, unlike Fig. 5-37(c)). The ○□ and □○ quartets of states in Fig. 5-42(b) are renumbered, and rate constant notation is included, in Fig. 5-43. The vertical transitions are conformational changes (via transient intermediates). The subscripts 1 and 2 on α and β refer, respectively, to conformations ○ (E) and □ (E*), *not* to the position in the dimer (left or right). The inverses of α_i and β_i are α_i' and β_i' (not included in the figure). All of these rate constants (in the enzymatic cycle) are assumed to be unperturbed; the "perturbation" (from interactions) is the exclusion of conformation pairs ○○ and □□. Because of symmetry, equivalent pairs of states are: 1 and 5; 2 and 7; 3 and 6; and 4 and 8.

The thermodynamic force in either two-state cycle is $X = \mu_S - \mu_P$. This is related to rate constants by

$$\frac{\alpha_1\beta_1}{\alpha_1'\beta_1'} = \frac{\alpha_2\beta_2}{\alpha_2'\beta_2'} = e^{X/kT}. \tag{19.1}$$

There are two additional relations among the rate constants that follow from Fig. 5-43:

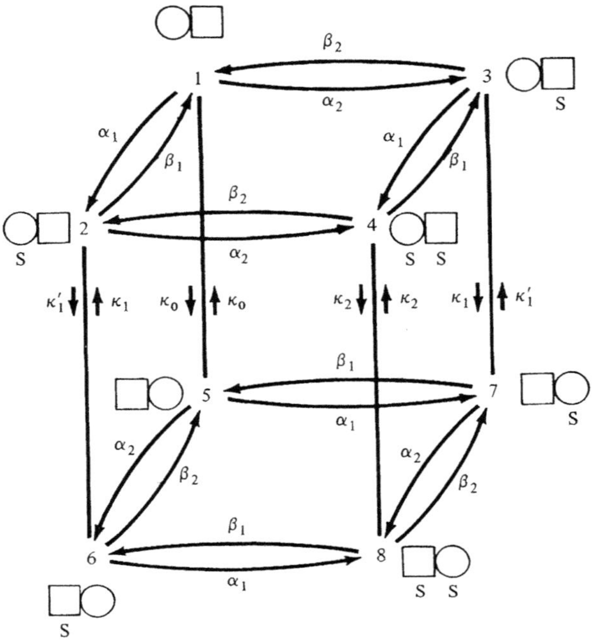

Fig. 5-43. Special case of Fig. 5-42(b) in which conformation pairs ○○ and □□ are unstable. See text for details.

$$\frac{\kappa_1}{\kappa_1'} = \frac{\beta_1'\beta_2}{\beta_1\beta_2'} = \frac{\alpha_1\alpha_2'}{\alpha_1'\alpha_2}. \tag{19.2}$$

In the remaining analysis we limit ourselves to one-way enzymatic cycles (α_i' and β_i' are neglected). The steady-state probabilities satisfy (see above)

$$\begin{gathered} p_1 = p_5, \qquad p_2 = p_7, \qquad p_3 = p_6, \qquad p_4 = p_8, \\ p_1 + p_2 + p_3 + p_4 = \tfrac{1}{2}. \end{gathered} \tag{19.3}$$

The steady-state linear equations in p_1 to p_4 [using Eqs. (19.3)] are:

$$\begin{aligned} (\alpha_1 + \alpha_2)p_1 &= \beta_1 p_2 + \beta_2 p_3, \\ (\beta_1 + \alpha_2 + \kappa_1')p_2 &= \alpha_1 p_1 + \beta_2 p_4 + \kappa_1 p_3, \\ (\alpha_1 + \beta_2 + \kappa_1)p_3 &= \alpha_2 p_1 + \beta_1 p_4 + \kappa_1' p_2, \\ (\beta_1 + \beta_2)p_4 &= a_2 p_2 + \alpha_1 p_3. \end{aligned} \tag{19.4}$$

Using any three of these equations and the conservation relation in Eqs. (19.3), we find

$$p_1 = p_5 = (\beta_1 + \beta_2)(\beta_2 C + \beta_1 D)/E,$$

$$\begin{aligned} p_2 = p_7 &= (\alpha_1 + \alpha_2)(\beta_1 + \beta_2)D/E, \\ p_3 = p_6 &= (\alpha_1 + \alpha_2)(\beta_1 + \beta_2)C/E, \\ p_4 = p_8 &= (\alpha_1 + \alpha_2)(\alpha_1 C + \alpha_2 D)/E, \end{aligned} \tag{19.5}$$

where

$$E = 2(AD + BC),$$

$$\begin{aligned} A &= \beta_1(\beta_1 + \beta_2) + \alpha_2(\alpha_1 + \alpha_2) + (\alpha_1 + \alpha_2)(\beta_1 + \beta_2), \\ B &= \beta_2(\beta_1 + \beta_2) + \alpha_1(\alpha_1 + \alpha_2) + (\alpha_1 + \alpha_2)(\beta_1 + \beta_2), \\ C &= (\alpha_1 + \alpha_2)(\beta_1 + \beta_2)\kappa_1' + \alpha_2\beta_1(\alpha_1 + \alpha_2 + \beta_1 + \beta_2), \\ D &= (\alpha_1 + \alpha_2)(\beta_1 + \beta_2)\kappa_1 + \alpha_1\beta_2(\alpha_1 + \alpha_2 + \beta_1 + \beta_2). \end{aligned} \tag{19.6}$$

The rate constants κ_0 and κ_2 do not appear in these expressions.

The S → P flux per subunit, J, must be the same for the two subunits, by symmetry. We calculate J using the α_1 and α_2 transitions of the left-hand subunit (Fig. 5-43):

$$\begin{aligned} J &= \alpha_1(p_1 + p_3) + \alpha_2(p_5 + p_7) \\ &= (\alpha_1 + \alpha_2)(\beta_1 + \beta_2)[(\alpha_1 + \beta_2)C + (\alpha_2 + \beta_1)D]/E. \end{aligned} \tag{19.7}$$

Finally, we consider an important special case of Fig. 5-43. Suppose that the enzyme operates only in conformation ○ = E (not in □ = E*). That is, $\alpha_2 = 0$ and $\beta_2 = 0$. Thus, in the dimer, only one subunit operates at a time because only one is in conformation ○ (Fig. 5-43). This is an example, then, of half-of-the-sites reactivity. In this case, Fig. 5-43 reduces to Fig. 5-44(a). In this example there are both missing states (○○, □□) and missing transitions ($\alpha_2 = 0$, $\beta_2 = 0$).

If we make use of the symmetry in Fig. 5-44(a) to combine states, we see that Fig. 5-44(a) becomes Fig. 5-44(b). The latter diagram is linear and has a quasiequilibrium p_i distribution at steady state:

$$p_1 = \frac{1}{2\xi}, \qquad p_2 = \frac{(\alpha_1/\beta_1)}{2\xi}, \qquad p_3 = \frac{(\alpha_1/\beta_1)(\kappa_1'/\kappa_1)}{2\xi}, \qquad p_4 = \frac{(\alpha_1/\beta_1)^2(\kappa_1'/\kappa_1)}{2\xi}, \tag{19.8}$$

where ξ is the sum of the four numerators. We then find, for the flux per subunit,

$$\begin{aligned} J &= \alpha_1(p_1 + p_3) \\ &= \frac{\alpha_1\beta_1}{2(\alpha_1 + \beta_1)}. \end{aligned} \tag{19.9}$$

This is half of the unperturbed flux [Eq. (17.1)], as expected.

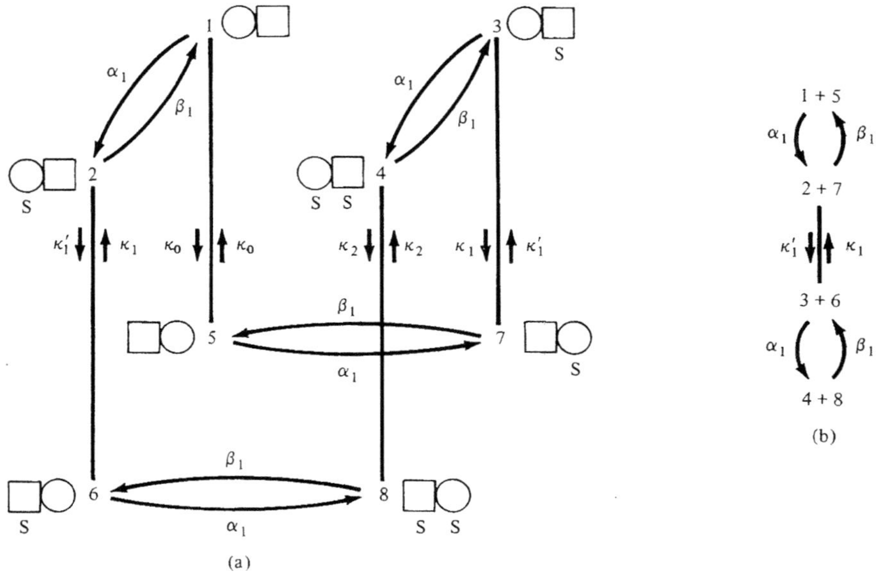

Fig. 5-44. (a) Special case of Fig. 5-43 in which the enzyme operates only in conformation ○ = E. (b) Equivalent diagram to (a), in view of symmetry.

Because of the linearity in Fig. 5-44, there is no difficulty in including the inverse rate constants α_i' and β_i' for this special case. In Eqs. (19.8), we make the change

$$\frac{\alpha_1}{\beta_1} \to \frac{\alpha_1 + \beta_1'}{\alpha_1' + \beta_1}.$$

Then, in place of Eq. (19.9),

$$\begin{aligned} J &= \alpha_1(p_1 + p_3) - \alpha_1'(p_2 + p_4) \\ &= \frac{\alpha_1\beta_1 - \alpha_1'\beta_1'}{2(\alpha_1 + \beta_1' + \alpha_1' + \beta_1)}, \end{aligned} \tag{19.10}$$

which is again half of the unperturbed flux [Eq. (17.1)].

20. Functional Interactions between Subunits

This section and the next are essentially appendices to this chapter. The principal concern of this book is the effect of intermolecular interactions on the steady-state and equilibrium behavior of enzyme or protein complexes or lattices. This section deals with neighbor interactions of a very different kind: one enzyme molecule of a complex transfers *directly* to another a small molecule, molecular fragment, electron, etc. The "interaction" between the two enzyme molecules is functional: they necessarily share a transition together; their kinetic cycles are interlocking (see below). Intermolecular interactions are introduced into these problems in Sections 27 and 28.

Another quite different type of interaction is the subject of Section 21: two different complexes in a closed vesicular membrane share a ligand or other small molecule via the vesicular interior. As a consequence, the enzymatic activities of the two complexes are coupled to each other.

These two special "interactions" are important and interesting, but because they are really outside of the scope of this book, they will merely be introduced here. Many more details and examples are available in Chapter 7 of Ref. 2 and in the General References.

Examples with $m = 2, 3, \rho = 2, 3$

We begin with Fig. 5-45(a). This represents the mechanism used by a complex of two three-state enzyme molecules E_1 and E_2. For simplicity, we employ one-way transitions in this figure and throughout the section. E_1 catalyzes $S \rightarrow P$ and transfers P directly to E_2; P is the substrate for E_2, which transforms P into Q. Thus the net reaction is $S \rightarrow Q$. The numbers in squares are shorthand state numbers for each enzyme. The two cycles are interlocked by the transfer reaction, $31 \rightarrow 12$. The two enzymes necessarily complete the same number of cycles; there is only one flux J (for $S \rightarrow Q$). The kinetic

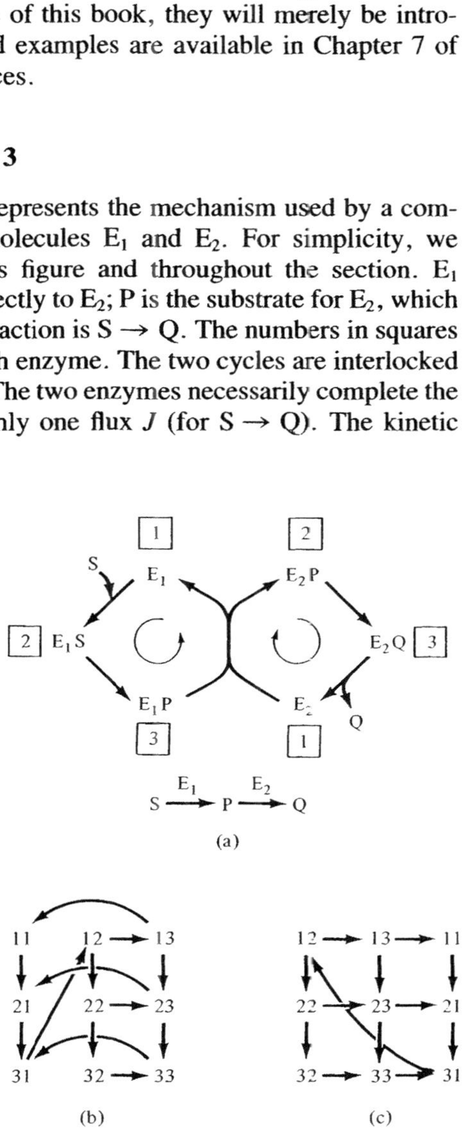

Fig. 5-45. (a) Reaction scheme for two three-state enzymes which, together, catalyze $S \rightarrow P \rightarrow Q$. See text for details. Numbers in squares are assigned state numbers. (b) Kinetic diagram for (a). All transitions are one way. State 12 is $E_1 \cdot E_2P$, etc. (c) Equivalent rearrangement of (b).

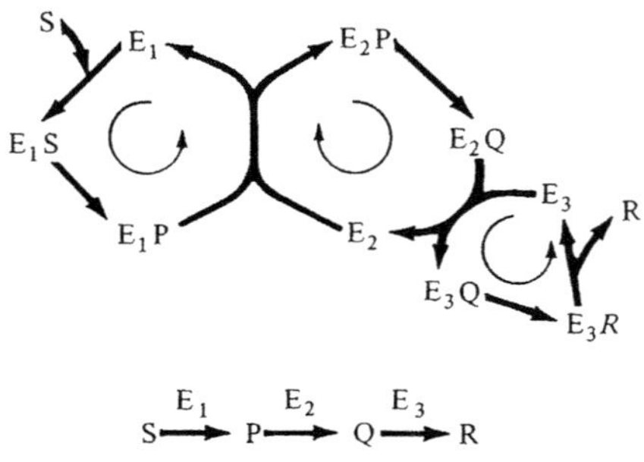

Fig. 5-46. Extension of Fig. 5-45 to three enzyme molecules. See text for explanation.

diagram is shown in Fig. 5-45(b) [compare Fig. 5-21(b)]. Actually the diagram has symmetry that is not evident in Fig. 5-45(b). This appears in Fig. 5-45(c), after a reordering of the states.

Figure 5-46 extends this example to a complex of three three-state enzyme molecules with two interlocking reactions. In Fig. 5-46, Q is transferred directly from E_2 to E_3, and is the substrate for E_3, which converts Q into R. Thus the net reaction catalyzed by this complex is $S \rightarrow R$. Again there is only one flux J (for $S \rightarrow R$); the interlocking reactions compel the three enzyme molecules to coordinate their cyclic activities.

We turn now to two-state cycles in order to be able to pursue the subject further with minimum algebraic complexity. Figures 5-47(a) and 5-48(a) give the two-state versions of Figs. 5-45(a) and 5-46. The species E_1P, E_2Q, and E_3R are all transient intermediates and do not appear in the two-state cycles. First-order rate constants (inverse transitions are neglected) are included in Figs. 5-47, 5-48(a), and 5-48(b); the kinetic diagrams appear in Figs. 5-47(b) and 5-48(b). Note that all vertical transitions in the two diagrams are of type α, all horizontal transitions are of type β, and diagonal transitions are transfer reactions (k, l).

In the two-enzyme ($m = 2$) complex case (Fig. 5-47), the steady-state probabilities of states can be determined from

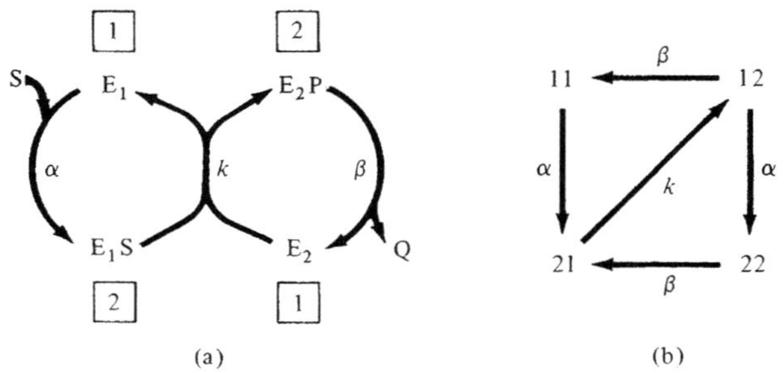

Fig. 5-47. (a) Two-state version of Fig. 5-45(a). See text for details. (b) Corresponding kinetic diagram.

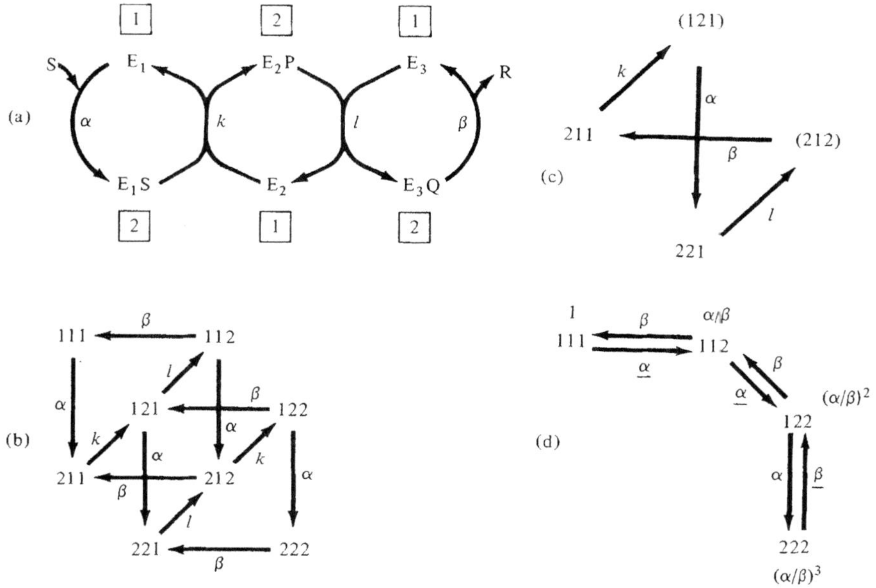

Fig. 5-48. (a) Two-state version of Fig. 5-46. See text for details. (b) Corresponding kinetic diagram. (c) Reduction of (b) when k and l are small. (d) Reduction of (b) when k and l are large. Relative probabilities of states are proportional to 1, α/β, etc.

$$\alpha p_{11} = \beta p_{12}, \qquad \beta p_{22} = \alpha p_{12},$$
$$\alpha p_{11} + \beta p_{22} = k p_{21}, \tag{20.1}$$
$$p_{11} + p_{12} + p_{21} + p_{22} = 1.$$

These give

$$p_{11} = \frac{k\beta^2}{\Sigma}, \qquad p_{12} = \frac{k\alpha\beta}{\Sigma},$$
$$p_{22} = \frac{k\alpha^2}{\Sigma}, \qquad p_{21} = \frac{\alpha\beta(\alpha + \beta)}{\Sigma}, \tag{20.2}$$
$$\Sigma = k(\alpha^2 + \alpha\beta + \beta^2) + \alpha\beta(\alpha + \beta).$$

Then the S $\rightarrow$ Q flux is

$$J = kp_{21} = \frac{k\alpha\beta(\alpha + \beta)}{\Sigma}. \tag{20.3}$$

If $\alpha = \beta = k$, Eq. (20.3) reduces to $J = 2k/5$. If k is much smaller than α and β, $J \rightarrow k$. That is, the rate-limiting transition is the transfer step $21 \rightarrow 12$. On the other hand, if α and β are relatively small,

$$J \rightarrow \frac{\alpha\beta(\alpha + \beta)}{\alpha^2 + \alpha\beta + \beta^2}. \tag{20.4}$$

This result can be obtained independently as follows. The mean lifetime of the 12 complex before either the α or the β transition occurs is $(\alpha + \beta)^{-1}$. The probability that this first transition is α is $\alpha/(\alpha + \beta)$. In this case, the mean time for the second transition (to state 21) is β^{-1}. The probability that the first transition is β is $\beta/(\alpha + \beta)$, and then the mean time for the second transition is α^{-1}. Hence the total mean time for $12 \rightarrow 21$ is

$$\frac{1}{\alpha + \beta} + \frac{\alpha}{(\alpha + \beta)\beta} + \frac{\beta}{(\alpha + \beta)\alpha}.$$

The reciprocal of this quantity is the effective rate constant for $12 \rightarrow 21$, or J (because the k transition is relatively fast).

The derivation of the state probabilities and flux at steady state for Figs. 5-48(a) and 5-48(b) is accomplished most easily (see Ref. 2, pp. 167–169) using the flux diagram method. We omit the details but give the final result for the flux:

$$J = \alpha\beta kl[\alpha^2 k(\alpha + \beta + l) + \beta^2 l\,(\alpha + \beta + k) + \alpha\beta(\alpha + \beta)(\alpha + \beta + k + l) + \alpha\beta kl]\,\Sigma^{-1}, \tag{20.5}$$

where

$$\Sigma = \alpha^2(a + \beta + l)[l(\alpha + \beta)(\beta^2 + \beta k + k^2) + a\beta k^2] + \beta^2(\alpha + \beta + k)[k(\alpha + \beta)(a^2 + al + l^2) + \alpha\beta l^2].$$

This simplifies, in three special cases, as follows:

$$J = \frac{5k}{14} \qquad (\alpha = \beta = k = l), \tag{20.6}$$

$$J = \frac{kl}{k + l} \qquad (k, l \ll \alpha, \beta), \tag{20.7}$$

$$J = \frac{\alpha\beta(\alpha^2 + \alpha\beta + \beta^2)}{(\alpha + \beta)(\alpha^2 + \beta^2)} \qquad (\alpha, \beta \ll k, l). \tag{20.8}$$

When k and l are small [Eq. (20.7)], Fig. 5-48(b) degenerates into Fig. 5-48(c), where states 121 and 212 are transient intermediates because $\alpha, \beta \gg k, l$. Thus this diagram is essentially a two-state (211, 221) cycle with rate constants k and l: Eq. (20.7) is the result for the flux.

When k and l are large [Eq. (20.8)], Fig. 5-48(b) simplifies, in effect, to Fig. 5-48(d), where the relative probabilities of states at quasiequilibrium are indicated above or beside the states. J can be found either from k transitions or from l transitions (embedded in some of the α, β transitions). For example, the underlined rate constants in Fig. 5-48(d) include k transitions. Then

$$J = \alpha(p_{111} + p_{112}) + \beta p_{222}. \tag{20.9}$$

If we use the probabilities from the figure, we recover Eq. (20.8).

Chain of Two-State Subunits with Arbitrary *m*

Figures 5-47 and 5-48 can be extended to a chain of two-state subunits of arbitrary size m. If there is no chemical transformation of the substrate, but the substrate is merely transferred from one subunit binding site to another along the chain, we have a model for the transport of a ligand L across a membrane or for the transport of an electron along the components of the respiratory chain in oxidative phosphorylation. A ligand molecule jumps from site to site ("transfer reaction"); any number of sites of the chain can be occupied by ligand at one time. This is illustrated in Figs. 5-49(a) and 5-49(b) (for $m = 3$), where L is the ligand, E is the subunit with the binding site for L, and all rate constants are taken equal to k (as an illustration, and to make the mathematics tractable for arbitrary m). For any subunit, E is state 1 and EL is state 2. If the ligand is an electron, E or state 1 is the oxidized state of a component and EL or state 2 is the reduced state.

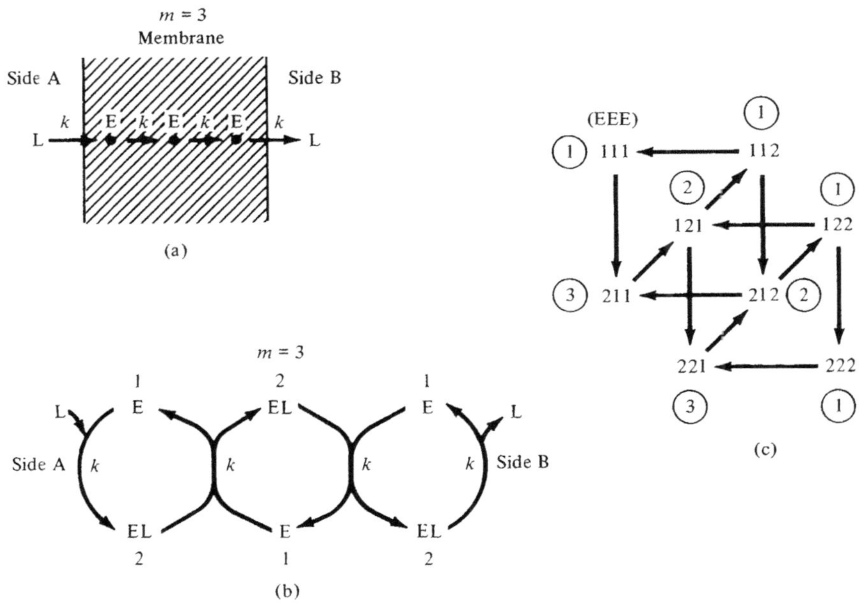

Fig. 5-49. (a) Simple membrane transport model with all transitions one way and all rate constants equal to k. See text for further details. (b) Another version of the same model. (c) Corresponding kinetic diagram. All rate constants are k. The numbers in circles are relative steady-state probabilities.

Figure 5-49(c) is a simplified version of Fig. 5-48(b) in which each arrow has the same rate constant k and the numbers in circles are relative state probabilities at steady state, found as in Figs. 5-24, Eq. (17.37), etc. The flux J (transport of L) in this special case ($m = 3$) can be found, for example, from all the vertical arrows [this is the rate at which L enters the chain from the left, side A, in Fig. 5-49(b); the left subunit undergoes the transition $1 \rightarrow 2$ for each vertical arrow]:

$$J = k(p_{111} + p_{121} + p_{112} + p_{122}) = \frac{k(1 + 2 + 1 + 1)}{14} = \frac{5k}{14}. \tag{20.10}$$

This same result was obtained in Eq. (20.6) [also, we found $J = 2k/5$ for $m = 2$ from Eq. (20.3)].

If we define $P_1^{(i)}$ as the probability that subunit i in the chain (counting from the left) is in state 1, then the flux can also be written as

$$J = kP_1^{(1)}. \tag{20.11}$$

That is, $P_1^{(1)}$ is the sum over all ps with subunit $i = 1$ in state 1. In other words, this is the probability that the left-hand subunit of the chain has an empty site in order to be able to accept an L from side A, irrespective of the states of subunits $i = 2, \ldots, m$. An alternative expression for J is obviously $J = kP_2^{(m)}$. This is the rate at which L enters side B. Hence we must have $P_1^{(1)} = P_2^{(m)}$ at steady state. In fact, from the relative probabilities in Fig. 5-49(c), we see that ($m = 3$)

$$P_1^{(1)} = \frac{5}{14}, \qquad P_1^{(2)} = \frac{7}{14}, \qquad P_1^{(3)} = \frac{9}{14}. \tag{20.12}$$

Of course

$$P_2^{(i)} = 1 - P_1^{(i)}. \tag{20.13}$$

The site occupation probability falls as i increases from $i = 1$ to $i = m = 3$.

Steady-state properties, as in the $m = 3$ case, were found by the author[5] for $m = 1, 2, \ldots, 7$. General formulas (arbitrary m) were then conjectured from these special cases. Rigorous proofs of the general results were provided later by Shapiro and Zeilberger.[6] The general formula for J turns out to be

$$J = \frac{(m + 2)k}{2(2m + 1)} = kP_1^{(1)} = k(1 - P_1^{(m)}). \tag{20.14}$$

Note that this gives $k/2$, $2k/5$, and $5k/14$, as we have already seen, for $m = 1, 2, 3$ [see Eq. (17.1) for $m = 1$]. In the limit $m \rightarrow \infty$, $J \rightarrow k/4$. Thus the range in J is from $k/2$ ($m = 1$) to $k/4$ ($m = \infty$). It may seem surprising that the flux of ligand through a long chain of subunits is not much less than through a short chain of subunits. However, in both cases (when all rate constants are k) the mean occupation of sites throughout the chain is $\frac{1}{2}$. Thus there is a ready supply of ligand molecules to enter side B, even for large m.

The general equation for the ps [chain states, as in Fig. 5-49(c)] is too complicated to repeat here. The formula for $P_1^{(i)}$ is somewhat simpler:

$$P_1^{(i)} = \frac{1}{C_{m+1}} \sum_{j=0}^{i-1} C_j C_{m-j}, \tag{20.15}$$

where the C_j are the Catalan numbers

$$C_j = \frac{(2j)!}{(j!)^2(j+1)}. \tag{20.16}$$

The first few of these are $C_0 = 1$, $C_1 = 1$, $C_2 = 2$, $C_3 = 5$, $C_4 = 14$. From Eq. (20.15), $P_1^{(1)}$ is C_m/C_{m+1}, which agrees with Eq. (20.14). The first few $P_1^{(i)}$ are:

$$\begin{array}{llllll}
m = 1\text{:} & P_1^{(1)} = 1/2 & & & & \\
m = 2\text{:} & P_1^{(1)} = 2/5 & P_1^{(2)} = 3/5 & & & \\
m = 3\text{:} & 5/14 & 7/14 & 9/14 & & \\
m = 4\text{:} & 14/42 & 19/42 & 23/42 & 28/42 & \\
m = 5\text{:} & 42/132 & 56/132 & 66/132 & 76/132 & 90/132
\end{array} \tag{20.17}$$

Note that the average in each row (i.e., for each m) is $\frac{1}{2}$.

21. Coupled Enzyme Systems in a Vesicular Membrane

If two or more different enzyme systems or complexes are present in the same closed vesicular membrane and, in their activity, share one or more small molecules (ligands, substrates, etc.) that are present in the interior of the vesicle, the enzyme complexes interact with each other by virtue of this sharing. The various systems of the mitochondrial inner membrane that participate in oxidative phosphorylation furnish an example that has been studied in considerable detail.[7–9] Here we merely introduce the subject by means of the simplest possible hypothetical prototype.[10]

Suppose we have a small vesicle of volume V, with two kinds of enzyme molecules (or complexes or units) embedded in the vesicular membrane. There are M_A molecules of type A and M_B of type B. Each of these units behaves independently of the rest, except for the fact that units of type A transport a ligand L from outside the vesicle to inside, whereas units of type B transport L from inside to outside. Because V is small and the inside ligand concentration is responsive to the ligand transport, the shared ligand has the effect of coupling the A and B systems to each other in a rather sensitive way. We examine this coupling here, assuming that both A and B units can be treated as two-state enzyme molecules.

Let us write the complete process catalyzed by A molecules as

$$L_{out} + P_A \rightarrow L_{in} + S_A;\ X_A \tag{21.1}$$

and by B molecules as

$$L_{in} + S_B \rightarrow L_{out} + P_B;\ X_B, \tag{21.2}$$

where S refers to substrate, P to product, and X to the net thermodynamic force. At steady state, when the inward ligand transport by A units is just balanced by the outward ligand transport by B units, the overall process is

$$S_B + P_A \rightarrow P_B + S_A;\ X_A + X_B. \tag{21.3}$$

To choose an explicit case (suggested by oxidative phosphorylation), we assume that the "downhill" L gradient out $\rightarrow$ in is large enough to drive the "backward" reaction $P_A \rightarrow S_A$ (the former has a positive thermodynamic force, the latter has a negative force, while the net force X_A is positive). Similarly, we assume that $S_B \rightarrow P_B$ (positive force) is able to drive L up its gradient (in $\rightarrow$ out; negative force) with a net positive force X_B. In the oxidative phosphorylation example (ignoring stoichiometry), the vesicle interior is the mitochondrial matrix, the L gradient is the proton electrochemical gradient, an A unit is the reverse ATPase complex (P_A is ADP, P_i; S_A is ATP), and a B unit is a "transducing site" region of the respiratory chain complex (S_B refers to electron donor and acceptor; P_B refers to oxidized donor and reduced acceptor).

Figures 5-50 and 5-51 should be examined together. Figure 5-50 shows several A and B enzyme units schematically (● and x, respectively), with only the ligand (○), not substrates and products, appearing explicitly in the figure. State 2 (for both A and B) has ligand bound; state 1 has ligand not bound. With these two-state cycles, it is to be understood that more complicated cycles have been "reduced" because of transient intermediates, etc.

The same rate constants appear in both Figs. 5-50 and 5-51. The associated physical processes (binding or release of ligand) are indicated in Fig. 5-50,

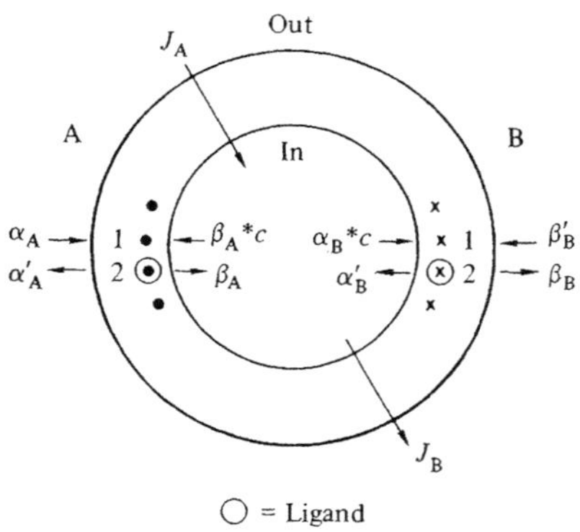

Fig. 5-50. Schematic vesicle with membrane. Units of type A(●) and B(x) transport a ligand L(○) across the vesicular membrane. Interior concentration of L is c. Arrows indicate binding or release of ligand. See text for details.

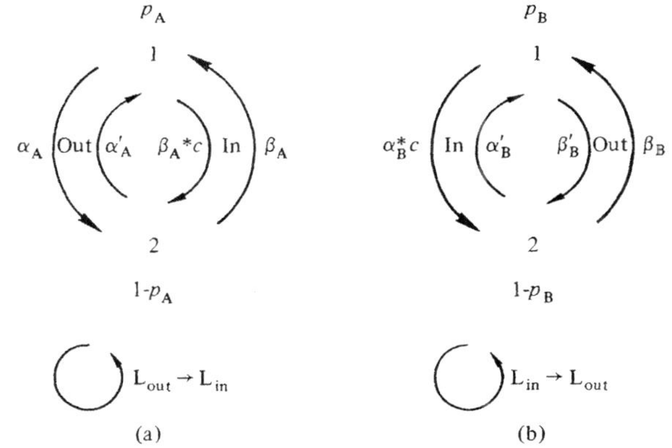

Fig. 5-51. Two-state cycles for A(a) and B(b) units, with first-order rate constant notation for transitions (arrows). Dominant cycle direction is counterclockwise in both cases. "Out" refers to binding and release of ligand to exterior, etc. See text for details.

whereas the kinetic cycles are shown in Fig. 5-51. Unprimed rate constants are usually the dominant ones (counterclockwise in Fig. 5-51); the smaller inverse rate constants are primed. An asterisk indicates a second-order rate constant; the others are first-order or pseudo-first-order rate constants. The interior concentration of ligand is $c = N/V$; because V is small, c cannot be treated as a constant. However, for simplicity, we assume that ligand is maintained at a constant *exterior* concentration over the time period of interest. Also, for clarity and simplicity, we assume that substrates and products are maintained at constant concentrations wherever they appear (inside or out). The membrane potential, if involved, is also assumed to be constant. In the reverse ATPase case (above), for example, the interior ADP, P_i, and ATP concentrations could be kept essentially constant by separate systems that are also present in the mitochondrial inner membrane. It is because of the above-mentioned simplifying assumptions that no other second-order rate constants are used in Figs. 5-50 and 5-51. The constant concentrations referred to above are "embedded" in pseudo-first-order rate constants. For example, α_A and β_B' would both be proportional to the constant outside ligand concentration.

The fraction of A units in state 1 is denoted by p_A, etc. (Fig. 5-51). The kinetic equations for the variables p_A, p_B, and c are

$$\frac{dp_A}{dt} = (\alpha_A' + \beta_A)(1 - p_A) - (\alpha_A + \beta_A^* c)p_A, \qquad (21.4)$$

$$\frac{dp_B}{dt} = (\alpha_B' + \beta_B)(1 - p_B) - (\alpha_B^* c + \beta_B')p_B, \qquad (21.5)$$

$$\frac{dc}{dt} = \frac{M_A}{V}[\beta_A(1 - p_A) - \beta_A^* c p_A] + \frac{M_B}{V}[\alpha_B'(1 - p_B) - \alpha_B^* c p_B]. \qquad (21.6)$$

We are interested, primarily, in the steady state. For this, let us consider first the *hypothetical separate* steady states of A and B units for an arbitrary value of c. We find, for the total fluxes J_A and J_B (Fig. 5-50), through the M_A and M_B units,

$$J_A(c) = \frac{M_A(\alpha_A\beta_A - \alpha'_A\beta^*_A c)}{\alpha'_A + \alpha_A + \beta_A + \beta^*_A c}, \tag{21.7}$$

$$J_B(c) = \frac{M_B(\alpha^*_B c\beta_B - \alpha'_B\beta'_B)}{\alpha'_B + \alpha^*_B c + \beta_B + \beta'_B}. \tag{21.8}$$

These are not only the ligand fluxes but also the fluxes of substrates and products involved in the two cycles. Incidentally, if only A units or only B units were present in the membrane, c would be free to adjust itself to the corresponding equilibrium value:

$$J_A(c^e_A) = 0, \qquad c^e_A = \alpha_A\beta_A/\alpha'_A\beta^*_A, \tag{21.9}$$

$$J_B(c^e_B) = 0, \qquad c^e_B = \alpha'_B\beta'_B/\alpha^*_B\beta_B. \tag{21.10}$$

Note, in Eqs. (21.7) and (21.8), that $J_A(c)$ starts ($c = 0$) from a positive value and ends ($c = \infty$) at a negative value, while $J_B(c)$ does the reverse. That is, J_A decreases and J_B increases with increasing c; the latter dependence on c is usually stronger because an unprimed rate constant is involved. The two curves necessarily cross each other. The intersection occurs at the steady-state value of c, c_∞, determined by $J_A(c_\infty) = J_B(c_\infty)$ [Eqs. (21.7) and (21.8)]. This is, in effect, the desired steady-state solution of Eqs. (21.4)–(21.6). $J_A = J_B$ leads to a quadratic equation in c_∞, which we omit. We denote the steady-state flux $J_A = J_B$ by J_∞. This flux will be positive if $c^e_A > c^e_B$ [the usual case; note the inverse rate constants in Eqs. (21.9) and (21.10)]. The following subsection provides an explicit numerical example. The A and B systems are coupled to each other, at steady state, through their common dependence on the value of c_∞ in the small volume V.

If we neglect the inverse rate constants in Fig. 5-51, the equation in c_∞ is linear:

$$J_A(c) = \frac{M_A\alpha_A\beta_A}{\alpha_A + \beta_A} = \text{independent of } c = J_A(c_\infty), \tag{21.11}$$

$$J_B(c) = \frac{M_B\alpha^*_B\beta_B c}{\beta_B + \alpha^*_B c}, \tag{21.12}$$

$$J_A(c_\infty) = J_B(c_\infty), \tag{21.13}$$

$$c_\infty = \frac{M_A\alpha_A\beta_A\beta_B}{\alpha^*_B[M_B\beta_B(\alpha_A + \beta_A) - M_A\alpha_A\beta_A]}. \tag{21.14}$$

The saturation ($c \to \infty$) value of $J_B(c)$, namely, $M_B\beta_B$, must be larger than the constant value of J_A in Eq. (21.11) in order for c_∞ to exist.

If the rate constants happen to be such [see Eqs. (21.9) and (21.10)] that

$c_A^e = c_B^e$, then this is also the value of c_∞, and $J_\infty = 0$. In this special case, the steady state is an equilibrium state (for both A and B).

The steady-state thermodynamic forces X_A and X_B are given by

$$e^{X_A/kT} = \alpha_A \beta_A / \alpha_A' \beta_A^* c_\infty, \qquad e^{X_B/kT} = \alpha_B^* c_\infty \beta_B / \alpha_B' \beta_B'. \quad (21.15)$$

Both X_A and X_B are usually positive; each is a combination of two subforces [Eqs. (21.1) and (21.2)]. Because $J_A = J_B$ at steady state, there is no net ligand transport in the coupled system (A plus B), and the sum $X_A + X_B$, determined by

$$e^{(X_A+X_B)/kT} = \alpha_A \beta_A \alpha_B^* \beta_B / \alpha_A' \beta_A^* \alpha_B' \beta_B', \quad (21.16)$$

has the physical significance indicated in Eq. (21.3). Note that c_∞ does not appear in Eq. (21.16). Operationally (i.e., viewed as a "black box"), the coupled A and B systems simulate a *single* kind of system with flux J_∞, force $X_A + X_B$, and net reaction given by Eq. (21.3). Even though $M_A \neq M_B$, the stoichiometry in Eq. (21.3) is precise because $J_A = J_B$. At this level, the role of the ligand does not appear explicitly. These comments seem pertinent to the history of the study of oxidative phosphorylation: until rather recently, perhaps because of the simulation of a single kind of system, ATP was generally thought to be synthesized by the respiratory chain, via an "active intermediate" (rather than by a separate reverse ATPase system).

Numerical Example

We now provide a numerical illustration of the steady-state properties of the coupled two-state systems in Fig. 5-51. For convenience, the rate constants and c are chosen to be dimensionless and of order unity, as shown in Fig. 5-52. Also, we take $M_A = M_B \equiv M$. The rate constant β_A is considered to be proportional to the fixed but adjustable concentration of P_A in Eq. (21.1) (e.g., ADP in the oxidative phosphorylation case).

For any choice of β_A, the steady state can be determined as in Fig. 5-53. Here we plot $J_A(c)/M$ and $J_B(c)/M$ from Eqs. (21.7) and (21.8), using the rate constants in Fig. 5-52 and several values of β_A. The points of intersection determine J_∞/M and c_∞. Although $J_B(c)$ itself is independent of β_A, the

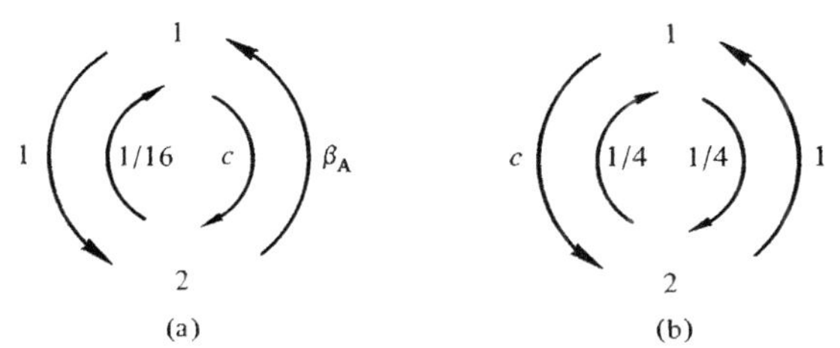

Fig. 5-52. Rate constant assignment for Fig. 5-51 in numerical example.

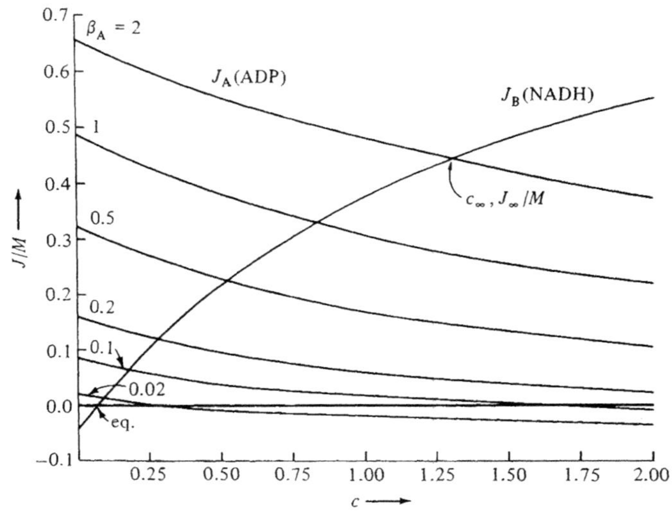

Fig. 5-53. Plot of J_A/M and J_B/M against c, for several values of β_A, for numerical example in Fig. 5-52. Intersection points give steady-state values of J_∞/M and c_∞. Equilibrium point is labeled "eq."

steady-state flux J_∞ of *both* systems (A and B) is clearly fairly sensitive to β_A because of the coupling of the two systems through the interior ligand concentration c. Figure 5-54 presents J_∞/M and c_∞ as functions of β_A (from the intersection points in Fig. 5-53). The flux $J_\infty(\beta_A)/M$ simulates that of a single system. For large β_A,

$$J_\infty/M \rightarrow a\beta_A^{1/2}/(1 + a\beta_A^{1/2}), \tag{21.17}$$

where $a = (2/3)^{1/2}$. The equilibrium point in Fig. 5-53 (labelled "eq.") has $\beta_A = \frac{1}{256}$, $c_\infty = \frac{1}{16}$, and $J_\infty = 0$.

For comparison, in Fig. 5-54, we include (dashed lines) the one-way special case (the clockwise rate constants in Fig. 5-52 are set equal to zero). Here $c_\infty = \beta_A$ and

$$J_\infty/M = \beta_A/(1 + \beta_A). \tag{21.18}$$

References

1. Hill, T. L. and Kirschner, M. W. (1982) Inter. Rev. Cytol. **78**, 1.
2. Hill, T. L. (1977) *Free Energy Transduction in Biology* (Academic, New York).
3. Gresser, M. J., Myers, J. A., and Boyer, P. D. (1982) J. Biol. Chem. **257**, 12030.
4. Stryer, L. (1981) *Biochemistry* (Freeman, San Francisco), Second Edition.

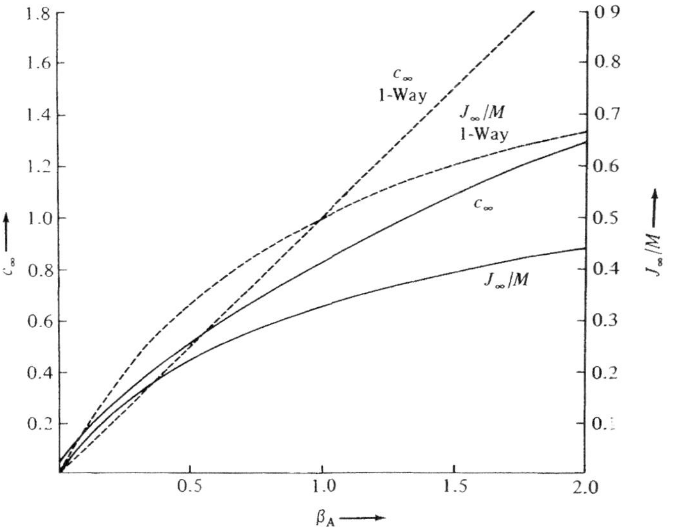

Fig. 5-54. Plots of J_∞/M and c_∞ against β_A, taken from intersection points in Fig. 5-53. The flux represents qualitatively the dependence of the rate of oxygen uptake and ATP synthesis on ADP concentration (β_A). Dashed curves are for one-way special cases.

5. Hill, T. L. (1977) in *Statistical Mechanics and Statistical Methods in Theory and Applications*, ed. Landman, U. (Plenum, New York).
6. Shapiro, L. W. and Zeilberger, D. (1982) J. Math. Biol. **15**, 351.
7. van Dam, K. and Westerhoff, H. V. (1977) in *Structure and Function of Energy-Transducing Membranes*, eds. van Dam, K. and van Gelder, B. F. (Elsevier North-Holland, Amsterdam).
8. Hill, T. L. (1980) Proc. Natl. Acad. Sci. USA **77**, 2681.
9. Walz, D. (1983) in *Biological Structures and Coupled Flows*, eds. Oplatka, A. and Balaban, M. (Academic, New York).
10. Hill, T. L. (1979) Proc. Natl. Acad. Sci. USA **76**, 232.

General References

Levitzki, A. (1978) *Quantitative Aspects of Allosteric Mechanisms* (Springer-Verlag, New York).

King, E. L. and Altman, C. (1956) J. Phys. Chem. **60**, 1375.

Hill, T. L. (1977) Proc. Natl. Acad. Sci. USA **74**, 3632.

Hill, T. L. (1977) Proc. Natl. Acad. Sci. USA **74**, 4111.

Kurganov, B. I. (1982) *Allosteric Enzymes* (Wiley, New York).

II

Small Systems or Complexes with Explicit Interactive Cooperativity

6

Equilibrium Small Systems with Explicit Interactions

The intent of this chapter is to discuss a sufficient number of simple but important models to illustrate the utility of the grand partition function for equilibrium small systems with explicit intermolecular interactions. This subject was introduced in Ref. 1. The treatment is meant to be of pedagogical value, primarily; it is *not* a comprehensive survey of systems of this type that have been discussed in the literature. In particular, cooperativity in hemoglobin is not treated explicitly because the application of the grand partition function method to hemoglobin has already been thoroughly covered in the literature.[2–4]

The emphasis is on explicit molecular models represented by explicit grand partition functions. This is in contrast to the use of empirical multiple equilibrium constants that can represent cooperativity in a formal way but provide little insight into molecular mechanisms.

This chapter generalizes some of the topics introduced in Chapter 4. Chapters 8 and 10 continue the discussion by a consideration of large one and two-dimensional lattices of units with interactions.

22. Simple Equilibrium Binding, with Interactions, on a Group of Sites

The primary topic in this section is the study of the binding of a single ligand on a group of sites, each with the same intrinsic binding constant for the ligand, but for cases in which neighboring bound ligand molecules interact with each other.[1] These interactions perturb the equilibrium properties of the system. In particular, positive cooperativity results if the ligand–ligand interactions are attractive and negative cooperativity occurs if the interactions are repulsive.

Most of the results obtained in this section are actually more general than the above-described model would suggest. They also apply, for example, to the binding of a ligand on the subunits of an aggregate or complex for those cases in which ligand binding induces a conformational change in a subunit and in which neighboring subunits interact with each other. This generalization is the subject of Section 23.

Two Different Sites, Two Ligands

To introduce interactions, we start with the model already considered in Eqs. (11.5)–(11.11). The protein P has two different binding sites, the first for ligand L_1 and the second for ligand L_2. As in Eq. (11.7), the grand partition function for one molecule of P, as an open system, is

$$\begin{array}{ccccccc} \xi = q_P & + & q_{L_1P}\lambda_1 & + & q_{PL_2}\lambda_2 & + & q_{L_1PL_2}\lambda_1\lambda_2 . \\ \boxed{} & & \boxed{1} & & \boxed{2} & & \boxed{1\ 2} \end{array} \tag{22.1}$$

In the present example, L_1 and L_2 do *not* bind independently on P (as in Section 11). That is, if we factor out q_P from the terms in ξ, and then omit q_p as usual [see Eq. (7.2)], we do not obtain

$$\xi = 1 + q_1\lambda_1 + q_2\lambda_2 + q_1q_2\lambda_1\lambda_2,$$

as in Eq. (11.9), but rather there is an extra factor in the last term that represents the lack of independence of the L_1 and L_2 binding when both are present on P. We write this as

$$\xi = 1 + q_1\lambda_1 + q_2\lambda_2 + q_1q_2e^{-w/kT}\lambda_1\lambda_2. \tag{22.2}$$

In the independent binding case, $w = 0$. Using the definitions of q_1 and q_2 in Eq. (11.8), we see from Eqs. (22.1) and (22.2) that w is defined by

$$e^{-w/kT} = \frac{q_{L_1PL_2}}{q_Pq_1q_2} = \frac{q_{L_1PL_2}q_P}{q_{L_1P}q_{PL_2}}. \tag{22.3}$$

If we write each q in the last expression as $e^{-A/kT}$, where A is a Helmholtz free energy, then, in obvious notation,

$$e^{-w/kT} = \frac{e^{-A_{12}/kT}e^{-A_0/kT}}{e^{-A_1/kT}e^{-A_2/kT}},$$

$$w = (A_{12} + A_0) - (A_1 + A_2). \tag{22.4}$$

Thus w has the physical significance of the extra free energy in L_1PL_2 owing to the simultaneous presence of L_1 and L_2 on P. We refer to w as the "interaction free energy"; w can be positive or negative; w is the isothermal reversible work required to carry out the process

$$\boxed{1} + \boxed{2} \rightarrow \boxed{1\ 2} + \boxed{}\,. \tag{22.5}$$

At one extreme, w could be associated entirely with the effective intermolecular interaction or force (see below) between the molecules L_1 and L_2 in their sites on P, in the presence of solvent. In this case, any perturbations of the structure of P itself resulting from the binding of L_1 and L_2 on the two sites are local, strictly additive, and make no contribution to w. At the other extreme, L_1 and L_2 might be such small molecules or the two binding sites on P could be so far apart that the effective interaction or force between L_1 and L_2 (above) does not contribute at all to w. Rather, w is a consequence of relatively long-range perturbations of the structure of P (e.g., a conformational change) by bound L_1 and L_2 that interfere with or reinforce each other. Intermediate cases are, of course, also possible. The remainder of this section uses the first extreme (for conceptual simplicity), the second extreme is implicit in the two-conformation model in Section 25, while the next section, Section 23, is not restricted in either of these ways.

In the first special case, above, in which w is due entirely to the effective intermolecular force between L_1 and L_2, it is possible to be more precise about the meaning of "effective." If the distance r between the centers of mass of L_1 and L_2, in their binding sites, is varied infinitesimally, the work required for this variation defines a mean force[5] at r. This mean force includes contributions from: (a) the direct intermolecular force between L_1 and L_2; (b) perturbation of solvent-L_1 and solvent-L_2 interactions owing to the proximity of L_1 and L_2 (at r compared to $r = \infty$); and (c) perturbation of solvent–solvent interactions owing to the proximity of L_1 and L_2. The mean force is calculated in principle by statistical mechanical averaging[5] over the rocking orientations of L_1 and L_2 in their sites and over all solvent molecule spatial positions and orientations. The potential function of this mean force is w ($\bar{F} = -dw/dr$). The zero of w must be chosen at $r = \infty$ in order to be consistent with Eqs. (22.4) and (22.5). In a gas (i.e., if no solvent were present), w would simply be the intermolecular pair potential between L_1 and L_2.

In the usual way [see Eq. (11.5)], Eq. (22.2) can be rewritten in the form

$$\xi = 1 + K_1c_1 + K_2c_2 + yK_1c_1K_2c_2, \tag{22.6}$$

where

$$y \equiv e^{-w/kT} \tag{22.7}$$

and K_1 and K_2 are the binding constants for L_1 and L_2 in the absence of the other ligand. If L_1 and L_2 interfere with each other when simultaneously bound, either directly or indirectly (e.g., through P), w will be positive and $0 < y < 1$. In the limit $y \to 0$, L_1 and L_2 cannot be bound at the same time. This case is indistinguishable from competitive binding of two ligands on a single site [Eq. (7.3)]. If the two ligands enhance each other's binding, i.e., L_1 and L_2 attract each other, directly or indirectly, w is negative and $y > 1$. If the two ligands bind independently of each other, $w = 0$ and $y = 1$.

The binding isotherm for L_1 on its site (L_2 is similar, by symmetry) follows immediately from the terms in ξ or from

$$\theta_1 = c_1 \frac{\partial \ln \xi}{\partial c_1}$$

$$= \frac{K_1 c_1 + y K_1 c_1 K_2 c_2}{\xi}. \tag{22.8}$$

This can be rearranged into the form

$$\theta_1 = \frac{\Phi K_1 c_1}{1 + \Phi K_1 c_1}, \tag{22.9}$$

where

$$\Phi = \frac{1 + y K_2 c_2}{1 + K_2 c_2}. \tag{22.10}$$

Thus the binding of L_1 has the simple Langmuir form, even in the presence of L_2 at concentration c_2, but the effective binding constant is ΦK_1. If $y = 1$, $\Phi = 1$ and the binding of L_1 is unperturbed by the binding of L_2. If $y > 1$, $\Phi > 1$ and the binding of L_1 is enhanced by the presence of L_2. If $y < 1$, $\Phi < 1$ and the binding of L_1 is diminished.

One Ligand on Two Sites ($m = 2$)

We now turn to a special case that is also used in the remaining subsections of this section (where $m = 3, 4, 6$). P has two sites for a single ligand L that is at concentration c; both sites have the same intrinsic or unperturbed binding constant K. To be explicit, we assume that ligand molecules in the two sites interact directly with each other (not through P) and that w is a potential of mean force, as described above. Operationally, K would be found from the initial (low c) binding of L on the pair of sites, where $\theta = Kc$. When c is larger, the interaction free energy w perturbs the binding. The grand partition function in Eq. (22.6) simplifies in the present case to

$$\xi = 1 + 2Kc + yK^2c^2. \tag{22.11}$$

· · ⊙· ⊙⊙
·⊙

The successive terms in ξ are for $s = 0$, $s = 1$, and $s = 2 = m$.

To simplify notation, we introduce the dimensionless quantity $x = Kc$ as the independent concentration variable instead of c. Then

$$\xi = 1 + 2x + yx^2. \tag{22.12}$$

In Section 15, for example in Eq. (15.9), we can then also use x in place of c. Thus

$$\xi = \alpha_0 + \alpha_1 x + \alpha_2 x^2, \tag{22.13}$$

where, in the present example,

$$\alpha_0 = 1, \qquad \alpha_1 = 2, \qquad \alpha_2 = y,$$
$$x_o = Kc_o = (\alpha_0/\alpha_2)^{1/2} = y^{-1/2}, \tag{22.14}$$
$$c_o = 1/Ky^{1/2}.$$

That is, $\theta = \frac{1}{2}$ when $x = x_o = y^{-1/2}$.

The explicit expression for $\theta(x)$ is

$$\theta(x) = \frac{\bar{s}}{2} = \frac{x}{2}\frac{\partial \ln \xi}{\partial x} = \frac{x + yx^2}{1 + 2x + yx^2}. \tag{22.15}$$

Note that this simplifies to $x/(1 + x)$ if $y = 1$. In this case, $\theta = \frac{1}{2}$ when $x = 1$ or $c = 1/K$. The function $\theta(\ln x)$ from Eq. (22.15) is plotted in Fig. 6-1 for four values of y. All four curves are symmetrical about $\theta = \frac{1}{2}$ (Section 15). Compared to the reference (unperturbed) $y = 1$ case, the $y = 20$ curve reaches $\theta = \frac{1}{2}$ at a smaller value of x ($x_o = 20^{-1/2} = 0.224$) and has a larger slope at $\theta = \frac{1}{2}$. With $y = 20$, we have $w = -1786$ cal mole^{-1} at 300 °K. This type of behavior (with $y > 1$) is referred to as "positive cooperativity": the favorable interaction between ligand molecules bound on neighboring sites enhances the ligand binding. The presence of one bound ligand molecule aids the binding of the second molecule. The $y = \frac{1}{20}$ case shows the opposite properties relative to the reference case ($w = +1786$ cal mole^{-1}); this is called "negative cooperativity."

The $y = \frac{1}{20}$ curve in Fig. 6-1 gives some indication of two-stage binding. This is much more obvious in the $y = \frac{1}{400}$ curve ($w = +3573$ cal mole^{-1} at 300 °K). The strong repulsion between the two ligands causes the system to avoid the doubly bound state as long as possible. When $y \to 0$, the binding

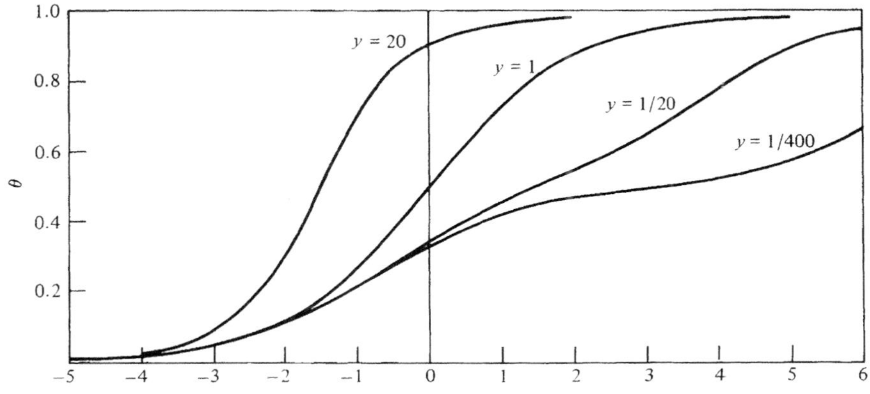

Fig. 6-1. Plot of θ against $\ln x$ [Eq. (22.15)] for the binding of one ligand on two sites, with ligand–ligand interaction parameter y.

on either site but not both is governed by $\xi = 1 + 2x$, from which

$$\theta = \frac{x}{1 + 2x}.$$

This takes θ from 0 to $\frac{1}{2}$. When one site is already occupied, the binding on the second site is determined by $\xi = 2x + yx^2$. From this ξ,

$$\theta = \frac{1 + yx}{2 + yx}.$$

This covers θ from $\frac{1}{2}$ to 1.

The slope at $\theta = \frac{1}{2}$, referred to above, is related to fluctuations in s at $\theta = \frac{1}{2}$ and to the Hill coefficient n (see Sections 11 and 15). The relations are

$$\sigma_{\frac{1}{2}}^2 = \overline{s^2} - \bar{s}^2 = \left(\frac{\partial \bar{s}}{\partial \ln x}\right)_{\theta = 1/2} = 2\left(\frac{\partial \theta}{\partial \ln x}\right)_{\theta = 1/2}, \tag{22.16}$$

$$n = 4\left(\frac{\partial \theta}{\partial \ln x}\right)_{\theta=1/2} = 2\sigma_{\frac{1}{2}}^2 = \frac{2y^{1/2}}{1 + y^{1/2}}. \tag{22.17}$$

The last expression follows from Eqs. (15.13) and (22.14). Of course $n = 1$ when $y = 1$. For $y = 20$ and $\frac{1}{20}$, in Fig. 6-1, the values of n are 1.63 and 0.365, respectively. The maximum possible n, $n = 2 = m$, occurs when $y \to \infty$. Thus the slope of $\theta(\ln x)$ at $\theta = \frac{1}{2}$ remains finite even when $w \to -\infty$. When $y \to 0$, $n \to 0$. Positive cooperativity is associated with $n > 1$, negative cooperativity with $n < 1$.

When y is very large, the term $2Kc$ in Eq. (22.11) is unimportant for finite c, but the leading term, unity, remains in the limit $c \to 0$. Thus ξ in Eq. (22.11) simplifies to $1 + yK^2c^2$ for very large y. This is essentially the A. V. Hill model of Eq. (11.30), for $m = 2$, with an effective binding constant [in Eq. (11.30)] of $y^{1/2}K$. This provides a physical basis for the A. V. Hill model (with $m = 2$) and shows explicitly that it is only a limiting special case (large y).

We digress from this $m = 2$ case to mention that the above argument holds also for any larger group of m sites (see $m = 3, 4, 6$ below) with nearest-neighbor interactions. The largest number of neighbor interactions between ligands occurs when the sites are *all* occupied by ligand molecules. That is, y will occur to its highest power in the final term (c^m) of ξ. Thus, when $y \to \infty$, ξ reduces to the form $1 + ac^m$ (we do not need an explicit expression for a). This has the same dependence on c as Eq. (11.30). It is easy to show (Section 11) that the Hill coefficient in this case ($\xi = 1 + ac^m$) is m. Thus we will find in all of the models of this section that $n \to m$ as $y \to \infty$. Hence the main legitimate use of the Hill equation, Eq. (11.30), is to furnish an upper limit for n in the case of a real system with m binding sites, a symmetrical $\theta(\ln c)$, and positive cooperativity.

It is instructive to compare $m = 2$ heterogeneous binding (sites different) in the *absence* of interactions [see Eqs. (15.15) and (15.16)] with homogeneous binding (sites equivalent) in the *presence* of interactions (as above).

The two grand partition functions are

$$\xi = 1 + (K_1 + K_2)c + K_1 K_2 c^2, \tag{22.18}$$

$$\xi = 1 + 2Kc + yK^2c^2. \tag{22.19}$$

As shown following Eq. (15.16), $n < 1$ for Eq. (22.18) if $K_1 \neq K_2$ (the heterogeneous case). The (well-known) point we wish to make here, using ξ, is that the above two cases are operationally indistinguishable if $y < 1$ (negative cooperativity; $n < 1$). This occurs in the present case because there are two parameters (K_1, K_2; K, y) in each ξ. For given K_1 and K_2, the corresponding K and y are, from Eqs. (22.18) and (22.19),

$$K = \frac{K_1 + K_2}{2}, \tag{22.20}$$

$$y = \frac{K_1 K_2}{K^2} = \frac{4K_1 K_2}{(K_1 + K_2)^2}. \tag{22.21}$$

If $K_1 \neq K_2$, necessarily $y < 1$ [Eq. (15.16)]. With these values of K and y, for the given K_1 and K_2, there is no way to distinguish between the two models on the basis of equilibrium binding measurements.

One Ligand on Three Sites ($m = 3$)

We first extend the above model to a ring or triangle (closed chain) of three equivalent binding sites for L, with an interaction free energy w for each neighbor pair of bound ligands in the ring. The grand partition function corresponding to Eqs. (22.11) and (22.12) is then

$$\begin{array}{rcccc} \xi = & 1 & + \; 3Kc & + \; 3yK^2c^2 & + \; y^3K^3c^3 \\ = & 1 & + \; 3x & + \; 3yx^2 & + \; y^3x^3. \\ & \cdot & \odot & \odot & \odot \\ & \cdot \;\; \cdot & \cdot \;\; \cdot & \odot \;\; \cdot & \odot\odot \\ & s = 0 & s = 1 & s = 2 & s = 3 \end{array} \tag{22.22}$$

The power of y in each term is the number of neighbor pairs (e.g., $y^3 = e^{-3w/kT}$). It should be mentioned that there is an implicit assumption involved in using $3w$ for the interaction free energy when there are three neighbor pairs ($s = 3$ term) if w is the interaction free energy when there is one neighbor pair ($s = 2$ term). The assumption is usually referred to as the "pairwise additivity" of the interaction free energy. It is probably quite accurate for van der Waals interactions in a vacuum (no solvent), but there may be an appreciable error in some cases when w is a potential of mean force[5] (solvent present).

Using

$$\alpha_0 = 1, \qquad \alpha_1 = 3, \qquad \alpha_2 = 3y, \qquad \alpha_3 = y^3 \tag{22.23}$$

from Eq. (22.22), Eqs. (15.19) show that $\theta(\ln x)$ is symmetrical and that $\theta = \frac{1}{2}$ at $x_0 = Kc_0 = y^{-1}$, or $c_0 = 1/Ky$. The expression for θ, from Eq. (22.22), is

$$\theta = \frac{\bar{s}}{3} = \frac{x + 2yx^2 + y^3x^3}{1 + 3x + 3yx^2 + y^3x^3}. \tag{22.24}$$

The function $\theta(\ln x)$ is plotted in Fig. 6-2 for the same four values of y. When $y = 1$, θ is the usual $x/(1 + x)$. The interaction effects are stronger in Fig. 6-2 than in Fig. 6-1, as one might expect, because here there are three possible pair interactions for three sites (i.e., one for one), whereas in Fig. 6-1 there is one possible interaction for two sites. The Hill coefficient is

$$n = 4\left(\frac{\partial \theta}{\partial \ln x}\right)_{\theta = 1/2} = \frac{4}{3}\sigma_3^2 = \frac{1 + 3y}{3 + y}, \tag{22.25}$$

having used Eq. (15.20), where $\phi = 3/y$. When $y = 1$, $n = 1$; when $y \to \infty$, $n \to 3 = m$; and when $y \to 0$, $n \to 1/3$ (not zero). For $y = 20$ and $\frac{1}{20}$, we find $n = 2.65$ and 0.377, respectively.

The $y = \frac{1}{400}$ curve in Fig. 6-2 (all curves are symmetrical about $\theta = \frac{1}{2}$) makes it clear that, when $y \to 0$, the binding occurs in three stages (avoiding multiple occupancy of sites as much as possible). The initial stage is based on $\xi = 1 + 3x$. Consequently,

$$\theta = \frac{x}{1 + 3x}.$$

This covers the range $\theta = 0$ to $\theta = \frac{1}{3}$. The slope of $\theta(\ln x)$ is finite at $\theta = \frac{1}{2}$, when $y \to 0$, because $\theta = \frac{1}{2}$ is in the middle of the second stage of binding.

The above example illustrates very well the general procedure to follow in equilibrium binding problems with interactions: select a model; enumerate all

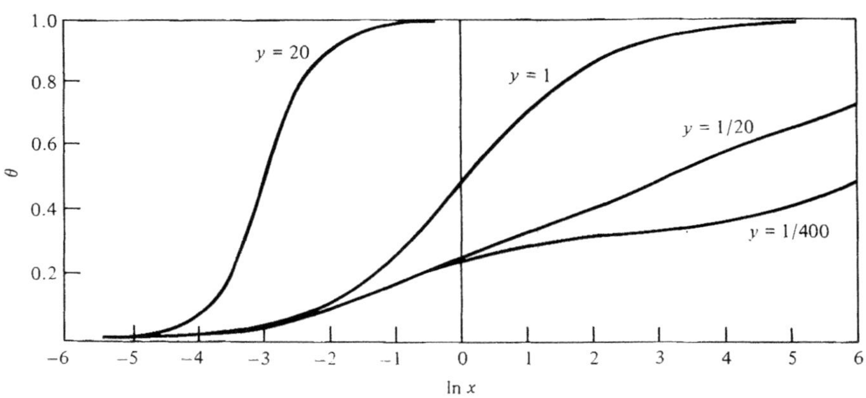

Fig. 6-2. Same as Fig. 6-1 but for the binding of a ligand on a triangle of three sites [Eq. (22.24)].

of the states, with pictures [Eq. (22.22)], so that interaction pairs may be counted; write out the terms of the grand partition function ξ, one term for each state, including in each term a factor for each bound ligand and another factor for each interacting neighbor pair; use the terms in ξ to write θ (or use $m\theta = \partial \ln \xi / \partial \ln c$).

It is again interesting to compare heterogeneous binding (three different sites) without interactions with the present model (homogeneous sites; pair interactions). The two grand partition functions are

$$\xi = (1 + K_1c)(1 + K_2c)(1 + K_3c) \tag{22.26}$$

$$= 1 + (K_1 + K_2 + K_3)c + (K_1K_2 + K_1K_3 + K_2K_3)c^2 + K_1K_2K_3c^3,$$

$$\xi = 1 + 3Kc + 3yK^2c^2 + y^3K^3c^3. \tag{22.27}$$

Are these two cases operationally indistinguishable, as for $m = 2$ (see above)? The answer is yes, if and only if the K_i in Eq. (22.26) are such as to give a symmetrical $\theta(\ln c)$ [recall that $\theta(\ln c)$ from Eq. (22.27) is always symmetrical]. The condition for symmetry in Eq. (22.26) is $\alpha_1^3\alpha_3 = \alpha_2^3\alpha_0$ [see Eq. (15.19)]. That is, the K_i must be related by

$$(K_1 + K_2 + K_3)^3K_1K_2K_3 = (K_1K_2 + K_1K_3 + K_2K_3)^3. \tag{22.28}$$

Assuming that this condition is satisfied, the parameters K and y in Eq. (22.27) can be related to the K_i by equating the coefficients of c and c^2 in Eqs. (22.26) and (22.27):

$$K = \frac{K_1 + K_2 + K_3}{3}, \tag{22.29}$$

$$y = \frac{K_1K_2 + K_1K_3 + K_2K_3}{3K^2} = \frac{3(K_1K_2 + K_1K_3 + K_2K_3)}{(K_1 + K_2 + K_3)^2}. \tag{22.30}$$

Using these relations, we then see that the two coefficients of c^3, y^3K^3 and $K_1K_2K_3$, are also equal if the symmetry condition, Eq. (22.28), is satisfied. It is not hard to show that the maximum value of y, as defined in Eq. (22.30), is $y = 1$, which occurs when $K_1 = K_2 = K_3$, but $y < 1$ (negative cooperativity) if the K_i are not all equal.

A second model with $m = 3$ is an *open or linear chain* of three equivalent binding sites. The grand partition function in this case is

$$\xi = 1 \quad + 3x \quad + 2yx^2 \quad + x^2 \quad + y^2x^3. \tag{22.31}$$

$$\cdot\cdot\cdot \qquad \odot\cdot\cdot \qquad \odot\odot\cdot \qquad \odot\cdot\odot \qquad \odot\odot\odot$$

The symmetry test in Eq. (15.19) is not satisfied, so $\theta(\ln x)$ from this ξ does not have the familiar symmetry about $\theta = \frac{1}{2}$. The equation for θ or $\bar{s}$ is

$$3\theta = \bar{s} = x\frac{\partial \ln \xi}{\partial x} = \frac{3x + 4yx^2 + 2x^2 + 3y^2x^3}{\xi}. \tag{22.32}$$

When $y = 1$, this reduces to the usual $\theta = x/(1 + x)$. Because of the lack of symmetry, the analytical properties of this model are not simple and we do not pursue them further.

Equation (22.31) contains two parameters, K and y; Eq. (22.26) contains three parameters. There is no symmetry condition. Hence there cannot be, in general, an operational equivalence between the open three-site model with interactions and the heterogeneous model without interactions.

Incidentally, the introduction of interaction free energy factors into grand partition functions is not restricted in any way, though by far the most common procedure is to include pairwise additive nearest-neighbor interactions only, as in the above examples. As a variation on these examples, suppose that there is a significant *second-neighbor* interaction free energy w_s between two ligands bound on the *outer* sites of the group of three sites in Eq. (22.31). If we introduce $y_s \equiv e^{-w_s/kT}$, then ξ becomes

$$\xi = 1 + 3x + 2yx^2 + y_s x^2 + y^2 y_s x^3. \tag{22.33}$$

Note that this is the same as Eq. (22.22) if $y_s = y$. Thus this second-neighbor model, with $1 < y_s < y$, would have properties intermediate between those determined by Eq. (22.31) (where $y_s = 1$) and by Eq. (22.22) (where $y_s = y$).

As a third and final example with $m = 3$, we mention the model introduced by Ackers *et al.*[6] for the *binding of repressor dimers* (the ligand) on three linear operator sites. The model is shown in Fig.6-3. Each of the three sites has its own K_i; in addition, there are neighbor interactions, designated w_{12} and w_{23} (both found to be about -2 kcal mole^{-1}), in states, 5, 7, and 8, as indicated schematically in the figure ($y_{12} \equiv e^{-w_{12}/kT}$, etc.). In state 8, the 12 interaction occurs rather than the 23 interaction. The grand partition function is the sum of the terms in Fig. 6-3:

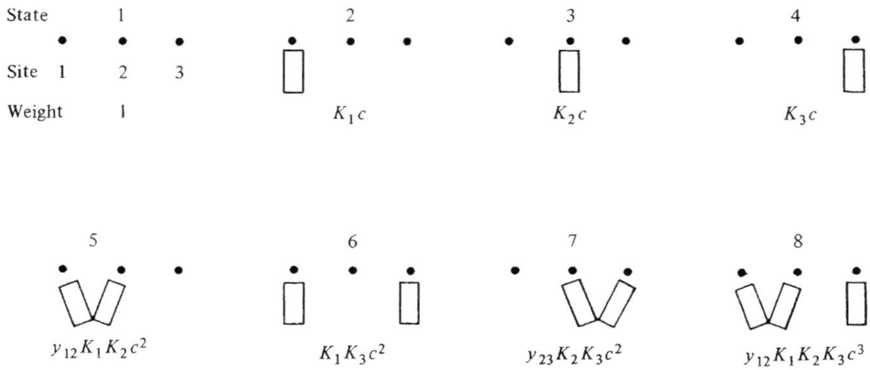

Fig. 6-3. Model for binding a ligand (repressor dimer) on three operator sites. See text for details.

$$\xi = 1 + (K_1 + K_2 + K_3)c + y_{12}K_1K_2c^2 + K_1K_3c^2 + y_{23}K_2K_3c^2 + y_{12}K_1K_2K_3c^3. \quad (22.34)$$

The probabilities of the eight states in the figure are given directly by the terms in ξ, as usual. For example, the probability of state 1 is $p_1^e = 1/\xi$. Important quantities are the probability that site 3 is occupied (required for shutoff of transcription at promoter P_{RM}),

$$f_{P_{RM}} = p_4^e + p_6^e + p_7^e + p_8^e \quad (22.35)$$

$$= \frac{K_3c + K_1K_3c^2 + y_{23}K_2K_3c^2 + y_{12}K_1K_2K_3c^3}{\xi}, \quad (22.36)$$

and the probability that at least one of sites 1 or 2 is occupied (required for shutoff of transcription from promotor P_R),

$$\begin{aligned} f_{P_R} &= p_2^e + p_3^e + p_5^e + p_6^e + p_7^e + p_8^e \\ &= 1 - p_1^e - p_4^e \\ &= 1 - \frac{1 + K_3c}{\xi}. \end{aligned} \quad (22.37)$$

The reader is referred to the original paper for further details, including numerical values for all parameters.

Ligand on Square of Four Sites ($m = 4$)

In this model, ligand molecules can bind, with interactions, on any of four equivalent sites arranged in a square. This is a closed ring with $m = 4$. The only interactions (to begin with) are nearest-neighbor interactions. There are a total of $2^4 = 16$ states. The grand partition function is

$$\xi = 1 \quad + 4x \quad + 4yx^2 + 2x^2 \quad + 4y^2x^3 + y^4x^4. \quad (22.38)$$

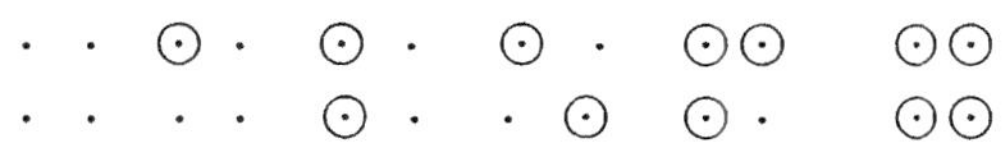

From Eqs. (15.25), we find that this ξ gives a $\theta(\ln c)$ that is symmetrical about $\theta = \frac{1}{2}$, and that $x_o = y^{-1}$ just as for a closed ring of three sites (above).

The binding isotherm is

$$\theta = (x + 2yx^2 + x^2 + 3y^2x^3 + y^4x^4)\xi^{-1}. \quad (22.39)$$

This reduces to $x/(1 + x)$ when $y = 1$. The function $\theta(\ln x)$ is plotted in Fig. 6-4 for the usual y values. On using Eqs. (15.26)–(15.28), we find that

$$n = 4\left(\frac{\partial \theta}{\partial \ln x}\right)_{\theta=1/2} = \sigma_4^2 = \frac{4y(1 + y)}{1 + 6y + y^2}. \quad (22.40)$$

This gives $n = 1$ when $y = 1$, $n = 4 = m$ when $y \to \infty$, and $n \to 0$ when

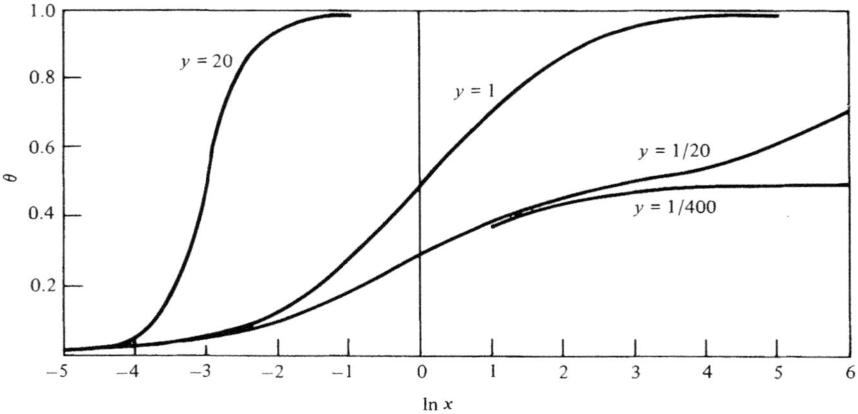

Fig. 6-4. Binding of a ligand on a square of four sites [Eq. (22.39)]. Compare Figs. 6-1 and 6-2.

$y \to 0$. For $y = 20$ and $\frac{1}{20}$, the values of n are 3.22 and 0.161, respectively.

The $y = \frac{1}{400}$ curve in Fig. 6-4 shows that there are two stages of binding when $y \to 0$, not four stages as might have been expected. This occurs because two ligand molecules can bind diagonally without a neighbor interaction (see the term $2x^2$ in ξ). When $y \to 0$, the first stage of binding is represented by

$$\xi = 1 + 4x + 2x^2,$$

$$\theta = \frac{x + x^2}{1 + 4x + 2x^2}.$$

This $\theta(x)$ starts at 0 and ends at $\frac{1}{2}$.

The heterogeneous binding model without interactions has four parameters $K_1, \ldots, K_4$. These are effectively reduced to three if we impose the symmetry condition, Eq. (15.25). However, ξ in Eq. (22.38) has only two parameters, K and y. Hence there cannot be a general operational equivalence between the two cases, even with symmetry imposed on the heterogeneous binding model.

A variation on the square model is to include the *diagonal second-neighbor interaction,* with interaction free energy designated w_s. If we define $y_s \equiv e^{-w_s/kT}$, the grand partition function becomes

$$\xi = 1 + 4x + 4yx^2 + 2y_s x^2 + 4y^2 y_s x^3 + y^4 y_s^2 x^4. \qquad (22.41)$$

In the previous model we took $w_s = 0$ and $y_s = 1$ implicitly. Equation (22.41) also gives a symmetrical $\theta(\ln x)$. The reader can easily write out $\theta(x)$, if desired. The value of x_0 is $y^{-1}y_s^{-1/2}$. The Hill coefficient ($\theta = \frac{1}{2}$) is [Eqs. (15.26)–(15.28)]

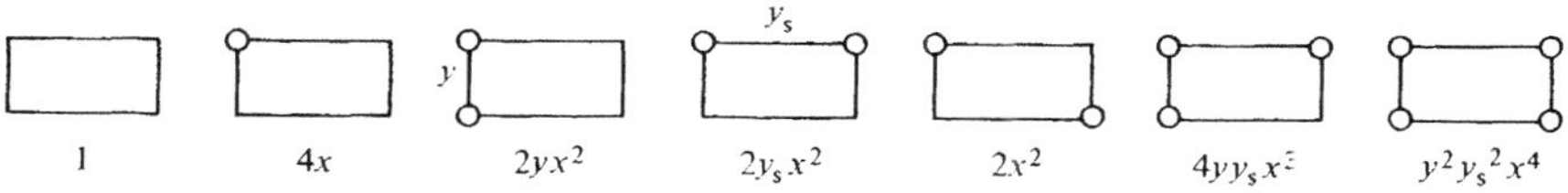

Fig. 6-5. Model for binding of a ligand on the sites at corners of a rectangle. Terms in the grand partition function [Eq. (22.43)] are included. See text for further details.

$$n = \frac{4yy_s^{1/2}(1 + yy_s^{1/2})}{y_s + 2y + 4yy_s^{1/2} + y^2y_s}. \tag{22.42}$$

Incidentally, Eqs. (22.41) and (22.42) reduce to the tetrahedral model (below) if we put $y_s = y$.

Because there is now an additional parameter in Eq. (22.41), namely, y_s, it is again possible in principle to establish a formal connection with the heterogeneous binding model without interactions, provided that symmetry is imposed. However, we do not pursue the details here.

Another variation on the square model is to convert it into a "*rectangular*" *model*: vertical interaction pairs (Fig. 6-5) have a free energy w and $y = e^{-w/kT}$, whereas horizontal pairs have w_s and $y_s = e^{-w_s/kT}$. States and terms in ξ are shown in Fig. 6-5. Thus

$$\xi = 1 + 4x + 2yx^2 + 2y_sx^2 + 2x^2 + 4yy_sx^3 + y^2y_s^2x^4. \tag{22.43}$$

This ξ also leads to a symmetrical $\theta(\ln c)$ function. Again we omit $\theta(x)$, which follows immediately from Eq. (22.43). The value of x_0 is $(yy_s)^{-1/2}$. The expression for n (at $\theta = \frac{1}{2}$, as usual) is found from Eqs. (15.26)–(15.28) to be

$$n = \frac{4(yy_s)^{1/2}[1 + (yy_s)^{1/2}]}{1 + y + y_s + 4(yy_s)^{1/2} + yy_s}. \tag{22.44}$$

Ligand on Tetrahedron of Four Sites ($m = 4$)

This model is shown in Fig. 6-6: there are four equivalent binding sites at the corners of a tetrahedron, with pairwise neighbor interactions, w. As usual, $y = e^{-w/kT}$. The grand partition function is

$$\xi = 1 + 4x + 6yx^2 + 4y^3x^3 + y^6x^4. \tag{22.45}$$

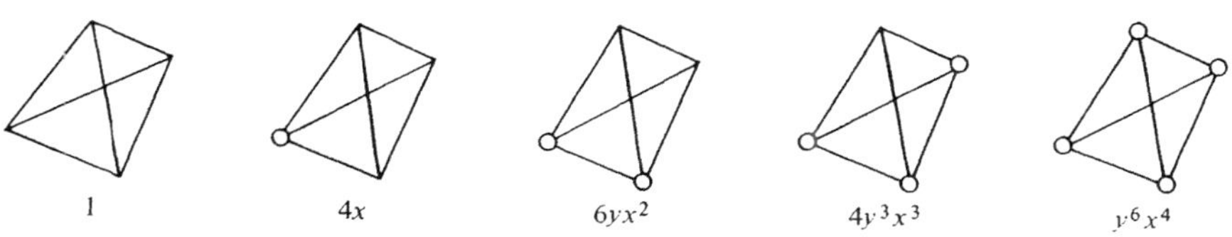

Fig. 6-6. Model for binding a ligand on tetrahedrally arranged sites. Terms in ξ [Eq. (22.45)] are included.

The "density" of interactions is larger in a tetrahedron than in a square: at complete saturation with ligand, there are six neighbor pairs compared to four (y^6x^4 in ξ compared to y^4x^4). Thus we expect stronger cooperativity here than in the square model.

Equation (15.25) shows that $\theta(\ln x)$ from ξ is symmetrical about $\theta = \frac{1}{2}$. The value of x_0 is $y^{-3/2}$; $c_0 = 1/Ky^{3/2}$. The equation for θ is

$$\theta = \frac{x + 3yx^2 + 3y^3x^3 + y^6x^4}{\xi}. \tag{22.46}$$

This becomes $x/(1 + x)$ when $y = 1$. The function $\theta(\ln x)$ is plotted in Fig. 6-7 for several values of y. Cooperative effects are clearly stronger than in Fig. 6-4 (square; $m = 4$).

Equations (15.26)–(15.28) lead to

$$n = 4\left(\frac{\partial \theta}{\partial \ln x}\right)_{\theta=1/2} = \sigma_4^2 = \frac{4y^{1/2}(1 + y^{3/2})}{3 + 4y^{1/2} + y^2}. \tag{22.47}$$

This gives $n = 1$ for $y = 1$, $n \to 4 = m$ for $y \to \infty$, and $n \to 0$ for $y \to 0$. For $y = 20$ and $\frac{1}{20}$, $n = 3.84$ and 0.232, respectively. The corresponding values for the square model are $n = 3.22$ and 0.161.

The $y = \frac{1}{400}$ curve in Fig. 6-7 (keeping symmetry about $\theta = \frac{1}{2}$ in mind) shows that there is four-stage binding in this case when $y \to 0$. In the first stage,

$$\xi = 1 + 4x,$$

$$\theta = \frac{x}{1 + 4x}.$$

This covers θ from 0 to $\frac{1}{4}$.

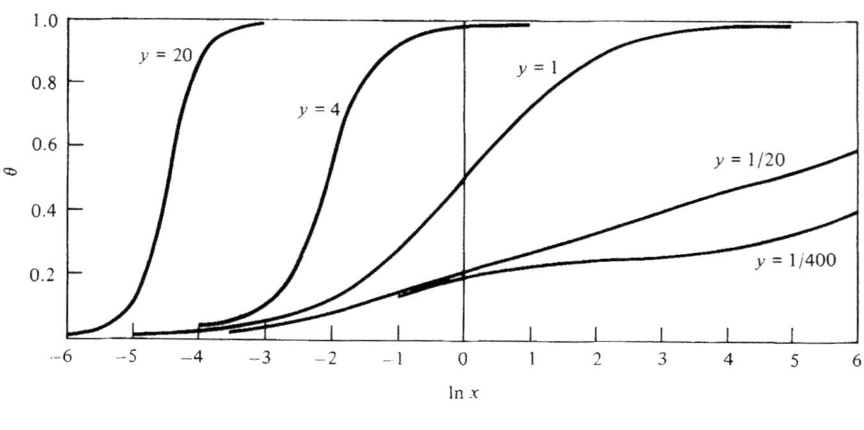

Fig. 6-7. Binding of a ligand on tetrahedrally arranged sites [Eq. (22.46)].

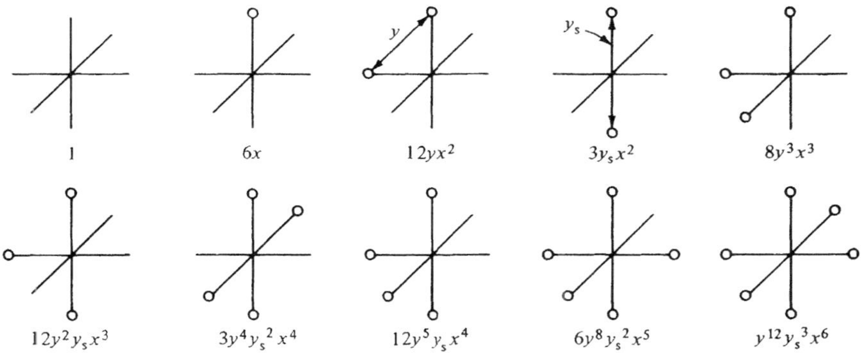

Fig. 6-8. Model for binding on octahedrally arranged sites. First (y) and second (y_s) neighbor interaction factors are included in the terms of ξ given.

Ligand on Octahedron of Six Sites ($m = 6$)

The model, states (a total of $2^6 = 64$), and terms in ξ are shown in Fig. 6-8. Second-neighbor interaction factors y_s are included in the figure but are not retained in the equations below, which refer to first-neighbor interactions only (interaction factor y).

The grand partition function is

$$\xi = 1 + 6x + 12yx^2 + 3x^2 + 8y^3x^3 + 12y^2x^3 + 3y^4x^4 + 12y^5x^4 + 6y^8x^5 + y^{12}x^6, \tag{22.48}$$

and, from this,

$$\theta = (x + 4yx^2 + x^2 + 4y^3x^3 + 6y^2x^3 + 2y^4x^4 + 8y^5x^4 + 5y^8x^5 + y^{12}x^6)\xi^{-1}. \tag{22.49}$$

When $y = 1$, $\theta = x/(1 + x)$. Equation (15.29) gives $x_o = y^{-2}$ (i.e., $c_o = 1/Ky^2$) and shows that $\theta(\ln x)$ is symmetrical about $\theta = \frac{1}{2}$. This symmetry is retained, incidentally, if y_s is included in ξ. There are two pair interactions per site in the ξ term $y^{12}x^6$. Consequently, we expect stronger interaction effects than in the other cases above. This is confirmed in Fig. 6-9, where $\theta(\ln x)$ is presented for several y values.

As is clear from the $y = \frac{1}{20}$ and $\frac{1}{400}$ curves, the binding takes place in three stages (not six) when $y \to 0$. This occurs because of the term $3x^2$ in ξ (compare the square, $m = 4$ case): it is possible to bind two ligand molecules without an interaction. The first-stage binding equation, when $y \to 0$, is

$$\theta = \frac{x + x^2}{1 + 6x + 3x^2},$$

which has limits $\theta = 0$ and $\theta = \frac{1}{3}$.

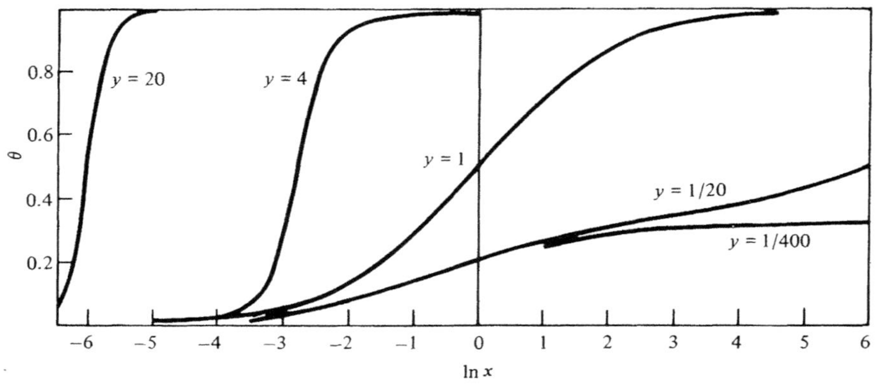

Fig. 6-9. Plots of θ against ln x for a ligand binding on octahedral sites with nearest-neighbor interactions only [Eq. (22.49)].

Equations (15.30) and (15.31) give, for the Hill coefficient,

$$n = 4\left(\frac{\partial \theta}{\partial \ln x}\right)_{\theta=1/2} = \frac{2}{3}\sigma_6^2 = \frac{2(1 + 4y + 8y^2 + 3y^4)}{9 + 16y + 6y^2 + y^4}. \tag{22.50}$$

When $y = 1$, $n = 1$; when $y = 0$, $n = \frac{2}{9}$; when $y \to \infty$, $n \to 6 = m$. The nonzero value of n at $y = 0$ arises because $\theta = \frac{1}{2}$ is in the middle of the second stage of binding. For $y = 20$ and $\frac{1}{20}$, in Fig. 6-9, $n = 5.940$ and 0.249, respectively. Even though $|w|$ is not large (1786 cal mole^{-1} at 300 °K), these values of n are rather close to the limits 6.0 and 0.222. This is a consequence of the high maximum neighbor pair density (two per site).

23. More General Models for Binding on Subunits or Sites

The preceding section dealt with the effects on binding of direct interactions between pairs of bound ligand molecules. The primary object of this section is to show that the results in Section 22 for symmetrical models are more general than they appear: the same grand partition functions (and properties derived from it) are obtained for a wider class of interactions than those referred to above.

Binding on Two Sites ($m = 2$)

We begin with a dimer (the protein P) comprised of two identical subunits. Each subunit has a site for binding L. A subunit without an L bound is in state 1; a subunit with an L bound is in state 2. At one extreme [Fig. 6-10(a)], we suppose that binding of L on a subunit induces a conformational change that alters the interaction between the subunits; the ligand and its site do not

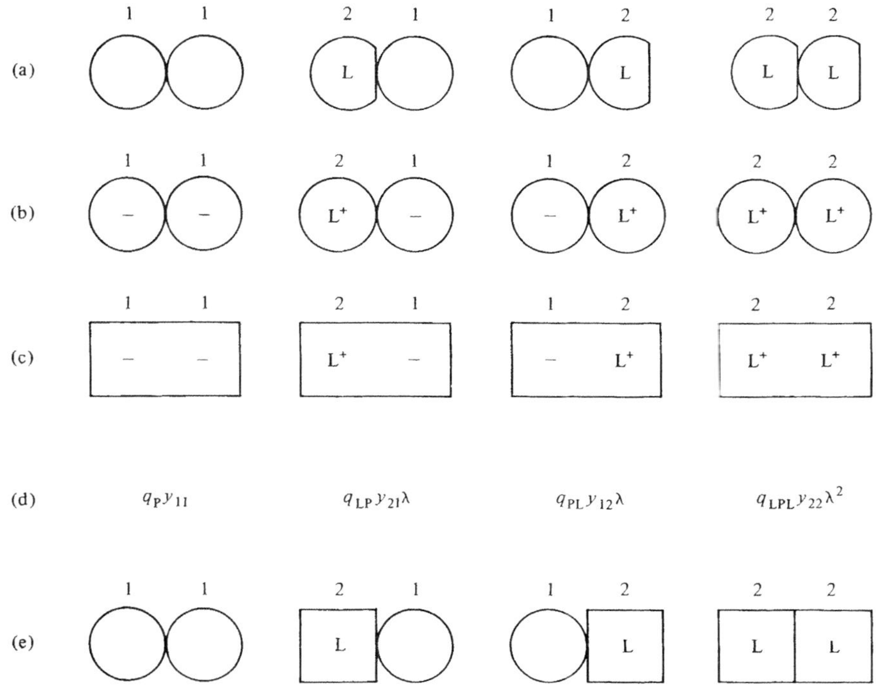

Fig. 6-10. (a) Binding of L induces conformational change in subunit. (b) Binding of L^{+2} on sites with negative charge. (c) Same as in (b) except that the protein P (rectangle) is not divided into subunits. (d) Separation of interaction factors y_{ij} from remainder of grand partition function terms [Eq. (23.1)]. (e) Symmetrical case of ligand-induced conformational change. See text for further details on all subfigures.

contribute at all to this interaction. At the other extreme, a subunit does not change conformation or shape when L is bound, but there are possible ligand–ligand (as in Section 22), ligand–opposite subunit, ligand–site, and site–site interactions. For example, a binding site may have a charge -1 and L may be an ion with charge $+2$ [Fig. 6-10(b)]. Intermediate cases, combining these two extremes, are of course also possible. For example, there may be a conformational change that alters the subunit–subunit interaction as well as ligand–ligand and ligand–opposite subunit interactions.

A third case, essentially equivalent to Fig. 6-10(b), is shown in Fig. 6-10(c): the protein P is not divided into subunits but, unlike the $m = 2$ model in Section 22, the sites may be involved in the interactions as well as the ligands.

In all of these cases an interaction free energy factor may be separated from the remainder of the partition function for P, as shown in Fig. 6-10(d). The factor y_{ij} $(i, j = 1, 2)$ is defined as $e^{-w_{ij}/kT}$, where w_{ij} is the potential of mean force (see Section 22) for bringing two subunits together from $r = \infty$, or two ligands, two charges on sites, etc. As will be seen below, the separate y_{ij} values are not so important as differences among the y_{ij} values. In all cases we have the grand partition function

$$\xi = q_P y_{11} + q_{LP} y_{21} \lambda + q_{PL} y_{12} \lambda + q_{LPL} y_{22} \lambda^2. \tag{23.1}$$

Because the interactions have been separated out and already contain any asymmetry [i.e., possibly $y_{21} \neq y_{12}$, as in Fig. 6-10(a)],

$$q_{LP} = q_{PL} \qquad \text{and} \qquad q_{LPL} = \frac{q_{LP} q_{PL}}{q_P}, \tag{23.2}$$

as in Eq. (11.10). If we factor q_P out of Eq. (23.1), and then drop q_P as usual,

$$\xi = y_{11} + y_{21} K_o c + y_{12} K_o c + y_{22} K_o^2 c^2, \tag{23.3}$$

where

$$\frac{q_{LP} \lambda}{q_P} = K_o c. \tag{23.4}$$

The physical significance of K_o defined here is that this is the unperturbed binding constant of L on an isolated subunit, or, if these are not separable subunits, K_o is the hypothetical binding constant for L if the interactions were not present. Note that if all $y_{ij} = 1$,

$$\xi = (1 + K_o c)^2. \tag{23.5}$$

In fact, if all y_{ij} are equal but $y_{ij} \neq 1$,

$$\xi = y_{11}(1 + K_o c)^2. \tag{23.6}$$

Both of these ξs give $\theta = K_o c/(1 + K_o c)$ and hence have the same operational K_o: interactions need not be absent to have K_o as binding constant; it suffices that no *change* in interaction parameter y_{ij} occurs when binding takes place.

In the important special case $y_{12} = y_{21}$, shown schematically in Fig. 6-10(e) [and including Figs. 6-10(b) and 6-10(c)],

$$\xi = y_{11} + 2y_{12} K_o c + y_{22} K_o^2 c^2. \tag{23.7}$$

If we factor out and then omit y_{11},

$$\xi = 1 + 2\frac{y_{12}}{y_{11}} K_o c + \frac{y_{22}}{y_{11}} K_o^2 c^2. \tag{23.8}$$

Initial binding (i.e., at small c) is governed by the effective constant

$$K = \frac{y_{12} K_o}{y_{11}}. \tag{23.9}$$

If we replace K_o in the c^2 term of ξ by $y_{11} K / y_{12}$, we arrive at

$$\xi = 1 + 2Kc + yK^2 c^2, \tag{23.10}$$

$$y \equiv \frac{y_{11}y_{22}}{y_{12}^2}. \tag{23.11}$$

The quantity y can be viewed as the equilibrium constant for the process

$$\begin{gathered}\text{state } 12 + 21 \rightarrow 11 + 22, \\ w \equiv w_{11} + w_{22} - w_{12} - w_{21}, \qquad y = e^{-w/kT}.\end{gathered} \tag{23.12}$$

The corresponding free energy change is w. When $y > 1$ (positive cooperativity), the combination 11 + 22 is more stable than 12 + 21. In very large two- or three-dimensional two-state systems, this effect can lead to the binary mixture (1, 2) dividing into two phases (almost all 1; almost all 2). When $y < 1$ (negative cooperativity), 12 + 21 is more stable than 11 + 22. Note that there need not be any actual repulsive interactions to obtain negative cooperativity. We might have [as in Fig. 6-10(e)] all $y_{ij} > 1$ (attractive interactions), but $y_{12}^2 > y_{11}y_{22}$. For example, $y_{12} = 20$, $y_{11} = y_{22} = 10$, and $y = \frac{1}{4}$. In the reverse case, $y_{12} = 10$, $y_{11} = y_{22} = 20$, and $y = 4$ (positive cooperativity).

If $y = 1$ in Eq. (23.10), we get

$$\xi = (1 + Kc)^2, \qquad \theta = Kc/(1 + Kc).$$

This shows that simple noncooperative binding will be observed if $y_{11}y_{22} = y_{12}^2$ (i.e., if $y = 1$). The requirement $y_{11} = y_{12} = y_{22}$ mentioned following Eq. (23.6) is seen here to be too restrictive; it suffices to have $y = 1$.

Equation (23.10) is identical to Eq. (22.11), but here y has a more general meaning. In Eq. (22.11), y referred to a ligand-ligand interaction only (i.e., $y_{11} = 1$, $y_{12} = 1$, $y = y_{22}$). Here an interaction may be present in states 11, 21, and 12 as well as in 22 (e.g., Fig. 6-10e): Eq. (22.11) is a special case of Eq. (23.10). All of the discussion of the $m = 2$ case following Eq. (22.11) applies to the more general model [Figs. 6-10(b), 6-10(c), and 6-10(e)].

Incidentally, the same formal type of procedure can be used in the more general $y_{12} \neq y_{21}$ case, but it is not really natural because there are *two* initial effective binding constants [linear terms in c, corresponding to states 21 and 12 in Fig. 6-10(a)]:

$$\xi = 1 + \frac{y_{21}}{y_{11}}K_o c + \frac{y_{12}}{y_{11}}K_o c + \frac{y_{22}}{y_{11}}K_o^2 c^2. \tag{23.13}$$

If we proceed, despite this objection, and define K by

$$K = \left(\frac{y_{21} + y_{12}}{2y_{11}}\right)K_o, \tag{23.14}$$

and then use this relation to replace K_o by K in the c^2 term of ξ, we again deduce Eq. (23.10) but with y defined by

$$y \equiv \frac{4y_{11}y_{22}}{(y_{21} + y_{12})^2}. \tag{23.15}$$

A more satisfactory alternative is to define the two initial constants

$$K_{21} \equiv \frac{y_{21}K_o}{y_{11}}, \qquad K_{12} \equiv \frac{y_{12}K_o}{y_{11}}, \tag{23.16}$$

and then find

$$\xi = 1 + K_{21}c + K_{12}c + yK_{21}K_{12}c^2,$$

$$y = \frac{y_{11}y_{22}}{y_{12}y_{21}}. \tag{23.17}$$

If $y = 1$,

$$\xi = (1 + K_{21}c)(1 + K_{12}c), \tag{23.18}$$

as in heterogeneous binding without interactions.

We return now to the symmetrical case $y_{12} = y_{21}$, and Eqs. (23.10) and (23.11). So far for this case, we have introduced the unperturbed binding constant K_o and the initial binding constant $K = y_{12}K_o/y_{11}$. A third binding constant is also very useful, especially in more complicated cases (below). This is the mean constant $\overline{K}$, defined in this case by [see the coefficient of c^2 in Eq. (23.10)]

$$\overline{K}^2 = yK^2, \qquad \overline{K} = y^{1/2}K. \tag{23.19}$$

If we replace K by $\overline{K}$ in Eq. (23.10), we obtain

$$\xi = 1 + 2y^{-1/2}\overline{K}c + \overline{K}^2c^2. \tag{23.20}$$

The physical significance of $\overline{K}^2$ is that this is the equilibrium constant for the over-all (double binding) process

$$\text{state } 11 + 2\text{L} \rightarrow \text{state } 22. \tag{23.21}$$

In terms of K_o, this equilibrium constant is $y_{22}K_o^2/y_{11}$ [see Eq. (23.8)]. The y_{ii} factors are accounted for by the loss of a 11 interaction and the gain of a 22 interaction. The constant $\overline{K}$ is an average binding constant per ligand when both ligands are bound. The main advantage of the use of $\overline{K}$, which becomes more obvious in more complicated symmetrical cases (see below), is that the coefficients of the powers of $\overline{K}c$ in ξ become symmetrical [for example, in Eq. (23.20), the coefficients of $(\overline{K}c)^0$ and $(\overline{K}c)^2$ are both unity].

In terms of K, the binding isotherm is [Eq. (22.15)]

$$\theta = \frac{Kc + yK^2c^2}{1 + 2Kc + yK^2c^2}. \tag{23.22}$$

The value $\theta = \frac{1}{2}$ is reached when $Kc = Kc_o = y^{-1/2}$. That is,

$$c_o = \frac{1}{Ky^{1/2}} = \left(\frac{y_{11}}{y_{22}}\right)^{1/2}\frac{1}{K_o}.$$

In terms of $\overline{K}$, from Eq. (23.20),

$$\theta = \frac{y^{-1/2}\overline{K}c + \overline{K}^2c^2}{1 + 2y^{-1/2}\overline{K}c + \overline{K}^2c^2}. \tag{23.23}$$

This gives $\theta = \frac{1}{2}$ when $\overline{K}c = \overline{K}c_o = 1$; that is, $c_o = 1/\overline{K}$. This simple result, $\overline{K}c_o = 1$, is a consequence of the symmetry in the coefficients of ξ, mentioned above. If we define $\overline{x} = \overline{K}c$ as a new concentration variable (reserving x for Kc),

$$\xi = 1 + 2y^{-1/2}\overline{x} + \overline{x}^2. \tag{23.24}$$

Then in Eq. (15.9),

$$\begin{aligned} &\alpha_0 = 1, \qquad \alpha_1 = 2y^{-1/2}, \qquad \alpha_2 = 1, \\ &\overline{x}_o = (\alpha_0/\alpha_2)^{1/2} = 1. \end{aligned} \tag{23.25}$$

When $y = 1$, Eq. (23.23) gives $\theta = \overline{K}c/(1 + \overline{K}c)$, just as for K, above. The reason for this agreement is that $\overline{K} = K$ when $y = 1$.

If Fig. 6-1 is replotted as θ against $\ln \overline{x}$, from Eq. (23.23), the curves do not change shape, but the three $y \neq 1$ curves are shifted horizontally so that all four curves pass through the point $\theta = \frac{1}{2}$, $\ln \overline{x} = 0$ ($\overline{x}_o = 1$ for any y). This of course obscures one of the significant effects of different y values, namely, the earlier (lower c) rise in θ for $y > 1$. This occurs because y is embedded in $\overline{x}$:

$$\overline{x} = \overline{K}c = y^{1/2}Kc = y^{1/2}x. \tag{23.26}$$

It should also be recalled that in the present more general interaction model, where we might well have $y_{12} \neq 1$ and $y_{11} \neq 1$, the additional interaction parameter y_{12}/y_{11} is embedded in K: $K = y_{12}K_o/y_{11}$. Hence a plot of θ vs $\ln Kc$ (i.e., $\ln x$) might involve some relative shifting of curves compared to a plot of θ vs $\ln K_oc$.

A "Ring" of Two Sites ($m = 2$, Closed)

This model (Fig. 6-11), with two identical subunits and *two* pair interactions, is worth treating separately [it can also be viewed as a special case of Fig. 6-10(e) and Eq. (23.7)], because it is then the first in the infinite sequence of

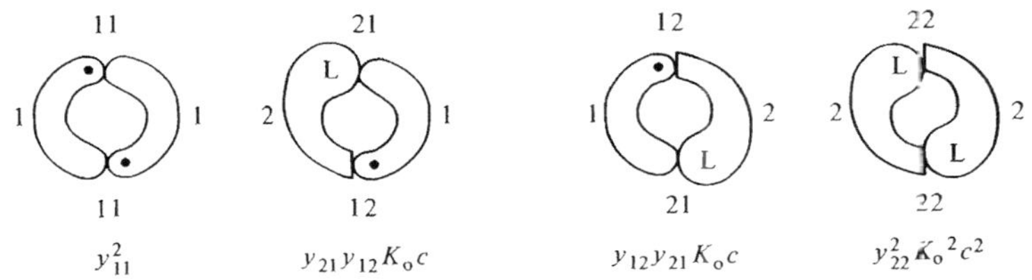

Fig. 6-11. Model for two identical subunits with *two* pair interactions. See text for details. Nearest-neighbor pairs are labelled in clockwise direction.

models with closed rings of sites or subunits, $m = 2$, 3 (triangle), 4 (square), ... The pair interactions (reading states in the clockwise direction) are specified at the top and bottom of each picture in Fig. 6-11. The grand partition function is

$$\xi = y_{11}^2 + 2y_{21}y_{12}K_o c + y_{22}^2 K_o^2 c^2. \tag{23.27}$$

If we factor out y_{11}^2, define the initial binding constant (low c)

$$K \equiv \left(\frac{y_{12}y_{21}}{y_{11}^2}\right)K_o, \tag{23.28}$$

replace K_o by K in Eq. (23.27), and finally define

$$y \equiv \frac{y_{11}y_{22}}{y_{12}y_{21}} \tag{23.29}$$

as in Eq. (23.17), the grand partition function becomes

$$\xi = 1 + 2Kc + y^2K^2c^2. \tag{23.30}$$

This is the same as, and has the same properties as, Eqs. (22.11) and (23.10) if y in these equations is replaced by y^2. For example, in place of Eq. (22.15), we have

$$\theta = \frac{Kc + y^2K^2c^2}{1 + 2Kc + y^2K^2c^2}, \tag{23.31}$$

and instead of Eq. (22.17), we have, for the Hill coefficient,

$$n = \frac{2y}{1 + y}. \tag{23.32}$$

The curves in Fig. 6-1 apply, for the present model, to $y^2 = 20, 1, \frac{1}{20}, \frac{1}{400}$, or to $y = 20^{1/2}$, etc. Smaller values of y suffice here to produce the same effects because there are two pair interactions in the dimer instead of one.

To put Eq. (23.30) in a symmetrical form, as in Eq. (23.20), we define

$$\overline{K}^2 = y^2K^2, \qquad \overline{K} = yK = y_{22}K_o/y_{11}. \tag{23.33}$$

Then, replacing K by $\overline{K}$ in ξ, we have

$$\xi = 1 + 2y^{-1}\overline{K}c + \overline{K}^2c^2. \tag{23.34}$$

The coefficients of $(\overline{K}c)^0$ and $(\overline{K}c)^2$ are both unity. Hence $\theta = \frac{1}{2}$ occurs at

$$\overline{K}c_o = 1, \qquad c_o = \frac{1}{\overline{K}} = \frac{1}{yK} = \frac{y_{11}}{y_{22}K_o}. \tag{23.35}$$

Another equivalent expression for θ, from Eq. (23.34), is

$$\theta = \frac{y^{-1}\overline{K}c + \overline{K}^2c^2}{1 + 2y^{-1}\overline{K}c + \overline{K}^2c^2}. \tag{23.36}$$

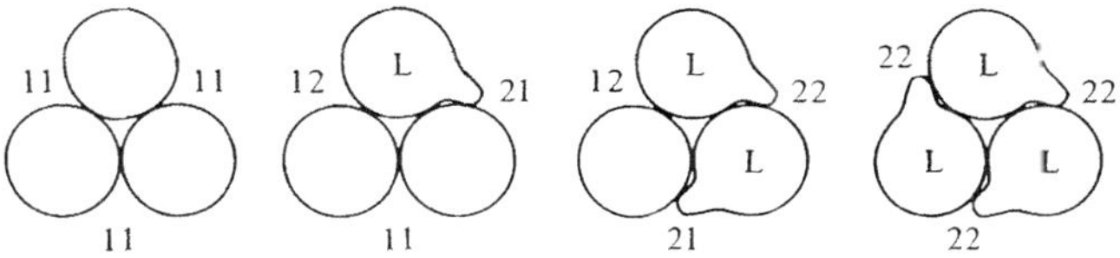

Fig. 6-12. Binding of L on a ring of three identical subunits. Nearest-neighbor pairs are labelled in the clockwise direction.

Binding on Three Sites ($m = 3$, Closed)

We consider binding of L on a ring of three identical subunits, each with one binding site. Figure 6-12 ($m = 3$) corresponds to Fig. 6-10(a) ($m = 2$). The cases in Figs. 6-10(b), 6-10(c), and 6-10(e) are also possible for $m = 3$, but they are not shown explicitly. The nearest-neighbor pairs are labelled in Fig. 6-12, arbitrarily choosing the clockwise direction. Corresponding to Eqs. (23.3) and (23.27), we have

$$\xi = y_{11}^3 + 3y_{11}y_{12}y_{21}K_oc + 3y_{12}y_{21}y_{22}K_o^2c^2 + y_{22}^3K_o^3c^3. \qquad (23.37)$$

There are three y_{ij} factors in each term because there are three neighbor pairs. If we remove the factor y_{11}^3 from ξ,

$$\xi = 1 + 3\left(\frac{y_{12}y_{21}}{y_{11}^2}\right)K_oc + 3\left(\frac{y_{12}y_{21}y_{22}}{y_{11}^3}\right)K_o^2c^2 + \left(\frac{y_{22}}{y_{11}}\right)^3K_o^3c^3. \qquad (23.38)$$

The initial binding constant (see the linear term in c) is defined as in Eq. (23.28). If we then replace K_o by K throughout ξ, we find

$$\xi = 1 + 3Kc + 3yK^2c^2 + y^3K^3c^3, \qquad (23.39)$$

where y is defined as in Eq. (23.29). Equation (23.39) is the same as Eq. (22.22), but again the model here is more general and y has a more general definition. The properties already deduced from Eq. (22.22) follow from Eq. (23.39) as well.

To obtain the symmetrical form for ξ, we define $\overline{K} = yK$; $\overline{K}^3 = (y_{22}/y_{11})^3K_o^3$ is the equilibrium constant for the overall process

$$\text{state } 1^11 + 3\text{L} \rightarrow \text{state } 2^22. \qquad (23.40)$$

In this process three 11 interactions are lost and three 22 interactions are gained. If we now replace K by $\overline{K}/y$ in Eq. (23.39), we obtain

$$\xi = 1 + 3y^{-1}\overline{K}c + 3y^{-1}\overline{K}^2c^2 + \overline{K}^3c^3. \qquad (23.41)$$

The coefficients of powers of $\overline{x} = \overline{K}c$ here are symmetrical:

$$\alpha_0 = 1, \qquad \alpha_1 = 3y^{-1}, \qquad \alpha_2 = 3y^{-1}, \qquad \alpha_3 = 1. \qquad (23.42)$$

As a consequence, Eq. (15.5) is satisfied for each r ($r = 0, 1$), with

$\bar{x}_o = \bar{K}c_o = 1$, or $c_o = 1/yK$, as in Eq. (23.35). Hence $\theta(\ln c)$ is symmetrical about $\theta = \frac{1}{2}$, which we already knew [Eq. (22.23)]. However, here the deduction of $\theta(\ln c)$ symmetry is immediate because of the symmetry in ξ coefficients [Eq. (23.42)]. That is, we can use Eq. (15.5) as the $\theta(\ln c)$ symmetry test on Eq. (23.41), but the less direct Eqs. (15.19) must be used on Eq. (22.22).

The binding isotherm, using $\bar{K}$, is

$$\theta = \frac{y^{-1}\bar{x} + 2y^{-1}\bar{x}^2 + \bar{x}^3}{\xi}. \tag{23.43}$$

Binding on a Square of Four Sites ($m = 4$)

Here we have binding on a ring of four sites, for example, four identical subunits arranged in a square, with one site per subunit. The possibilities are $m = 4$ versions of Figs. 6-10(a), 6-10(b), 6-10(c), and 6-10(e). In Fig. 6-13, state 1 is an empty site and state 2 is an occupied site, as usual. The interaction factors follow from the numbering of the four successive pairs, going around the square in, say, the clockwise direction (Fig. 6-13). Hence,

$$\xi = y_{11}^4 + 4y_{11}^2 y_{12} y_{21} K_o c + 4y_{11} y_{12} y_{21} y_{22} K_o^2 c^2 + 2y_{12}^2 y_{21}^2 K_o^2 c^2 + 4y_{12} y_{21} y_{22}^2 K_o^3 c^3 + y_{22}^4 K_o^4 c^4. \tag{23.44}$$

If we first divide by y_{11}^4, then define the initial binding constant K by Eq. (23.28) and y by Eq. (23.29), we obtain again Eq. (22.38) for ξ, but now with a more general model and meaning for y. The properties of Eq. (22.38) apply here as well.

Because the last term in Eq. (22.38) is $y^4K^4c^4$, we again introduce $\bar{K} = yK$, where now $\bar{K}^4 = (y_{22}/y_{11})^4 K_o^4$ is the equilibrium constant for

$$\begin{array}{ccc} 1\;1 & & 2\;2 \\ \text{state } 1\;1 + 4\text{L} & \rightarrow & \text{state } 2\;2. \end{array} \tag{23.45}$$

On replacing K by $\bar{K}$ throughout Eq. (22.38), we find

$$\xi = 1 + 4y^{-1}\bar{K}c + 4y^{-1}\bar{K}^2c^2 + 2y^{-2}\bar{K}^2c^2 + 4y^{-1}\bar{K}^3c^3 + \bar{K}^4c^4. \tag{23.46}$$

There is again symmetry in the coefficients (with $\bar{x} = \bar{K}c$ as concentration variable):

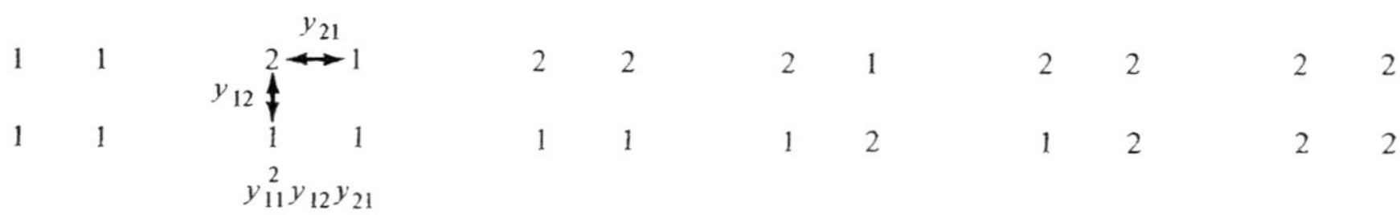

Fig. 6-13. Schematic extension of Fig. 6-12 to a ring (square) of four sites. State 2 = L bound; state 1 = site empty. The y_{ij} are labelled in the clockwise direction.

$$\alpha_0 = \alpha_4 = 1, \qquad \alpha_1 = \alpha_3 = 4y^{-1}. \tag{23.47}$$

Hence $\theta(\ln c)$ is symmetrical about $\theta = \frac{1}{2}$, which occurs at $\bar{x}_o = \bar{K}c_o = 1$, or $c_o = 1/yK$, as in Eq. (23.35). The expression for θ is Eq. (22.39) or, alternatively,

$$\theta = \frac{y^{-1}\bar{x} + 2y^{-1}\bar{x}^2 + y^{-2}\bar{x}^2 + 3y^{-1}\bar{x}^3 + \bar{x}^4}{\xi}. \tag{23.48}$$

Tetrahedron and Octahedron ($m = 4, 6$)

The above approach can be applied to tetrahedral or octahedral groups of identical subunits or sites. It is left to the reader to fill in the details in these cases. In these models there is no distinction between y_{12} and y_{21} ($y_{12} = y_{21}$).

The important relations for the tetrahedral case are:

$$\xi = y_{11}^6 + 4y_{11}^3 y_{12}^3 K_o c + 6y_{11} y_{12}^4 y_{22} K_o^2 c^2 + 4y_{12}^3 y_{22}^3 K_o^3 c^3 + y_{22}^6 K_o^4 c^4, \tag{23.49}$$

$$K = \left(\frac{y_{12}}{y_{11}}\right)^3 K_o, \qquad y = \frac{y_{11} y_{22}}{y_{12}^2}, \tag{23.50}$$

$$\xi = 1 + 4Kc + 6yK^2c^2 + 4y^3K^3c^3 + y^6K^4c^4, \tag{23.51}$$

$$\bar{K}^4 = y^6 K^4, \qquad \bar{K} = y^{3/2} K, \tag{23.52}$$

$$\xi = 1 + 4y^{-3/2}\bar{K}c + 6y^{-2}\bar{K}^2c^2 + 4y^{-3/2}\bar{K}^3c^3 + \bar{K}^4c^4. \tag{23.53}$$

Equation (23.51) has the same appearance as, and is a generalization of, Eq. (22.45). Powers of $\bar{K}c$ in Eq. (23.53) have symmetrical coefficients; at $\theta = \frac{1}{2}$

$$\bar{K}c_o = 1, \qquad c_o = \frac{1}{y^{3/2}K} = \left(\frac{y_{11}}{y_{22}}\right)^{3/2} \frac{1}{K_o};$$

$\theta(\ln c)$ is symmetrical about $\theta = \frac{1}{2}$.

The corresponding results for the octahedral model are:

$$\begin{aligned}\xi = {} & y_{11}^{12} + 6y_{11}^8 y_{12}^4 K_o c + 12y_{11}^5 y_{12}^6 y_{22} K_o^2 c^2 + 3y_{11}^4 y_{12}^8 K_o^2 c^2 \\ & + 8y_{11}^3 y_{12}^6 y_{22}^3 K_o^3 c^3 + 12y_{11}^2 y_{12}^8 y_{22}^2 K_o^3 c^3 + 3y_{12}^8 y_{22}^4 K_o^4 c^4 \\ & + 12y_{11} y_{12}^6 y_{22}^5 K_o^4 c^4 + 6y_{12}^4 y_{22}^8 K_o^5 c^5 + y_{22}^{12} K_o^6 c^6,\end{aligned} \tag{23.54}$$

$$K = \left(\frac{y_{12}}{y_{11}}\right)^4 K_o, \qquad y = \frac{y_{11} y_{22}}{y_{12}^2}, \tag{23.55}$$

$$\xi = 1 + 6Kc + \cdots \qquad \text{as in Eq. (22.48)},$$

$$\bar{K}^6 = y^{12} K^6, \qquad \bar{K} = y^2 K, \tag{23.56}$$

$$\begin{aligned}\xi = {} & 1 + 6y^{-2}\bar{K}c + 12y^{-3}\bar{K}^2c^2 + 3y^{-4}\bar{K}^2c^2 + 8y^{-3}\bar{K}^3c^3 \\ & + 12y^{-4}\bar{K}^3c^3 + 3y^{-4}\bar{K}^4c^4 + 12y^{-3}\bar{K}^4c^4 + 6y^{-2}\bar{K}^5c^5 + \bar{K}^6c^6.\end{aligned} \tag{23.57}$$

The powers of $\overline{K}c$ have symmetrical coefficients; at $\theta = \frac{1}{2}$,

$$\overline{K}c_o = 1, \qquad c_o = \frac{1}{y^2K} = \left(\frac{y_{11}}{y_{22}}\right)^2 \frac{1}{K_o};$$

$\theta(\ln c)$ is symmetrical about $\theta = \frac{1}{2}$.

Symmetry in Ring with Arbitrary *m*

Among the models above, we have considered closed rings of subunits or sites with $m = 2, 3, 4$. In all of these cases the coefficients of powers of $\overline{K}c$ in ξ are symmetrical; hence, from Eq. (15.5), there is symmetry in $\theta(\ln c)$ about $\theta = \frac{1}{2}$, with $\theta = \frac{1}{2}$ at $c = c_o$, where $\overline{K}c_o = 1$. In this subsection we show that this symmetry property holds for a ring of m subunits with m arbitrary ($m \to \infty$ is the classical one-dimensional Ising problem, which is discussed in Chapter 8).

Our object is to show that any single term (i.e., with *numerical* coefficient unity) in ξ, representing a single state of the ring with r subunits in state 2, $r < m/2$, when ξ is written in powers of $\overline{K}c$, has a companion term in ξ, representing a single state of the ring with $m - r$ subunits in state 2, *with the same (nonnumerical) coefficient*. Because of the equality of the nonnumerical coefficients of *all* pairs of companion terms in $(\overline{K}c)^r$ and $(\overline{K}c)^{m-r}$, Eq. (15.5) applies with $\overline{K}c_o = 1$ at $\theta = \frac{1}{2}$. Hence $\theta(\ln c)$ is symmetrical about $\theta = \frac{1}{2}$, with $\theta = \frac{1}{2}$ at $c_o = 1/\overline{K}$.

We start, as usual, with the grand partition function in the form

$$\xi = y_{11}^m + \cdots + y_{22}^m K_o^m c^m. \tag{23.58}$$

The first and last terms are easy to write for any m. We represent a single arbitrary term in this ξ, with $r < m/2$, where r is the number of subunits in state 2 (i.e., r ligand molecules are bound), by

$$y_{11}^{\alpha}(y_{12}y_{21})^{\beta} y_{22}^{\gamma}(K_o c)^r. \tag{23.59}$$

The numerical coefficient is unity. That is, this is merely one term in ξ out of a total of 2^m. An example of a particular state or configuration (corresponding to a single term in ξ) of a ring of $m = 12$ subunits is shown in Fig. 6-14. Pairs are counted, say, in the clockwise direction. Strings of 1s begin and end with a 21 and a 12 pair, respectively; strings of 2s begin with a 12 pair and end with a 21. Thus the total number of 12 pairs must equal the total number of 21 pairs. Both numbers are designated β in Eq. (23.59). Also, there are two conservation relations:

$$m = \alpha + 2\beta + \gamma, \tag{23.60}$$

$$r = \beta + \gamma \qquad \text{or} \qquad m - r = \alpha + \beta. \tag{23.61}$$

That is, all state 2 subunits can be accounted for by counting 12 and 22 pairs (or 21 and 22 pairs). These relations are illustrated in Fig. 6-14. Using the conservation relations to eliminate β and γ, Eq. (23.59) becomes

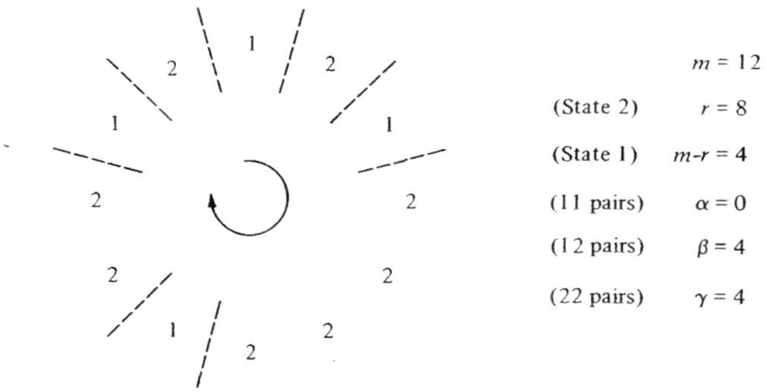

Fig. 6-14. Ring of $m = 12$ subunits in states 1 or 2. Dashed lines are placed at 1, 2 or 2, 1 intersections. Pairs are designated in the clockwise direction. See text for further details.

$$y_{11}^{\alpha}(y_{12}y_{21})^{m-r-\alpha}y_{22}^{2r+\alpha-m}(K_o c)^r. \tag{23.62}$$

From Eq. (23.58), upon dividing by y_{11}^m,

$$\overline{K} = \left(\frac{y_{22}}{y_{11}}\right)K_o, \tag{23.63}$$

as already found for $m = 2, 3, 4$. If we now divide Eq. (23.62) by y_{11}^m, so that ξ begins with $1 + \cdots$, as usual, and replace K_o by $\overline{K}$, Eq. (23.62) simplifies to

$$y^{r+\alpha-m}(\overline{K}c)^r, \tag{23.64}$$

where

$$y \equiv \frac{y_{11}y_{22}}{y_{12}y_{21}}, \tag{23.65}$$

as in Eq. (23.29).

To find the companion term to Eq. (23.62), each state 1 in the corresponding configuration (see Fig. 6-14, for example) is replaced by state 2, and each state 2 by state 1. The number of bound ligands (state 2) is now $m - r$, each 11 pair becomes a 22 pair, and each 22 becomes a 11. There is no change in the numbers of 12 or 21 pairs. Hence the companion term to Eq. (23.62) is

$$y_{11}^{2r+\alpha-m}(y_{12}y_{21})^{m-r-\alpha}y_{22}^{\alpha}(K_o c)^{m-r}. \tag{23.66}$$

Finally, we replace K_o by $\overline{K}$ [Eq. (23.62)] and divide by y_{11}^m, as before. Then Eq. (23.66) reduces to

$$y^{r+\alpha-m}(\overline{K}c)^{m-r}. \tag{23.67}$$

The coefficients in the companion pair Eqs. (23.64) and (23.67) are the same, as was to be proved.

24. Binding on Subunits with Two Conformations

This section is devoted to one of the more important classes of small cooperative systems in biology. The model has been applied extensively to hemoglobin,[2–4] for example. We have a complex of m identical subunits, each with one binding site for a ligand L. Each subunit can exist in two conformations or states, 1 and 2. The intermolecular interactions between nearest-neighbor subunit pairs depend on the conformational state of the subunits but not on whether a ligand is bound or not. This is in distinct contrast with Section 23. That is, there are no direct ligand–ligand interactions, and the binding of a ligand on a subunit does not perturb the subunit structure in a way that affects its interactions with neighboring subunits. However, ligand binding will influence the equilibrium between the two conformations because the binding constant for L is K_1 on a subunit in state 1 and is K_2 on a state 2 subunit. If $K_1 \neq K_2$, binding of L will shift the $1 \rightleftarrows 2$ equilibrium toward the better binding state. In fact, to be explicit, we arbitrarily choose state 2 as the better binding state. That is, $K_2 \geq K_1$.

This is the same model as in Section 14, except that here there are intersubunit interactions. Two variations on the above model will be considered at the end of the section: (a) each subunit has binding sites for two different ligands; and (b) instead of the $1 \rightleftarrows 2$ equilibrium being modulated by ligand binding, it is altered by an electric field (e.g., in a membrane).

Complex with Two Subunits ($m = 2$)

We shall introduce most of the fundamental topics by treating this important but relatively simple special case. The analysis of more complicated models then follows easily by analogy with the dimer.

The 16 possible states of a symmetrical dimer are shown in Figs. 4-6 and 6-15. Now, however, we introduce the intersubunit interaction parameter $y_{ij} = e^{-w_{ij}/kT}$, as in Section 23, for a dimer with conformational states ij (irrespective of whether L is bound). For this symmetrical dimer, we take $y_{12} = y_{21}$ [as in Eq. (23.7)].

In the absence of ligand ($c = 0$), to begin with, the grand partition function of the dimer is (see Sections 12 and 14)

$$\xi = j_1^2 y_{11} + 2j_1 j_2 y_{12} + j_2^2 y_{22}, \tag{24.1}$$

○○ ○□, □○ □□

where j_1 and j_2 are partition functions for isolated subunits in states 1 and 2. The y_{ij} factors are contributions associated with bringing two subunits together from $r = \infty$ to form the dimer (see Section 23); w_{ij} is a potential of mean force relative to a zero at $r = \infty$. If we factor out $j_1^2 y_{11}$,

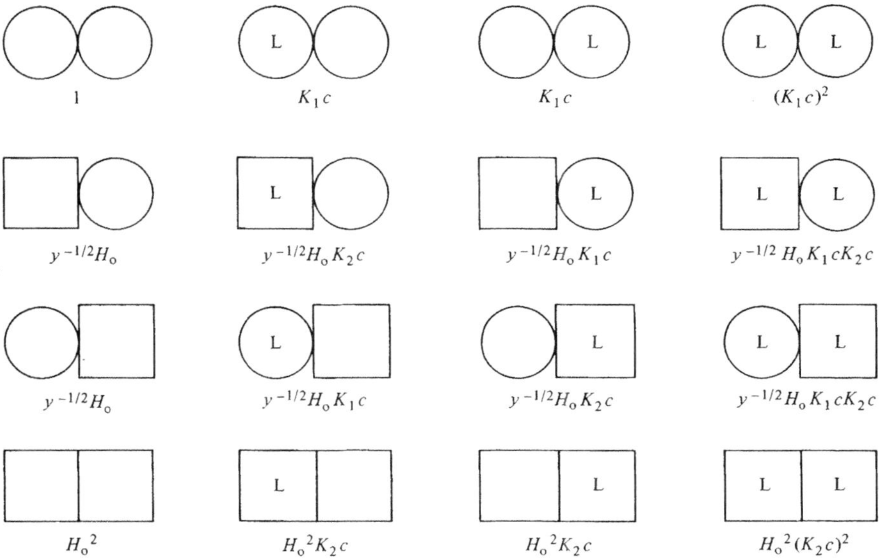

Fig. 6-15. States and grand partition function terms [Eq. (24.5)] for binding on a symmetrical dimer, each subunit of which has two possible conformations.

$$\begin{aligned} \xi &= 1 + 2\frac{j_2}{j_1}\cdot\frac{y_{12}}{y_{11}} + \left(\frac{j_2}{j_1}\right)^2\frac{y_{22}}{y_{11}} \\ &= 1 \quad + 2y^{-1/2}H_o \quad + H_o^2, \end{aligned} \tag{24.2}$$

○○ ○□, □○ □□

where we have defined

$$y \equiv \frac{y_{11}y_{22}}{y_{12}^2} \qquad \text{and} \qquad H_o \equiv \left(\frac{y_{22}}{y_{11}}\right)^{1/2}\frac{j_2}{j_1}. \tag{24.3}$$

Equation (23.12) should be noted in connection with the definition of y. Whereas j_2/j_1 is the equilibrium constant for $1 \rightleftarrows 2$ with isolated subunits, H_o^2 is the equilibrium constant for the process

$$\underset{1\ 1}{○○} \rightarrow \underset{2\ 2}{□□} \tag{24.4}$$

in the dimer, including interaction free energies, but in the absence of ligand.

Note the formal resemblance between Eqs. (23.2C, and (24.2). In Eq. (23.20), the states 11, 12, 21, 22 are directly associated with ligand binding (see Fig. 6-10); in Eq. (24.2), on the other hand, there is a distribution among these states that depends on *intrinsic* properties of the subunits [see Eq.

(24.1)] in the *absence* of ligand. The subunits are involved in an internal equilibrium among conformational states that is modified by nearest-neighbor interactions.

In this model, as already mentioned, the binding constant for a state 1 subunit is K_1 irrespective of whether the neighboring subunit is in state 1 or state 2. The same is true for state 2 and K_2. Thus, the grand partition functions for the *ligand substates* of the several conformational states of the dimer are:

○○ $(1 + K_1c)^2$,

○□, □○ $(1 + K_1c)(1 + K_2c)$,

□□ $(1 + K_2c)^2$.

Consequently, the generalization of Eq. (24.2), in the presence of ligand at concentration c, is

$$\xi = \underbrace{(1 + K_1c)^2}_{11} + \underbrace{2y^{-1/2}H_o(1 + K_1c)(1 + K_2c)}_{12,\ 21} + \underbrace{H_o^2(1 + K_2c)^2}_{22}. \tag{24.5}$$

This now includes all $4^2 = 16$ states of the dimer: each subunit has four possible states,

○ Ⓛ □ 🄻.

Correspondingly, Eq. (24.5) has 16 terms. The association between states and terms in ξ is included in Fig. 6-15. These terms give the relative weights of the 16 states at equilibrium. Thus any equilibrium average of interest, over these states, can be calculated easily.

Although it might appear in Eq. (24.5) and Fig. 6-15 that interactions effects are included only in conformational states 12 and 21 (because of the factor $y^{-1/2}$), but not in 11 and 22, this is definitely *not* the case. The terms in ξ represent *relative* probabilities of states, not absolute probabilities. The factor y_{11} has disappeared from state 11 terms because of normalization [Eq. (24.1) → Eq. (24.2)]; and y_{22} is missing from 22 terms because of the introduction of H_o [Eq. (24.3)]. Indeed, $y^{-1/2}$ appears in the 12 and 21 terms, and not y_{12}, because of both of the above steps as well as the introduction of the composite parameter y [Eq. (24.3)].

If $y = 1$ (no cooperativity) in Eq. (24.5), we have

$$\xi = [(1 + K_1c) + H_o(1 + K_2c)]^2, \tag{24.6}$$

which is formally the same as Eq. (14.1). Note that $y = 1$ does not mean that the subunits of a dimer do not interact with each other. Rather, the interaction factors y_{ij} happen to be related to each other in such a way that $y_{12}^2 = y_{11}y_{22}$. That is [Eq. (23.12)],

$$2w_{12} = w_{11} + w_{22}.$$

When $y > 1$ (positive cooperativity), the 11 and 22 terms in Eq. (24.5) become relatively more important than the 12, 21 term; the opposite is true when $y < 1$ (negative cooperativity).

Equation (24.5) is needed to obtain ligand binding properties formally (see below). However, to find the conformational distribution (i.e., between states 1 and 2) and related properties, it is simpler to normalize one more time: we divide Eq. (24.5) by $(1 + K_1c)^2$ and find

$$\xi' = 1 + 2y^{-1/2}H + H^2, \tag{24.7}$$

where

$$H \equiv \left(\frac{1 + K_2c}{1 + K_1c}\right)H_o. \tag{24.8}$$

Note again the formal resemblance to Eq. (23.20). Ligand binding is involved in Eq. (24.7) via Eq. (24.8); but except in special cases, the role of ligand binding here is as a modulator; the effect of ligand in Eq. (23.20) is more direct. H^2 is the equilibrium constant for $11 \rightarrow 22$ in the presence of L at concentration c. That is, there is equilibrium binding of L on both 11 and 22, and this influences the $11 \rightleftarrows 22$ equilibrium. Because $K_2 \geq K_1$ by definition, $H \geq H_o$. Equation (24.7) simplifies to Eq. (24.2) if $c = 0$ or if $K_1 = K_2$. In the latter case, binding of L has no effect on the subunit equilibrium.

The prime is placed on ξ in Eq. (24.7) as a reminder that this is not a proper ξ insofar as binding of ligand is concerned. Thus $\partial \ln \xi'/\partial \ln c$ cannot be used to find θ, the fractional binding of L. To obtain Eq. (24.7), we divided by a function of c, not by a constant. Hence ξ and ξ' are not equivalent in all respects.

If we let p_r^e be the probability that r subunits are in state (conformation) 2, then the terms in Eq. (24.7) give immediately

$$p_0^e = \frac{1}{\xi'}, \qquad p_1^e = \frac{2y^{-1/2}H}{\xi'}, \qquad p_2^e = \frac{H^2}{\xi'}. \tag{24.9}$$

Also, the fraction of subunits in state 2 is (see Section 12)

$$\delta = \frac{1}{2}H\frac{\partial \ln \xi}{\partial H} = \frac{y^{-1/2}H + H^2}{\xi'} = \frac{1}{2}p_1^e + p_2^e. \tag{24.10}$$

The relation between δ and the p_r^e is physically obvious. The derivative relationship between δ and H follows by analogy with $m\theta = \partial \ln \xi/\partial \ln c$.

Actually, we need not rely on an analogy to establish this latter relationship. The fact that $p_r^e \propto H^r$ for all r is the essential point. For, whatever the meaning of ξ', r, and H, if $0 \leqslant r \leqslant m$ and

$$\xi' = \sum_{r=0}^{m} A_rH^r, \qquad p_r^e = \frac{A_rH^r}{\xi'} \tag{24.11}$$

(ξ' is a normalization sum), then

$$H\frac{\partial \ln \xi'}{\partial H} = \frac{1}{\xi'}\sum_r rA_rH^r = \sum_r rp_r^e = \bar{r}, \tag{24.12}$$

and, in addition, a little further algebra shows that

$$H\frac{\partial \bar{r}}{\partial H} = \overline{r^2} - \bar{r}^2. \tag{24.13}$$

In the context of this section, δ and $\bar{r}$ are related ($m\delta = \bar{r}$) in the same way that θ and $\bar{s}$ are related ($m\theta = \bar{s}$). That is, we use $r = 0, 1, 2, \ldots, m$ to count subunits in state 2 (δ is the fraction of subunits in state 2) and we use $s = 0, 1, 2, \ldots, m$ to count bound ligand molecules (θ is the fraction of sites occupied).

Because of the formal resemblance of Eq. (24.7) to Eq. (23.20) already mentioned, a plot of δ against $\ln H$ would be symmetrical about $\delta = \frac{1}{2}$. Furthermore, $\delta = \frac{1}{2}$ occurs at $H = 1$, $\ln H = 0$, as can be verified from Eq. (24.10). Operationally, the only way to vary H is by varying c. When $c \to 0$, $H \to H_o$. When $c \to \infty$, $H \to K_2H_o/K_1$. Thus the range in H is from H_o to K_2H_o/K_1 (recall that $K_2 \geq K_1$). Because $\delta = \frac{1}{2}$ occurs when $H = 1$, this value of δ is achievable, by varying c, if $H_o < 1$ and $K_2H_o/K_1 > 1$ (i.e., if $K_2H_o > K_1$). In such a case, state 1 is intrinsically more stable in the absence of ligand ($\delta < \frac{1}{2}$), but ligand binding shifts the equilibrium to favor state 2 ($\delta > \frac{1}{2}$). If we denote the value of c at $\delta = \frac{1}{2}$ by c', c' is determined (when it exists) by

$$\left(\frac{1 + K_2c'}{1 + K_1c'}\right)H_o = 1, \tag{24.14}$$

or by

$$c' = \frac{1 - H_o}{K_2H_o - K_1}. \tag{24.15}$$

Note that c' does not depend on y, but it does depend on y_{22}/y_{11}, through Eq. (24.3).

As mentioned above, $\delta(\ln H)$ is a symmetrical function. It is from a plot of δ against $\ln H$ (when c' exists) that the Hill coefficient n should be calculated for this kind of model because it is only this plot that has a fundamental connection with the cooperativity parameter y analogous to that in $\theta(\ln c)$ in Section 22. That is, we take it as axiomatic that the Hill coefficient is designed to be an operational index of cooperativity, whatever its origin. In the present type of model, the cooperativity does not originate with binding. Of course y itself is an even more direct index, but it is not an operational quantity. From Eq. (22.17),

$$n = 4\left(\frac{\partial \delta}{\partial \ln H}\right)_{\delta=1/2} = 2(\overline{r^2} - \bar{r}^2) = \frac{2y^{1/2}}{1 + y^{1/2}}. \tag{24.16}$$

We can simply rewrite Eq. (22.17) in this way because of the formal connections, already pointed out, among Eqs. (22.11), (23.10), (23.20), and (24.7). Alternatively, Eq. (24.16) follows directly on differentiating $\delta(H)$ in Eq. (24.10).

Application of Eq. (24.16) may not be practical in most cases because δ is not ordinarily measured directly (but see Ref. 7). Despite this difficulty, Eq. (24.16) *is* the fundamental equation for n for this model ($m = 2$). Of course it is not necessary to use H itself in Eq. (24.16): $(1 + K_2c)/(1 + K_1c)$ would suffice in its place.

The plots $\theta(\ln x)$ in Fig. 6-1 are also plots of $\delta(\ln H)$, except that the $y \neq 1$ curves must be shifted horizontally to give $\delta = \frac{1}{2}$ at $\ln H = 0$ for all values of y.

We turn now to a consideration of the binding isotherm $\theta(c)$. When the dimer is in state $r = 0$ (probability p_0^e), the binding isotherm is $\theta_1(c)$, where

$$\theta_1 \equiv \frac{K_1 c}{1 + K_1 c}. \tag{24.17}$$

When the dimer is in state $r = 2$ (probability p_2^e), the isotherm is $\theta_2(c)$:

$$\theta_2 \equiv \frac{K_2 c}{1 + K_2 c}. \tag{24.18}$$

Similarly, in state $r = 1$, the isotherm is $\frac{1}{2}\theta_1 + \frac{1}{2}\theta_2$. Thus the overall binding isotherm is

$$\theta = p_0^e \theta_1 + \tfrac{1}{2} p_1^e (\theta_1 + \theta_2) + p_2^e \theta_2 \tag{24.19}$$

$$= (1 - \delta)\theta_1 + \delta\theta_2. \tag{24.20}$$

Equation (24.20) follows from Eq. (24.10) and

$$1 - \delta = p_0^e + \frac{1}{2} p_1^e, \tag{24.21}$$

which is the fraction of subunits in state 1.

Equations (24.19) and (24.20) are physically obvious for this model, but they can also be obtained by differentiation of Eq. (24.5):

$$2\theta = \bar{s} = c\frac{\partial \ln \xi}{\partial c}. \tag{24.22}$$

Although δ is a symmetrical function of $\ln H$, δ is not a symmetrical function of $\ln c$. Because $m = 2$, θ is necessarily a symmetrical function of $\ln c$ about $\theta = \frac{1}{2}$ [Eq. (15.9)]. If Eq. (24.5) is rearranged as a polynomial in c, the value of c at which $\theta = \frac{1}{2}$ is found to be

$$c_o = \left(\frac{\alpha_0}{\alpha_2}\right)^{1/2} = \left(\frac{1 + 2y^{-1/2}H_o + H_o^2}{K_1^2 + 2y^{-1/2}H_o K_1 K_2 + H_o^2 K_2^2}\right)^{1/2}. \tag{24.23}$$

When $y = 1$, this simplifies to $(1 + H_o)/(K_1 + H_o K_2)$. If we use Eq. (24.19) for θ and obtain the p_r^e from Eq. (24.5) [rather than from Eq. (24.9)], it is not difficult to deduce

$$\frac{\partial \theta}{\partial \ln c} = \frac{1}{2}\sigma_2^2 = \frac{1}{2}(\overline{s^2} - \bar{s}^2)$$

$$= \theta(1 - 2\theta) + p_0^e \theta_1^2 + p_1^e \theta_1 \theta_2 + p_2^e \theta_2^2. \tag{24.24}$$

At $\theta = \frac{1}{2}$ and $c = c_o$ [Eq. (24.23)], this becomes

$$\left(\frac{\partial \theta}{\partial \ln c}\right)_{\theta=1/2}$$

$$= \frac{(K_1 c_o)^2 + 2y^{-1/2} H_o K_1 c_o K_2 c_o + H_o^2 (K_2 c_o)^2}{(1 + K_1 c_o)^2 + 2y^{-1/2} H_o (1 + K_1 c_o)(1 + K_2 c_o) + H_o^2 (1 + K_2 c_o)^2}. \tag{24.25}$$

When $y = 1$, this simplifies to

$$\left(\frac{\partial \theta}{\partial \ln c}\right)_{\theta=1/2} = \frac{1}{4}, \tag{24.26}$$

but, for arbitrary y, Eq. (24.25) [including Eq. (24.23) for c_o] depends on y in a very complicated way. This is in contrast with the simplicity of Eq. (24.16).

In the special case $y = 1$ (noncooperative), we obtain results for δ and θ that are actually valid for any m:

$$\xi' = (1 + H)^2, \qquad \delta = \frac{H}{1 + H} = \frac{H_o(1 + K_2 c)}{(1 + K_1 c) + H_o(1 + K_2 c)}, \tag{24.27}$$

$$\begin{gathered} \theta = (1 - \delta)\theta_1 + \delta\theta_2, \\ p_0^e = (1 - \delta)^2, \qquad p_1^e = 2\delta(1 - \delta), \qquad p_2^e = \delta^2, \end{gathered} \tag{24.28}$$

$$\frac{\partial \theta}{\partial \ln c} = \theta(1 - 2\theta) + \theta^2 = \theta(1 - \theta). \tag{24.29}$$

Further details are included in Section 14.

In the special case that there is no binding of L on the state 1 subunit ($K_1 = 0$), we have

$$H = (1 + K_2 c)\, H_o, \qquad c' = \frac{1 - H_o}{K_2 H_o}, \qquad \theta_1 = 0, \tag{24.30}$$

$$\theta = \delta\theta_2, \; \frac{\partial \theta}{\partial \ln c} = \theta(1 - 2\theta) + p_2^e \theta_2^2. \tag{24.31}$$

A more severe special case than this shows how the *induced conformation model* of Sections 14 (where $y = 1$) and 23 is related to the present formulation. In addition to $K_1 = 0$, we also assume that: (a) $H_o \to 0$ (state 1 completely dominates over state 2 in the absence of ligand); (b) $K_2 \to \infty$ (binding of L on state 2 is very strong, so that $K_2 c \gg 1$); and (c) $H_o K_2$ is finite. In this case the surviving terms in the ligand grand partition function are

$(1 + K_1 c)^2 \to 1$ ○○

$(1 + K_1 c)(1 + K_2 c) \to K_2 c$ ○[L], [L]○

$$(1 + K_2c)^2 \rightarrow (K_2c)^2 \qquad \boxed{\text{L}}\,\boxed{\text{L}}$$

as in Fig. 6-10e. Thus $H \rightarrow H_oK_2c$ and Eq. (24.5) reduces to

$$\xi = 1 + 2y^{-1/2}(H_oK_2)c + (H_oK_2)^2c^2, \tag{24.32}$$

which is the same as Eq. (23.20) but with the effective binding constant H_oK_2 in place of $\bar{K}$. Only the diagonal states, starting from the upper left, remain in Fig. 6-15. In this special case we have $\theta_1 = 0$, $\theta_2 = 1$, and $\vartheta = \delta$ (ligand binding and conformational change are tightly coupled):

$$\theta = \delta = \frac{y^{-1/2}(H_oK_2)c + (H_oK_2)^2c^2}{1 + 2y^{-1/2}(H_oK_2)c + (H_oK_2)^2c^2}, \tag{24.33}$$

as in Eq. (23.23). Both θ and δ are symmetrical functions of ln c about $\theta = \delta = \frac{1}{2}$. At $\theta = \delta = \frac{1}{2}$ [Eq. (24.15)],

$$c_o = c' = 1/H_oK_2. \tag{24.34}$$

Equation (24.31) simplifies to

$$\frac{\partial \theta}{\partial \ln c} = \theta(1 - 2\theta) + p_2^e, \tag{24.35}$$

$$p_2^e = \frac{(H_oK_2)^2c^2}{\xi}. \tag{24.36}$$

Equation (24.35) leads to the Hill coefficient n in Eq. (22.17) at $\theta = \frac{1}{2}$.

Another important special case is the MWC *concerted model* already discussed in Sections 12 and 14. In the present context, this is the case $y \rightarrow \infty$ (very strong positive cooperativity): the 12 and 21 states are very unstable compared to 11 and 22 so that 12 and 21 are used only as transient intermediates. The only important states in Fig. 6-15 are those in the top and bottom rows. We give a summary of a few properties of the MWC model in the present notation:

$$\xi = (1 + K_1c)^2 + H_o^2(1 + K_2c)^2, \tag{24.37}$$

$$\xi' = 1 + H^2, \qquad p_0^e = \frac{1}{\xi'}, \qquad p_1^e = 0, \qquad p_2^e = \frac{H^2}{\xi'}, \tag{24.38}$$

$$\delta = p_2^e = \frac{H^2}{\xi'} = \frac{H_o^2(1 + K_2c)^2}{\xi}, \tag{24.39}$$

$$\theta = (1 - \delta)\theta_1 + \delta\theta_2, \tag{24.40}$$

$$\frac{\partial \theta}{\partial \ln c} = \theta(1 - 2\theta) + \frac{(\theta_1^2 + H^2\theta_2^2)}{1 + H^2}, \tag{24.41}$$

$$c_o = \left(\frac{1 + H_o^2}{K_1^2 + H_o^2K_2^2}\right)^{1/2}, \tag{24.42}$$

$$\left(\frac{\partial \theta}{\partial \ln c}\right)_{\theta=1/2} = \frac{(K_1 c_o)^2 + H_o^2 (K_2 c_o)^2}{(1 + K_1 c_o)^2 + H_o^2 (1 + K_2 c_o)^2}, \tag{24.43}$$

$$n = 4\left(\frac{\partial \delta}{\partial \ln H}\right)_{\delta=1/2} = 2. \tag{24.44}$$

The fact that $4(\partial\theta/\partial \ln c)$ at $\theta = \frac{1}{2}$ does not give $2(=m)$ when $y \to \infty$ shows that this is not a suitable index to measure the true degree of cooperativity in this model.

A numerical illustration of the general equations above, for $m = 2$, is given in Fig. 6-16. The parameter choices are $H_o = 0.55$, $K_1/K_2 = 1/10$, and $y = 1$ and 400 [in Eq. (23.12), $w = -3573$ cal mole^{-1} at 25°C]. The concentration variable chosen is $K_2 c$. In the main figure, θ_1, θ_2, θ, and δ are plotted against $\ln K_2 c$. The θ_1 and θ_2 curves are independent of the value of y. With the above choices of H_o and K_1/K_2, the value of $K_2 c'$ [Eq. (24.15)] is 1. Consequently $\delta = \frac{1}{2}$ at $\ln K_2 c = 0$ at any y. It will be noted that δ is an unsymmetrical function of $\ln K_2 c$; also, the $y = 400$ curve for δ is somewhat steeper than the $y = 1$ curve. The θ curves are linear combinations of θ_1 and θ_2, with $1 - \delta$ and δ as weights [Eq. (24.20)]. Because binding of L is only indirectly associated with cooperativity in this model, the steepness of the θ curve is not very sensitive to y [Eq. (24.25)]. The slope $\partial\theta/\partial \ln c$ at $\theta = \frac{1}{2}$ is 0.2500 at $y = 1$ and it is 0.3015 at $y = 400$. If we multiply the latter slope by 4, we obtain 1.206 as a pseudo Hill coefficient. However, this does not

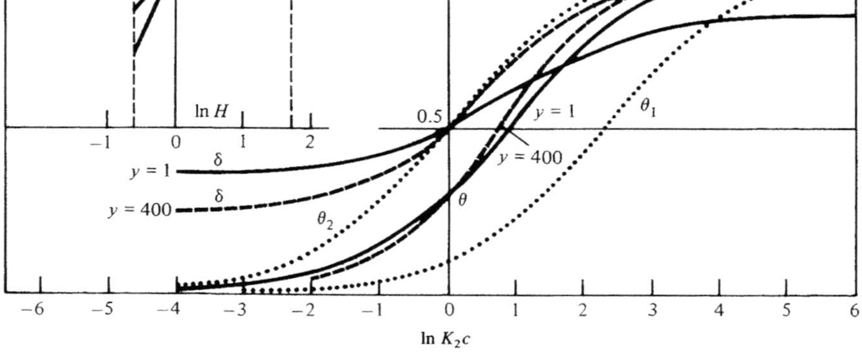

Fig. 6-16. Numerical example of binding (θ, θ_1, θ_2) and conformational equilibrium (δ) for model in Fig. 6-15, with $H_o = 0.55$ and $K_1/K_2 = 1/10$. See text for details.

reflect the actual strength of the positive cooperativity in this model and should not be used as the Hill coefficient.

For this same example, δ is plotted against $\ln H$ for $y = 1$ and 400 in the inset of Fig. 6-16. The possible range in H is from $H_o = 0.55$ to $H_oK_2/K_1 = 5.5$. The two slopes at $\delta = \frac{1}{2}$ (the point of symmetry) are 0.2500 ($y = 1$) and 0.4762 ($y = 400$). The corresponding n values [Eq. (24.16)] are 1 and 1.905 (rather close to the limit $2 = m$).

Ligand Interactions in a Dimer ($m = 2$)

In the model of the previous subsection, binding of L has no direct effect on the intermolecular interaction parameters of the model, y_{ij}. One of many possible more general models arises if we suppose that, in addition to the intersubunit interactions already included, there is a direct ligand–ligand interaction free energy w' ($y' \equiv e^{-w'/kT}$), when *both* sites of the dimer are occupied (whatever the conformational states). This type of interaction was introduced in Section 22 (with parameters w and y). We merely indicate the change required in ξ and leave further details to the interested reader. Corresponding to Eq. (22.11), Eq. (24.5) is modified to

$$\begin{aligned}\xi = {} & (1 + 2K_1c + y'K_1^2c^2) \\ & + 2y^{-1/2}H_o(1 + K_1c + K_2c + y'K_1K_2c^2) \\ & + H_o^2(1 + 2K_2c + y'K_2^2c^2).\end{aligned} \tag{24.45}$$

Triangular Complex ($m = 3$)

There are three identical subunits in an equilateral triangle. Each subunit can be in two states or conformations, 1 and 2. We allow for the possibility that $y_{12} \neq y_{21}$, as in Section 23. Because of ligand binding, as in the $m = 2$ case, there are now a total of $4^3 = 64$ possible states of the trimer. The interaction parameters are the same as in Eq. (23.37).

In the absence of ligand binding ($c = 0$), the grand partition function is [compare Eqs. (23.37) and (24.1)]

$$\xi = j_1^3y_{11}^3 + 3j_1^2j_2y_{11}y_{12}y_{21} + 3j_1j_2^2y_{12}y_{21}y_{22} + j_2^3y_{22}^3. \tag{24.46}$$

We follow the procedure used in the $m = 2$ case. On dividing by $j_1^3y_{11}^3$ and defining

$$y \equiv \frac{y_{11}y_{22}}{y_{12}y_{21}}, \qquad H_o \equiv \frac{y_{22}}{y_{11}} \cdot \frac{j_2}{j_1}, \tag{24.47}$$

we have

$$\xi = 1 + 3y^{-1}H_o + 3y^{-1}H_o^2 + H_o^3. \tag{24.48}$$

This resembles Eq. (23.41). H_o^3 is the equilibrium constant for $1^11 \rightarrow 2^22$, without ligand.

With ligand binding included,

$$\xi = (1 + K_1c)^3 + 3y^{-1}H_o(1 + K_1c)^2(1 + K_2c) + 3y^{-1} H_o^2(1 + K_1c)(1 + K_2c)^2 + H_o^3(1 + K_2c)^3, \tag{24.49}$$

$$\xi' = 1 + 3y^{-1}H + 3y^{-1}H^2 + H^3, \tag{24.50}$$

with H defined as in Eq. (24.8). H^3 is the equilibrium constant for $1^1 1 \to 2^2 2$ in the presence of L at concentration c. Note that there are $4^3 = 64$ terms in Eq. (24.49). These terms determine the equilibrium distribution among the 64 states.

All of the models in this section ($m = 2, 3, 4, 6$) are formally similar to the corresponding models in Section 23 (they, in turn, are related to models in Section 22). Hence there are many properties carried over to the present section. For the cases below with $m = 4$ and 6, we can start immediately with the analogues of Eq. (24.50), using Section 23 as a guide.

The main properties of interest for the trimer are

$$p_0^e = \frac{1}{\xi'}, \qquad p_1^e = \frac{3y^{-1}H}{\xi'}, \qquad p_2^e = \frac{3y^{-1}H^2}{\xi'}, \qquad p_3^e = \frac{H^3}{\xi'}, \tag{24.51}$$

$$\delta = \tfrac{1}{3}p_1^e + \tfrac{2}{3}p_2^e + p_3^e, \tag{24.52}$$

$$1 - \delta = p_0^e + \tfrac{2}{3}p_1^e + \tfrac{1}{3}p_2^e, \tag{24.53}$$

$$\theta = p_0^e\theta_1 + p_1^e(\tfrac{2}{3}\theta_1 + \tfrac{1}{3}\theta_2) + p_2^e(\tfrac{1}{3}\theta_1 + \tfrac{2}{3}\theta_2) + p_3^e\theta_2, \tag{24.54}$$

$$= (1 - \delta)\theta_1 + \delta\theta_2. \tag{24.55}$$

Because of the symmetry of coefficients in Eq. (24.50), δ is a symmetrical function of $\ln H$ with $\delta = \frac{1}{2}$ at $H = 1$. The value of c at $\delta = \frac{1}{2}$, $c = c'$, is given by Eq. (24.15). The range in H is again from H_o to K_2H_o/K_1. The curves in Fig. 6-2 apply to $\delta(\ln H)$ if the $y \neq 1$ curves are shifted horizontally to pass through $\delta = \frac{1}{2}$, $\ln H = 0$. The Hill coefficient is [Eq. (22.25)]

$$n = 4\left(\frac{\partial \delta}{\partial \ln H}\right)_{\delta=1/2} = \frac{4}{3}(\overline{r^2} - \bar{r}^2) = \frac{1 + 3y}{3 + y}. \tag{24.56}$$

If Eq. (24.49) is rearranged as a polynomial in c, one finds that in general the condition $\alpha_1^3\alpha_3 = \alpha_2^3\alpha_0$ in Eq. (15.19) is not satisfied. In fact, for this purpose it suffices to examine the simple special case $K_1 = 0$, $y \to \infty$. Hence θ is not usually a symmetrical function of $\ln c$. From Eqs. (24.49) and (24.54), we find

$$\frac{\partial \theta}{\partial \ln c} = \frac{1}{3}\sigma_3^2 = \frac{1}{3}(\overline{s^2} - \bar{s}^2)$$

$$= \theta - 3\theta^2 + 2p_0^e\theta_1^2 + p_1^e\left(\frac{2}{3}\theta_1^2 + \frac{4}{3}\theta_1\theta_2\right) + p_2^e\left(\frac{4}{3}\theta_1\theta_2 + \frac{2}{3}\theta_2^2\right) + 2p_3^e\theta_2^2. \tag{24.57}$$

In the *induced conformation* special case ($K_1 = 0, K_2 \to \infty, H_o \to 0, H_o K_2$ finite), Eq. (24.49) simplifies to

$$\xi = 1 + 3y^{-1}(H_o K_2)c + 3y^{-1}(H_o K_2)^2 c^2 + (H_o K_2)^3 c^3. \quad (24.58)$$

This has the same properties as Eq. (23.41), with $H_o K_2$ in place of $\overline{K}$. Because $\theta = \delta$ and $H = H_o K_2 c$, θ is now a symmetrical function of ln c.

In the *concerted model,* we put $y \to \infty$ or $y^{-1} \to 0$ in Eqs. (24.48)–(24.56). A few basic properties are

$$\xi = (1 + K_1 c)^3 + H_o^3(1 + K_2 c)^3, \quad (24.59)$$

$$\xi' = 1 + H^3, \quad (24.60)$$

$$p_1^e = p_2^e = 0, \qquad p_0^e = \frac{1}{\xi'}, \qquad p_3^e = \frac{H^3}{\xi'}. \quad (24.61)$$

Further details are in Sections 12 and 14. The function $\theta(\ln c)$ is not symmetrical about $\theta = \frac{1}{2}$ [Eq. (15.19)].

It should be noted from the $m = 2$ and $m = 3$ examples that the induced conformation case is model dependent (as in Section 23), but the concerted case is essentially model independent: $\xi' = 1 + H^m$, whatever the other details; the concerted case depends only on m.

Arbitrary Model with m Identical Subunits

The above results for $m = 2$ and $m = 3$ suggest a digression to obtain certain general properties for *any* model with m identical subunits, each of which has two states or conformations, 1 and 2, in equilibrium with each other, and corresponding binding constants K_1 and K_2, as above. We arrange the grand partition function in order of increasing r, from $r = 0$ to $r = m$, where r is the number of subunits in state 2. Then

$$\xi = \sum_{r=0}^{m} a_r(1 + K_1 c)^{m-r}(1 + K_2 c)^r, \quad (24.62)$$

where $a_0 = 1$ and the other a_r depend on the details of the particular model [see Eq. (24.5) for an example]. Then the equilibrium probability of r is

$$p_r^e = \frac{a_r(1 + K_1 c)^{m-r}(1 + K_2 c)^r}{\xi}. \quad (24.63)$$

The fractions of subunits in states 1 and 2 are

$$1 - \delta = \sum_r \left(\frac{m - r}{m}\right) p_r^e, \quad (24.64)$$

$$\delta = \sum_r \frac{r}{m} p_r^e. \quad (24.65)$$

The fractional binding is

$$\frac{1}{m} \cdot c \frac{\partial \ln \xi}{\partial c} = \theta = \sum_r \left[\left(\frac{m - r}{m} \right) \theta_1 + \frac{r}{m} \theta_2 \right] p_r^{\mathrm{e}} \tag{24.66}$$

$$= (1 - \delta)\theta_1 + \delta\theta_2. \tag{24.67}$$

From Eq. (24.66),

$$\frac{\partial \theta}{\partial \ln c} = \theta - m\theta^2 + \sum_r \left[\frac{(m - r)(m - r - 1)}{m} \theta_1^2 + \frac{2(m - r)r}{m} \theta_1 \theta_2 + \frac{r(r - 1)}{m} \theta_2^2 \right] p_r^{\mathrm{e}}. \tag{24.68}$$

From Eq. (24.65),

$$\frac{\partial \delta}{\partial \ln c} = -m\theta\delta + \sum_r \left[\frac{(m - r)r}{m} \theta_1 + \frac{r^2}{m} \theta_2 \right] p_r^{\mathrm{e}}. \tag{24.69}$$

Similar expressions for $\partial^2\theta/\partial \ln c^2$ and $\partial^2\delta/\partial \ln c^2$ are not difficult to derive, but they are omitted because of their length. We have already seen examples of Eqs. (24.62)–(24.68), above, for $m = 2$ and $m = 3$.

Complexes with $m = 4$ and 6

Because of the availability of the above general results, we can give a very brief discussion of the present cases.

For a *square complex* of four subunits, from Eq. (23.46),

$$\xi' = 1 + 4y^{-1}H + 4y^{-1}H^2 + 2y^{-2}H^2 + 4y^{-1}H^3 + H^4, \tag{24.70}$$

where H, y, and H_o are defined as in Eqs. (24.8) and (24.47). Thus

$$p_0^{\mathrm{e}} = \frac{1}{\xi'}, \quad p_1^{\mathrm{e}} = \frac{4y^{-1}H}{\xi'}, \quad p_2^{\mathrm{e}} = \frac{(4y^{-1} + 2y^{-2})H^2}{\xi'}, \quad p_3^{\mathrm{e}} = \frac{4y^{-1}H^3}{\xi'}, \quad p_4^{\mathrm{e}} = \frac{H^4}{\xi'}. \tag{24.71}$$

Then δ and θ are given in Eqs. (24.65) and (24.67). Because of the symmetry in Eq. (24.70), $\delta(\ln H)$ is symmetrical about $\delta = \frac{1}{2}$ (see Fig. 6-4). However, $\theta(\ln c)$ is not symmetrical in general. This can be seen very easily from the special case $K_1 = 0$, $y \to \infty$ using Eq. (15.25). Because the definition of H [Eq. (24.8)] is the same in all cases, c' and the range in H are as for $m = 2$ and $m = 3$ above. The Hill coefficient is [Eq. (22.40)]

$$n = 4\left(\frac{\partial \delta}{\partial \ln H}\right)_{\delta=1/2} = \overline{r^2} - \bar{r}^2 = \frac{4y(1+y)}{1+6y+y^2}. \qquad (24.72)$$

The induced conformation special case has $H_o K_2 c$ in place of H in Eq. (24.70), and thus is essentially the same as Eq. (23.46). In the MWC (concerted) special case, $y^{-1} \to 0$ and $p_1^e = p_2^e = p_3^e = 0$. Hence $\xi' = 1 + H^4$.

Turning now to a *tetrahedral complex* of four subunits, Eq. (23.53) leads to

$$\xi' = 1 + 4y^{-3/2}H + 6y^{-2}H^2 + 4y^{-3/2}H^3 + H^4, \qquad (24.73)$$

where H is defined as usual, but y is given by Eq. (24.3) (because $y_{12} = y_{21}$) and

$$H_o \equiv \left(\frac{y_{22}}{y_{11}}\right)^{3/2} \frac{j_2}{j_1}. \qquad (24.74)$$

From Eq. (24.73),

$$p_0^e = \frac{1}{\xi'}, \qquad p_1^e = \frac{4y^{-3/2}H}{\xi'}, \qquad p_2^e = \frac{6y^{-2}H^2}{\xi'},$$
$$p_3^e = \frac{4y^{-3/2}H^3}{\xi'}, \qquad p_4^e = \frac{H^4}{\xi'}. \qquad (24.75)$$

Then θ and δ follow from Eqs. (24.65) and (24.67). Again $\delta(\ln H)$ is symmetrical (Fig. 6-7) and $\theta(\ln c)$ is unsymmetrical. The Hill coefficient is [Eq. (22.47)]

$$n = 4\left(\frac{\partial \delta}{\partial \ln H}\right)_{\delta=1/2} = \overline{r^2} - \bar{r}^2 = \frac{4y^{1/2}(1+y^{3/2})}{3+4y^{1/2}+y^2}. \qquad (24.76)$$

In the induced conformation special case, replace H by $H_o K_2 c$ in Eq. (24.73) [compare Eq. (23.53)]. In the MWC special case, put $y^{-1} = 0$. Then $\xi' = 1 + H^4$.

The tetrahedral model is illustrated in Fig. 6-17, where the basic parameter choices are the same as in Fig. 6-16: $H_o = 0.55$, $K_1/K_2 = 1/10$, $K_2 c' = 1$, and the concentration variable is $K_2 c$. Also, we use $y = 1$ and 20 (interactions are more dense for a tetrahedron than for a dimer, so we take a smaller value of y). The θ_1, θ_2, $\delta(y = 1)$ and $\theta(y = 1)$ curves are the same as in Fig. 6-16; these are all independent of the value of m. With $y = 20$, $\delta(\ln c)$ and $\theta(\ln c)$ are significantly steeper than for $y = 1$. Both curves are unsymmetrical; this is not obvious in the $\theta(\ln c)$ case, but it can be verified numerically. The inset shows the true strength of the interactive cooperativity in this example, not disguised by ligand binding. These two curves, $\delta(\ln H)$, are the same as in Fig. 6-7 for $y = 1$ and 20. The Hill coefficient [Eq. (24.76)] for $y = 20$ is $n = 3.84$. Incidentally, if we mistakenly used the slope (0.384) for the $\theta(\ln c)$ curve for $y = 20$, at $\theta = \frac{1}{2}$, to obtain a Hill coefficient, we would find $4 \times 0.384 = 1.54$, which is very misleading.

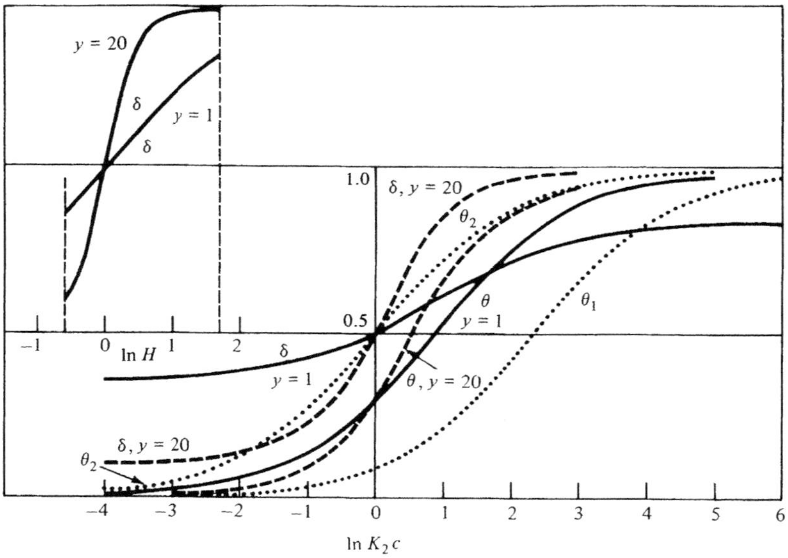

Fig. 6-17. Same as Fig. 6-16, except that this is a tetrahedral model. Again $H_o = 0.55$ and $K_1/K_2 = 1/10$.

Figure 6-18 presents a contrasting example, also for the tetrahedral model with $y = 1$ and 20. However, this example approximates the induced conformation special case: we take $K_1 = 0$ and $H_o = 0.1$. That is, L does not bind at all on state 1 subunits, and state 1 subunits are much more stable than state 2 subunits in the absence of L. The midpoint of the δ curve is at $K_2c' = 9$ whereas the midpoint of the θ_2 curve is at $K_2c = 1$. Hence most state 2 sites are occupied by L in the rising part of the δ curve. Consequently, because $\theta = \delta\theta_2$ in this case, θ and δ are not very different. Furthermore, these curves do not differ much from the inset (the same curves as in Fig. 6-17) because

$$H = (1 + K_2c)H_o$$

and $K_2c \gg 1$. The inset is shifted horizontally, relative to the main figure, by approximately $\ln H_o$.

Finally, we consider an *octahedral model, m* = 6. The expression for ξ' is the same as in Eq. (23.57), but with $\overline{K}c$ replaced by H. H has the usual definition; y is given by Eq. (24.3); and

$$H_o = \left(\frac{y_{22}}{y_{11}}\right)^2 \frac{j_2}{j_1}. \tag{24.77}$$

From Eq. (23.57), for ξ',

$$p_0^e = \frac{1}{\xi'}, \qquad p_1^e = \frac{6y^{-2}H}{\xi'}, \qquad p_2^e = \frac{(12y^{-3} + 3y^{-4})H^2}{\xi'}, \tag{24.78}$$

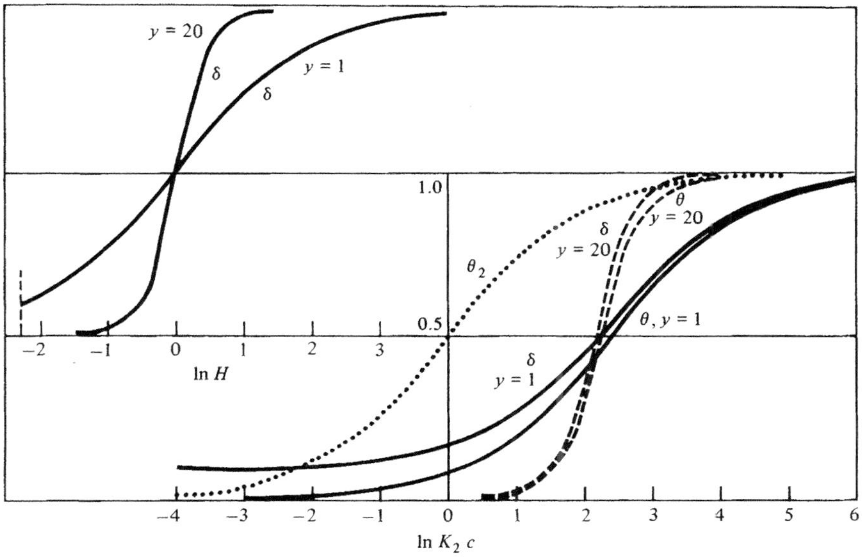

Fig. 6-18. A second numerical example (see Fig. 6-17) for a tetrahedral model. Here $K_1 = 0$ and $H_o = 0.1$. See text for details.

$$p_3^e = \frac{(8y^{-3} + 12y^{-4})H^3}{\xi'}, \qquad p_4^e = \frac{(3y^{-4} + 12y^{-3})H^4}{\xi'},$$
$$p_5^e = \frac{6y^{-2}H^5}{\xi'}, \qquad p_6^e = \frac{H^6}{\xi'}. \tag{24.78}$$

Expressions for δ and θ then follow from Eqs. (24.65) and (24.67). The function $\delta(\ln H)$ is symmetrical (Fig. 6-9) because of the symmetry in ξ'; $\theta(\ln c)$ is not symmetrical. The Hill coefficient $n(y)$ from $\delta(\ln H)$ at $\delta = \frac{1}{2}$ is given in Eq. (22.50).

In the induced conformation special case, $\overline{K}$ in Eq. (23.57) is replaced by H_oK_2. In the MWC case, $y^{-1} = 0$ so that

$$p_1^e = p_2^e = \cdots = p_5^e = 0, \qquad \xi' = 1 + H^6. \tag{24.79}$$

More Complicated Binding on Subunits

It is easy to generalize the above treatment to any model with more complicated binding on the subunits providing that, as above, the binding parameters for each subunit state or conformation (1 or 2) do not depend on r, the number of subunits out of m that are in state 2. In particular, the binding parameters cannot depend on intersubunit interactions, nor can the intersubunit interactions depend on the degree of binding [as they do in Eq. (24.45)].

To be more general, we rewrite Eq. (24.62) as

$$\xi = \sum_{r=0}^{m} a_r \xi_1^{m-r} \xi_2^r, \tag{24.80}$$

where $a_0 = 1$ and ξ_i is the *binding* grand partition for a subunit in state i ($i = 1, 2$). For example, in Eq. (24.62),

$$\xi_1 = 1 + K_1 c, \qquad \xi_2 = 1 + K_2 c. \tag{24.81}$$

The a_r are the terms of ξ *in the absence of ligands* in the various cases we have considered above. For example, for a tetrahedron [Eq. (24.73)],

$$\begin{gathered} a_0 = 1, \qquad a_1 = 4y^{-3/2}H_o, \qquad a_2 = 6y^{-2}H_o^2, \\ a_3 = 4y^{-3/2}H_o^3, \qquad a_4 = H_o^4. \end{gathered} \tag{24.82}$$

No matter how complicated are ξ_1 and ξ_2 (see below for *not* very complicated examples), we have the general properties

$$p_r^e = \frac{a_r \xi_1^{m-r} \xi_2^r}{\xi}, \tag{24.83}$$

$$1 - \delta = \sum_r \left(\frac{m-r}{m}\right) p_r^e, \tag{24-84}$$

$$\delta = \sum_r \frac{r}{m} p_r^e, \tag{24.85}$$

$$H \equiv \frac{\xi_2}{\xi_1} H_o. \tag{24.86}$$

Equations like Eqs. (24.73) and (24.75) (for a tetrahedron) are unchanged, but H has a more general meaning.

The most important special case [see Eqs. (14.19)–(14.27), where $y = 1$] is probably the following: each subunit has two independent sites, one for binding ligand A (concentration c_A), and one for binding ligand B (concentration c_B). The binding constants are K_{A1} and K_{B1} on a state 1 subunit and K_{A2} and K_{B2} on a state 2 subunit. In this case, then,

$$\xi_1 = (1 + K_{A1} c_A)(1 + K_{B1} c_B), \tag{24.87}$$

$$\xi_2 = (1 + K_{A2} c_A)(1 + K_{B2} c_B). \tag{24.88}$$

The amount of binding of A and B are then

$$\begin{aligned} \theta_A &= \frac{\bar{s}_A}{m} = \frac{1}{m} c_A \frac{\partial \ln \xi}{\partial c_A} \\ &= (1 - \delta)\theta_{A1} + \delta \theta_{A2}, \end{aligned} \tag{24.89}$$

$$\theta_{A1} = \frac{K_{A1} c_A}{1 + K_{A1} c_A}, \tag{24.90}$$

$$\theta_{A2} = \frac{K_{A2}c_A}{1 + K_{A2}c_A}, \tag{24.91}$$

with δ given by Eq. (24.85). There are analogous equations for the binding of B, with subscript A above replaced by B. For example, for a dimer, Eqs. (24.7), (24.9), and (24.10) all hold, but

$$H = \frac{\xi_2}{\xi_1}H_o = \frac{(1 + K_{A2}c_A)(1 + K_{B2}c_B)}{(1 + K_{A1}c_A)(1 + K_{B1}c_B)}H_o. \tag{24.92}$$

With δ determined by Eq. (24.10), Eq. (24.89) gives θ_A.

As Eq. (24.92) shows, the $1 \rightleftarrows 2$ equilibrium is shifted toward state 2 by binding of A or B on state 2 (K_{A2}, K_{B2}) and it is shifted toward state 1 by binding of A or B on state 1 (K_{A1}, K_{B1}). An important special case, mentioned in Section 14, arises when, say, A binds only on state 1 subunits ($K_{A2} = 0$) and B binds only on state 2 subunits ($K_{B1} = 0$). Then the direction of the $1 \rightleftarrows 2$ equilibrium can be regulated by varying c_A and c_B: A binding shifts towards 1 and B binding shifts towards 2. The expression for H in this case is

$$H = \frac{(1 + K_{B2}c_B)}{(1 + K_{A1}c_A)}H_o. \tag{24.93}$$

The sensitivity of $1 \rightleftarrows 2$ to changes in c_A and c_B is, of course, enhanced here (compared to Section 14, where $y = 1$) by the steepened $\delta(\ln H)$ curves when $y > 1$ (positive cooperativity), as seen in the insets of Figs. 6-16, 6-17, and 6-18.

Competitive binding of two ligands, A and B, on a single site on each subunit state is a closely related example [see Eqs. (11.3) and (11.4)]. In this case

$$\xi_1 = 1 + K_{A1}c_A + K_{B1}c_B, \tag{24.94}$$

$$\xi_2 = 1 + K_{A2}c_A + K_{B2}c_B. \tag{24.95}$$

The general Eqs. (24-83)–(24.86) apply, as does Eq. (24.89) for θ_A (and similarly for θ_B), but Eqs. (24.90) and (24.91) are replaced by

$$\theta_{A1} = \frac{K_{A1}c_A}{1 + K_{A1}c_A + K_{B1}c_B}, \tag{24.96}$$

$$\theta_{A2} = \frac{K_{A2}c_A}{1 + K_{A2}c_A + K_{B2}c_B} \tag{24.97}$$

(and similarly for θ_{B1} and θ_{B2}). In the special case $K_{A2} = 0$, $K_{B1} = 0$ (see above), this model [Eqs. (24.94) and (24.95)] becomes the same as in the previous case [Eqs. (24.87) and (24.88)]; in fact, the binding is no longer competitive.

A more general hybrid case encompasses both of the above two special cases (A and B each have a site; A and B compete for the same site). In the hybrid case, there are two sites on a subunit, near each other, one for A and one for B; when both sites are occupied, the ligand molecules A and B interact with each other. In state 1, the interaction parameter for the A–B interaction is $y_1 \equiv e^{-w_1/kT}$ and, in state 2, it is $y_2 \equiv e^{-w_2/kT}$. As in Eq.(22.6), the binding grand partition functions are

$$\xi_1 = 1 + K_{A1}c_A + K_{B1}c_B + y_1 K_{A1}c_A K_{B1}c_B , \tag{24.98}$$

$$\xi_2 = 1 + K_{A2}c_A + K_{B2}c_B + y_2 K_{A2}c_A K_{B2}c_B . \tag{24.99}$$

The two special cases above correspond first, to $y_1 = y_2 = 1$ (independent binding on the two sites) and, second, to $y_1 = y_2 = 0$ (strong repulsion between A and B allows only one of the two sites to be occupied at a time). The second case ($y_1 = y_2 = 0$) is operationally indistinguishable from competitive binding of A and B on a single site.

The basic equations are again Eqs. (24.83)–(24.86) and (24.89) (similarly for θ_B), but θ_{A1}, etc., are different. As in Eq. (22.8),

$$\theta_{A1} = \frac{K_{A1}c_A + y_1 K_{A1}c_A K_{B1}c_B}{\xi_1}, \tag{24.100}$$

$$\theta_{A2} = \frac{K_{A2}c_A + y_2 K_{A2}c_A K_{B2}c_B}{\xi_2}, \tag{24.101}$$

$$\theta_{B1} = \frac{K_{B1}c_B + y_1 K_{A1}c_A K_{B1}c_B}{\xi_1}, \tag{24.102}$$

$$\theta_{B2} = \frac{K_{B2}c_B + y_2 K_{A2}c_A K_{B2}c_B}{\xi_2}. \tag{24.103}$$

All four of these can be put in simple Langmuir form, as in Eq. (22.9) (there are four different Φs).

Complexes Modulated by an Electric Field

Here we generalize Section 13 by including the possibility of intersubunit interactions, with resultant cooperativity. We consider the same class of protein complexes, with m two-conformation subunits, as previously treated in this section. Now, however, a given complex is in a membrane and the $1 \rightleftarrows 2$ equilibrium of the subunits of the complex can be shifted by variations in the membrane potential; also, the equilibrium may be modified by intersubunit interactions. The treatment is formally the same as above, in this section, except that the ligand grand partition functions ξ_1 and ξ_2 [see Eqs. (24.80)–(24.86)] are replaced by corresponding Boltzmann factors that take into account the dependence of the subunit free energy on membrane potential.

Let us choose a tetrahedral model ($m = 4$) to illustrate the needed modifications. In the absence of ligand binding and at zero membrane potential ($\psi \equiv \psi_{in} - \psi_{out} = 0$, where the symbol ψ represents the electrostatic potential), we have for the grand partition function of the complex,

$$\xi = y_{11}^6 j_1^4 + 4y_{11}^3 y_{12}^3 j_1^3 j_2 + 6y_{11} y_{12}^4 y_{22} j_1^2 j_2^2 + 4y_{12}^3 y_{22}^3 j_1 j_2^3 + y_{22}^6 j_2^4, \tag{24.104}$$

or, upon factoring out $y_{11}^6 j_1^4$,

$$\xi = 1 + 4y^{-3/2}H_o + 6y^{-2}H_o^2 + 4y^{-3/2}H_o^3 + H_o^4, \tag{24.105}$$

with y defined by Eq. (24.3) and H_o by Eq. (24.74). This is the familiar type of starting point used repeatedly above for several values of m.

In Section 13 it was shown that a subunit in state i ($i = 1, 2$), with partition function j_i in the absence of an electric field, would have a partition function $j_i e^{\alpha_i \mathscr{E}^2/2kT}$ in the presence of a field $\mathscr{E}$ if the polarizability of the subunit is α_i. The quantity $-\alpha_i \mathscr{E}^2/2$ is a contribution to the energy of the subunit. Here we take a more general and somewhat empirical approach in order to deal with an arbitrary type of subunit: we replace $\alpha_i \mathscr{E}^2/2kT$ above by

$$\epsilon_i \psi + \gamma_i \psi^2 \qquad (i = 1, 2), \tag{24.106}$$

where ψ is the membrane potential and ϵ_i and γ_i are constants characteristic of state i of the subunit. In the simplest case, the electrostatic potential varies linearly across the membrane and $\psi \sim \mathscr{E}$. Thus the ψ^2 term in Eq. (24.106) would include the polarizability effect already mentioned. The linear term in ψ could arise, for example, from a fixed charge on the subunit in state i. A more detailed discussion of the source, at the molecular level, of possible contributions to both terms is outside the scope of this book but is available elsewhere.[8]

When $\psi \neq 0$, we replace j_1 and j_2 in Eq. (24.104) by

$$j_1 e^{\epsilon_1 \psi + \gamma_1 \psi^2} \qquad \text{and} \qquad j_2 e^{\epsilon_2 \psi + \gamma_2 \psi^2}, \tag{24.107}$$

where we have now included the new Boltzmann factors associated with ψ. The expressions in Eq. (24.107) are the analogues of $j_1(1 + K_1 c)$ and $j_2(1 + K_2 c)$, used previously in this section. If we include these ψ factors in Eq. (24.104), and then factor out

$$y_{11}^6 (j_1 e^{\epsilon_1 \psi + \gamma_1 \psi^2})^4,$$

we obtain

$$\xi' = 1 + 4y^{-3/2}H + 6y^{-2}H^2 + 4y^{-3/2}H^3 + H^4, \tag{24.108}$$

just as in Eq. (24.73), but here H is defined by

$$H = \left(\frac{e^{\epsilon_2 \psi + \gamma_2 \psi^2}}{e^{\epsilon_1 \psi + \gamma_1 \psi^2}} \right) H_o$$

$$= e^{\epsilon\psi + \gamma\psi^2} H_o, \tag{24.109}$$

$$\epsilon \equiv \epsilon_2 - \epsilon_1, \qquad \gamma \equiv \gamma_2 - \gamma_1. \tag{24.110}$$

It is clear from Eq. (24.109) that the significant quantities are the *differences* ϵ and γ, not the separate values ϵ_i and γ_i. The above definition of H would in fact be the same for a model with any m (however, H_o depends on the model). For any m, H^m is the equilibrium constant for the process all-1 $\rightleftarrows$ all-2 in the presence of a membrane potential ψ.

Whereas previously in this section the 1 $\rightleftarrows$ 2 subunit equilibrium was modulated by the ligand concentration, here the equilibrium is modulated by the membrane potential. The fraction δ of subunits in state 2 is given by [Eq. (24.65)]

$$\delta = \frac{1}{4}p_1^e + \frac{1}{2}p_2^e + \frac{3}{4}p_3^e + p_4^e$$
$$= \frac{y^{-3/2}H + 3y^{-2}H^2 + 3y^{-3/2}H^3 + H^4}{\xi'}. \tag{24.111}$$

This is, of course, the same mathematical function $\delta(H)$ as above for a tetrahedron [Eqs. (24.75) and (24.76)]; however, the definition and physical significance of H have changed.

In the special case $y = 1$,

$$\xi' = (1 + H)^4, \qquad \delta = \frac{H}{1+H}, \qquad 1 - \delta = \frac{1}{1+H}, \tag{24.112}$$

$$p_0^e = (1-\delta)^4, \quad p_1^e = 4\delta(1-\delta)^3, \quad p_2^e = 6\delta^2(1-\delta)^2,$$
$$p_3^e = 4\delta^3(1-\delta), \quad p_4^e = \delta^4. \tag{24.113}$$

Of course, with $y = 1$, any model with four identical subunits (square, rectangle, tetrahedron, etc.) would have these same properties. Such models would differ, though, in the way in which H_o is defined with respect to subunit interactions [compare, for example, Eqs. (24.47) for a square with Eq. (24.74) for a tetrahedron].

As a numerical example, we consider the potassium channel of the giant squid axon membrane. This channel probably contains four protein subunits. Significant cooperativity among these subunits is apparently ruled out by certain transient experiments that we shall not discuss here. Hence, for this example, we take $y = 1$ and use Eqs. (24.112) and (24.113). Because $y = 1$, we need not commit ourselves to any particular geometry for the tetramer. The K^+ channel is open for transport of ions only when all four subunits are in state 2; that is, at equilibrium, the probability of the channel being open is $p_4^e = \delta^4$.

If we reinterpret the experimental observations of Hodgkin and Huxley[9] in terms of the present model with four protein subunits (as first suggested[10] in 1970) the fraction δ of subunits in state 2, as a function of ψ (in mV), is as shown in Fig. 6-19. The rise in δ with ψ is fairly steep, but this is not due to cooperativity ($y = 1$). Then, using

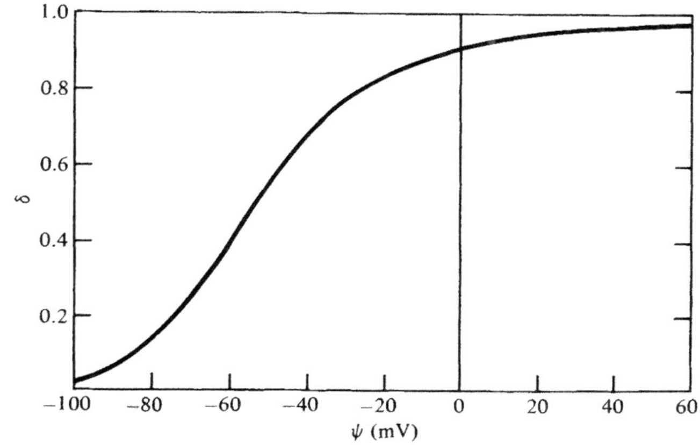

Fig. 6-19. Fraction of subunits in K^+ channel in state 2, from Hodgkin and Huxley[9] (see text), as function of membrane potential.

$$\frac{\delta}{1-\delta} = H = e^{\epsilon\psi + \gamma\psi^2} H_0, \tag{24.114}$$

we can obtain ln H as a function of ψ, as shown in Fig. 6-20. This shows how the equilibrium constant H must depend on ψ in order to account for $\delta(\psi)$ in Fig. 6-19. The curve in Fig. 6-20 can be fitted quite closely by the quadratic function

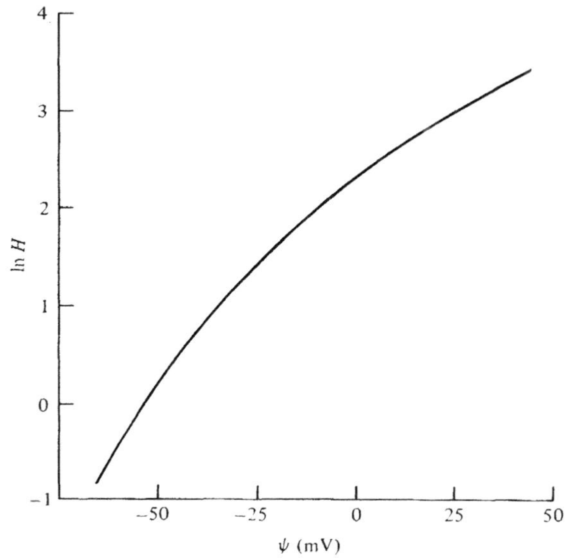

Fig. 6-20. Dependence of equilibrium constant H, for subunit conformational change, on membrane potential. This curve is a consequence of Fig. 6-19 [see Eq. (24.114)].

$$\ln H = \ln H_o + \epsilon\psi + \gamma\psi^2 \tag{24.115}$$

with parameter values

$$\begin{aligned} \ln H_o &= 2.33, \qquad H_o = 10.3, \\ \epsilon &= 3.24 \times 10^{-2}\,(\text{mV})^{-1}, \qquad \gamma = -2.08 \times 10^{-4}\,(\text{mV})^{-2}. \end{aligned} \tag{24.116}$$

$H_o = 10.3$ is the equilibrium constant for $1 \rightleftarrows 2$ when $\psi = 0$; this large value favors state 2 at $\psi = 0$ ($\delta = 0.91$). At the rest potential, $\psi = -65$ mV, Eq. (24.115) gives $\ln H = -0.65$, $H = 0.52$, and $\delta = 0.34$. The values of $p_4^c = \delta^4$ (channel open) corresponding to $\delta = 0.91$ and $\delta = 0.34$ are 0.69 and 0.014, respectively.

Incidentally, if there is ligand binding (e.g., Ca^{+2}) as well as the membrane potential effects just examined, H is generalized in an obvious way:

$$H = e^{\epsilon\psi + \gamma\psi^2}\left(\frac{1 + K_2 c}{1 + K_1 c}\right) H_o. \tag{24.117}$$

This assumes that K_1 and K_2 are independent of ψ. There are now two ways to shift the values of H and δ. The calculation of δ proceeds as before [Eq. (24.111)]. The amount of ligand binding is

$$\theta = (1 - \delta)\theta_1 + \delta\theta_2, \tag{24.118}$$

as usual.

25. Binding of Two Ligands with Induced Conformational Changes

This section is a generalization of Section 23. We study a complex of m identical subunits, each of which has one site for the competitive binding of either L_1 (concentration c_1) or L_2 (concentration c_2). Alternatively, a subunit may have a site for each of L_1 and L_2, but only one site can be occupied at a time. The subunit is in conformation or state 0 when the site is empty, in state 1 when the site is occupied by L_1, and in state 2 when the site is occupied by L_2. These state changes, $0 \to 1$ and $0 \to 2$, are immediately induced by binding L_1 or L_2, respectively, but otherwise the state changes do not occur. The inverses ($1 \to 0$, $2 \to 0$) are also assumed to be effectively single-step processes. There are intersubunit nearest-neighbor pair interactions in the complex that depend on the states (conformations) of the subunits in the pairs. A pair ij ($i, j = 0, 1, 2$) in a complex has an interaction free energy (potential of mean force relative to infinite separation) w_{ij}. We define $y_{ij} = e^{-w_{ij}/kT}$, as usual. We assume $y_{ij} = y_{ji}$ for simplicity. The new feature here is that each subunit has three possible states: $i, j = 0, 1, 2$. There are six kinds of pair interactions: 00, 01, 02, 11, 12, 22. The total number of states for a complex with m subunits is 3^m. This is also the number of terms in the grand partition function.

We illustrate this problem with two special cases: a dimer ($m = 2$); and a square tetramer ($m = 4$). The latter model has been used to interpret the binding of Ca^{2+} and H^+ on the Ca-ATPase of sarcoplasmic reticulum.[11] This will provide us with a realistic numerical example.

Binding of Two Ligands on a Dimer ($m = 2$)

We begin as in Eq. (23.7). The grand partition function is

$$\xi = y_{00} + 2y_{01}K_1^\circ c_1 + 2y_{02}K_2^\circ c_2 + 2y_{12}K_1^\circ c_1 K_2^\circ c_2 + y_{11}(K_1^\circ c_1)^2 + y_{22}(K_2^\circ c_2)^2. \tag{25.1}$$

There are $3^2 = 9$ terms in ξ. The nature of the different states of the dimer represented here is obvious from the subscripts on the y_{ij} factors. The constants K_1° and K_2° are unperturbed binding constants for L_1 and L_2 on an isolated subunit [see Eq. (23.4)]. The associated standard free energy changes relate to the induced conformational changes as well as to the binding. If we factor out y_{00} from Eq. (25.1), we obtain

$$\xi = 1 + 2K_1c_1 + 2K_2c_2 + 2\left(\frac{yz}{u}\right)^{1/2} K_1c_1K_2c_2 + y(K_1c_1)^2 + z(K_2c_2)^2, \tag{25.2}$$

where

$$K_1 = \frac{y_{01}K_1^\circ}{y_{00}}, \qquad K_2 = \frac{y_{02}K_2^\circ}{y_{00}}, \tag{25.3}$$

$$y \equiv \frac{y_{00}y_{11}}{y_{01}^2}, \qquad z \equiv \frac{y_{00}y_{22}}{y_{02}^2}, \qquad u = \frac{y_{11}y_{22}}{y_{12}^2}. \tag{25.4}$$

The constants K_1 and K_2 are operational initial binding constants, as in Eq. (23.9). Three interaction parameters [Eq. (25.4)] are needed here. These are equilibrium constants for the processes

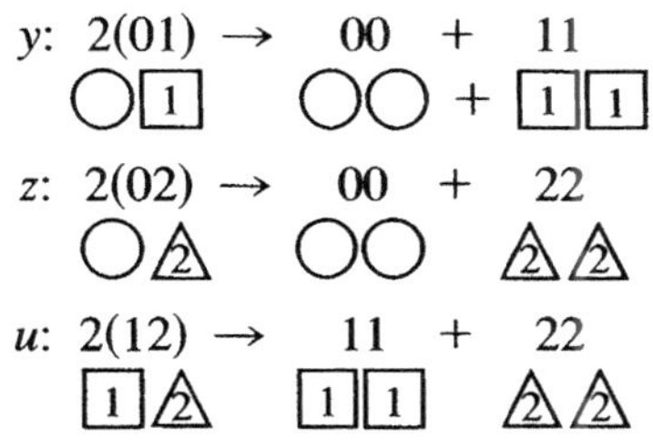

(25.5)

as in Eq. (23.12). Here we have represented the molecular shapes of states 0, 1, 2 schematically by circle, square, triangle. Values of $y, z, u > 1$ would indicate greater stability of "pure" dimer states compared to mixed states; the opposite is true if $y, z, u < 1$. However, there is no reason why all three parameters need be of the same type (>1 or <1).

The next step is analogous to Eqs. (23.19) and (23.20). We define new binding constants $\overline{K}_1$ and $\overline{K}_2$ by

$$\overline{K}_1^2 \equiv yK_1^2, \qquad \overline{K}_2^2 \equiv zK_2^2. \tag{25.6}$$

The two quantities in Eq. (25.6) are also operational equilibrium constants, in this case for the processes

$$\begin{aligned} &\overline{K}_1^2\text{: } 00 + 2\text{L}_1 \rightarrow 11, \\ &\overline{K}_2^2\text{: } 00 + 2\text{L}_2 \rightarrow 22. \end{aligned} \tag{25.7}$$

Using the new constants, ξ becomes

$$\begin{aligned} \xi = &\underset{00}{1} + \underset{01}{2y^{-1/2}\overline{K}_1c_1} + \underset{02}{2z^{-1/2}\overline{K}_2c_2} + \underset{12}{2u^{-1/2}\overline{K}_1c_1\overline{K}_2c_2} \\ &+ \underset{11}{(\overline{K}_1c_1)^2} + \underset{22}{(\overline{K}_2c_2)^2}. \end{aligned} \tag{25.8}$$

If y, z, u are all very large, the mixed states have relatively small populations. Then ξ degenerates into

$$\xi = 1 + (\overline{K}_1c_1)^2 + (\overline{K}_2c_2)^2, \tag{25.9}$$

which is another example of an A. V. Hill binding model.

To simplify the notation, we rewrite Eq. (25.8) as

$$\xi = \underset{00}{1} + \underset{01}{2y^{-1/2}a} + \underset{02}{2z^{-1/2}b} + \underset{12}{2u^{-1/2}ab} + \underset{11}{a^2} + \underset{22}{b^2}, \tag{25.10}$$

where the new dimensionless concentration variables are

$$a \equiv \overline{K}_1c_1, \qquad b \equiv \overline{K}_2c_2. \tag{25.11}$$

These variables are the analogues of $\bar{x}$ in Eq. (23.24). If $y = z = u = 1$ (no cooperativity),

$$\xi = (1 + a + b)^2. \tag{25.12}$$

There is simple competitive binding of L_1 and L_2 on each subunit in this case. The same form of ξ follows from Eq. (25.2). This agreement occurs because $\overline{K}_1 = K_1$ and $\overline{K}_2 = K_2$ when $y = z = u = 1$.

The probabilities of the various states, at equilibrium, are determined by the terms in ξ. From these, the fraction of subunits with L_1 bound (i.e., the fraction of subunits in state 1) is

$$\theta_1 = \frac{1}{2}\frac{\partial \ln \xi}{\partial \ln a} = \frac{a(y^{-1/2} + u^{-1/2}b + a)}{\xi}. \tag{25.13}$$

Similarly, for L_2 bound,

$$\theta_2 = \frac{1}{2}\frac{\partial \ln \xi}{\partial \ln b} = \frac{b(z^{-1/2} + u^{-1/2}a + b)}{\xi}. \tag{25.14}$$

The fraction of subunits in state 0 (site empty) is $1 - \theta_1 - \theta_2$. When $y = z = u = 1$, these simplify to

$$\theta_1 = \frac{a}{1 + a + b}, \qquad \theta_2 = \frac{b}{1 + a + b}, \tag{25.15}$$

as in Eq. (11.4).

Binding of Two Ligands on a Square ($m = 4$)

The grand partition function in this case has $3^4 = 81$ terms:

$$\begin{aligned} \xi = {} & \underset{\substack{0\ \ 0\\ 0\ \ 0}}{y_{00}^4} + \underset{\substack{1\ \ 0\\ 0\ \ 0}}{4y_{00}^2 y_{01}^2 K_1^\circ c_1} + \underset{\substack{2\ \ 0\\ 0\ \ 0}}{4y_{00}^2 y_{02}^2 K_2^\circ c_2} \\ & + \underset{\substack{1\ \ 2\\ 0\ \ 0}}{8y_{00}y_{01}y_{02}y_{12} K_1^\circ c_1 K_2^\circ c_2} + \cdots + \underset{\substack{1\ \ 1\\ 1\ \ 1}}{y_{11}^4 (K_1^\circ c_1)^4} + \underset{\substack{2\ \ 2\\ 2\ \ 2}}{y_{22}^4 (K_2^\circ c_2)^4}. \end{aligned} \tag{25.16}$$

Only 19 of the 81 terms are shown explicitly here. The interested reader should work with the complete equation. If we divide each term by y_{00}^4 and define the initial binding constants

$$K_1 \equiv \frac{y_{01}^2 K_1^\circ}{y_{00}^2}, \qquad K_2 \equiv \frac{y_{02}^2 K_2^\circ}{y_{00}^2}, \tag{25.17}$$

then

$$\xi = 1 + 4K_1c_1 + 4K_2c_2 + \cdots + y^4(K_1c_1)^4 + z^4(K_2c_2)^4. \tag{25.18}$$

The cooperativity parameters are defined in Eq. (25.4). Finally, we define

$$\begin{aligned} \overline{K}_1 &\equiv yK_1, & \overline{K}_2 &\equiv zK_2, \\ a &\equiv \overline{K}_1c_1, & b &\equiv \overline{K}_2c_2. \end{aligned} \tag{25.19}$$

$\overline{K}_1^4$ is the equilibrium constant for all-0 + $4L_1 \rightarrow$ all-1, and similarly for $\overline{K}_2^4$. Then the complete ξ, the analogue of Eq. (25.10) for a dimer, is

$$\xi = A_0 + 4y^{-1}A_1a + 4y^{-1}A_2a^2 + 2y^{-2}A_3a^2 + 4y^{-1}A_4a^3 + a^4, \tag{25.20}$$

where

$$A_0 = 1 + 4z^{-1}b + 4z^{-1}b^2 + 2z^{-2}b^2 + 4z^{-1}b^3 + b^4 \tag{25.21}$$

$$= (1 + z^{-1}b)^2 + 2z^{-1}b(1 + b)^2 + z^{-2}b^2(1 + zb)^2, \tag{25.22}$$

$$A_1 = \left[1 + \left(\frac{y}{zu}\right)^{1/2} b\right]^2 + z^{-1}b\left[1 + \left(\frac{yz}{u}\right)^{1/2} b\right]^2, \tag{25.23}$$

$$A_2 = \left[1 + \left(\frac{y}{zu}\right)^{1/2} b\right] + \left(\frac{y}{zu}\right)^{1/2} b\left[1 + \left(\frac{yz}{u}\right)^{1/2} b\right], \tag{25.24}$$

$$A_3 = (1 + yu^{-1}b)^2, \qquad A_4 = 1 + yu^{-1}b. \tag{25.25}$$

Equation (25.20) has been written in a form that emphasizes the binding of L_1 (even though the roles of L_1 and L_2 in the model and in ξ are equivalent). That is, Eq. (25.20) is a polynomial in a with coefficients that are functions of b. This is especially convenient if, for example, one is interested in calculating $\theta_1(c_1)$ or $\theta_1(\ln c_1)$ for L_1 binding at a fixed concentration c_2 of L_2. This procedure is, in fact, used in the next subsection. An equivalent expression for ξ as a polynomial in b is obtained simply by exchanging a and b and, also, y and z throughout Eqs. (25.20)–(25.25).

It should be noted that Eqs. (25.20) and (25.21) are very closely related to Eq. (23.46). If $c_2 = 0$ (i.e., $b = 0$), all of the $A_i = 1$ and Eqs. (23.46) and (25.20) become identical. If $c_1 = 0$ (i.e., $a = 0$), one has the same relation expressed in terms of b and z: $\xi = A_0$. If $y = z = u = 1$ (no cooperativity),

$$\begin{aligned} A_0 &= (1 + b)^4, \qquad A_1 = (1 + b)^3, \\ A_2 &= A_3 = (1 + b)^2, \qquad A_4 = 1 + b, \end{aligned} \tag{25.26}$$

and then $\xi = (1 + a + b)^4$. Note that when $y = z = u = 1$, there is no longer a distinction between $\overline{K}_1$ and K_1 and between $\overline{K}_2$ and K_2.

Another special case is important. If $c_2 = \text{constant} \to \infty$ (i.e., $b \to \infty$), the terms of highest power in b give for the A_i:

$$\begin{aligned} A_0 &= b^4, \qquad A_1 = \frac{yb^3}{u}, \qquad A_2 = \frac{yb^2}{u}, \\ A_3 &= \frac{y^2b^2}{u^2}, \qquad A_4 = \frac{yb^4}{u}. \end{aligned} \tag{25.27}$$

Then Eq. (25.20) becomes

$$\begin{aligned} \xi = b^4 &+ 4u^{-1}b^3a + 4u^{-1}b^2a^2 + 2u^{-2}b^2a^2 \\ &+ 4u^{-1}ba^3 + a^4. \end{aligned} \tag{25.28}$$

This is again essentially Eq. (23.46) but with $a/b = (\overline{K}_1/\overline{K}_2c_2)c_1$ (c_2 = constant) in place of $\overline{K}c$ and u in place of y. Physically, Eq. (25.28) relates to the binding of L_1 on complexes that are initially saturated with L_2. As c_1 and a/b increase (c_2 = constant), L_2 is displaced by L_1. Hence u [Eq. (25.4)] is the natural cooperativity parameter.

From Eq. (25.20), the probability, p_s^e that a complex has s molecules of L_1 bound (i.e., the complex has s subunits in state 1) is

$$\begin{aligned} p_0^e &= \frac{A_0}{\xi}, \qquad p_1^e = \frac{4y^{-1}A_1a}{\xi}, \qquad p_2^e = \frac{4y^{-1}A_2a^2 + 2y^{-2}A_3a^2}{\xi}, \\ p_3^e &= \frac{4y^{-1}A_4a^3}{\xi}, \qquad p_4^e = \frac{a^4}{\xi}. \end{aligned} \tag{25.29}$$

Then the fraction of all subunits with L_1 bound is

$$\theta_1 = \frac{1}{4}\frac{\partial \ln \xi}{\partial \ln a} = \frac{1}{4}p_1^e + \frac{1}{2}p_2^e + \frac{3}{4}p_3^e + p_4^e. \tag{25.30}$$

The corresponding expression for θ_2 is obtained by exchanging a and b and, also, y and z in the θ_1 equation.

In general, $\theta_1(\ln c_1)$ at constant c_2, or $\theta_2(\ln c_2)$ at constant c_1, from Eq. (25.30), are not symmetrical functions. They are symmetrical, however, in the special cases mentioned above: $b = 0$; $a = 0$; $b \to \infty$; and, also, by symmetry, $a \to \infty$.

Numerical Example

This model has been used successfully to fit equilibrium data on the binding of Ca^{2+} (L_1 at c_1), at constant pH (H^+ is L_2 at c_2), on the Ca-ATPase complex of sarcoplasmic reticulum.[11] That is, in using this model, Ca^{2+} and H^+ are assumed to compete for a binding site on each subunit of a square complex that has nearest-neighbor interactions between subunits.

We shall not repeat details of the parameter selection process. Units used are c_1 and c_2 in M, $\overline{K}_1$ and $\overline{K}_2$ in M^{-1}. The final values chosen were

$$\begin{gathered} \log \overline{K}_1 = 6.95, \qquad \log \overline{K}_2 = 7.86, \\ y = 25, \qquad z = \tfrac{1}{5}, \qquad u = 5. \end{gathered} \tag{25.31}$$

Note that proton binding occurs with negative cooperativity ($z < 1$). Using Eqs. (25.19), the initial binding constants are

$$\log K_1 = 5.55, \qquad \log K_2 = 8.56. \tag{25.32}$$

The Ca^{2+} binding data, θ_1 as a function of pCa $= -\log c_1$ at constant pH $= -\log c_2$, are presented in Fig. 6-21. The data appear to be essentially symmetrical about $\theta_1 = \frac{1}{2}$. Although θ_1, as calculated from Eq. (25.30), is not strictly symmetrical, for the parameters in Eq. (25.31) calculated curves are practically symmetrical: deviations (of order 0.3% or less) can be detected only numerically. Therefore the data are analyzed below as if $\theta_1 = \frac{1}{2}$ is a center of symmetry.

If we put $\theta_1 = \frac{1}{2}$ in Eq. (25.30), and call a_o the value of a at which $\theta_1 = \frac{1}{2}$ for a given b, then

$$a_o^4 + 2y^{-1}A_4 a_o^3 - 2y^{-1}A_1 a_o - A_0 = 0 \tag{25.33}$$

is a quartic equation that determines $a_o(b)$ for given y, z, u. Using the definitions of a and b in Eq. (25.19) and the parameter values in Eq. (25.31), the function $a_o(b)$ from Eq. (25.33) can be plotted in the form pCa as a function of pH at $\theta_1 = \frac{1}{2}$. This is the solid curve in Fig. 6-22. The solid straight lines in this figure correspond to the asymptotic solutions of Eq. (25.33) in the two limits $c_2 \to 0$ (large pH) and $c_2 \to \infty$ (small pH). These are the special cases $b \to 0$ and $b \to \infty$ already mentioned above. In these

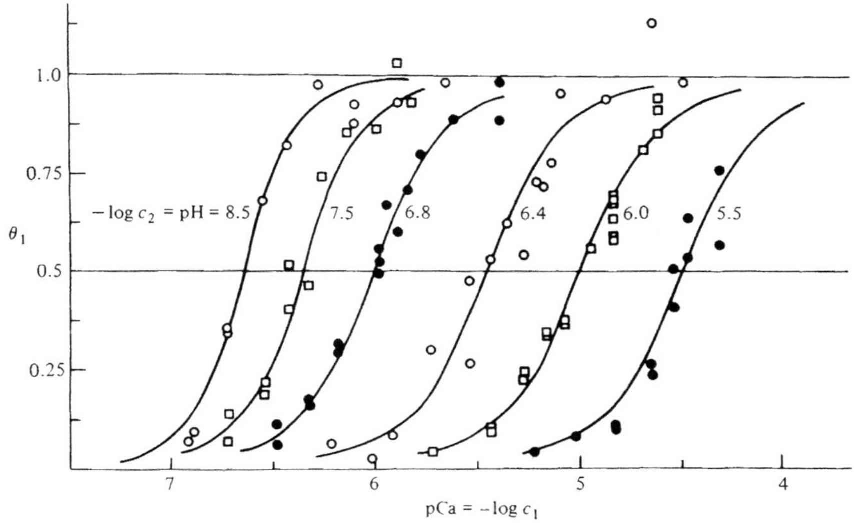

Fig. 6.21. Experimental data for binding of Ca^{2+} on the Ca-ATPase complex of sarcoplasmic reticulum, at different pH values.

two special cases, the solutions of Eq. (25.33) (for $\theta_1 = \frac{1}{2}$) are $a_o = 1$ and $a_o = b$. These two lines intersect at $a_o = 1$, $b = 1$, or

$$\begin{aligned} \overline{K}_1 c_1 &= 1, \qquad \log \overline{K}_1 = -\log c_1 = \text{pCa} = 6.95, \\ \overline{K}_2 c_2 &= 1, \qquad \log \overline{K}_2 = -\log c_2 = \text{pH} = 7.86. \end{aligned} \tag{25.34}$$

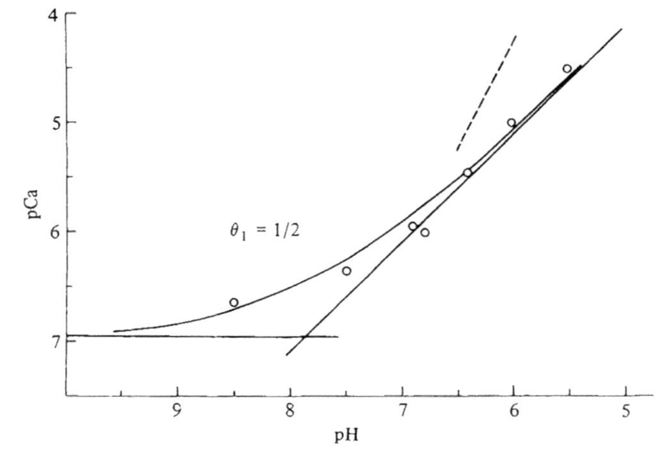

Fig. 6-22. The solid curve is a plot of pCa as a function of pH at half-saturation with Ca^{2+} ($\theta_1 = \frac{1}{2}$), from Eq. (25.33). The solid lines show the asymptotic behavior of this function. The points are experimental. The dashed line shows the asymptotic slope at low pH if two bound H^+ were replaced by one Ca^{2+}. See text for further details.

The slope of unity at low pH in Fig. 6-22 is a consequence of the assumed 1 : 1 replacement of bound H^+ by Ca^{2+}. If two bound H^+ were replaced by one Ca^{2+}, the corresponding slope would be 2, as shown by the dashed line in Fig. 6-22.

A smooth curve (not shown) drawn through the data in Fig. 6-21, near $\theta_1 = \frac{1}{2}$, for each pH leads to an experimental value of pCa at $\theta_1 = \frac{1}{2}$ for that pH. These pCa values appear as points in Fig. 6-22 and are included in Table 6-1 (column 2). (The extra point at pH 6.9 in Fig. 6-22 was obtained from another source.[11]) Column 3 in Table 6-1 gives the corresponding theoretical pCa values from the solid curve in Fig. 6-22 [based on Eq. (25.33)]. The points in Fig. 6-22 at low pH confirm the assumed 1 : 1 competition between Ca^{2+} and H^+.

Using the parameters in Eq. (25.31), theoretical $\theta_1(c_1)$ curves can be calculated from Eq. (25.30) for the pH values in Fig. 6-21. These are the smooth curves in Fig. 6-21, all of which show positive cooperativity. However, to observe the fit in slope at $\theta_1 = \frac{1}{2}$, these curves have been shifted slightly, horizontally, to give $\theta_1 = \frac{1}{2}$ at the *experimental* pCa value in Table 6-1. The amount of the shift at each pH can be seen by comparing columns 2 and 3 in Table 6-1 (or by comparing points with the solid curve in Fig. 6-22). The fit of slopes at $\theta_1 = \frac{1}{2}$ in Fig. 6-21 is seen to be excellent. The theoretical slopes, $\partial\theta_1/\partial \ln c_1$ at $\theta_1 = \frac{1}{2}$, are included in Table 6-1 (column 4). These can also be considered to be the experimental slopes because of the fit just mentioned. Multiplication by 4 gives the Hill coefficient (n) values in column 5. The n values are all greater than 2, which shows that a dimer could not have been used as a model for this cooperative system. Even infinitely strong positive cooperativity cannot give $n > 2$ for a dimer.

It has already been mentioned that at very high and low pH ($b \to 0$ or $b \to \infty$, respectively), Eq. (25.20) has the same form as Eq. (23.46) but with a in place of $\overline{K}c$ in the former case and with a/b and u in place of $\overline{K}c$ and y in the latter case. Because Eq. (23.46) produces a symmetrical $\theta(\ln c)$ function, limiting high and low pH curves from Eq. (25.20) will be symmetrical. Hence it is not surprising that *all* of the curves in Fig. 6-21 are almost

Table 6-1. Properties at $\theta_1 = \frac{1}{2}$ and Cooperativity Parameter η

1 pH	2 Exptl. pCa	3 Theoret. pCa	4 Theoret. Slope	5 Theoret. n	6 η
8.5	6.64	6.706	0.815	3.26	21.36
7.5	6.36	6.244	0.715	2.86	11.95
6.8	6.01	5.764	0.617	2.47	7.48
6.4	5.45	5.430	0.577	2.31	6.18
6.0	4.99	5.064	0.555	2.22	5.51
5.5	4.48	4.581	0.542	2.17	5.17

symmetrical. Furthermore, the near symmetry suggests the use of an *approximate* ξ at any pH with the form of Eq. (23.46):

$$\xi = 1 + 4\eta^{-1}\alpha + 4\eta^{-1}\alpha^2 + 2\eta^{-2}\alpha^2 + 4\eta^{-1}\alpha^3 + \alpha^4, \quad (25.35)$$

where η is a single composite cooperativity parameter and α is a Ca^{2+} concentration variable. If we divide all of the terms in Eq. (25.20) by A_0, comparison of a^4/A_0 with α^4 in Eq. (25.35) suggests that α be defined as

$$\alpha = \frac{a}{A_0^{1/4}}. \quad (25.36)$$

Furthermore, if we equate the second terms in the two ξs,

$$4\eta^{-1}\alpha = \frac{4y^{-1}A_1 a}{A_0}.$$

If we now use Eq. (25.36) to eliminate a, we find

$$\eta = \frac{A_0^{3/4}y}{A_1}. \quad (25.37)$$

With these definitions of α and η, numerical calculations of $\theta_1(c_1)$ show that Eq. (25.35) is indeed a very good approximation to Eq. (25.20). Values of η, calculated from Eq. (25.37), are included in Table 6-1 (column 6). The limiting values of η are $y = 25$ at high pH and $u = 5$ at low pH. The values of η in Table 6-1 show directly the degree of positive cooperativity at the various pH values; this is the main point of introducing the approximation. The n values in Table 6-1 are related to the η values by Eq. (22.40), to a close approximation. That is, n and η are alternative cooperativity parameters.

By using the parameters in Eq. (25.31) and Eq. (25.30), with the roles of H^+ and Ca^{2+} reversed, the fractional proton binding θ_2 as a function of pH can be calculated at $\theta_1 = 0$ (no Ca^{2+}) and at $\theta_1 = \frac{1}{2}$ (half-saturation by Ca^{2+}). These curves are presented in Fig. 6-23. The $\theta_1 = 0$ curve shows the ex-

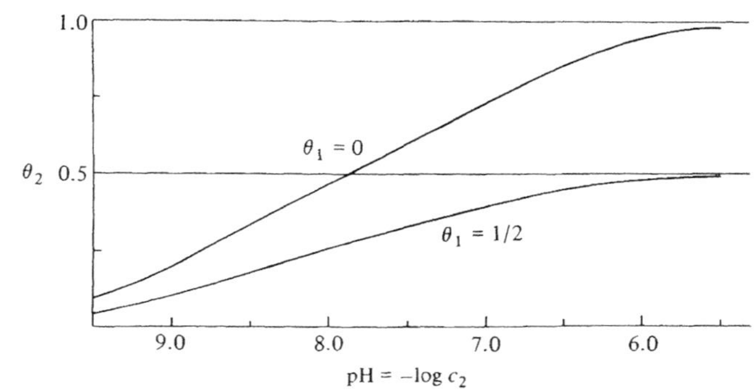

Fig. 6-23. Fractional proton binding as a function of pH, at $\theta_1 = 0$ and $\theta_1 = \frac{1}{2}$ (half-saturation with Ca^{2+}), as calculated from the model.

pected negative cooperativity (because $z = \frac{1}{5}$) and gives the total amount of H^+ (per site) displaced by Ca^{2+} binding (when $c_1 = 0 \rightarrow c_1 = \infty$). At low pH, the sites are almost saturated with H^+ when $\theta_1 = 0$; when $\theta_1 = \frac{1}{2}$, $\theta_2 \cong \frac{1}{2}$ (i.e., almost no sites are empty in the course of the Ca^{2+} binding from $\theta_1 = 0$ to $\theta_1 = \frac{1}{2}$).

26. Rate Constant Perturbations in Equilibrium Systems

Cooperative systems at steady state (see Chapters 7 and 9) cannot be analyzed without employing rate constants. Hence the perturbation of rate constants (see Section 5) by intermolecular interactions will play a significant role in the discussion of these steady-state systems. On the other hand, as amply illustrated in this chapter, no mention of rate constants need be made in the treatment of cooperative *equilibrium* systems. However, rate constants and rate constant perturbations would appear in any analysis of transients that such equilibrium systems might undergo (e.g., in passing from one possible equilibrium state to another). Although, to limit the length of this book, discussion of transients leading to either steady state or equilibrium is excluded, in this section we present a few examples of rate constant perturbations in equilibrium systems. Considerations of this type would be needed as background by any reader who wished to pursue equilibrium-transient problems.

Possible Cooperativity in the Potassium Channel

We return to the four-subunit tetrahedral model represented by Eq. (24.108). Transients in this model have been used to test for the possible existence of cooperativity in the potassium channel of the giant squid axon (the conclusion was negative). We present the elements of the analysis here, for pedagogical purposes, but not the numerical results (which are available in the literature[12, 13]).

At an arbitrary membrane potential ψ, we take $\alpha(\psi)$ and $\beta(\psi)$ to be the first-order rate constants for the transitions $1 \rightarrow 2$ and $2 \rightarrow 1$, respectively, in a single subunit of a complex with $y = 1$. Then the rate constants between states of the complete complex with different r values (r is the number of subunits in state 2), still with $y = 1$, are

$$[0] \underset{\beta}{\overset{4\alpha}{\rightleftarrows}} [1] \underset{2\beta}{\overset{3\alpha}{\rightleftarrows}} [2] \underset{3\beta}{\overset{2\alpha}{\rightleftarrows}} [3] \underset{4\beta}{\overset{\alpha}{\rightleftarrows}} [4]. \tag{26.1}$$

For example, the first rate constant is 4α because state $r = 0$ has four subunits in state 1, any one of which can undergo $1 \rightarrow 2$ with rate constant α. The corresponding probabilities p_r^e for the complex at equilibrium are related by

$$\frac{p_1^e}{p_0^e} = \frac{4\alpha}{\beta} = 4H, \qquad \frac{p_2^e}{p_1^e} = \frac{3\alpha}{2\beta} = \frac{3H}{2},$$
$$\frac{p_3^e}{p_2^e} = \frac{2\alpha}{3\beta} = \frac{2H}{3}, \qquad \frac{p_4^e}{p_3^e} = \frac{\alpha}{4\beta} = \frac{H}{4}. \tag{26.2}$$

These are essentially detailed balance relations at equilibrium (e.g., $\beta p_1^e = 4\alpha p_0^e$). The relations to H follow from Eq. (24.108) with $y = 1$; the p_r^e are proportional to the successive terms in ξ',

$$\xi' = 1 + 4H + 6H^2 + 4H^3 + H^4. \tag{26.3}$$
$$r = 0 \quad 1 \quad 2 \quad 3 \quad 4$$

Obviously, then, $H = \alpha/\beta$. We can regard the ratio α/β as an unperturbed equilibrium constant.

We turn now to the more general situation with arbitrary y and pursue the above argument in reverse. The cooperativity parameter y perturbs the equilibria and rate constants. The expression for ξ' is [Eq. (24.108)]

$$\xi' = 1 + 4y^{-3/2}H + 6y^{-2}H^2 + 4y^{-3/2}H^3 + H^4. \tag{26.4}$$

At equilibrium, the probability ratios are

$$\frac{p_1^e}{p_0^e} = 4y^{-3/2}H = \frac{4\alpha y^{-3/2}}{\beta}, \qquad \frac{p_2^e}{p_1^e} = \frac{3y^{-1/2}H}{2} = \frac{3\alpha y^{-1/2}}{2\beta},$$
$$\frac{p_3^e}{p_2^e} = \frac{2y^{-1/2}H}{3} = \frac{2\alpha y^{1/2}}{3\beta}, \qquad \frac{p_4^e}{p_3^e} = \frac{y^{3/2}H}{4} = \frac{\alpha y^{3/2}}{4\beta}. \tag{26.5}$$

Finally, we need to include y factors in Eq. (26.1). In the equilibrium $[0] \rightleftarrows [1]$, the unperturbed equilibrium constant $4\alpha/\beta$ in Eq. (26.2) has been modified by the perturbation to $4\alpha y^{-3/2}/\beta$ in Eq. (26.5). This is a special case of the general relation, Eq. (5.3). The split of the factor $y^{-3/2}$ between forward and backward rate constants is handled by introducing the parameter f, as in Eqs. (5.6) and (5.8). This gives $4\alpha y^{-3f/2}$ for the forward rate constant and $\beta y^{3(1-f)/2}$ for the backward rate constant. If we assume that the same f applies for transitions at any r, which is very plausible because the same conformational change $1 \rightleftarrows 2$ is involved, the generalization of Eq. (26.1) is

$$[0] \underset{\beta y^{3(1-f)/2}}{\overset{4\alpha y^{-3f/2}}{\rightleftarrows}} [1] \underset{2\beta y^{(1-f)/2}}{\overset{3\alpha y^{-f/2}}{\rightleftarrows}} [2] \underset{3\beta y^{(f-1)/2}}{\overset{2\alpha y^{f/2}}{\rightleftarrows}}$$
$$[3] \underset{4\beta y^{3(f-1)/2}}{\overset{\alpha y^{3f/2}}{\rightleftarrows}} [4]. \tag{26.6}$$

It is easy to see that these rate constants are consistent with Eqs. (26.5). An obvious first choice for f, in the absence of other information, would be $f = \frac{1}{2}$.

The reaction scheme in Eq. (26.6) determines a set of five linear first-order differential equations in the p_r (one equation is redundant because of normalization):

$$\frac{dp_0}{dt} = \beta y^{3(1-f)/2} p_1 - 4\alpha y^{-3f/2} p_0, \tag{26.7}$$

etc. Such equations are easy to solve numerically by computer, given the values of the p_r at $t = 0$. The final values of the p_r, at $t = \infty$, will satisfy Eqs. (26.5). The complete solutions $p_r(t)$ will obviously depend on the parameter choices y and f, that is, on the interactive perturbation and on the way in which the perturbation alters the rate constants.

The reader can easily verify, using

$$\begin{aligned} \delta &= \frac{1}{4}p_1 + \frac{1}{2}p_2 + \frac{3}{4}p_3 + p_4, \\ 1 - \delta &= p_0 + \frac{3}{4}p_1 + \frac{1}{2}p_2 + \frac{1}{4}p_3, \end{aligned} \tag{26.8}$$

that the complete set of Eqs. (26.7) simplifies, when $y = 1$, to the single differential equation

$$\frac{d\delta}{dt} = \alpha(1 - \delta) - \beta\delta. \tag{26.9}$$

This equation is physically obvious: it follows the kinetics of individual independent subunits. The solution of Eq. (26.9) is

$$\begin{aligned} \delta(t) &= \delta_\infty + (\delta_0 - \delta_\infty)\, e^{-(\alpha + \beta)t}, \\ \delta_\infty &= \frac{\alpha}{\alpha + \beta} = \frac{H}{1 + H}, \end{aligned} \tag{26.10}$$

where δ_0 is the initial value of δ and δ_∞ the final equilibrium value.

A comparison of the solutions of Eqs. (26.7) with experimental curves has led to the conclusion[12,13] that $y \cong 1$.

Membrane Potential as a Perturbation

We now assume $y = 1$, as in Eqs. (24.112)–(24.116) and (26.8)–(26.10). The rate constants $\alpha(\psi)$ and $\beta(\psi)$ can be found, as functions of ψ, from experimentally derived values of $\delta_\infty(\psi)$ and $\alpha(\psi) + \beta(\psi)$ [see Eq. (26.10)]. This was done in the classical work of Hodgkin and Huxley.[9] In this subsection, we adopt the point of view that the membrane potential is a perturbation of the $\psi = 0$ rate constants α_o and β_o and of the $\psi = 0$ equilibrium constant $\alpha_o/\beta_o = H_o$. Thus we write

$$\frac{\alpha(\psi)}{\beta(\psi)} = H(\psi) = H_o \cdot \frac{H(\psi)}{H_o} = \frac{\alpha_o}{\beta_o} \cdot \frac{H(\psi)}{H_o}. \tag{26.11}$$

This is another illustration of Eq. (5.3). The perturbation factor (on α_o/β_o) is $H(\psi)/H_o$, which corresponds to $e^{(w_1-w_2)/kT}$ in Eq. (5.3). A very good approximation to $H(\psi)/H_o$ is $e^{\epsilon\psi + \gamma\psi^2}$, as in Eqs. (24.114)–(24.116). However, there is no need to use this approximation here: the perturbation factor $H(\psi)/H_o$ can be found from the experimentally derived $\alpha(\psi)$ and $\beta(\psi)$,

$$\frac{H(\psi)}{H_o} = \frac{\alpha(\psi)/\beta(\psi)}{\alpha_o/\beta_o}. \tag{26.12}$$

The separate rate constants can then be written [see Eqs. (5.6) and (5.8)]

$$\alpha(\psi) = \alpha_o\left[\frac{H(\psi)}{H_o}\right]^{f(\psi)}, \tag{26.13}$$

$$\beta(\psi) = \beta_o\left[\frac{H(\psi)}{H_o}\right]^{f(\psi)-1}, \tag{26.14}$$

where $f(\psi)$ is the "splitting" factor that divides $H(\psi)/H_o$ between α and β. An interesting feature of this example is that $\alpha(\psi)$ and $\beta(\psi)$ are available from experimental data so that $f(\psi)$ can actually be calculated at each ψ from Eq. (26.13) or (26.14) instead of merely being speculated about.

The calculated function $f(\psi)$ from the Hodgkin–Huxley squid data[9] is shown in Fig. 6-24. Also included is $f(\psi)$ for K^+ channels in the frog node, from work by Dodge[14] and Hille.[15] In neither case would $f = \frac{1}{2}$, or even f = constant, have been a good guess.

Thermodynamic Perturbation: Binding on Two Sites

It was pointed out in Section 5 that the fundamental way to treat kinetic perturbations is at the microscopic level: perturbations of the rate constants

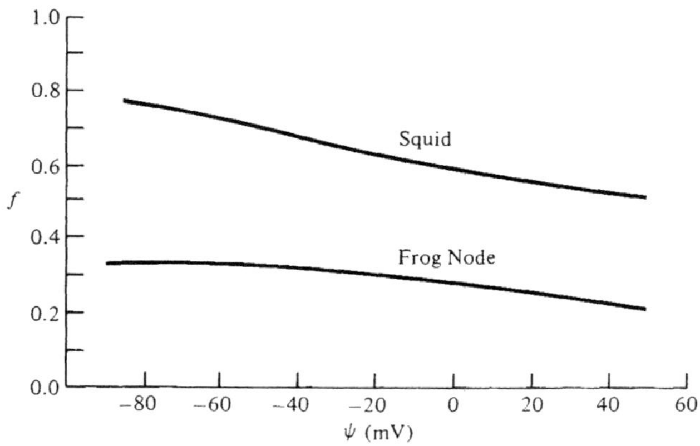

Fig. 6-24. Calculation of the kinetic splitting factor $f(\psi)$ from experimental values of $\alpha(\psi)$ and $\beta(\psi)$.

for elementary processes are used, as in Eqs. (5.6) and (5.8). However, if for some reason (e.g., ignorance of microscopic details) kinetic perturbations are introduced at a more macroscopic level, the question arises as to the connection between the microscopic and macroscopic approaches. In this subsection, we work out the details both ways, in a very simple example, and then establish the connection between the two treatments. A slightly more complicated case is considered in the following subsection. This subject is continued in Section 34.

We examine the model in Eq. (22.11). A protein P has two equivalent binding sites for a ligand L at concentration c. When both sites are occupied by L there is a ligand–ligand interaction that introduces a factor $y = e^{-w/kT}$ in ξ:

$$\xi = 1 + 2Kc + yK^2c^2. \tag{26.15}$$

First, in the *microscopic approach,* where y is the perturbation, the unperturbed first-order "on" rate constant is α^*c and the "off" rate constant is β. Then $K = \alpha^*/\beta$. The unperturbed rate constants apply in the first step of binding, onto either site, when the other site is not occupied by L. The initial binding constant is K. In the second step of binding, when the other site *is* occupied by L, the effective binding constant is yK and the rate constants become α^*y^fc and βy^{f-1}, as in Eqs. (5.6) and (5.8). The full microscopic kinetic scheme is then

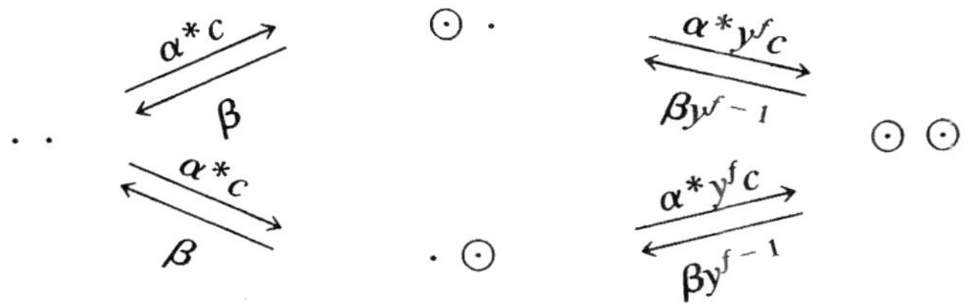

(26.16)

Incidentally, this scheme would be used to follow a transient in this system. If we denote the four state probabilities at equilibrium by (from the left) p^e_{00}, p^e_{10}, p^e_{01}, p^e_{11}, then the four detailed balance relations from Eq. (26.16), for example,

$$\alpha^*cp^e_{00} = \beta p^e_{10}, \tag{26.17}$$

are seen (using $\alpha^*/\beta = K$) to be consistent with

$$p^e_{00} = 1/\xi, \qquad p^e_{10} = p^e_{01} = Kc/\xi, \qquad p^e_{11} = yK^2c^2/\xi \tag{26.18}$$

from Eq. (26.15).

The *macroscopic approach* has already been outlined in Eqs. (5.9)–(5.14). We consider average on and off rates for one particular (either) site. In the unperturbed situation (i.e., $y = 1$), the ratio of unperturbed first-order rate constants is related to site occupation by

$$\frac{\alpha^* c}{\beta} = Kc = \frac{\theta}{1 - \theta}, \tag{26.19}$$

where θ is the probability either site is occupied by ligand. Because [Eq. (6.12)]

$$Kc = q\lambda = qe^{\mu/kT}, \tag{26.20}$$

the chemical potential of bound ligand is, as in Eq. (6.10),

$$\mu = -kT \ln q + kT \ln\left(\frac{\theta}{1 - \theta}\right). \tag{26.21}$$

This is the same as the chemical potential of an "ideal lattice gas"; there is no interaction between bound molecules. The detailed balance relation at equilibrium is, from Eq. (26.19),

$$\alpha^* c(1 - \theta) = \beta\theta. \tag{26.22}$$

This equates the average on and off rates for either site.

In the presence of the interaction ($y \neq 1$), the binding is no longer ideal. As is customary in thermodynamics, this is taken care of in Eq. (26.21) by the introduction of an activity coefficient γ,

$$\mu = -kT \ln q + kT \ln\left(\frac{\theta\gamma}{1 - \theta}\right), \tag{26.23}$$

where γ is a function of θ (or, indirectly, of c). Correspondingly, $\theta/(1 - \theta)$ in Eq. (26.19), which is the analogue of p_2^e/p_1^e in Eq. (5.3), becomes

$$\frac{\theta}{1 - \theta} = \frac{\alpha^* c}{\beta} \cdot \frac{1}{\gamma}. \tag{26.24}$$

The perturbation factor for the rate constant ratio is seen to be γ^{-1}. If we split this factor in the usual way, the on rate constant is $\alpha^*\gamma^{-\bar{f}}c$ and the off rate constant is $\beta\gamma^{1-\bar{f}}$. These are average, not microscopic, rate constants that pertain to on and off transitions from either site. The bar on $\bar{f}$ reminds us that a single average $\bar{f}$ is being used. The precise nature of the averaging will be made clear below. The (averaged) detailed balance relation for on-off transitions from either site is then, from Eq. (26.24),

$$\alpha^*\gamma^{-\bar{f}}c(1 - \theta) = \beta\gamma^{1-\bar{f}}\theta. \tag{26.25}$$

Note that there is only one (averaged) detailed balance relation here compared to four in the microscopic treatment [one for each transition pair in Eq. (26.16); Eq. (26.17) is an example].

Finally, in this subsection, we return to the microscopic treatment and recast it in the form of the averages referred to in the macroscopic case. In this way we can see the interconnections and find the molecular interpretation of γ and $\bar{f}$ for this particular model.

For comparison with Eq. (26.25), we need to calculate the *mean* on and off rates for either site at equilibrium, say, for the right-hand site, using the details implicit in Eq. (26.16). There are two possibilities that need be analyzed, and then averaged: the left-hand site is either empty or it is occupied. The probability that the left-hand site is empty is [Eq. (26.18)]

$$p_{00}^{e} + p_{01}^{e} = \frac{1 + Kc}{\xi}. \tag{26.26}$$

When the left-hand site is empty, the on-off detailed balance relation for the right-hand site is, from Eq. (26.16),

$$\alpha^{*}c(1 - \theta_0) = \beta\theta_0, \tag{26.27}$$

where θ_0 is the probability that the right-hand site is occupied if the left-hand site is empty. Obviously,

$$\theta_0 = \frac{p_{01}^{e}}{p_{00}^{e} + p_{01}^{e}} = \frac{Kc}{1 + Kc}. \tag{26.28}$$

The probability that the left-hand site is occupied is

$$p_{10}^{e} + p_{11}^{e} = \frac{Kc(1 + yKc)}{\xi}. \tag{26.29}$$

When the left-hand site is occupied, the on–off detailed balance relation for the right-hand site is [Eq. (26.16)]

$$\alpha^{*}y^{f}c(1 - \theta_1) = \beta y^{f-1}\theta_1, \tag{26.30}$$

where

$$\theta_1 = \frac{p_{11}^{e}}{p_{10}^{e} + p_{11}^{e}} = \frac{yKc}{1 + yKc}. \tag{26.31}$$

This is the probability that the right-hand site is occupied if the left-hand site is occupied.

The *average* on rate for the right-hand site is then

$$\begin{aligned} on &= (p_{00}^{e} + p_{01}^{e})\,\alpha^{*}c(1 - \theta_0) + (p_{10}^{e} + p_{11}^{e})\,\alpha^{*}y^{f}c(1 - \theta_1) \\ &= \frac{\alpha^{*}c(1 + y^{f}Kc)}{\xi}. \end{aligned} \tag{26.32}$$

The average off rate is

$$\begin{aligned} off &= (p_{00}^{e} + p_{01}^{e})\,\beta\theta_0 + (p_{10}^{e} + p_{11}^{e})\,\beta y^{f-1}\theta_1 \\ &= \frac{\beta Kc(1 + y^{f}Kc)}{\xi}. \end{aligned} \tag{26.33}$$

The self-consistency of Eqs. (26.32) and (26.33) is evident from the result $\alpha^* = \beta K$ when the two expressions are equated.

Our objective is to equate Eqs. (26.32) and (26.33) with the left-hand and right-hand sides of Eq. (26.25), respectively, and thus to deduce the relation between $\bar{f}$ and f. First, though, we need to find θ and γ from Eq. (26.15). As in Eq. (22.15),

$$\theta = \frac{Kc(1 + yKc)}{\xi}, \qquad 1 - \theta = \frac{(1 + Kc)}{\xi}. \tag{26.34}$$

From Eq. (26.23),

$$q\lambda = Kc = \frac{\theta\gamma}{1 - \theta} = \frac{Kc(1 + yKc)\gamma}{1 + Kc}. \tag{26.35}$$

Thus we obtain

$$\gamma = \frac{1 + Kc}{1 + yKc}. \tag{26.36}$$

Note that $\gamma \rightarrow 1$ as $c \rightarrow 0$ and that $\gamma = 1$ if $y = 1$. If we now equate Eq. (26.32) and the left-hand side of Eq. (26.25), we find

$$\left(\frac{1 + yKc}{1 + Kc}\right)^{\bar{f}} = \frac{1 + y^f Kc}{1 + Kc}. \tag{26.37}$$

The same result is found using Eq. (26.33) and the right-hand side of Eq. (26.25). By taking logarithms, we can write an explicit equation for $\bar{f}$, if desired. On the right-hand side of Eq. (26.37) (with the microscopic f), there is averaging that involves y^f. On the left-hand side, there is similar averaging that involves y, with the macroscopic $\bar{f}$ entering *after* the averaging. Although f is presumed to be a constant, $\bar{f}$ is clearly a function of c.

Thermodynamic Perturbation: Binding on Three Sites

The above type of analysis can be extended quite easily to binding on a ring of three sites, as in Eq. (22.22). We present the main results without intervening details. The expression for θ is Eq. (22.24). Then from $Kc = \theta\gamma/(1 - \theta)$, we find

$$\gamma = \frac{1 + 2Kc + yK^2c^2}{1 + 2yKc + y^3K^2c^2}. \tag{26.38}$$

Note again that $\gamma = 1$ if $y = 1$ and that $\gamma \rightarrow 1$ if $c \rightarrow 0$. Typical transitions in the kinetic scheme, with rate constants, are

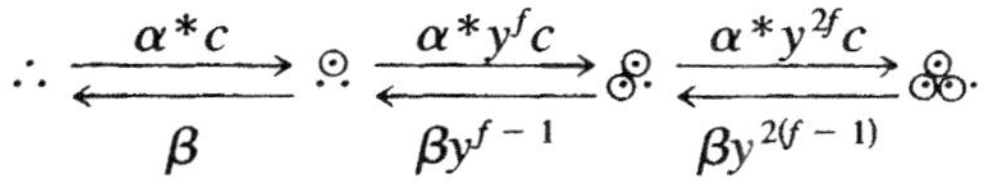 (26.39)

This includes only four out of eight states. For comparison with the macroscopic approach, the on-off detailed balance is examined for any one particular site in three cases: the other two sites are empty; one of them is occupied; and both of them are occupied. The respective probabilities for the three types of occupation of the other two sites are

$$\frac{1 + Kc}{\xi}, \qquad \frac{2Kc(1 + yKc)}{\xi}, \qquad \frac{yK^2c^2(1 + y^2Kc)}{\xi}. \tag{26.40}$$

The three detailed balance equations are [see Eq. (26.39)]

$$\begin{aligned} \alpha^* c(1 - \theta_0) &= \beta\theta_0, \\ \alpha^* c y^f(1 - \theta_1) &= \beta y^{f-1}\theta_1, \\ \alpha^* c y^{2f}(1 - \theta_2) &= \beta y^{2(f-1)}\theta_2, \end{aligned} \tag{26.41}$$

where

$$\theta_0 = \frac{Kc}{1 + Kc}, \qquad \theta_1 = \frac{yKc}{1 + yKc}, \qquad \theta_2 = \frac{y^2Kc}{1 + y^2Kc}. \tag{26.42}$$

The meaning of θ_i is that this is the probability that the one particular site is occupied if i of its neighbors are occupied. We then find, using Eqs. (26.40)–(26.42), that the *average* on and off rates for the particular site are

$$\begin{aligned} on &= \frac{\alpha^* c(1 + 2y^fKc + y^{2f+1}K^2c^2)}{\xi}, \\ off &= \frac{\beta Kc(1 + 2y^fKc + y^{2f+1}K^2c^2)}{\xi}. \end{aligned} \tag{26.43}$$

These are obviously self-consistent ($\alpha^* = \beta K$ if $on = off$). Comparison with Eq. (26.25) then yields, for the final result relating $\bar{f}$ to f,

$$\left(\frac{1 + 2yKc + y^3K^2c^2}{1 + 2Kc + yK^2c^2}\right)^{\bar{f}} = \frac{1 + 2y^fKc + y^{2f+1}K^2c^2}{1 + 2Kc + yK^2c^2}. \tag{26.44}$$

Upon taking logarithms, we obtain an explicit expression for $\bar{f}$ as a function of f. These two expressions [Eq. (26.44)] differ with respect to averaging in essentially the same way as in Eq. (26.37).

In summary, because the above two examples can be worked out easily in microscopic detail, they afford an opportunity to see what the precise meaning of $\bar{f}$ would be if these models had been handled, kinetically, at the macroscopic (thermodynamic) level. See Section 34 for another example.

References

1. Hill, T. L. (1960) *Introduction to Statistical Thermodynamics* (Addison-Wesley, Reading, MA).
2. Szabo, A. and Karplus, M. (1972) J. Molec. Biol. **72,** 163.
3. Szabo, A. and Karplus, M. (1976) Biochemistry **15,** 2869.
4. Chay, T. R. and Brillhart, D. K. (1974) Biochemistry **13,** 4579.
5. Hill, T. L. (1956) *Statistical Mechanics* (McGraw-Hill, New York), p. 276.
6. Ackers, G. K., Johnson, A. D., and Shea, M. A. (1982) Proc. Natl. Acad. Sci. USA **79,** 1129.
7. Inesi, G., Kurzmack, M., Coan, C., and Lewis, D. (1980) J. Biol. Chem. **255,** 3025.
8. Hill, T. L. and Chen, Y. (1972) Proc. Natl. Acad. Sci. USA **69,** 1723.
9. Hodgkin, A. L. and Huxley, A. F. (1952) J. Physiol. **117,** 500.
10. Hill, T. L. (1972) in *Perspectives in Membrane Biophysics,* ed. Agin, D. (Gordon and Breach, New York).
11. Hill, T. L. and Inesi, G. (1982) Proc. Natl. Acad. Sci. USA **79,** 3978.
12. Hill, T. L. and Chen, Y. (1971) Proc. Natl. Acad. Sci. USA **68,** 1711.
13. Hill, T. L. and Chen, Y. (1971) Proc. Natl. Acad. Sci. USA **68,** 2488.
14. Dodge, F. A. (1961) in *Biophysics of Physiological and Pharmacological Actions,* ed. Shanes, A. M. (Amer. Assoc. Adv. Sci., Washington, D.C.).
15. Hille, B. (1967) Ph.D. Thesis, Rockefeller University, New York, NY.

General References

Poland, D. (1978) *Cooperative Equilibria in Physical Biochemistry* (Clarendon Press, Oxford).

Koshland, D. E., Nemethy, G., and Filmer, D. (1966) Biochemistry **5,** 365.

Hess, V. L. and Szabo, A. (1979) J. Chem. Educ. **56,** 289.

Levitzki, A. (1978) *Quantitative Aspects of Allosteric Mechanisms* (Springer-Verlag, New York).

Wyman, J. (1968) Q. Revs. Biophys. **1,** 1.

Weber, G. (1972) Biochemistry **11,** 864.

7

Steady-State Small Systems with Explicit Interactions

Having introduced explicit interactions into a number of equilibrium models of small systems in Chapter 6, we now extend this treatment to similar steady-state systems. The basic theoretical approach, however, has to be quite different: there is no steady-state analogue of the grand partition function from which to derive all of the important steady-state properties. Instead, we have to introduce interaction effects explicitly into the rate constants, and consequently into the rate equations that apply in each particular problem, and then solve the rate equations under steady-state conditions ($dp_i/dt = 0$). We are thus led, mathematically, to the solution of a set of linear algebraic equations in the state probabilities p_i. The related transient problem, which we do not consider here, would require the solution of the corresponding set of linear differential equations in the dp_i/dt. This is easy to do, numerically, by computer.

The above paragraph outlines the formal or mathematical approach that has to be taken for problems treated in this chapter. Physically, all of the models are of the following type, a class of systems of great importance in biochemistry: several enzymatic or transport units are clustered or complexed together (dimer, trimer, etc.); because of the proximity, the units interact with each other; these interactions alter rate constants as well as equilibrium constants; consequently, the rate constants applicable for the possible transitions of any given unit of the complex at any given time depend on the particular states of the neighbors of the given unit at that time. Because of this interdependence, it is necessary to treat the entire cluster or complex as a single kinetic entity: the kinetic diagram that must be used is the diagram of the complete complex, not separate diagrams for each of the individual units.

The models that we use in this chapter to illustrate the above introductory comments are related to those in Chapters 5 (steady state) and 6 (equilibrium). However, they are generally somewhat simpler because explicit steady-state problems, with interactions included, readily get out of hand algebraically. Only the simplest can be treated analytically, usually by the diagram method. At the next stage in complexity (roughly from about five or six states in the diagram up to approximately 50 states), the linear steady-state algebraic equations can be written out in detail and then solved numerically by computer. Because our primary emphasis in this chapter is on *how* to set up and solve problems rather than in actual properties of particular models, in many cases neither analytical nor numerical final results will be obtained.

For still larger systems, the complexity of the diagram, of the algebraic equations, and of the related computer programming become prohibitive. In this case, one can turn to Monte Carlo computer calculations without explicit introduction of a kinetic diagram (e.g., Hill and Chay[1] studied a kinetic model of the complete respiratory chain with 8520 states). The computer keeps track of the state of each unit (and of its neighbors, for interactions) as transitions occur stochastically in the collection of units, controlled by a random number generator (see Chapters 8 and 9).

It should be emphasized that the treatment of steady-state cooperative systems in this chapter (and in Chapter 9) differs from the conventional approach in enzymology in that the interactive effects are introduced into the rate constants of the steady-state problem from the outset, as they should be; an equilibrium distribution among the states is *not* assumed (conventionally, the steady-state flux is calculated from an assumed equilibrium distribution).

One additional general comment should be made. In this chapter and in Chapter 9, we investigate the effects of intermolecular interactions among macromolecular subunits in systems operating at steady state. Any steady-state system necessarily possesses one or more nonzero thermodynamic forces that maintain the steady state. The inclusion of intermolecular interactions in the kinetic models will have no effect on these forces but will alter fluxes. In fact, our main objective is to study or to show how to study the effect of these interactions on fluxes. The reason the forces are not affected is that forces are determined not by intrinsic properties of the macromolecular subunits, including possible intermolecular interactions, but rather by concentrations of small molecules or ions in solution (and possibly the membrane potential) that interact with the macromolecular subunits of the system via binding, chemical reaction while bound, etc. Each force is a net free energy change for a net small-molecule process catalyzed or promoted by the macromolecular subunits.

27. Simple Membrane Transport Models with Interactions

Although most of the examples in this chapter relate to enzyme systems, we begin with three illustrative membrane transport models in this section. These models have intrinsic interest; they are quite easy to visualize physically; and, furthermore, they provide simple introductions to Sections 28, 29, and 32.

Transport Across a Two-Site Channel

This is a generalization of a model first introduced in Fig. 3-6. In Fig. 7-1, we have a symmetrical two-site channel for the steady-state transport of an

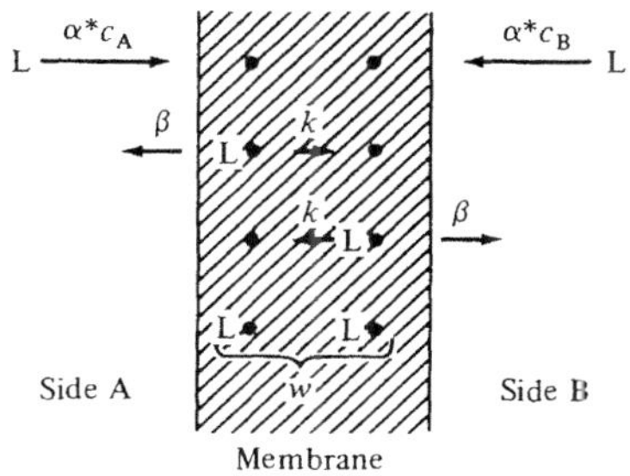

State	Equil.
11	1
21	Kc
12	Kc
22	yK^2c^2

Fig. 7-1. Four possible states of a symmetrical two-site channel for the transport of L across a membrane. The right-hand column gives the relative population of states at equilibrium ($c_A = c_B = c$), as in Eq. (27.2).

uncharged ligand L, across a membrane, from side A (with L at concentration c_A) to side B (with L at concentration c_B). The direction A → B is arbitrarily chosen as positive for the flux J and thermodynamic force [Eq. (8.3)]

$$X = \mu_A - \mu_B = k_B T \ln(c_A/c_B). \tag{27.1}$$

We use k_B here (and occasionally elsewhere) for the Boltzmann constant, to avoid confusion with the rate constant k (for the jump of L from one site to the other). When both sites are occupied, the two ligand molecules interact with each other (free energy of interaction w, relative to infinite separation; $y \equiv e^{-w/k_BT}$). A repulsion corresponds to $w > 0$, $y < 1$; an attraction to $w < 0$, $y > 1$. This interaction introduces cooperative effects into the transport accomplished by the channel. At equilibrium, $c_A = c_B = c$ and $X = 0$. In this case, the relative populations of the states (note the labels for the states) are as given on the right-hand side of Fig. 7-1, where $K = \alpha^*/\beta$ is the initial ($c \to 0$) binding constant of L from either side of the membrane. These relative populations are the terms in the grand partition function

$$\xi = 1 + 2Kc + yK^2c^2, \tag{27.2}$$

as in Eqs. (22.11) and (26.15).

The kinetic diagram for one channel (the channels are independent of each other in this model) is given in Fig. 7-2. This should be compared with Eq. (26.16), which is closely related. The interaction free energy w occurs in state 22 only; it influences the rate constants for the transitions $12 \rightleftarrows 22$ and $21 \rightleftarrows 22$, but no other rate constants. The same f is used for both transition pairs (because of the assumed symmetry of the channel) to split w between forward and backward transitions. Note that if we put $c_A = c_B = c$ in Fig. 7-2, the rate constants are consistent (via detailed balance) with the equilibrium terms in Fig. 7-1. For example,

$$\frac{p_{22}^e}{p_{21}^e} = \frac{p_{22}^e}{p_{12}^e} = \frac{\alpha^* y^f c}{\beta y^{f-1}} = \frac{\alpha^* yc}{\beta} = yKc. \tag{27.3}$$

The equations determining the state probabilities at steady state can be read off of Fig. 7-2:

$$\frac{dp_{11}}{dt} = 0 = \beta(p_{12} + p_{21}) - (\alpha^* c_A + \alpha^* c_B)p_{11}, \tag{27.4}$$

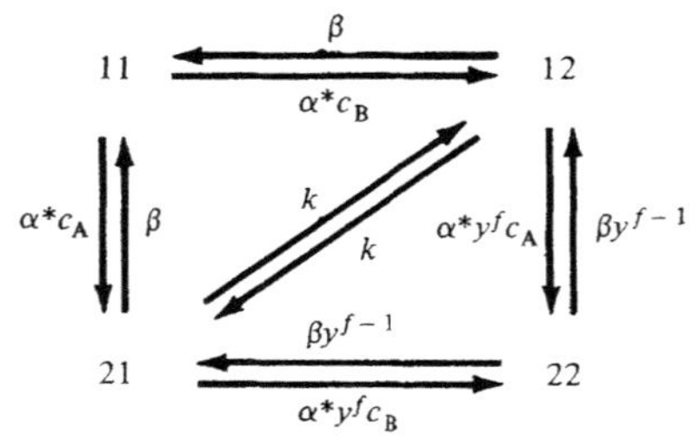

Fig. 7-2. Kinetic diagram for the model in Fig. 7-1. See text for further details.

etc., together with

$$p_{11} + p_{12} + p_{21} + p_{22} = 1. \tag{27.5}$$

The solution of these linear algebraic equations is found most easily in this case from the diagram method.[2] The partial diagrams are shown in Fig. 7-3. From these we find

$$p_{11} = [2\beta^3 y^{f-1} + \alpha\beta^2 y^{2f-1}(c_A + c_B) + 4\beta^2 k y^{f-1}]\Sigma^{-1}, \tag{27.6}$$

etc., where Σ is the sum of the four numerators. If we divide the four numerators and Σ by $\beta^3 y^{f-1}$, the equations simplify to

$$\begin{aligned}
p_{11} &= [2 + y^f(x_A + x_B) + 4(k/\beta)]\Sigma'^{-1}, \\
p_{12} &= [2x_B + y^f x_B(x_A + x_B) + 2(x_A + x_B)(k/\beta)]\Sigma'^{-1}, \\
p_{21} &= [2x_A + y^f x_A(x_A + x_B) + 2(x_A + x_B)(k/\beta)]\Sigma'^{-1}, \\
p_{22} &= [2x_A x_B + y^f x_A x_B(x_A + x_B) + (x_A + x_B)^2 (k/\beta)] y\Sigma'^{-1}, \\
p_2 &= \tfrac{1}{2}(p_{12} + p_{21}) + p_{22},
\end{aligned} \tag{27.7}$$

where Σ' is the sum of the four new numerators, p_2 is the fractional occupation of sites, and

$$x_A \equiv Kc_A, \qquad x_B \equiv Kc_B. \tag{27.8}$$

Note that, at equilibrium ($x_A = x_B \equiv x = Kc$), the four numerators in Eqs. (27.7) have the relative magnitudes 1, x, x, yx^2, as required by Eq. (27.2). It is impressive to observe, beginning with this first example, how much more complicated are the steady-state p_{ij} than the equilibrium p_{ij}^e (e.g., $p_{11}^e = 1/\xi$).

The diagram has three cycles (Fig. 7-4), labelled a, b, and c. The cycle forces (in the directions shown for a and b in the figure) are

$$X_a = X, \qquad X_b = X, \qquad X_c = 0. \tag{27.9}$$

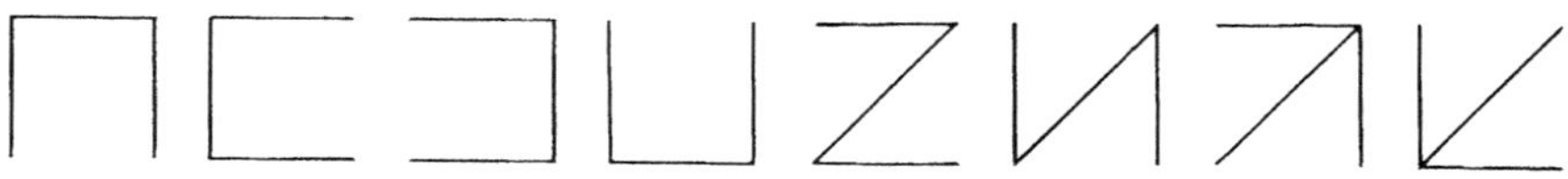

Fig. 7-3. Partial diagrams for the kinetic diagram in Fig. 7-2.

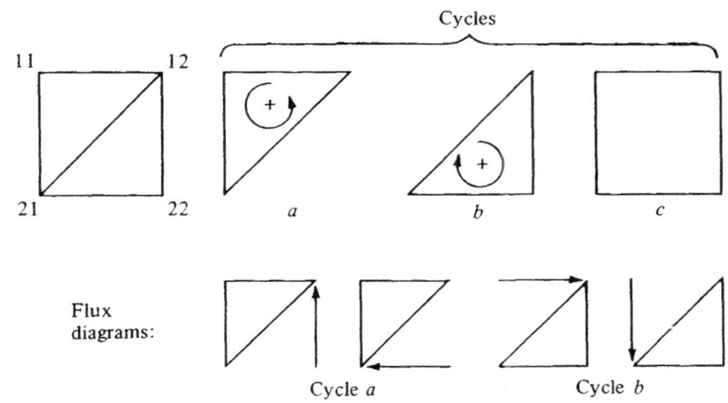

Fig. 7-4. Cycles and flux diagrams for the kinetic diagram in Fig. 7-2.

These are found, as in Eqs. (8.7)–(8.9), from

$$e^{X_a/kT} = \frac{\Pi_{a+}}{\Pi_{a-}} = \frac{\alpha^* c_A k\beta}{\alpha^* c_B k\beta} = \frac{c_A}{c_B}, \tag{27.10}$$

etc. Because $X_c = 0$, the flux in this cycle is $J_c = 0$. The total flux J (i.e., rate of transport of L from A to B, per channel) is then $J = J_a + J_b$, where J_a and J_b are determined[2] by the flux diagrams in Fig. 7-4:

$$J_a = (\Pi_{a^+} - \Pi_{a^-})(\beta y^{f-1} + \beta y^{f-1})\Sigma^{-1}, \tag{27.11}$$

$$J_b = (\Pi_{b^+} - \Pi_{b^-})(\alpha^* c_B + \alpha^* c_A)\Sigma^{-1}. \tag{27.12}$$

Upon dividing numerators and Σ by $\beta^3 y^{f-1}$, as in Eqs. (27.7), we find

$$J = J_a + J_b = k(x_A - x_B)[2 + y^f(x_A + x_B)]\Sigma'^{-1}, \tag{27.13}$$

where Σ' is defined in Eq. (27.7). The term 2 in [ ] in Eq. (27.13) is the J_a contribution. Because all of the net transport of L at steady state, from A to B, has to pass from the left site of the channel to the right site, the same expression for J is easily found from

$$J = k(p_{21} - p_{12}), \tag{27.14}$$

using Eqs. (27.7).

If there is strong repulsion between two bound ligand molecules (e.g., the two sites are close together relative to the size of L), state 22 becomes unstable and only cycle a is used. Mathematically, we let $y \to 0$ in Eq. (27.13) and

$$J \to \frac{k(x_A - x_B)}{[1 + 2(k/\beta)](1 + x_A + x_B)} \qquad (y \to 0). \tag{27.15}$$

This is the flux for a two-site channel that can accomodate only one L at a time (as in Fig. 3-4).

If there is a strong attraction between ligand molecules ($y \to \infty$), and if we assume $0 < f < 1$, Eq. (27.13) becomes

$$J \to \frac{k(x_A - x_B)}{y x_A x_B} \qquad (y \to \infty). \tag{27.16}$$

The flux approaches zero because the system becomes "frozen" in state 22; positive cooperativity ($w < 0, y > 1$) is a disadvantage. Transport across the membrane by either cycle a or b requires use of states 21 and 12 [Eq. (27.14)].

A numerical illustration of Eq. (27.13) is presented in Fig. 7-5. In this figure, J/k is shown as a function of $\ln x_A$ at constant $x_B = 1$. That is, c_B is held constant at the value $1/K$ and c_A is varied. Equilibrium occurs at $\ln x_A = 0$ or $x_A = 1$. We choose, arbitrarily, $k/\beta = 2$ and $f = \frac{1}{2}$ (this implies that L passes over some free energy barrier in the membrane on its way from either solution, A or B, to a binding site). Curves are given for $y = \frac{1}{400}$, 1, and 20. The main effect of the L–L interaction in this case is at high c_A, where large y (attraction between Ls) inhibits transport and small y (repulsion between Ls) enhances it (provided y is not too small: at $\ln x_A = 6$, the

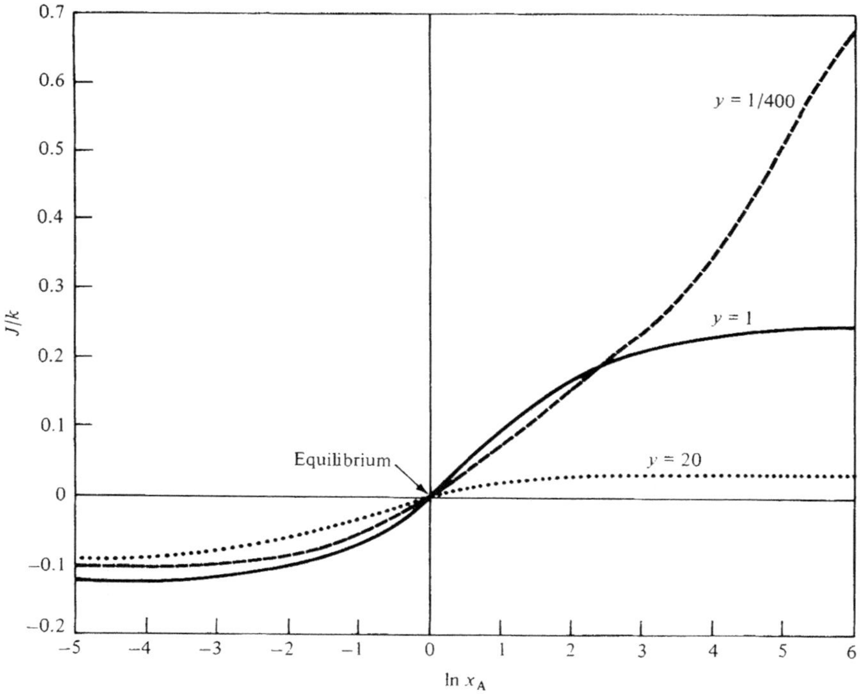

Fig. 7-5. Numerical illustration of Eq. (27.13): J/k as a function of $\ln x_A$, with $x_B = 1, f = \frac{1}{2}, k/\beta = 2$.

Table 7-1. State Probabilities in a Numerical Example (at $\ln x_A = 6$, $x_B = 1$)

State	$y = 0$	$y = 1/400$	$y = 1$	$y = 20$
11	0.0025	0.0023	0.0006	0.0001
12	0.3995	0.1253	0.0031	0.0002
21	0.5980	0.8083	0.2512	0.0334
22	0.0000	0.0642	0.7451	0.9663

maximum J/k is 0.716 at $y \cong 1/121$; at $y = 0$, $J/k = 0.198$ at this value of x_A). The numerical values of the p_{ij} at $\ln x_A = 6$ are informative and are given in Table 7-1. The flux J/k is simply $p_{21} - p_{12}$.

Near equilibrium, where $c_A - c_B$ is a small difference around a concentration c ($x \equiv Kc$),

$$J \to \frac{k(x_A - x_B)(1 + y^f x)}{[1 + y^f x + 2(k/\beta)]\xi}. \tag{27.17}$$

The thermodynamic force near equilibrium is

$$X/k_B T = (c_A - c_B)/c. \tag{27.18}$$

Hence, if we write Eq. (27.17) in the conventional form $J = LX$, the coefficient L is

$$L = \frac{kx(1 + y^f x)}{k_B T[1 + y^f x + 2(k/\beta)]\xi}. \tag{27.19}$$

The physical significance of L is

$$L = (J^e_{a\pm} + J^e_{b\pm})/k_B T, \tag{27.20}$$

where $J^e_{\kappa\pm}$ is the one-way cycle κ flux ($\kappa = a, b$), at equilibrium [see Eq. (9.12) for a simpler case]. For example, from Eq. (27.11),

$$J^e_{a\pm} = \left(\frac{\Pi_{a+} \cdot 2\beta y^{f-1}}{\Sigma}\right)_{c_A = c_B = c} = \frac{kx}{[1 + y^f x + 2(k/\beta)]\xi}. \tag{27.21}$$

There is a contribution in Eq. (27.20) from each cycle that contributes to the flux J.

If we extend the model in Fig. 7-1 to a linear three-site channel, with equivalent sites, site-to-site jump constant k, and nearest-neighbor interactions, the diagram becomes that shown in Fig. 7-6 (compare Fig. 5-49). Incidentally, this is typical of a diagram that is too large to treat, with reasonable effort, algebraically (i.e., at steady state) but would be simple to handle numerically by computer. One has to solve (in this case) eight linear algebraic equations in the eight p_{ijk}. At equilibrium ($c_A = c_B = c$), the corresponding grand partition function terms, which determine the relatively simple p^e_{ijk}, are given in Fig. 7-7. That is, the terms in Fig. 7-7 are the solution

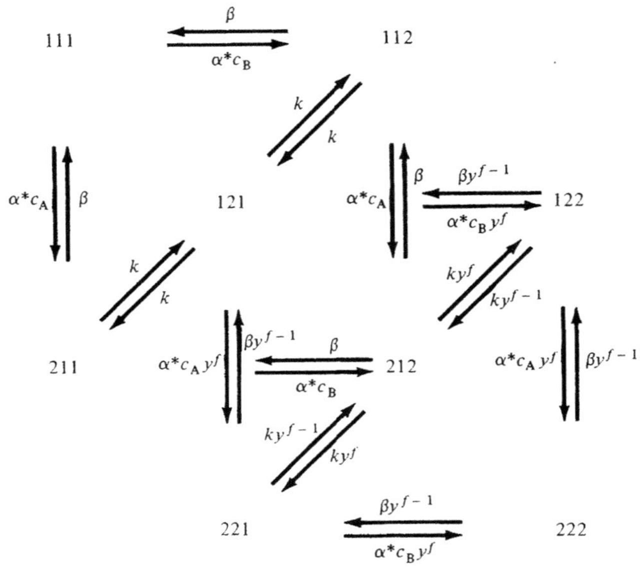

Fig. 7-6. Diagram for a linear three-site model, an extension of the model in Fig. 7-1. See text for details.

of the steady-state algebraic equations that follow from Fig. 7-6 in the special case $c_A = c_B = c$. The grand partition function itself has already been encountered in Eq. (22.31). We do not pursue this case any further.

Transport by a Two-Subunit Complex

In the previous model, two ligand molecules, bound to sites in the same channel, interact. Here transport is accomplished by a symmetrical dimer in a symmetrical membrane (the symmetry is invoked to keep the algebra simple). Each subunit of the dimer can act as a one-site "channel" (compare Figs.

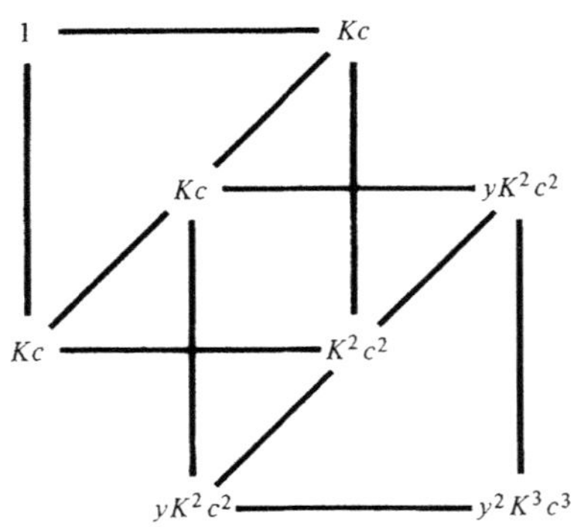

Fig. 7-7. Grand partition function terms for the various states in Fig. 7-6, at equilibrium ($c_A = c_B = c$). See text.

3-1 and 3-2) for transport of L. However, the two subunits or "channels" interact with each other because of their physical proximity: the rate constants of one subunit depend on the state of the other subunit. This example serves as a simple prototype for most of the models in this chapter: in a complex of interacting enzyme molecules, the rate constants of any one enzyme molecule depend on the (fluctuating) states of its neighbors in the complex.

Figure 7-8(a) shows the states and rate constants for an isolated subunit (i.e., separated from its neighbor in the dimer). These are the unperturbed rate constants. The unperturbed binding constant of L on a subunit is $K = \alpha^*/\beta$; this is the same constant called K_o in Eqs. (23.4) and (23.7). Binding of L includes an induced conformational change in the subunit, indicated schematically in Fig. 7-8 by ○ → □.

The two-state transport cycle for an isolated subunit is given in Fig. 7-8(b) (compare Fig. 3-2). For example, one-counterclockwise circuit 1 → 2 → 1 transports one L from side A to side B. Note that a bound L (state 2) is equally likely to depart (rate constant β) in either direction because of the membrane symmetry.

Figure 7-8(c) shows the four possible states of the dimer; the y_{ij} are the usual interaction free energy factors $y_{ij} \equiv e^{-w_{ij}/kT}$. The y_{ij} may be different because of the different shapes (conformations) of the subunits in states 1 and

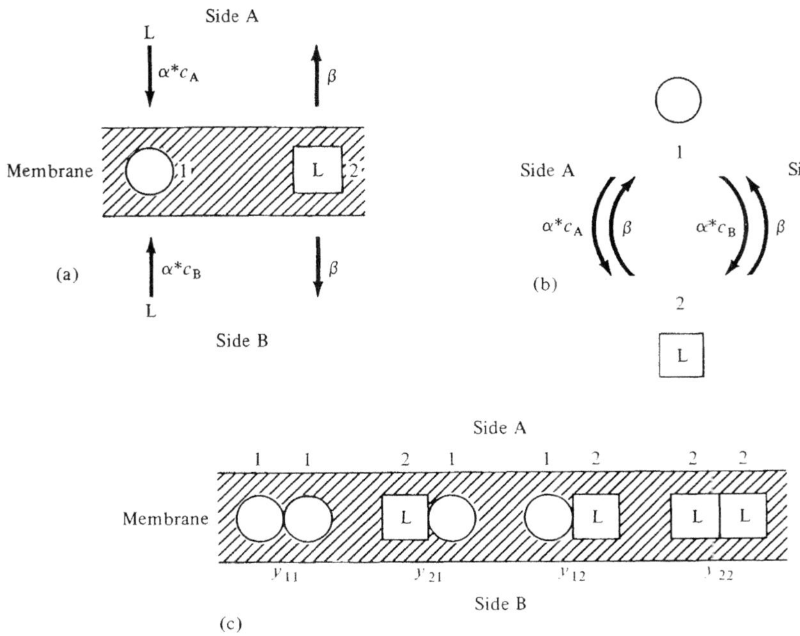

Fig. 7-8. (a) Two states, including conformational change on binding of L, of a single subunit in a symmetrical membrane. (b) Corresponding kinetic diagram (single subunit). (c) States and interaction parameters for a pair of such subunits.

2, and also possibly because of a direct ligand–ligand interaction in state 22 (if L is a large molecule). The symmetry of the membrane is reflected in the use of the same α^* and β for both sides of the membrane, and the symmetry of the dimer makes states 12 and 21 equivalent, including $y_{12} = y_{21}$.

The kinetic diagram for the dimer is presented in Fig. 7-9. The rate constants α^* and β in the unperturbed cycle of Fig. 7-8(b) are altered here, depending on the state of the neighbor in the dimer: the rate constants are designated α_1 and β_1 if the neighbor is in state 1 and they are designated α_2 and β_2 if the neighbor is in state 2. The perturbation of the rate constants has its origin in the interaction free energies w_{ij} in the dimer. The three relevant equilibrium (binding) constants are [see Eq. (5.3)]:

$$\text{Unperturbed: } K = \frac{\alpha^*}{\beta},$$

$$\text{Neighbor in state 1: } \frac{\alpha_1}{\beta_1} = \frac{\alpha^*}{\beta} \cdot \frac{y_{12}}{y_{11}}, \tag{27.22}$$

$$\text{Neighbor in state 2: } \frac{\alpha_2}{\beta_2} = \frac{\alpha^*}{\beta} \cdot \frac{y_{22}}{y_{12}}. \tag{27.23}$$

The split of the free energy factor y_{12}/y_{11} between α_1 and β_1 is accomplished via the constant f, as shown in Fig. 7-9 [see Eqs. (5.6) and (5.8)]. The same f is used for side A and side B transitions because of the assumed symmetry of the membrane. We also use this same f for the split of y_{22}/y_{12} between α_2 and β_2 (Fig. 7-9) because the same process (binding of L, and its inverse) is involved in the two cases (α_1/β_1 and α_2/β_2); only the "environment" is different (neighbor in state 1 or in state 2). This is a very plausible assumption, but it is not necessary. With the explicit assignment of rate constants in

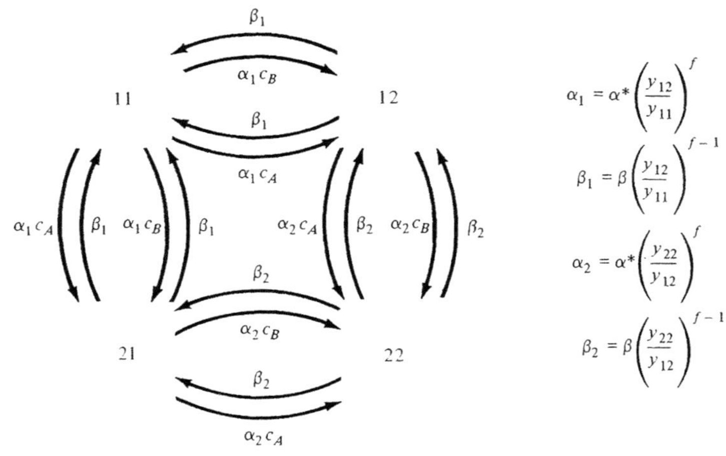

Fig. 7-9. Kinetic diagram for Fig. 7-8(c). Details are given in the text.

Fig. 7-9, the model is complete. The steady-state properties of the model can then be deduced by solving the four steady-state rate equations in the p_{ij}, which follow immediately from the diagram. Fortunately, in the present example, there is a very simple quasiequilibrium solution of the steady-state equations.

We begin with the unperturbed case, Fig. 7-8(b). As in Section 9, at steady state,

$$\frac{p_2}{p_1} = \frac{\alpha^*(c_A + c_B)}{2\beta} = \frac{\alpha^*\bar{c}}{\beta} = K\bar{c}, \tag{27.24}$$

where

$$\bar{c} = \frac{c_A + c_B}{2} \tag{27.25}$$

is the mean concentration of L in the two baths. Equation (27.24) has the form of an equilibrium detailed balance relation for the effective inverse transitions

$$1 \underset{2\beta}{\overset{2\alpha^*\bar{c}}{\rightleftarrows}} 2. \tag{27.26}$$

Actually, there are two pairs of inverse transitions, but they can be treated formally as one pair. The amount of binding of L at steady state is the same as it would be at equilibrium with $c = \bar{c}$:

$$p_2 = \theta = \frac{K\bar{c}}{1 + K\bar{c}}. \tag{27.27}$$

The thermodynamic force is given by Eq. (27.1). The unperturbed flux (rate of transport per subunit), from Eq. (9.5) and Fig. 7-8(b), is

$$\begin{aligned} J_o &= \frac{\alpha^* c_A \beta - \alpha^* c_B \beta}{\alpha^*(c_A + c_B) + 2\beta} \\ &= \frac{\beta K(c_A - c_B)}{2(1 + K\bar{c})}. \end{aligned} \tag{27.28}$$

We turn now to the dimer and Fig. 7-9. In this particular case, an effective detailed balance can be satisfied simultaneously in all four pairs of states (the sides of the square in Fig. 7-9), even at steady state. Each effective detailed balance involves two transition pairs, as in Eq. (27.26). A true detailed balance concerns only one transition pair. The p_{ij} are related by

$$\frac{p_{12}}{p_{11}} = \frac{p_{21}}{p_{11}} = \frac{\alpha_1(c_A + c_B)}{2\beta_1} = \frac{y_{12}}{y_{11}} K\bar{c}, \tag{27.29}$$

$$\frac{p_{22}}{p_{12}} = \frac{p_{22}}{p_{21}} = \frac{\alpha_2(c_A + c_B)}{2\beta_2} = \frac{y_{22}}{y_{12}} K\bar{c}. \tag{27.30}$$

These equations provide the steady-state solution of the rate equations. This is quite unusual. In most steady-state systems, there is *not* detailed balance between all pairs of interreacting states but rather only zero net flux out of each individual state. Furthermore, Eqs. (27.29) and (27.30) have a quasi-equilibrium form: they are exactly what one would have for an equilibrium dimer with $c = \overline{c}$. (It is possible to have an effective detailed balance without a quasiequilibrium, as will be seen at the end of this section.) Consequently, for this quasiequilibrium system, the relative state populations at steady state are represented by the terms in a grand partition function:

$$\xi = 1 + 2\frac{y_{12}}{y_{11}}K\overline{c} + \frac{y_{22}}{y_{11}}K^2\overline{c}^2, \tag{27.31}$$

with

$$p_{11} = \frac{1}{\xi}, \qquad p_{12} = p_{21} = \frac{y_{12}}{y_{11}}\frac{K\overline{c}}{\xi}, \qquad p_{22} = \frac{y_{22}}{y_{11}}\frac{K^2\overline{c}^2}{\xi}, \tag{27.32}$$
$$p_2 = \tfrac{1}{2}(p_{12} + p_{21}) + p_{22},$$

where p_2 is the fractional binding of L. Equation (27.31) has the same form as Eq. (23.8).

The steady-state flux or transport of L, per dimer, in the direction A → B can be obtained most easily in this model as the net flow either into side B or out of side A. For example, using the four transition pairs in Fig. 7-9 relating to side B,

$$J = 2[(\beta_2 p_{22} - \alpha_2 c_B p_{12}) + (\beta_1 p_{12} - \alpha_1 c_B p_{11})]. \tag{27.33}$$

On introducing the expressions for the rate constants (Fig. 7-9) and the p_{ij}, we find for the rate of transport by the dimer,

$$J = \frac{\beta K\,(c_A - c_B)}{\xi}\left[\left(\frac{y_{12}}{y_{11}}\right)^f + \frac{y_{12}}{y_{11}}\left(\frac{y_{22}}{y_{12}}\right)^f K\overline{c}\right]. \tag{27.34}$$

There are two independent interaction parameters in J, for example, y_{11}/y_{12} and y_{22}/y_{12}. In the absence of interaction effects ($y_{11} = y_{12} = y_{22}$), this simplifies to

$$J = \frac{\beta K\,(c_A - c_B)}{1 + K\overline{c}} = 2J_o, \tag{27.35}$$

as expected. That is, the two channels are independent of each other in this case.

If there is no potential barrier involved in the binding of L (e.g., the binding rate constant α^* is diffusion controlled, as in Section 4), then intersubunit interactions do not alter α^*. That is, $f = 0$. In this special case,

$$J = \frac{\beta K\,(c_A - c_B)}{\xi}\left(1 + \frac{y_{12}}{y_{11}}K\overline{c}\right) \qquad (f = 0). \tag{27.36}$$

The further special case $y_{11} = y_{12} = 1$ was treated by Hill and Kedem (Ref. 3, Eq. 51).

To provide a plausible numerical example, we assume, in Eq. (27.36), that $y_{11} = y_{22}$ (i.e., the interaction free energy is the same in the two symmetrical forms of the dimer). Then, if we introduce the usual parameter y,

$$y = \frac{y_{11}y_{22}}{y_{12}^2}, \qquad y^{1/2} = \frac{y_{11}}{y_{12}} = \frac{y_{22}}{y_{12}}, \tag{27.37}$$

Eq. (27.36) for the transport of L by the dimer becomes

$$J = \frac{\beta K(c_A - c_B)(1 + y^{-1/2}K\bar{c})}{1 + 2y^{-1/2}K\bar{c} + K^2\bar{c}^2} \qquad (f = 0,\ y_{11} = y_{22}). \tag{27.38}$$

In this special case, the rate constants in Fig. 7-9 are

$$\begin{aligned} \alpha_1 &= \alpha^*, & \beta_1 &= \beta y^{1/2}, \\ \alpha_2 &= \alpha^*, & \beta_2 &= \beta y^{-1/2}. \end{aligned} \tag{27.39}$$

For very small y (negative cooperativity), Eq. (27.38) becomes

$$J \to \frac{\beta K(c_A - c_B)}{2} \qquad (y \to 0). \tag{27.40}$$

Here states 12 and 21 dominate; the transition from 12 or 21 back to state 11 is very slow; the effective cycles use state 22 and 12 or 21. For very large y (positive cooperativity),

$$J \to \frac{\beta K(c_A - c_B)}{1 + K^2\bar{c}^2} \qquad (y \to \infty). \tag{27.41}$$

Here states 11 and 22 dominate; state 22 returns very slowly to 12 or 21; the effective cycles are between 11 and 12 or 21. If we hold c_B constant and vary c_A, $J \propto c_A$ at large c_A when $y \to 0$, but $J \propto 1/c_A$ at large c_A when $y \to \infty$. That is, $J \to 0$. For a finite y, in Eq. (27.38), when c_A is very large,

$$J \to 2\beta y^{-1/2}. \tag{27.42}$$

This asymptotic value of J is relatively large for negative cooperativity ($y < 1$) and relatively small for positive cooperativity ($y > 1$). As in the previous model, we observe that positive cooperativity is a disadvantage.

Figure 7-10 shows plots of J/β as a function of ln x_A ($x_A = Kc_A$), from Eq. (27.38), with $x_B = Kc_B$ held constant at $x_B = 1$. The three y values used are $y = \frac{1}{400}$, 1, and 20, as in Fig. 7-5. The respective asymptotic values of J/β, at large x_A, are [Eq. (27.42)] 40, 2, and 0.45. The details of Figs. 7-5 and 7-10 are of course quite different, but the main qualitative effects are similar.

Transport by a Closed Three-Subunit Complex

This is a direct extension of the previous model to a symmetrical triangular (closed) trimer of identical subunits, each of which can transport L from A

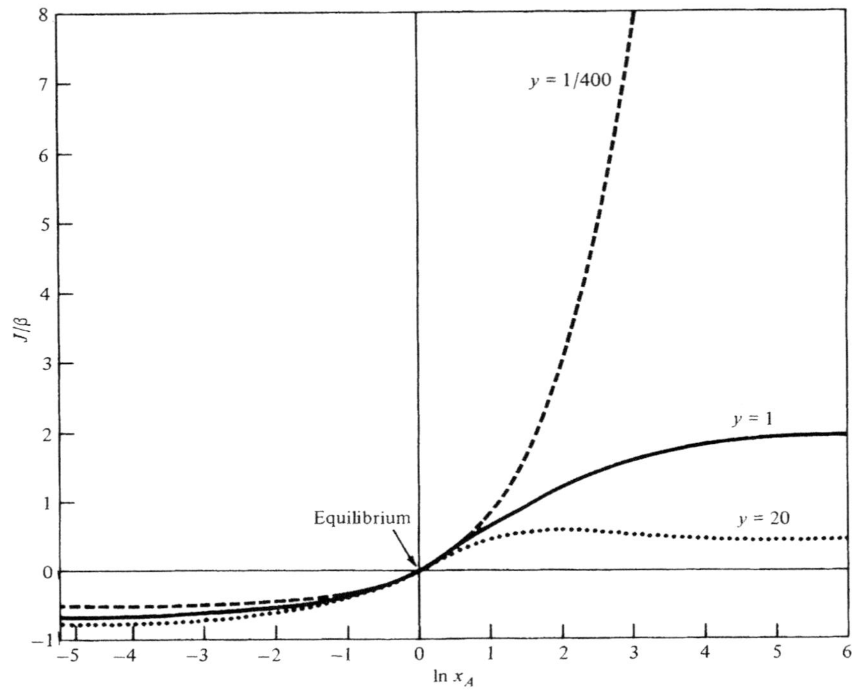

Fig. 7-10. Flux J/β as a function of ln x_A, from Eq. (27.38), with $x_B = 1$. See text for further details.

to B via a one-site "channel." The plane of the triangle is in the plane of the membrane. The three subunits interact with each other as in Eq. (23.38) (except that here $y_{12} = y_{21}$). Incidentally, in the two-dimensional hexagonal lattice of bacteriorhodopsin molecules, there are triangles of close neighbors that possibly interact with each other. These molecules transport protons across a membrane, using photons as an energy source. The transport cycle involves about six significant states.

The kinetic diagram for the trimer is shown rather schematically in Fig. 7-11; Fig. 5-31 is similar. The cycles (pairs of curved lines) in Fig. 7-11 are of the same type as in Fig. 7-9. The labels (0, 1, 2) on the cycles specify the number of neighbors of the "reacting" subunit that are in state 2. In this model a reacting subunit has an "environment" of two other subunits; the rate constants of the reacting subunit depend on the states of these two neighbors.

The interaction parameters for representative states of the trimer are:

$$\begin{array}{cccc} 1 & 2 & 2 & 2 \\ 1\;\;1 & 1\;\;1 & 1\;\;2 & 2\;\;2. \\ y_{11}^3 & y_{11}y_{12}^2 & y_{12}^2 y_{22} & y_{22}^3 \end{array} \qquad (27.43)$$

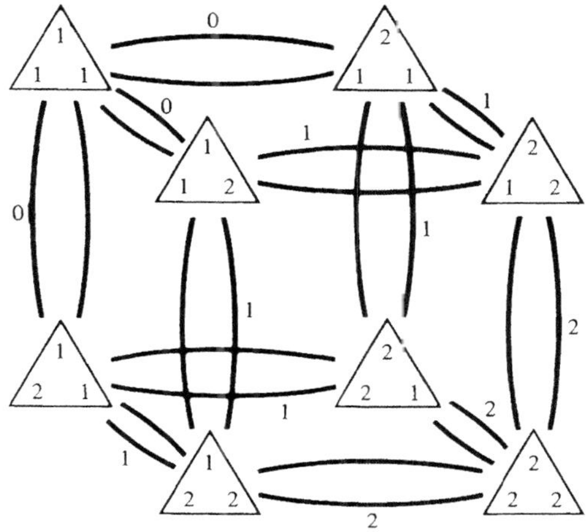

Fig. 7-11. Schematic kinetic diagram for a triangular trimer of identical interacting subunits, each of which can transport L as in Fig. 7-8(a). The numbers on the cycles give the number of neighbors of the reacting subunit that are in state 2.

These are in the order $r = 0, 1, 2, 3$, where r is the number of subunits in state 2. Ratios of successive pairs of these expressions are:

$$(y_{12}/y_{11})^2, \qquad y_{22}/y_{11}, \qquad (y_{22}/y_{12})^2. \tag{27.44}$$

These are the free energy factors we need in order to correct rate constant ratios for interaction effects, as in Eqs. (27.22) and (27.23):

$$\begin{aligned} &\text{Unperturbed: } K = \alpha^*/\beta, \\ &\text{Neighbors in states 11: } \frac{\alpha_0}{\beta_0} = \frac{\alpha^*}{\beta}\left(\frac{y_{12}}{y_{11}}\right)^2, \\ &\text{Neighbors in states 12: } \frac{\alpha_1}{\beta_1} = \frac{\alpha^*}{\beta}\cdot\frac{y_{22}}{y_{11}}, \\ &\text{Neighbors in states 22: } \frac{\alpha_2}{\beta_2} = \frac{\alpha^*}{\beta}\left(\frac{y_{22}}{y_{12}}\right)^2. \end{aligned} \tag{27.45}$$

The subscripts here on α and β refer to the labels in Fig. 7-11.

In writing individual rate constants, we use a single f for all transitions, for the reasons given following Eqs. (27.22) and (27.23) (however, this is not essential). Consequently (compare Fig. 7-9),

$$\begin{aligned} \alpha_0 &= \alpha^*\left(\frac{y_{12}}{y_{11}}\right)^{2f}, & \beta_0 &= \beta\left(\frac{y_{12}}{y_{11}}\right)^{2f-2}, \\ \alpha_1 &= \alpha^*\left(\frac{y_{22}}{y_{11}}\right)^{f}, & \beta_1 &= \beta\left(\frac{y_{22}}{y_{11}}\right)^{f-1}, \\ \alpha_2 &= \alpha^*\left(\frac{y_{22}}{y_{12}}\right)^{2f}, & \beta_2 &= \beta\left(\frac{y_{22}}{y_{12}}\right)^{2f-2}. \end{aligned} \tag{27.46}$$

If the above rate constants are introduced into Fig. 7-11 (just as in Fig. 7-9), there is again a quasiequilibrium solution of the steady-state linear kinetic equations (the reason for this is discussed in the next subsection). As a result, there is an effective grand partition function whose separate terms give the state probabilities at steady state:

$$\xi = 1 + 3\left(\frac{y_{12}}{y_{11}}\right)^2 K\bar{c} + 3\left(\frac{y_{12}}{y_{11}}\right)^2 \frac{y_{22}}{y_{11}} K^2\bar{c}^2 + \left(\frac{y_{22}}{y_{11}}\right)^3 K^3\bar{c}^3, \tag{27.47}$$

$$p_{111} = \frac{1}{\xi}, \qquad p_{121} = \left(\frac{y_{12}}{y_{11}}\right)^2 \frac{K\bar{c}}{\xi}, \qquad \text{etc.}$$
$$p_2 = p_{121} + 2p_{122} + p_{222}, \tag{27.48}$$

where the subscript 121 refers to the state $1\overset{2}{}1$, etc. The interaction parameter factors in Eq. (27.47) follow from the terms in Eq. (27.43), on dividing by y_{11}^3. Equations (27.43) and (27.47) should be compared with Eqs. (23.37) and (23.38).

To obtain the steady-state flux of L from side A to side B, per trimer, we use the 12 transition pairs in Fig. 7-11 that relate to side B [compare Eq. (27.33)]:

$$J = 3(\beta_2 p_{222} - \alpha_2 c_B p_{122}) + 6(\beta_1 p_{122} - \alpha_1 c_B p_{121}) + 3(\beta_0 p_{121} - \alpha_0 c_B p_{111}) \tag{27.49}$$

$$= \frac{3\beta K(c_A - c_B)}{2\xi}\left[\left(\frac{y_{12}}{y_{11}}\right)^{2f} + 2\left(\frac{y_{12}}{y_{11}}\right)^2\left(\frac{y_{22}}{y_{11}}\right)^f K\bar{c} + \left(\frac{y_{22}}{y_{11}}\right)^3\left(\frac{y_{22}}{y_{12}}\right)^{2f-2} K^2\bar{c}^2\right]. \tag{27.50}$$

The second form [Eq. (27.50)] follows upon introduction of Eqs. (27.46) and (27.48). There are again two independent interaction parameters in J, say, y_{11}/y_{12} and y_{22}/y_{12}. When $y_{11} = y_{12} = y_{22}$, this simplifies to $J = 3J_o$ as it should, where J_o is given in Eq. (27.28).

In the special case $f = 0$, $y_{11} = y_{22}$ discussed in the previous subsection, Eq. (27.50) reduces to

$$J = \frac{3\beta K(c_A - c_B)(1 + 2y^{-1}K\bar{c} + y^{-1}K^2\bar{c}^2}{2(1 + 3y^{-1}K\bar{c} + 3y^{-1}K^2\bar{c}^2 + K^3\bar{c}^3)}. \tag{27.51}$$

This equation has properties that are qualitatively similar to Eq. (27.38), as one might expect. The details are left to the interested reader. See, also, Fig. 7-42.

Steady States that Resemble Equilibrium

Both of the preceding two models possess a quasiequilibrium state distribution at steady state. This naturally raises the question as to the conditions under which equilibrium-like solutions of the steady-state rate equations might be expected for interacting complexes of enzyme molecules. It can be stated at the outset that a necessary condition is that *all* members of the complex must be adequately represented by *two-state cycles*, not larger. There are then three types of cases.

The *first case* is a true quasiequilibrium. To begin with, consider a complex or lattice (of any size) of *identical* two-state enzyme molecules. Each member of the complex has its own environment of neighbors with which it interacts. For the particular member of the complex on which we fix attention, we designate this environment by e (e will not be used to refer to equilibrium here).

The unperturbed (i.e., in the absence of e) first-order rate constants of the two-state cycle are shown in Fig. 7-12(a). Usually α and β are the more important rate constants; counterclockwise is usually the more important cycle direction. The ratio of the two state probabilities, in this figure, at steady state is

$$\frac{p_2}{p_1} = \frac{\alpha + \beta'}{\beta + \alpha'}. \tag{27.52}$$

This has a quasiequilibrium form, as suggested by the effective inverse transitions of Fig. 7-12(b).

Returning now to the complex, let w_{1e} be the interaction free energy between the particular molecule of interest, in state 1, and its environment e. Similarly, w_{2e} applies when the molecule is in state 2. The definition of e must, of course, include a specification of the particular states (1 or 2) of each of the neighbors whose interactions are included in w_{1e} and w_{2e}. We designate the perturbed rate constants (i.e., perturbed by e) by a subscript e. Then the two perturbed rate constant ratios, in the counterclockwise cycle direction [Fig. 7-12(a)] are

$$\begin{aligned} \frac{\alpha_e}{\alpha_e'} &= \frac{\alpha}{\alpha'} \cdot \frac{e^{-w_{2e}/kT}}{e^{-w_{1e}/kT}} = \frac{\alpha}{\alpha'} \cdot \frac{y_{2e}}{y_{1e}}, \\ \frac{\beta_e}{\beta_e'} &= \frac{\beta}{\beta'} \cdot \frac{e^{-w_{1e}/kT}}{e^{-w_{2e}/kT}} = \frac{\beta}{\beta'} \cdot \frac{y_{1e}}{y_{2e}}, \end{aligned} \tag{27.53}$$

where $y_{ie} \equiv e^{-w_{ie}/kT}$. These are special cases of Eq. (5.3).

In general the α and β processes in the cycle may be quite different. Hence, we use the parameter f_α for the α process and f_β for the β process, both defined in the main cycle direction (this is a matter of convention). We assume that f_α and f_β retain the same values for any e because the α and β processes are

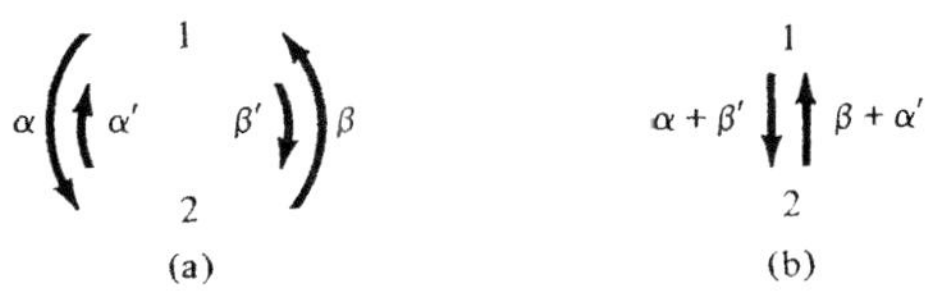

Fig. 7-12. (a) Unperturbed two-state cycle. (b) Quasiequilibrium version of the cycle in (a).

the same for any e; only the environment changes. Then the separate rate constants, as perturbed by e, are

$$\alpha_e = \alpha\left(\frac{y_{2e}}{y_{1e}}\right)^{f_\alpha}, \qquad \alpha'_e = \alpha'\left(\frac{y_{2e}}{y_{1e}}\right)^{f_\alpha - 1},$$
$$\beta_e = \beta\left(\frac{y_{1e}}{y_{2e}}\right)^{f_\beta}, \qquad \beta'_e = \beta'\left(\frac{y_{1e}}{y_{2e}}\right)^{f_\beta - 1}. \tag{27.54}$$

If we now form the quotient

$$\frac{\alpha_e + \beta'_e}{\beta_e + \alpha'_e} = \frac{\alpha(y_{2e}/y_{1e})^{f_\alpha} + \beta'(y_{2e}/y_{1e})^{1-f_\beta}}{\beta(y_{2e}/y_{1e})^{-f_\beta} + \alpha'(y_{2e}/y_{1e})^{f_\alpha - 1}}, \tag{27.55}$$

as in Eq. (27.52), we note that if and only if $f_\alpha + f_\beta = 1$, Eq. (27.55) simplifies to

$$\frac{\alpha_e + \beta'_e}{\beta_e + \alpha'_e} = \frac{\alpha + \beta'}{\beta + \alpha'} \cdot \frac{y_{2e}}{y_{1e}}. \tag{27.56}$$

This equation shows the same effect of interactions as for the truly inverse transition pairs in Eq. (27.53), even though we are using composite rate constants. Thus, when $f_\alpha + f_\beta = 1$, the perturbed *combined* transitions simulate equilibrium behavior.

The quasiequilibrium result in Eq. (27.56) holds for any e and for any molecule in the complex or lattice of identical two-state molecules, if $f_\alpha + f_\beta = 1$. As a result, there will be a "detailed balance" relation

$$\frac{p_{2e}}{p_{1e}} = \frac{\alpha_e + \beta'_e}{\beta_e + \alpha'_e} = \frac{\alpha + \beta'}{\beta + \alpha'} \cdot \frac{y_{2e}}{y_{1e}} \tag{27.57}$$

for every molecule in the complex, whatever its fluctuating environment e, that treats interaction effects just as for an *equilibrium* system. This will then provide a solution to the steady-state rate equations: if there is a detailed balance within every pair of interreacting states in the full diagram of the complex, there will automatically be zero net flux out of each state in the diagram, as required for the steady-state solution. (Ordinarily the latter condition, zero net flux, has to be satisfied *without* benefit of the former, detailed balance.)

Conclusion: the state distribution at steady state for any model in this class is the same as for a corresponding equilibrium system. This is a great advantage, mathematically, because interactions make equilibrium systems complicated and steady-state systems even more so (usually).

If we now reexamine the rate constants and f used in the preceding two models, we note the following relations to the present notation:

$$\alpha \leftrightarrow \alpha^* c_A, \qquad \alpha' \leftrightarrow \beta, \qquad f_\alpha \leftrightarrow f$$
$$\beta \leftrightarrow \beta, \qquad \beta' \leftrightarrow \alpha^* c_B, \qquad f_\beta \leftrightarrow 1 - f. \tag{27.58}$$

Thus we see that $f_\alpha + f_\beta = 1$ was satisfied in both of these models, and hence a quasiequilibrium solution to the steady-state rate equations could have been anticipated. Incidentally, Eqs. (27.22), (27.23), and (27.45) provide two simple examples of complete listings of possible environments, e.

If the complex or lattice is comprised of more than one kind of two-state enzyme, for example, molecules with rate constant notation α, β (as above); γ, δ; etc., then the above argument shows that a quasiequilibrium will arise if and only if $f_\alpha + f_\beta = 1, f_\gamma + f_\delta = 1$, etc., for each of the enzyme types.

The *second case* relates to one-way two-state cycles. These systems do not exhibit a true quasiequilibrium, but rather show a "false equilibrium." Consider first a complex or lattice of *identical* two-state enzyme molecules whose inverse transitions can be neglected [i.e., $\alpha' = 0$ and $\beta' = 0$ in Fig. 7-12(a)]. In this case, Eq. (27.55) reduces to

$$\frac{\alpha_e}{\beta_e} = \frac{\alpha}{\beta}\left(\frac{y_{2e}}{y_{1e}}\right)^{f_\alpha + f_\beta}. \tag{27.59}$$

The y expression here is not that of a true equilibrium, Eq. (27.53), because of the exponent $f_\alpha + f_\beta$. However, the steady-state population of states will still have an equilibrium form; the "equilibrium" distribution of states will not be that characteristic of the true interaction free energies w_{ij} between members of the complex but rather that of a "false equilibrium" system with interaction free energies $w_{ij}(f_\alpha + f_\beta)$. The interaction free energies are scaled by the factor $f_\alpha + f_\beta$, and then the steady-state distribution of states can be treated like an equilibrium. Examples will be found in Sections 37 and 44.

If the complex contains two or more different kinds of two-state one-way enzyme molecules with rate constants α, β; γ, δ; etc., then a "false equilibrium" solution of the steady-state rate equations will exist only if $f_\alpha + f_\beta = f_\gamma + f_\delta = \ldots$ (for all enzyme types). This is not a very likely coincidence.

The *third case* is concerned with systems whose kinetic diagrams are linear or can be put in linear form, as in Fig. 7-13(b). The rate equations at steady state for a linear diagram of this type, even though successive states are connected by cycles, can be solved one step at a time, starting from either end, as if it were an equilibrium system. There will be an effective detailed balance between each pair of successive states. The most important type of example is a small complex of identical two-state enzyme molecules, for which one-way cycles are not adequate and also for which $f_\alpha + f_\beta \neq 1$. Consequently, a quasiequilibrium or false equilibrium solution of the steady-state rate equations does not exist. However, if a linear diagram applies, there is an effective detailed balance at steady state to make the solution easy. Figure 7-13(a) is a schematic example for a trimer. If the grouped states in Fig. 7-13(a) are *identical,* the diagram can be collapsed into the form shown in Fig. 7-13(b). Whatever the rate constants (which have to include proper statistical factors), an effective detailed balance solution can be found for the

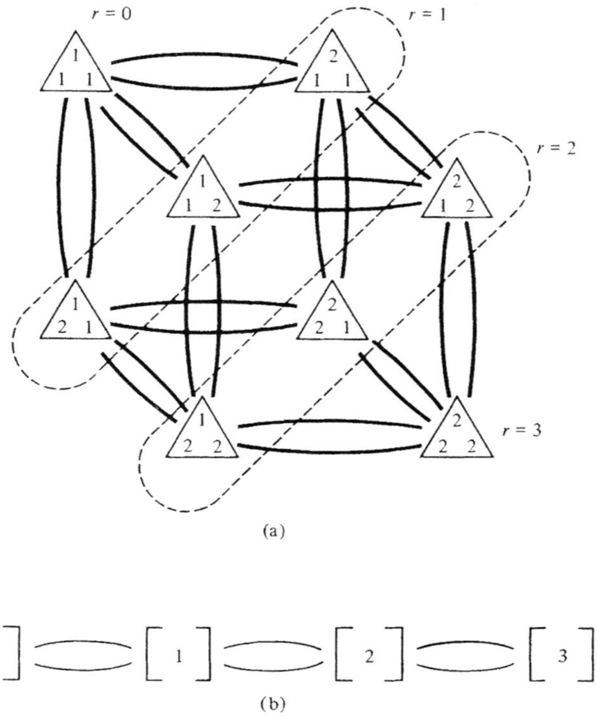

Fig. 7-13. (a) Symmetrical case in which grouped states (dashed lines) are identical. The r value is the number of subunits in state 2. (b) Linearized version of (a), using grouped states. The r values are in brackets. See text for further details.

steady-state rate equations by the procedure already mentioned. Examples will be encountered in Figs. 7-20 and 7-41. See also Fig. 7-11.

28. Simple Complexes with Direct Transfer of Small Molecules

The first part of Section 20 provides an introduction to this section. We consider simple enzymatic complexes in which small molecules, molecular fragments, electrons, etc., are transferred *directly* from one subunit to another. Figures 5-45 through 5-48 provide prototypes that we shall refer to here for concreteness. In Fig. 5-47, for example, we have a two-subunit complex of two-state enzymes (E_1, E_2) that catalyzes $S \rightarrow Q$, in the course of which the intermediate product P ($S \rightarrow P$ is catalyzed by E_1) is transferred directly from E_1 to E_2. The new feature that we introduce in this section is the possibility that intersubunit interactions may perturb the rate constants of the participating enzyme molecules. For example, in Fig. 5-47, the rate constant

β for $E_2P \rightarrow E_2 + Q$ might have different values depending on whether its companion in the complex is in state E_1 or in state E_1S. There is certainly an attractive interaction between E_1 and E_2 that is responsible for the existence of the complex in the first place. We are interested in cases in which the interaction free energy changes with the states of the subunits. We discuss three examples.

Dimer with Two-State Cycles

Here we consider the model in Fig. 5-47, but we now include inverse rate constants (α', k', β') and also intersubunit interaction effects. The numbering of the states in Fig. 5-47 should be noted. If intersubunit interaction effects are present, they will be reflected in different rate constants α, α' and β, β', depending on the state of the neighbor. We shall use the notation α, α' and β, β' for the unperturbed rate constants, and, for perturbed rate constants, we introduce a subscript 1 if the neighbor is in state 1 (E_1 or E_2) and a subscript 2 if the neighbor is in state 2 (E_1S or E_2P). The intermolecular interaction parameters are designated $y_{ij} = e^{-w_{ij}/k_BT}$, as usual, where w_{ij} is the free energy of the dimer in state ij relative to infinite separation of the two subunits in states i and j. Because the two subunits are different, $y_{12} \neq y_{21}$ in general.

The new kinetic diagram [compare Fig. 5-47(b)] is shown in Fig. 7-14. All rate constants are first order. The constants κ, κ' are the perturbed versions of the unperturbed k, k'. Because the nature of the reactions α, β, k are all quite different, we need to introduce three different parameters f_α, f_β, and f_k for the kinetic split of the interaction free energy differences. Then, following our usual convention,

$$\begin{aligned} \alpha_1 &= \alpha\left(\frac{y_{21}}{y_{11}}\right)^{f_\alpha}, & \alpha_1' &= \alpha'\left(\frac{y_{21}}{y_{11}}\right)^{f_\alpha - 1}, \\ \alpha_2 &= \alpha\left(\frac{y_{22}}{y_{12}}\right)^{f_\alpha}, & \alpha_2' &= \alpha'\left(\frac{y_{22}}{y_{12}}\right)^{f_\alpha - 1}, \end{aligned} \tag{28.1}$$

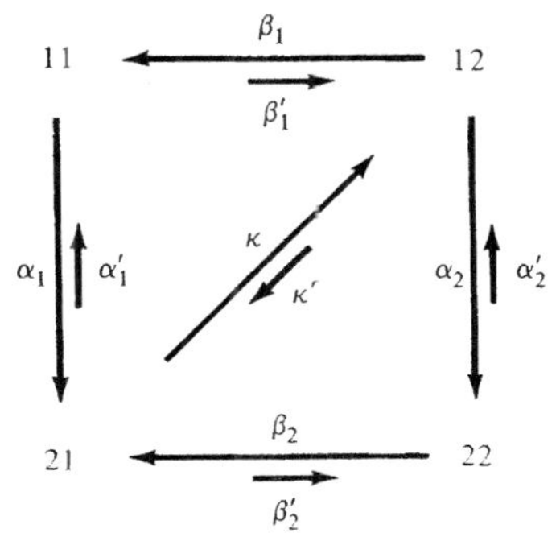

Fig. 7-14. Diagram for a generalization of the model in Fig. 5-47. See text for details.

$$\beta_1 = \beta\left(\frac{y_{11}}{y_{12}}\right)^{f_\beta}, \qquad \beta_1' = \beta'\left(\frac{y_{11}}{y_{12}}\right)^{f_\beta - 1},$$
$$\beta_2 = \beta\left(\frac{y_{21}}{y_{22}}\right)^{f_\beta}, \qquad \beta_2' = \beta'\left(\frac{y_{21}}{y_{22}}\right)^{f_\beta - 1}, \tag{28.2}$$

$$\kappa = k\left(\frac{y_{12}}{y_{21}}\right)^{f_k}, \qquad \kappa' = k'\left(\frac{y_{12}}{y_{21}}\right)^{f_k - 1}. \tag{28.3}$$

These rate constants, and the steady-state flux J, depend on three independent interaction parameters, for example, y_{11}/y_{12}, y_{22}/y_{12}, and y_{12}/y_{21}. With the rate constants in the kinetic diagram (Fig. 7-14) completely specified, the steady-state problem is now solved in principle. That is, four linear rate equations in the p_{ij} can be written and solved, for example, by the diagram method (Fig. 7-3), and then

$$J = \kappa p_{21} - \kappa' p_{12} \tag{28.4}$$

is the steady-state flux (in S $\to$ Q).

Before turning to some of the details, we make the general comment that Fig. 7-14 and Eqs. (28.1)–(28.3) are not restricted to the particular type of mechanism used in Fig. 5-47. These relations apply to any two-subunit complex with two-state cycles and an exchange transition. In fact, even though the model is quite different, Fig. 7-2 is a formal special case. In the Fig. 5-47 case, α is proportional to c_S and β' is proportional to c_Q.

The cycles are shown in Fig. 7-4. The cycle forces are [compare Eqs. (27.9) and (27.10)]

$$e^{X/k_BT} = e^{X_a/k_BT} = e^{X_b/k_BT}$$
$$= \frac{\kappa\alpha_1\beta_1}{\kappa'\alpha_1'\beta_1'} = \frac{\kappa\alpha_2\beta_2}{\kappa'\alpha_2'\beta_2'} = \frac{\kappa\alpha\beta}{\kappa'\alpha'\beta'}. \tag{28.5}$$

The cancellation of all the interaction parameters in the force expression is a self-consistency check on Eqs. (28.1)–(28.3). The force should be independent of intersubunit interactions.

The cycle fluxes are

$$J_a = \frac{(\kappa\alpha_1\beta_1 - \kappa'\alpha_1'\beta_1')(\alpha_2' + \beta_2)}{\Sigma}$$
$$= \frac{(k\alpha\beta - k'\alpha'\beta')(\alpha_2' + \beta_2)}{\Sigma}\left(\frac{y_{12}}{y_{21}}\right)^{f_k}\left(\frac{y_{21}}{y_{11}}\right)^{f_\alpha}\left(\frac{y_{11}}{y_{12}}\right)^{f_\beta}, \tag{28.6}$$

$$J_b = \frac{(\kappa\alpha_2\beta_2 - \kappa'\alpha_2'\beta_2')(\alpha_1' + \beta_1')}{\Sigma}$$
$$= \frac{(k\alpha\beta - k'\alpha'\beta')(\alpha_1' + \beta_1')}{\Sigma}\left(\frac{y_{12}}{y_{21}}\right)^{f_k}\left(\frac{y_{22}}{y_{12}}\right)^{f_\alpha}\left(\frac{y_{21}}{y_{22}}\right)^{f_\beta}, \tag{28.7}$$

where Σ is the sum of the 32 directional diagrams (see below). The total flux is $J = J_a + J_b$. Equations (28.6) and (28.7) confirm that $J \to 0$ when $X \to 0$ [Eq. (28.5)], as expected.

The state probabilities follow from the partial diagrams in Fig. 7-3. For example,

$$p_{11} = (\alpha_1'\alpha_2'\beta_1 + \alpha_1'\beta_1\beta_2 + \alpha_2'\beta_1\beta_2' + \alpha_1'\alpha_2\beta_2 + \kappa\beta_1\beta_2 + \kappa'\alpha_1'\alpha_2' + \kappa\alpha_2'\beta_1 + \kappa'\alpha_1'\beta_2)\Sigma^{-1}, \tag{28.8}$$

and similarly for p_{12}, p_{21}, and p_{22}. The sum of the four numerators (32 terms) is Σ. J is given by Eq. (28.4) or by $J_a + J_b$.

To avoid algebraic complexity, we now turn to the special case of one-way transitions. That is, we assume $\alpha' = 0$, $\beta' = 0$, $k' = 0$. This is a very plausible possibility. In this case, only five out of 32 of the directional diagrams are nonzero:

$$p_{11} = \frac{\kappa\beta_1\beta_2}{\Sigma} = \frac{k\beta^2}{\Sigma}\left(\frac{y_{12}}{y_{21}}\right)^{f_k}\left(\frac{y_{11}y_{21}}{y_{22}y_{12}}\right)^{f_\beta}, \tag{28.9}$$

$$p_{12} = \frac{\kappa\alpha_1\beta_2}{\Sigma} = \frac{k\alpha\beta}{\Sigma}\left(\frac{y_{12}}{y_{21}}\right)^{f_k}\left(\frac{y_{21}}{y_{11}}\right)^{f_\alpha}\left(\frac{y_{21}}{y_{22}}\right)^{f_\beta}, \tag{28.10}$$

$$p_{21} = \frac{\alpha_1\beta_2(\alpha_2 + \beta_1)}{\Sigma} = \frac{\alpha\beta^2}{\Sigma}\left(\frac{y_{21}}{y_{11}}\right)^{f_\alpha}\left(\frac{y_{11}y_{21}}{y_{22}y_{12}}\right)^{f_\beta} + \frac{\alpha^2\beta}{\Sigma}\left(\frac{y_{22}y_{21}}{y_{11}y_{12}}\right)^{f_\alpha}\left(\frac{y_{21}}{y_{22}}\right)^{f_\beta}, \tag{28.11}$$

$$p_{22} = \frac{\kappa\alpha_1\alpha_2}{\Sigma} = \frac{k\alpha^2}{\Sigma}\left(\frac{y_{22}y_{21}}{y_{11}y_{12}}\right)^{f_\alpha}\left(\frac{y_{12}}{y_{21}}\right)^{f_k}, \tag{28.12}$$

where Σ is the sum of the four numerators (five terms). The flux is then

$$J = \kappa p_{21} = k\left(\frac{y_{12}}{y_{21}}\right)^{f_k} p_{21}, \tag{28.13}$$

where p_{21} is given by Eq. (28.11). These results are generalizations of Eqs. (20.2) and (20.3). A wide variety of interaction effects on J is possible, depending on interaction parameter choices.

We proceed further with one particular special case of Eqs. (28.9)–(28.13). In Fig. 5-47, we take $f_\alpha = 0$ (diffusion-controlled binding of S on E_1) and $f_k = f_\beta = \frac{1}{2}$ (the rate-determining steps are, say, the chemical transformations $S \to P$ on E_1 and $P \to Q$ on E_2). We assume further that the binding of S on E_1 induces a conformational change in E_1 (but there is no such effect of P on E_2) so that the intermolecular interaction of E_1S (state 2) with E_2 and E_2P is different than that of E_1 (state 1) with E_2 and E_2P. For example, E_1S (state 2) may not fit as well against E_2, E_2P as does E_1 (state 1). Thus we assume

$$\begin{aligned} &\mathrm{E_1(state\ 1)}: && y_{11} = y_{12} \equiv y_1, \\ &\mathrm{E_1\,S(state\ 2)}: && y_{21} = y_{22} \equiv y_2, \end{aligned} \tag{28.14}$$

where the subscript on y_i refers to the state of E_1. We might note that the conventional equilibrium interaction parameter (Chapter 6)

$$y = y_{11}y_{22}/y_{12}y_{21}$$

is unity in this case. If there is a relatively poor fit in state 2 (see above), $y_2 < y_1$. If we now substitute Eqs. (28.14) and $f_\alpha = 0, f_k = f_\beta = \frac{1}{2}$ into Eqs. (28.1)–(28.3) and (28.9)–(28.13), the result is quite simple (especially because of $f_\alpha = 0$). Figure 7-14 simplifies to Fig. 7-15 and

$$\begin{aligned} p_{11} &= \frac{k\beta^2 Y}{\Sigma}, \qquad p_{12} = \frac{k\alpha\beta Y}{\Sigma}, \\ p_{21} &= \frac{\alpha\beta(\alpha+\beta)}{\Sigma}, \qquad p_{22} = \frac{k\alpha^2 Y}{\Sigma}, \end{aligned} \tag{28.15}$$

$$\Sigma = kY(\alpha^2 + \alpha\beta + \beta^2) + \alpha\beta(\alpha+\beta),$$

$$J = \frac{kY\alpha\beta(\alpha+\beta)}{kY(\alpha^2 + \alpha\beta + \beta^2) + \alpha\beta(\alpha+\beta)}, \tag{28.16}$$

where

$$Y \equiv \left(\frac{y_1}{y_2}\right)^{1/2}. \tag{28.17}$$

The dependence of J on Y in Eq. (28.16) is quite simple; J increases with Y. If E_1S makes a relatively poor fit with E_2, E_2S, as suggested above, then $y_2 < y_1$ and $Y > 1$. If the opposite is true, $y_2 > y_1$ and $Y < 1$. If $Y = 1$, Eq. (28.16) reduces to Eq. (20.3).

The dependence of J on α is of interest because α is proportional to c_S (e.g., c_Q is held constant and c_S is varied). When α is small, $J \rightarrow \alpha$. When α is large, the saturation value of J is

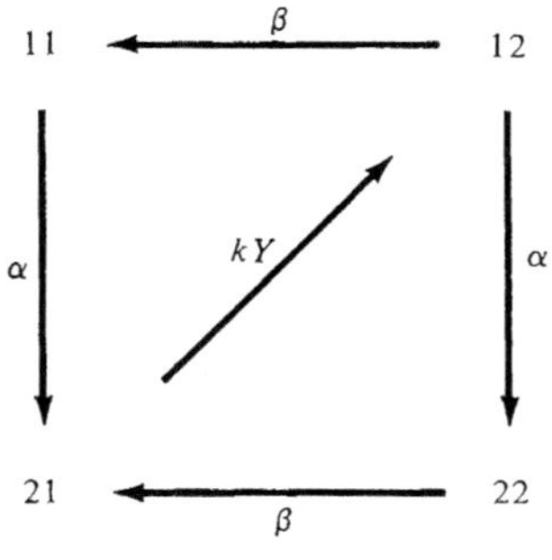

Fig. 7-15. Special case of Fig. 7-14 discussed in text [see Eq. (28.16)]. $Y \equiv (y_1/y_2)^{1/2}$.

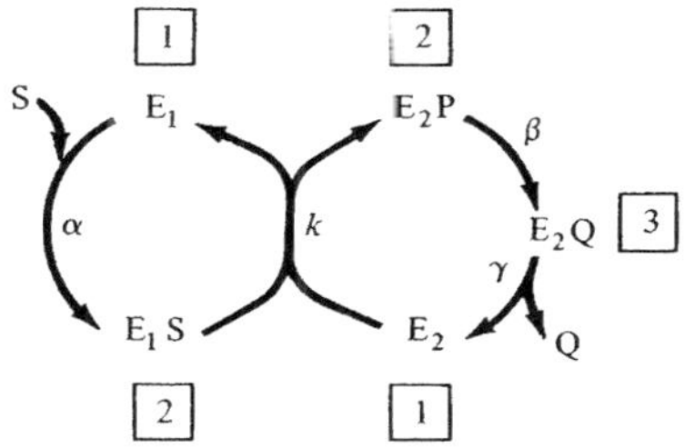

Fig. 7-16. Model of a dimer in which E_1 has a two-state cycle and E_2 has a three-state cycle. The dimer catalyzes $S \to Q$. Numbers in boxes are state numbers.

$$J \to \frac{kY\beta}{kY + \beta} \qquad (\alpha \to \infty). \tag{28.18}$$

If Y is very large, this saturation value of J is β; if Y is very small, it is $kY \to 0$. Note, from Eqs. (28.15), that p_{21} is relatively small if Y is large.

Dimer with a Two-State and a Three-State Cycle

This model (Fig. 7-16) is an extension of the previous one. In the E_1, E_2 complex (which again catalyzes $S \to Q$), E_2 has a three-state cycle here (i.e., E_2Q can no longer be treated as a transient intermediate). The unperturbed rate constants are included in Fig. 7-16. We consider one-way transitions only.

Intersubunit interactions may perturb the rate constants. That is, α may be different for each of the three states of E_2 and β, γ may be different for the two states of E_1. The kinetic diagram is shown in Fig. 7-17(a). Subscripts on rate constants refer to the state of the neighboring subunit.

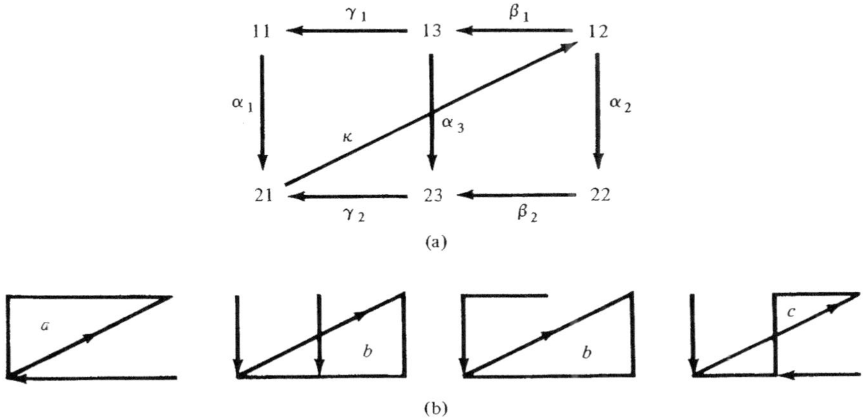

Fig. 7-17. (a) Kinetic diagram for Fig. 7-16 with same assignment of state numbers. Subscripts refer to the state of the nonreacting neighbor. (b) Flux diagrams for the cycles *a*, *b*, *c*.

The interaction parameters are $y_{ij} = e^{-w_{ij}/k_BT}$, as before, but here $i = 1, 2$ and $j = 1, 2, 3$. Again, following Eqs. (5.6) and (5.8), the explicit expressions for the possibly perturbed rate constants appearing in Fig. 7-17(a) are

$$\begin{aligned} \alpha_1 &= \alpha\left(\frac{y_{21}}{y_{11}}\right)^{f_\alpha}, & \alpha_2 &= \alpha\left(\frac{y_{22}}{y_{12}}\right)^{f_\alpha}, \\ \alpha_3 &= \alpha\left(\frac{y_{23}}{y_{13}}\right)^{f_\alpha}, & \kappa &= k\left(\frac{y_{12}}{y_{21}}\right)^{f_k}, \\ \beta_1 &= \beta\left(\frac{y_{13}}{y_{12}}\right)^{f_\beta}, & \beta_2 &= \beta\left(\frac{y_{23}}{y_{22}}\right)^{f_\beta}, \\ \gamma_1 &= \gamma\left(\frac{y_{11}}{y_{13}}\right)^{f_\gamma}, & \gamma_2 &= \gamma\left(\frac{y_{21}}{y_{23}}\right)^{f_\gamma}. \end{aligned} \tag{28.19}$$

If inverse rate constants are included, each inverse rate constant has the same interaction parameter factor but with the appropriate f replaced by $f - 1$ [see Eqs. (28.1)–(28.3)].

This model affords an opportunity to introduce an alternative and quite simple way of finding the p_{ij}, especially when only one-way transitions are used in a diagram [this method was mentioned in relation to Eq. (20.5)]. The procedure is: first, write the cycle fluxes, leaving Σ (sum of directional diagrams) unspecified; second, express the transition fluxes of the diagram in terms of both the cycle fluxes and the p_{ij}; third, obtain from these relations expressions for the p_{ij}; and fourth, find Σ by normalization. The method is advantageous basically because it is so much easier to enumerate and evaluate flux diagrams than directional diagrams.

The flux diagrams for the three possible cycles (a, b, c) with nonzero force (all have the same force) are shown in Fig. 7-17(b). From these,

$$\begin{aligned} J_a &= \frac{\kappa\alpha_1\beta_1\gamma_1 \cdot \beta_2\gamma_2}{\Sigma}, \qquad J_c = \frac{\kappa\alpha_3\beta_1\gamma_2 \cdot \alpha_1\beta_2}{\Sigma}, \\ J_b &= \frac{\kappa\alpha_2\beta_2\gamma_2(\alpha_1\alpha_3 + \alpha_1\gamma_1)}{\Sigma}. \end{aligned} \tag{28.20}$$

A transition flux[2] is equal to the sum of cycle fluxes for those cycles that use the transition. There are eight transitions and transition fluxes in Fig. 7-17a:

$$\begin{aligned} &\gamma_1 p_{13} = \alpha_1 p_{11} = J_a, \qquad \alpha_2 p_{12} = \beta_2 p_{22} = J_b, \\ &\alpha_3 p_{13} = J_c, \qquad \beta_1 p_{12} = J_a + J_c, \\ &\gamma_2 p_{23} = J_b + J_c, \qquad \kappa p_{21} = J_a + J_b + J_c. \end{aligned} \tag{28.21}$$

The six p_{ij} then follow easily (there are two redundancies). For example, p_{11} is J_a/α_1. Thus

$$\begin{aligned}
p_{11} &= \kappa\beta_1\beta_2\gamma_1\gamma_2\Sigma^{-1}, \qquad p_{13} = \kappa\alpha_1\beta_1\beta_2\gamma_2\Sigma^{-1},\\
p_{12} &= \kappa\alpha_1\beta_2\gamma_2(\alpha_3 + \gamma_1)\Sigma^{-1}, \qquad p_{22} = \kappa\alpha_1\alpha_2\gamma_2(\alpha_3 + \gamma_1)\Sigma^{-1},\\
p_{23} &= \kappa\alpha_1\beta_2(\alpha_2\alpha_3 + \alpha_2\gamma_1 + \alpha_3\beta_1)\Sigma^{-1},\\
p_{21} &= \alpha_1\beta_2\gamma_2(\beta_1\gamma_1 + \alpha_2\alpha_3 + \alpha_2\gamma_1 + \alpha_3\beta_1)\Sigma^{-1}.
\end{aligned} \tag{28.22}$$

Σ is then the sum of the six numerators, and the total flux (in S $\rightarrow$ Q) is

$$J = J_a + J_b + J_c = \kappa p_{21}. \tag{28.23}$$

As a more explicit example, consider the following. Because P is transferred directly from E_1 to E_2, the site for S and P on E_1 must be close in the complex to the site for P and Q on E_2. If S, P, and Q are fairly large molecules, they may interact with each other when both E_1 and E_2 have occupied sites. So we suppose that the intersubunit interaction between E_1 and E_2 is "normal" except when *both* sites are occupied. Thus we take (normal interaction)

$$y_{11} = y_{12} = y_{13} = y_{21} \equiv y_o. \tag{28.24}$$

However, y_o differs from y_{22} (S, P bound) and from y_{23} (S, Q bound). We define

$$Z \equiv y_o/y_{23}, \qquad z \equiv y_{23}/y_{22}. \tag{28.25}$$

We would expect the main effect of interactions to be from Z. If P and Q do not differ much in structure, $z \cong 1$. If S and Q on adjacent sites interfere with each other, $y_{23} < y_o$ and $Z > 1$. If S and Q attract each other, $Z < 1$.

A plausible set of f values in this example would be $f_\alpha = 0$, $f_\gamma = 1$ (diffusion-controlled binding of both S and Q) and $f_k = f_\beta = \frac{1}{2}$. Using these and Eqs. (28.24) and (28.25), we find from Eqs. (28.19) that all rate constants have their unperturbed values except

$$\beta_2 = \beta z^{1/2}, \qquad \gamma_2 = \gamma Z.$$

Consequently, Eqs. (28.22) simplify to

$$\begin{aligned}
p_{11} &= k\beta^2\gamma^2 z^{1/2} Z\Sigma^{-1}, \qquad p_{13} = k\alpha\beta^2\gamma z^{1/2} Z\Sigma^{-1},\\
p_{12} &= k\alpha\beta\gamma(\alpha + \gamma)z^{1/2} Z\Sigma^{-1},\\
p_{22} &= k\alpha^2\gamma(\alpha + \gamma)Z\Sigma^{-1},\\
p_{23} &= k\alpha^2\beta(\alpha + \gamma + \beta)z^{1/2}\Sigma^{-1},\\
p_{21} &= \alpha\beta\gamma(\beta\gamma + \alpha^2 + \alpha\gamma + \alpha\beta)z^{1/2} Z\Sigma^{-1},
\end{aligned} \tag{28.26}$$

where Σ is the sum of the six numerators. Also, $J = kp_{21}$. The dependence of J on Z has the form

$$J = AZ/(B + CZ). \tag{28.27}$$

For small and large α (recall that $\alpha \propto c_S$),

$$J \to \alpha \qquad (\alpha \to 0),$$
$$J \to \frac{k\beta\gamma z^{1/2}Z}{\beta\gamma z^{1/2}Z + k\beta z^{1/2} + k\gamma Z} \qquad (\alpha \to \infty). \tag{28.28}$$

Trimer with Two-State Cycles

This final example in this section extends the first example by adding a third two-state enzyme, making a linear complex of three subunits. The model, with unperturbed rate constants, is shown in Fig. 5-48. The net reaction is $S \to R$. P is transferred from E_1 to E_2 and Q from E_2 to E_3. We use one-way transitions only, to make the algebra tractable. The formula for the unperturbed flux (of $S \to R$) has already been given in Eq. (20.5). If two-way transitions are included below, the initial formalities are hardly more complicated [Eqs. (28.29) are matched by a corresponding set of inverse rate constants, with each f replaced by an $f - 1$], but the steady-state probabilities would have to be found by computer (from eight linear algebraic equations).

In the present treatment of the model, we assume that there are nearest-neighbor interactions between subunits that may perturb the rate constants. That is, E_1 and E_2 interact (interaction parameters y_{ij}) and E_2 and E_3 interact (interaction parameters z_{ij}), but E_1 and E_3 do not interact (the complex is linear). The diagram with perturbed rate constant notation is given in Fig. 7-18(a). The subscript on α_i refers to the state of E_2 (the neighbor of E_1); the subscript on β_i also refers to the state of E_2 (the neighbor of E_3); the subscript on k_i refers to the state of E_3; and the subscript on l_i refers to the state of E_1. The explicit expressions for the perturbed rate constants, found in each case by noting the pair interactions in final and initial states, are

$$\begin{aligned}
\alpha_1 &= \alpha\left(\frac{y_{21}}{y_{11}}\right)^{f_\alpha}, & \alpha_2 &= \alpha\left(\frac{y_{22}}{y_{12}}\right)^{f_\alpha},\\
\beta_1 &= \beta\left(\frac{z_{11}}{z_{12}}\right)^{f_\beta}, & \beta_2 &= \beta\left(\frac{z_{21}}{z_{22}}\right)^{f_\beta},\\
k_1 &= k\left(\frac{y_{12}z_{21}}{y_{21}z_{11}}\right)^{f_k}, & k_2 &= k\left(\frac{y_{12}z_{22}}{y_{21}z_{12}}\right)^{f_k},\\
l_1 &= l\left(\frac{y_{11}z_{12}}{y_{12}z_{21}}\right)^{f_l}, & l_2 &= l\left(\frac{y_{21}z_{12}}{y_{22}z_{21}}\right)^{f_l}.
\end{aligned} \tag{28.29}$$

The steady-state properties are found most easily from the transition flux method introduced in the preceding subsection. The six cycles with nonzero force (all have the same force) are shown in Fig. 7-18(b). The numbers in parentheses in this figure are the number of flux diagrams in each case. We leave the drawing of these diagrams as an exercise for the interested reader.

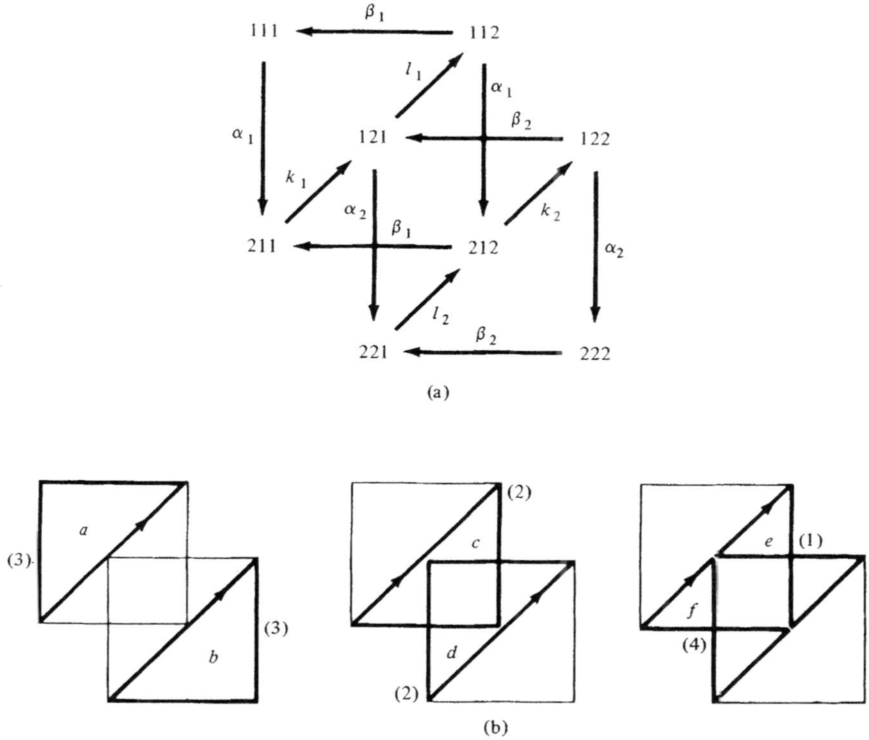

Fig. 7-18. (a) Kinetic diagram for a linear trimer with two-state cycles (see Fig. 5-48). The subscripts are explained in the text. (b) The six cycles belonging to the kinetic diagram. The numbers in parentheses are the number of flux diagrams in each case.

$$
\begin{aligned}
J_a &= k_1 l_1 \alpha_1 \beta_1 \cdot l_2 \beta_2 (k_2 \beta_2 + \alpha_2 \beta_1 + \beta_1 \beta_2) \Sigma^{-1}, \\
J_b &= k_2 l_2 \alpha_2 \beta_2 \cdot k_1 \alpha_1 (l_1 \alpha_1 + \alpha_1 \alpha_2 + \alpha_2 \beta_1) \Sigma^{-1}, \\
J_c &= k_1 l_1 \alpha_1 \beta_1 \cdot l_2 \alpha_1 \beta_2 (\alpha_2 + \beta_2) \Sigma^{-1}, \\
J_d &= k_2 l_2 \alpha_2 \beta_2 \cdot k_1 \alpha_1 \beta_2 (\alpha_1 + \beta_1) \Sigma^{-1}, \\
J_e &= k_2 l_1 \alpha_1 \beta_2 \cdot k_1 l_2 \alpha_1 \beta_2 \Sigma^{-1}, \\
J_f &= k_1 l_2 \alpha_2 \beta_1 \cdot \alpha_1 \beta_2 (\alpha_1 \alpha_2 + \alpha_1 \beta_2 + \alpha_2 \beta_1 + \beta_1 \beta_2) \Sigma^{-1}.
\end{aligned}
\tag{28.30}
$$

Figure 7-19 lists, on each transition line, those cycles that contribute to the particular transition flux [see Fig. 7-18(b)]. Twelve transition flux equations can then be read off of Fig. 7-19, and these, with Eqs. (28.30), immediately determine the eight p_{ijk} (with four redundancies). As before, Σ is found by

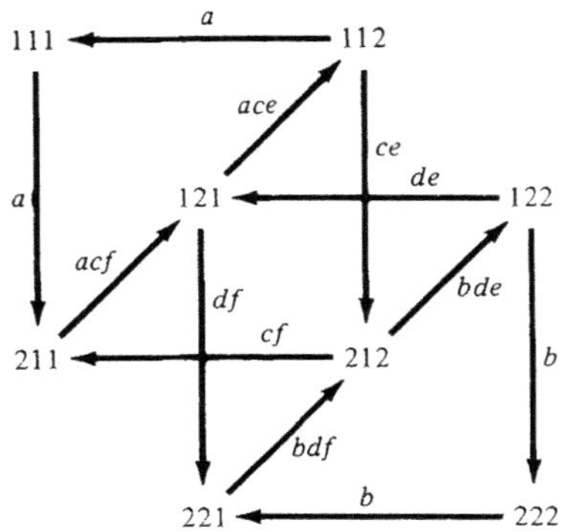

Fig. 7-19. Listing of cycles [Fig. 7-18(b)] that contribute to each transition flux in the diagram of Fig. 7-18.

normalizing the p_{ijk} to unity. Examples of transition flux equations, from Figs. 7-18(a) and 7-19, are

$$\begin{aligned} \beta_1 p_{112} &= \alpha_1 p_{111} = J_a, \\ \alpha_1 p_{112} &= J_c + J_e, \qquad l_1 p_{121} = J_a + J_c + J_e. \end{aligned} \tag{28.31}$$

Then $p_{111} = J_a/\alpha_1$, etc. The explicit writing out of the p_{ijk} is left to the reader.

The S → R flux is then

$$\begin{aligned} J &= k_1 p_{211} + k_2 p_{212} = l_1 p_{121} + l_2 p_{221} \\ &= (J_a + J_c + J_f) + (J_b + J_d + J_e) \\ &= (J_a + J_c + J_e) + (J_b + J_d + J_f). \end{aligned} \tag{28.32}$$

The terms in the second and third lines correspond to those in the first line, as can be seen from Fig. 7-19.

29. Complexes with Two One-Conformation Subunits

This section includes the treatment of the most basic case of intersubunit interaction effects on steady-state enzyme activity: two two-state enzyme molecules (e.g., of the Michaelis–Menten type), each capable of carrying out its own enzymatic activity, form a complex (dimer); because of intersubunit interactions, the rate constants in the enzymatic cycle of each subunit depend on the state of the other subunit.

We treat three examples in this section: the two two-state molecules of the dimer are identical and equivalent (isologous); the two two-state molecules are different; and the subunits of the isologous dimer of the first case have three-state cycles.

Two Identical Two-State Subunits[4]

Here we study a dimer of identical two-state molecules, each subunit of which, when by itself (unperturbed), has the kinetic diagram shown in Fig.

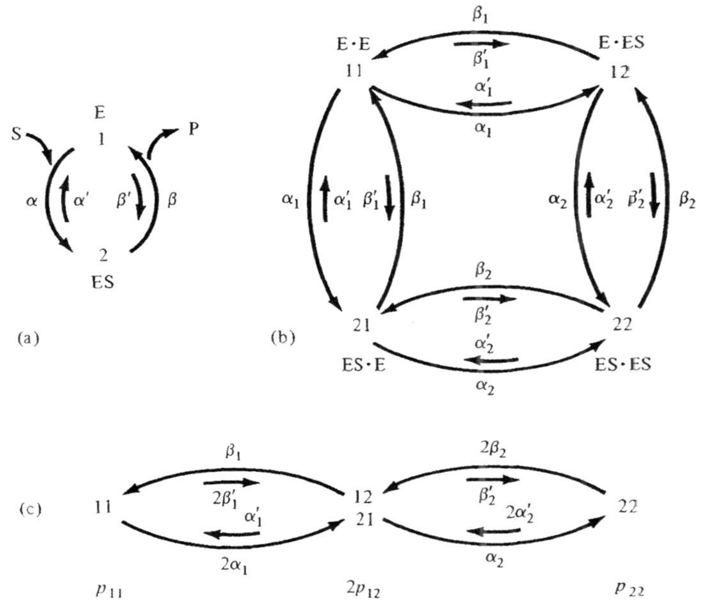

Fig. 7-20. (a) Kinetic diagram of a single subunit. E catalyzes S → P. (b) Diagram for an isologous dimer of two identical subunits. Subscripts refer to the state of the nonreacting neighbor. (c) Because of symmetry, (b) can be collapsed into a linear form. Steady-state probability notation is included.

7-20(a). The first-order rate constants $\alpha, \alpha', \beta, \beta'$ can apply to any two-state cyclic process, but, for concreteness, we shall consider that E catalyzes the reaction S → P, with EP a transient intermediate. Then state 1 is E and state 2 is ES. The thermodynamic force in this case is $X = \mu_S - \mu_P$, but in any case (Section 9), X is related to the rate constants of the cycle by

$$e^{X/kT} = \frac{\alpha\beta}{\alpha'\beta'}. \tag{29.1}$$

The unperturbed flux for one subunit is

$$J_o = \frac{\alpha\beta - \alpha'\beta'}{\alpha + \alpha' + \beta + \beta'}. \tag{29.2}$$

The dimer is symmetrical (isologous). That is, state 12 of the dimer is equivalent to state 21. The kinetic diagram of the dimer is given in Fig. 7-20(b). The subscripts on the rate constants refer to the state of the neighbor: the rate constants in the cycle of the reacting subunit are generally different, depending on whether the neighboring subunit is in state 1 or 2. The most obvious origin of this effect is that the conformation is different in states 1 and 2. The intersubunit interaction free energy is w_{ij}, relative to infinite separation of the subunits. Then $y_{ij} \equiv e^{-w_{ij}/kT}$ and $y_{12} = y_{21}$ (because of symmetry). We then have (Section 5)

$$\frac{\alpha_1}{\alpha_1'} = \frac{\alpha}{\alpha'} \cdot \frac{y_{12}}{y_{11}}, \qquad \frac{\alpha_2}{\alpha_2'} = \frac{\alpha}{\alpha'} \cdot \frac{y_{22}}{y_{12}},$$
$$\frac{\beta_1}{\beta_1'} = \frac{\beta}{\beta'} \cdot \frac{y_{11}}{y_{12}}, \qquad \frac{\beta_2}{\beta_2'} = \frac{\beta}{\beta'} \cdot \frac{y_{12}}{y_{22}}, \tag{29.3}$$

for the equilibrium constants of the dimer, as perturbed by interaction effects. The ratios y_{12}/y_{11} and y_{22}/y_{12} recur frequently, so we define

$$y_1 \equiv y_{12}/y_{11} \qquad \text{and} \qquad y_2 \equiv y_{22}/y_{12}. \tag{29.4}$$

The familiar interaction parameter y is then related to y_1 and y_2 by

$$y = y_{11}y_{22}/y_{12}^2 = y_2/y_1. \tag{29.5}$$

If we now introduce the parameters f_α and f_β (Section 5) in order to split the free energy effects in Eqs. (29.3) between forward and backward rate constants, we have

$$\alpha_1 = \alpha y_1^{f_\alpha}, \qquad \alpha_1' = \alpha' y_1^{f_\alpha - 1}, \qquad \alpha_2 = \alpha y_2^{f_\alpha}, \qquad \alpha_2' = \alpha' y_2^{f_\alpha - 1},$$
$$\beta_1 = \beta y_1^{-f_\beta}, \qquad \beta_1' = \beta' y_1^{1-f_\beta}, \qquad \beta_2 = \beta y_2^{-f_\beta}, \qquad \beta_2' = \beta' y_2^{1-f_\beta}. \tag{29.6}$$

Because of the symmetry in Fig. 7-20(b) (states 12 and 21 have the same properties), we can simplify the kinetic diagram, as in Fig. 7-20(c). That is, states 12 and 21 are combined and treated as a single state (with state probability, at steady state, designated $2p_{12}$). Note, in Fig. 7-20(c), the factors of 2 in the rate constants for transitions out of states 11 and 22, because of the double degeneracy. The linearity of the diagram in Fig. 7-20(c) allows a nonequilibrium detailed balance solution of the steady-state rate equations. That is,

$$\frac{2p_{12}}{p_{11}} = \frac{2(\alpha_1 + \beta_1')}{\beta_1 + \alpha_1'} = 2x_1, \tag{29.7}$$

$$\frac{p_{22}}{2p_{12}} = \frac{\alpha_2 + \beta_2'}{2(\beta_2 + \alpha_2')} = \frac{x_2}{2}, \tag{29.8}$$

where

$$x_1 \equiv \frac{\alpha_1 + \beta_1'}{\beta_1 + \alpha_1'} = \frac{\alpha y_1^{f_\alpha} + \beta' y_1^{1-f_\beta}}{\beta y_1^{-f_\beta} + \alpha' y_1^{f_\alpha - 1}}, \tag{29.9}$$

$$x_2 \equiv \frac{\alpha_2 + \beta_2'}{\beta_2 + \alpha_2'} = \frac{\alpha y_2^{f_\alpha} + \beta' y_2^{1-f_\beta}}{\beta y_2^{-f_\beta} + \alpha' y_2^{f_\alpha - 1}}. \tag{29.10}$$

Parameters like x_i were first introduced in Eqs. (16.52), (16.60), (17.3), and (17.5). Here they include interaction effects. In the S → P interpretation of Fig. 7-20(a), α is proportional to c_S and β' is proportional to c_P. If $\beta' = 0$ (Michaelis–Menten enzyme), x_1 and x_2 are proportional to c_S. The x_i are steady-state analogues of $x = Kc$, used in equilibrium problems (Chapter 6).

Equations (29.7) and (29.8) lead to the steady-state populations of the states:

$$p_{11} = \frac{1}{\xi}, \qquad p_{12} = p_{21} = \frac{x_1}{\xi}, \qquad p_{22} = \frac{x_1 x_2}{\xi},$$
$$p_2 = \frac{1}{2}(p_{12} + p_{21}) + p_{22} = \frac{x_1(1 + x_2)}{\xi}, \tag{29.11}$$

$$\xi = 1 + 2x_1 + x_1 x_2, \tag{29.12}$$

where p_2 is the fractional binding of S and ξ is the analogue of a grand partition function, which arises because of the detailed balance type of solution of the steady-state rate equations.

The total S → P flux of the dimer follows from Fig. 7-20(c), using either α transitions or β transitions (the two fluxes must be equal at steady state):

$$J = 2[(\alpha_1 p_{11} - \alpha_1' p_{12}) + (\alpha_2 p_{12} - \alpha_2' p_{22})], \tag{29.13}$$
$$= 2[(\beta_2 p_{22} - \beta_2' p_{12}) + (\beta_1 p_{12} - \beta_1' p_{11})]. \tag{29.14}$$

On substituting Eqs. (29.6) and (29.11), we find for the dimer flux

$$J = \frac{2(\alpha\beta - \alpha'\beta')(y_1^{f_\alpha - f_\beta} + y_2^{f_\alpha - f_\beta} x_{12})}{(\beta y_1^{-f_\beta} + \alpha' y_1^{f_\alpha - 1})(1 + 2x_1 + x_1 x_2)}, \tag{29.15}$$

where

$$x_{12} \equiv \frac{\alpha y_1^{f_\alpha} + \beta' y_1^{1-f_\beta}}{\beta y_2^{-f_\beta} + \alpha' y_2^{f_\alpha - 1}}. \tag{29.16}$$

It will be seen from Eqs. (29.9) and (29.10) that x_{12} is a hybrid between x_1 and x_2.

Equation (29.15) gives an explicit expression for the flux of the dimer, as perturbed by arbitrary interactions between the two subunits. The parameters that enter the equation are: unperturbed rate constants (α, α', β, β'); subunit interaction free energy differences (in the form $y_1 = y_{12}/y_{11}$ and $y_2 = y_{22}/y_{12}$); and free energy "splitting" parameters (f_α, f_β).

If there are no differential subunit interaction effects, that is, if

$$w_{11} = w_{12} = w_{22}, \qquad y_{11} = y_{12} = y_{22}, \qquad y_1 = y_2 = 1, \tag{29.17}$$

then Eq. (29.15) reduces, as expected, to $J = 2J_o$, where J_o, the unperturbed flux for one subunit, is given in Eq. (29.2).

Another important special case is $f_\alpha + f_\beta = 1$. As pointed out in the last subsection of Section 27, when this relation holds, the steady state is a quasiequilibrium state. For the present model, the two most likely combinations that would produce $f_\alpha + f_\beta = 1$ are $f_\alpha = 0, f_\beta = 1$ and $f_\alpha = f_\beta = \frac{1}{2}$. The former combination could arise in the S → P example if the binding of both c_S and c_P is diffusion controlled. In this case, interaction effects have no effect on the binding rate constants: $\alpha_1 = \alpha_2 = \alpha$ and $\beta_1' = \beta_2' = \beta'$ in Eqs. (29.6). (Interaction effects would be confined to the inverse rate constants.) This requires $f_\alpha = 0$ and $f_\beta = 1$. The latter combination ($f_\alpha = f_\beta = \frac{1}{2}$) would be possible, again in the S → P example, if the binding of S induces a conformational change in E, E + S → ES → E*S, and the conformational change is rate determining, and if, furthermore, in E*S → E*P → E + P, the transformation S → P is rate determining.

For any combination $f_\alpha + f_\beta = 1$, Eqs. (29.9) and (29.10) simplify to the equilibrium-like expressions

$$x_1 = y_1 x, \qquad x_2 = y_2 x, \tag{29.18}$$

where

$$x \equiv \frac{\alpha + \beta'}{\beta + \alpha'}. \tag{29.19}$$

Then, from Eq. (29.12),

$$\xi = 1 + 2y_1 x + y_1 y_2 x^2 = 1 + 2\frac{y_{12}}{y_{11}}x + \frac{y_{22}}{y_{11}}x^2, \tag{29.20}$$

which has the same form as the equilibrium relation, Eq. (23.8). The separate terms in ξ determine the p_{ij} [Eq. (29.11)]. The expression for the flux, Eq. (29.15), simplifies to (we put $f_\beta = 1 - f_\alpha$)

$$J = \frac{2(\alpha\beta - \alpha'\beta')(y_1^{f_\alpha} + y_1 y_2^{f_\alpha} x)}{(\beta + \alpha')(1 + 2y_1 x + y_1 y_2 x^2)}. \tag{29.21}$$

Incidentally, Eq. (27.34) is a special case of Eq. (29.21) in which

$$f_\alpha = f, \qquad x = K\bar{c}, \qquad \frac{\alpha\beta - \alpha'\beta'}{\beta + \alpha'} = \frac{\beta K(c_A - c_B)}{2}. \tag{29.22}$$

Also, we note that if state 22 is relatively unstable, then y_2 is small and Eq. (29.21) reduces to Eq. (17.12) (if we also take $y_{12} = y_{11}$ or $y_1 = 1$). This is an example of half-of-the-sites reactivity because, with state 22 eliminated from Fig. 7-20(b), only one subunit can cycle at a time. The other subunit is always in state 1. Half-of-the-sites reactivity also occurs if state 11 is unstable and is excluded from Fig. 7-20(b). For this case, y_{11} is relatively small, y_1 is large, and ξ in the denominator of Eq. (29.21) reduces to

$$\xi = y_1 x(2 + y_2 x).$$

That is, $p_{11} \to 0$. Then, if $0 < f_\alpha < 1$, Eq. (29.21) becomes

$$J = \frac{2(\alpha\beta - \alpha'\beta')y_2^{f_\alpha}}{(\beta + \alpha')(2 + y_2 x)}. \tag{29.23}$$

In this case, the noncycling subunit is always in state 2.

Example with States 11 and 22 as Transient Intermediates

In the discussion following Eq. (17.13), mention was made of the case in which states 11 and 22 are both unstable so that the dimer populates states 12 and 21 and, in its enzymatic cycling, uses states 11 and 22 essentially as transient intermediates. Here we consider an explicit example of this type. The more important three-state trimer (see Fig. 5-33), with analogous properties, will be discussed in Section 32.

In this example, we take $f_\alpha = f_\beta = \frac{1}{2}$ (see above) and $y_{11} = y_{22}$. That is, the two symmetrical forms of the dimer have the same intersubunit interaction free energy. The case of primary interest is $y \to 0$, where [see Eq. (27.37)]

$$y^{1/2} = \frac{y_{11}}{y_{12}} = \frac{y_{22}}{y_{12}} = \frac{1}{y_1} = y_2. \tag{29.24}$$

In this special case (arbitrary y to begin with), we put $f_\alpha = \frac{1}{2}$, $y_1 = y^{-1/2}$, and $y_2 = y^{1/2}$ in Eq. (29.21), and define, for convenience, $Y = y^{1/4}$. Then we have

$$J = \frac{2(\alpha\beta - \alpha'\beta')Y^{-1}(1 + x)}{(\beta + \alpha')(1 + 2Y^{-2}x + x^2)}. \tag{29.25}$$

The state populations are

$$p_{11} = \frac{1}{\xi}, \qquad p_{12} = p_{21} = \frac{Y^{-2}x}{\xi}, \qquad p_{22} = \frac{x^2}{\xi}, \tag{29.26}$$

$$\xi = 1 + 2Y^{-2}x + x^2. \tag{29.27}$$

The rate constants are shown in Fig. 7-21. When $Y \to 0$ (the primary case),

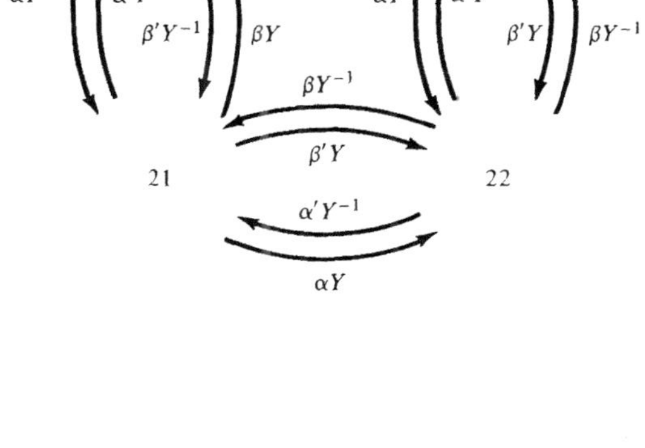

Fig. 7-21. Special case of Fig. 7-20(b) in which states 11 and 22 are transient intermediates ($Y \to 0$). See text for further details.

$$J = \frac{(\alpha\beta - \alpha'\beta')\,Y\,(1 + x)}{(\beta + \alpha')x} \qquad (Y \rightarrow 0), \qquad (29.28)$$

$$p_{12} = p_{21} \cong \frac{1}{2}, \qquad p_{11} = \frac{Y^2}{2x}, \qquad p_{22} = \frac{xY^2}{2}. \qquad (29.29)$$

The flux is small because all of the rate constants out of states 12 and 21 are proportional to Y (Fig. 7-21). It can be seen from Fig. 7-21 that only half of the completed cycles are of the type $12 \rightarrow 11 \rightarrow 21$ (there are four possibilities of this type). The other half are of the type $12 \rightarrow 11 \rightarrow 12$ (again there are four possibilities).

Incidentally, if Y is large (positive cooperativity) in Eq. (29.25),

$$J = \frac{2(\alpha\beta - \alpha'\beta')(1 + x)}{(\beta + \alpha')\,Y\,(1 + x^2)} \qquad (Y \rightarrow \infty). \qquad (29.30)$$

Again J is small (here $p_{12} = p_{21} \rightarrow 0$). The rate constants out of the populated states 11 and 22 are all proportional to $1/Y$ (Fig. 7-21). Thus both small and large Y lead to small J. At a given x, the maximum J occurs when $Y^2 = 2x/(1 + x^2)$ [Eq. (29.25)].

Michaelis–Menten Dimer[4]

We return now to Eqs. (29.6)–(29.16) for arbitrary f_α and f_β. In the important Michaelis–Menten special case, the catalyzed reaction is S $\rightarrow$ P, and we put $\beta' = 0$ in all of these equations (that is, far from equilibrium, cycles are completed essentially in only one direction). Also, we can write $\alpha = \alpha^* c_S$; x_1, x_2, and x_{12} are all proportional to c_S. The principal relations are

$$J = \frac{2\alpha\beta(y_1^{f_\alpha - f_\beta} + y_2^{f_\alpha - f_\beta} x_{12})}{(\beta y_1^{-f_\beta} + \alpha' y_1^{f_\alpha - 1})(1 + 2x_1 + x_1 x_2)}, \qquad (29.31)$$

$$x_1 = \frac{\alpha y_1^{f_\alpha}}{\beta y_1^{-f_\beta} + \alpha' y_1^{f_\alpha - 1}}, \qquad (29.32)$$

$$x_2 = \frac{\alpha y_2^{f_\alpha}}{\beta y_2^{-f_\beta} + \alpha' y_2^{f_\alpha - 1}}, \qquad (29.33)$$

$$x_{12} = \frac{\alpha y_1^{f_\alpha}}{\beta y_2^{-f_\beta} + \alpha' y_2^{f_\alpha - 1}}. \qquad (29.34)$$

The state populations are given by Eq. (29.11).

To provide a more explicit example, we take $f_\alpha = 0$ (diffusion-controlled binding of S), but leave f_β arbitrary. In this case, the flux is

$$J = \frac{2\beta x_1(y_1^{-f_\beta} + y_2^{-f_\beta} x_2)}{1 + 2x_1 + x_1 x_2}, \qquad (29.35)$$

$$x_1 = \frac{\alpha}{\beta y_1^{-f_\beta} + \alpha' y_1^{-1}}, \tag{29.36}$$

$$x_2 = \frac{\alpha}{\beta y_2^{-f_\beta} + \alpha' y_2^{-1}}, \tag{29.37}$$

where we have used $x_{12} = x_2$ (because $f_\alpha = 0$). The fractional binding of S at steady state is

$$p_2 = \frac{1}{2}(p_{12} + p_{21}) + p_{22} = p_{12} + p_{22}$$
$$= \frac{x_1(1 + x_2)}{1 + 2x_1 + x_1 x_2}. \tag{29.38}$$

Note that this expression has a form similar to but not the same as J [Eq. (29.35)].

We shall consider this case only briefly below because a very detailed discussion with numerical examples is easily accessible elsewhere.[4] As a convenience to the reader who wishes to pursue the more detailed analysis, we give here the correspondences in notation:

$$\alpha \to k_1, \quad \alpha' \to k_2, \quad \beta \to k_3, \quad f_\beta \to g, \quad J \to 2V,$$
$$y_1^{-1} \to y, \quad y_2^{-1} \to z, \quad x_1 \to Y, \quad x_2 \to Z, \quad \xi \to \Sigma, \tag{29.39}$$
$$\beta y_1^{-f_\beta} + \alpha' y_1^{-1} \to F_y, \quad \beta y_2^{-f_\beta} + \alpha' y_2^{-1} \to F_z.$$

Because x_1 and x_2 are both proportional to c_S, when c_S is small

$$J \to 2\beta x_1 y_1^{-f_\beta} \qquad (c_S \to 0), \tag{29.40}$$

and the asymptotic flux when c_S is large is

$$J \to 2\beta y_2^{-f_\beta} = 2\beta\left(\frac{y_{12}}{y_{22}}\right)^{f_\beta} \qquad (c_S \to \infty). \tag{29.41}$$

This is usually the maximum in the function $J(c_S)$ but not always. Upon examining $\partial J/\partial\alpha$ in Eq. (29.35), one finds[4] that J passes through a maximum at a finite c_S if and only if $y^{f_\beta} > 2$. With positive f_β (as would be expected), this can occur in positive cooperativity ($y > 1$) cases only, provided that y is sufficiently large (e.g., $y > 4$ if $f_\beta = \frac{1}{2}$).

Equation (29.41) shows that the asymptotic ($c_S \to \infty$) flux is decreased by a relatively stable interaction in state 22 (y_{22}) and increased by relatively stable interactions in states 12 and 21 (y_{12}).

When there are no intersubunit interaction effects ($y_1 = y_2 = 1$),

$$J = 2J_o = \frac{2\alpha\beta}{\alpha + \alpha' + \beta}. \tag{29.42}$$

The fractional binding of S in this case is

$$p_2 = \frac{\alpha}{\alpha + \alpha' + \beta}. \tag{29.43}$$

This has the same dependence on α (or c_S) as J_o because $J_o = \beta p_2$. This is a major difference between the unperturbed (or $y_1 = y_2 = 1$) and perturbed systems: p_2 and J in the perturbed system do not usually have the same dependence on α (see the examples below). It should be mentioned that J_o is proportional to p_2 in the unperturbed case only for a Michaelis–Menten enzyme ($\beta' = 0$). In the more general case, the expression for p_2 corresponding to Eq. (29.2) is

$$p_2 = \frac{\alpha + \beta'}{\alpha + \alpha' + \beta + \beta'}. \tag{29.44}$$

Figure 7-22 presents a numerical example with negative cooperativity and half-of-the-sites reactivity. The abscissa is α/β, which is proportional to c_S. The choice of parameters is $f_\beta = 0.8$, $y_1 = 1.0$, $y_2 = 0.01$, and $\alpha'/\beta = 1.0$. Then $y = y_2/y_1 = 0.01$ (negative cooperativity). The intersubunit interaction is the same in states 11, 12, and 21 ($y_{11} = y_{12} = y_{21}$), but the interaction in state 22 is relatively unfavorable ($y_{22} = 0.01y_{12}$). This decreases the stability of state 22 but increases the rate constants out of state 22. The net result, as can be seen in Fig. 7-22, is: p_2 reaches $\frac{1}{2}$ and J_o/β approaches its limit, 1.0,

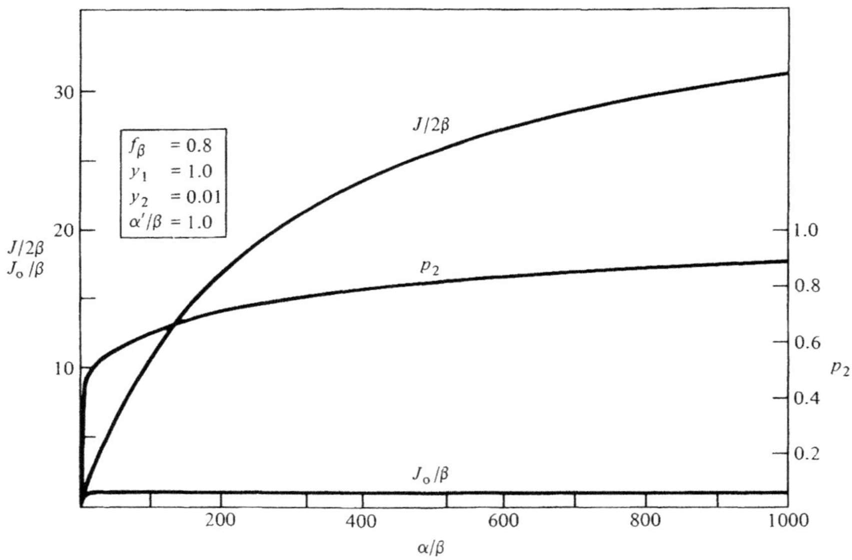

Fig. 7-22. Numerical illustration, for a Michaelis–Menten dimer ($\beta' = 0$), of Eqs. (29.35), (29.38), and (29.42). The abscissa, α/β, is proportional to c_S. Parameter choices are in inset. These correspond to negative cooperativity and half-of-the-sites reactivity. See text for details.

at relatively small α/β (i.e., small c_S); the second stage of p_2 ($\frac{1}{2} < p_2 < 1$) and $J/2\beta$ increase much more slowly with α/β; but in due course the flux $J/2\beta$ is considerably enhanced (over J_o/β) by the negative cooperativity (the asymptotic value of $J/2\beta$ is 39.8 when $\alpha/\beta \to \infty$). $J/2\beta$ reaches half of its maximum value at $\alpha/\beta = 273.6$. Except when α/β is very small, state 11 is hardly used (e.g., p_{11} is 0.045 at $\alpha/\beta = 20$ and 0.0073 at $\alpha/\beta = 100$). Thus only one subunit cycles at a time; the other is in state 2.

An example with positive cooperativity is given in Fig. 7-23. The parameters are the same as in Fig. 7-22, except that y_1 and y_2 are reversed: $y_1 = 0.01$ and $y_2 = 1.0$. Thus $y = y_2/y_1 = 100$ (positive cooperativity). States 12, 21, and 22 have equal intersubunit interactions ($y_{12} = y_{22}$), but state 11 has an unusually favorable interaction ($y_{11} = 100y_{12}$). This excess stability of state 11 delays the binding of S on the dimer; the shape of $p_2(c_S)$ is typical for positive cooperativity. The reference curve J_o/β is the same as in Fig. 7-22, but both scales are different. The peak in $J/2\beta$ is only 2.63 (at $\alpha/\beta = 21.2$), compared to 39.8 in Fig. 7-22. The asymptotic value of $J/2\beta$ at $\alpha/\beta \to \infty$ is 1.0.

Note, in both Figs. 7-22 and 7-23, that the p_2 and $J/2\beta$ curves have quite different shapes.

Two Different Two-State Subunits

We now consider a complex of two different subunits, each of which carries out its own cyclic activity by means of a two-state cycle. Because the subunits

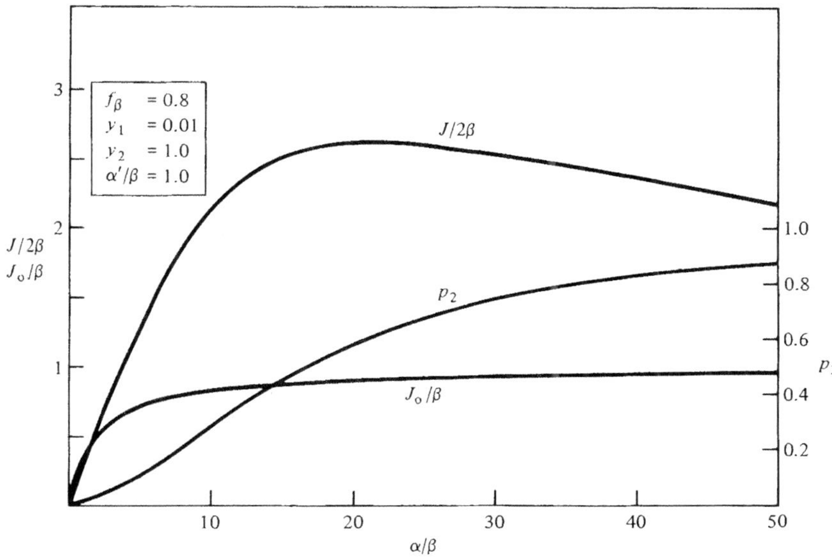

Fig. 7-23. Same as Fig. 7-22 except that the values of y_1 and y_2 are reversed (see inset) This corresponds to positive cooperativity. See text for further details.

are attached to each other, as a complex, the rate constants of one may depend on the state of the other. An example would be E_1 catalyzing $S_1 \rightarrow P_1$ and E_2 catalyzing $S_2 \rightarrow P_2$, as in Fig. 5-20. The unperturbed fluxes are denoted J_{o1} and J_{o2} and the perturbed fluxes J_1 and J_2. Equations (17.6)–(17.11) refer to a special case.

The interaction parameters are the usual $y_{ij} \equiv e^{-w_{ij}/kT}$, where $y_{12} \neq y_{21}$ here, in general, because the subunits are different. The rate constant notation that we use is shown in Fig. 7-24. Figure 7-24(a) relates to the unperturbed (separated) enzyme molecules, whereas Fig. 7-24(b) relates to the dimer. The subscripts in Fig. 7-24(b) refer to the state of the other (nonreacting) subunit. Inverse rate constants have primes, but they are omitted from the figure for simplicity and because we shall carry out the algebra only for the one-way cycle special case.

The dependence of rate constants on interaction parameters is similar to that in Eqs. (29.6):

$$\begin{aligned} \alpha_1 &= \alpha\left(\frac{y_{21}}{y_{11}}\right)^{f_\alpha}, \qquad \beta_1 = \beta\left(\frac{y_{21}}{y_{11}}\right)^{-f_\beta}, \\ \alpha_2 &= \alpha\left(\frac{y_{22}}{y_{12}}\right)^{f_\alpha}, \qquad \beta_2 = \beta\left(\frac{y_{22}}{y_{12}}\right)^{-f_\beta}, \end{aligned} \tag{29.45}$$

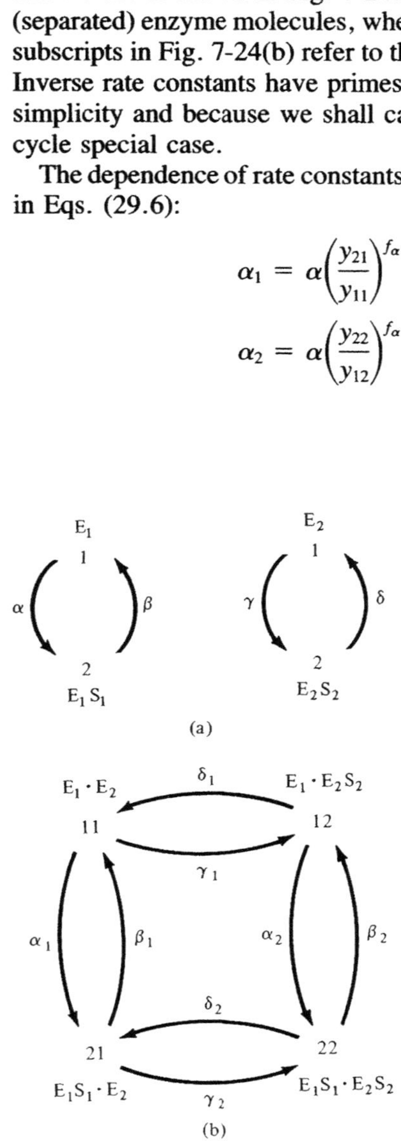

Fig. 7-24. (a) Unperturbed cycles for two different two-state subunits in a complex (dimer). (b) Diagram for the dimer. Subscripts refer to the state of the nonreacting neighbor. Inverse rate constants (omitted) have primes, in both (a) and (b).

$$\begin{aligned} \gamma_1 &= \gamma\left(\frac{y_{12}}{y_{11}}\right)^{f_\gamma}, & \delta_1 &= \delta\left(\frac{y_{12}}{y_{11}}\right)^{-f_\delta}, \\ \gamma_2 &= \gamma\left(\frac{y_{22}}{y_{21}}\right)^{f_\gamma}, & \delta_2 &= \delta\left(\frac{y_{22}}{y_{21}}\right)^{-f_\delta}. \end{aligned} \tag{29.45}$$

Examples of inverse rate constants are [see Eqs. (29.6)]

$$\alpha_1' = \alpha'\left(\frac{y_{21}}{y_{11}}\right)^{f_\alpha - 1}, \qquad \beta_1' = \beta'\left(\frac{y_{21}}{y_{11}}\right)^{1-f_\beta}. \tag{29.46}$$

With all the rate constants specified as above, for Fig. 7-24b, the steady-state rate equations follow from the diagram:

$$\begin{aligned} \frac{dp_{11}}{dt} &= 0 \\ &= (\delta_1 + \gamma_1')p_{12} + (\beta_1 + \alpha_1')p_{21} - (\gamma_1 + \delta_1' + \alpha_1 + \beta_1')p_{11}, \end{aligned} \tag{29.47}$$

etc. Four linear algebraic equations in the p_{ij} (including normalization) can then be solved to give the p_{ij} as functions of the rate constants (including interaction parameters). Then the two fluxes are

$$\begin{aligned} J_1 &= (\alpha_1 p_{11} - \alpha_1' p_{21}) + (\alpha_2 p_{12} - \alpha_2' p_{22}), \\ J_2 &= (\gamma_1 p_{11} - \gamma_1' p_{12}) + (\gamma_2 p_{21} - \gamma_2' p_{22}). \end{aligned} \tag{29.48}$$

The above is the way to proceed in principle to find the general steady-state properties. In practice, a numerical solution of the p_{ij} equations would probably be used even for this relatively simple model, because of burdensome algebra. As an alternative, we give the rather simple algebra below for the special case in which all inverse rate constants can be neglected.

The simplest procedure in this special case is again via the expression of transition fluxes in terms of cycle fluxes. With one-way transitions, the diagram in Fig. 7-24(b) has six cycles, $a, \ldots, f$, which are shown in Fig. 7-25(a). The transition fluxes are then composites of the cycle fluxes, as indicated in Fig. 7-25(b).

Each of cycles a, b, c, and d has three flux diagrams; each of cycles e and f has only one flux diagram. Thus the cycle fluxes are

$$\begin{aligned} J_a &= \alpha_1\beta_1(\alpha_2\delta_2 + \beta_2\delta_1 + \delta_1\delta_2)\Sigma^{-1}, \\ J_b &= \gamma_2\delta_2(\alpha_1\delta_1 + \alpha_2\gamma_1 + \alpha_1\alpha_2)\Sigma^{-1}, \\ J_c &= \alpha_2\beta_2(\alpha_1\gamma_2 + \beta_1\gamma_1 + \gamma_1\gamma_2)\Sigma^{-1} \\ J_d &= \gamma_1\delta_1(\beta_1\delta_2 + \beta_2\gamma_2 + \beta_1\beta_2)\Sigma^{-1}, \\ J_e &= \alpha_1\beta_2\gamma_2\delta_1\Sigma^{-1}, \qquad J_f = \alpha_2\beta_1\gamma_1\delta_2\Sigma^{-1}, \end{aligned} \tag{29.49}$$

where Σ is to be determined below. We can then find the p_{ij} from, for example,

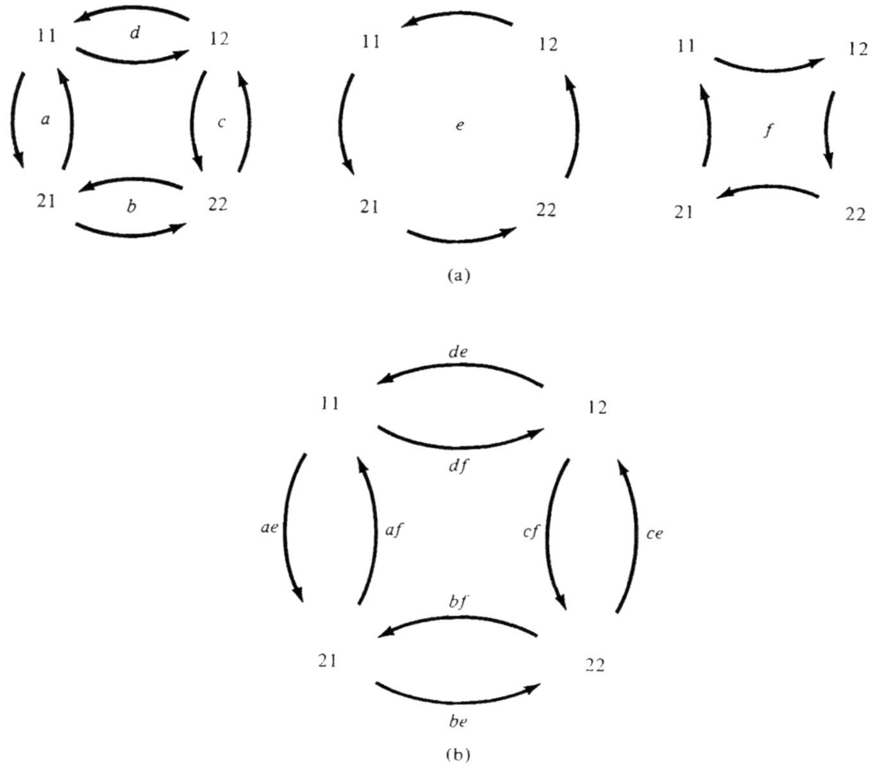

Fig. 7-25. (a) Six cycles in Fig. 7-24(b), with one-way transitions only. (b) Cycle contributions to transition fluxes in the same special case.

$$\begin{aligned} \alpha_1 p_{11} &= J_a + J_e, & \alpha_2 p_{12} &= J_c + J_f, \\ \beta_1 p_{21} &= J_a + J_f, & \beta_2 p_{22} &= J_c + J_e. \end{aligned} \tag{29.50}$$

Upon combining Eqs. (29.49) and (29.50), we have

$$\begin{aligned} p_{11} &= [\beta_1 \delta_2(\alpha_2 + \delta_1) + \beta_2 \delta_1(\beta_1 + \gamma_2)]\Sigma^{-1}, \\ p_{12} &= [\beta_2 \gamma_2(\alpha_1 + \gamma_1) + \beta_1 \gamma_1(\beta_2 + \delta_2)]\Sigma^{-1}, \\ p_{21} &= [\alpha_1 \delta_1(\beta_2 + \delta_2) + \alpha_2 \delta_2(\alpha_1 + \gamma_1)]\Sigma^{-1}, \\ p_{22} &= [\alpha_2 \gamma_1(\beta_1 + \gamma_2) + \alpha_1 \gamma_2(\alpha_2 + \delta_1)]\Sigma^{-1}, \end{aligned} \tag{29.51}$$

where Σ is the sum of the four numerators. Finally, the two fluxes are

$$\begin{aligned} J_1 &= \alpha_1 p_{11} + \alpha_2 p_{12} = \beta_1 p_{21} + \beta_2 p_{22} = J_a + J_c + J_e + J_f, \\ J_2 &= \gamma_1 p_{11} + \gamma_2 p_{21} = \delta_1 p_{12} + \delta_2 p_{22} = J_b + J_d + J_e + J_f. \end{aligned} \tag{29.52}$$

When there are no interaction effects ($y_{11} = y_{12} = y_{21} = y_{22}$), Eqs. (29.51) and (29.52) simplify to

$$p_{11} = \frac{\beta\delta}{\Sigma}, \qquad p_{12} = \frac{\beta\gamma}{\Sigma},$$
$$p_{21} = \frac{\alpha\delta}{\Sigma}, \qquad p_{22} = \frac{\alpha\gamma}{\Sigma}, \tag{29.53}$$
$$\Sigma = (\alpha + \beta)(\gamma + \delta),$$

$$J_{o1} = \frac{\alpha\beta}{\alpha + \beta}, \qquad J_{o2} = \frac{\gamma\delta}{\gamma + \delta}. \tag{29.54}$$

Equations (29.51) and (29.52) contain the required effects of interactions on the fluxes. However, the results are much more transparent in special cases. For example, let us assume that only the 22 interaction is unusual:

$$y_{11} = y_{12} = y_{21} \neq y_{22}. \tag{29.55}$$

Then

$$y = \frac{y_{11}y_{22}}{y_{12}y_{21}} = \frac{y_{22}}{y_{ij}}, \tag{29.56}$$

where ij = 11, 12, or 21. Thus there is only one independent interaction parameter, y. Also, let us take all four fs = $\frac{1}{2}$ and define $Y \equiv y^{1/2}$. Then

$$\begin{aligned} \alpha_1 &= \alpha, & \beta_1 &= \beta, & \gamma_1 &= \gamma, & \delta_1 &= \delta, \\ \alpha_2 &= \alpha Y, & \beta_2 &= \beta Y^{-1}, & \gamma_2 &= \gamma Y, & \delta_2 &= \delta Y^{-1}. \end{aligned} \tag{29.57}$$

From Fig. 7-24(b), we see that the rate constants into state 22 (α_2, γ_2) are proportional to Y and the rate constants out of state 22 (β_2, δ_2) are proportional to Y^{-1}. If state 22 is destabilized by the intersubunit interaction (negative cooperativity), $Y < 1$; if state 22 is stabilized (positive cooperativity), $Y > 1$.

Upon substituting Eqs. (29.57) into Eqs. (29.51) and (29.52), we find

$$\begin{aligned} p_{11} &= \beta\delta Y^{-1}(\alpha Y + \gamma Y + \beta + \delta)\Sigma^{-1}, \\ p_{12} &= \beta\gamma(\beta Y^{-1} + \delta Y^{-1} + \alpha + \gamma)\Sigma^{-1}, \\ p_{21} &= \alpha\delta(\beta Y^{-1} + \delta Y^{-1} + \alpha + \gamma)\Sigma^{-1}, \\ p_{22} &= \alpha\gamma Y(\alpha Y + \gamma Y + \beta + \delta)\Sigma^{-1}, \end{aligned} \tag{29.58}$$

$$\begin{aligned} J_1 &= \alpha p_{11} + \alpha Y p_{12}, \\ J_2 &= \gamma p_{11} + \gamma Y p_{21}, \end{aligned} \tag{29.59}$$

where Σ is the sum of the four numerators in Eqs. (29.58). It should be recalled that α is proportional to c_{S_1} and γ is proportional to c_{S_2}. Thus J_1 depends on c_{S_2} as well as on c_{S_1}, and J_2 depends on c_{S_1} as well as on c_{S_2}. This coupling between E_1 and E_2 is a consequence of the interaction $Y \neq 1$. If $Y = 1$ [Eq. (29.54)], the two molecules are kinetically independent of each other.

Note that if Y is small, p_{22} is the smallest of the p_{ij}; if Y is large, p_{22} is the largest of the p_{ij}. In the former case, in the limit $Y \to 0$,

$$p_{11} = \frac{\beta\delta}{\Sigma}, \qquad p_{12} = \frac{\beta\gamma}{\Sigma},$$
$$p_{21} = \frac{\alpha\delta}{\Sigma}, \qquad p_{22} \to 0, \tag{29.60}$$
$$\Sigma = \alpha\delta + \beta\delta + \beta\gamma,$$

$$J_1 = \frac{\alpha\beta\delta}{\alpha\delta + \beta\delta + \beta\gamma},$$
$$J_2 = \frac{\beta\gamma\delta}{\alpha\delta + \beta\delta + \beta\gamma}. \tag{29.61}$$

These results agree with Eqs. (17.7)–(17.9) if the inverse rate constants are dropped in the latter equations and the notation is adjusted. In this extreme case ($p_{22} \to 0$), only one enzyme can cycle at a time (the other is always in state 1; substrate is not bound). The concentration dependences in J_1 and J_2 [Eqs. (29.61)] are quite simple (put $\alpha = \alpha^* c_{S_1}$ and $\gamma = \gamma^* c_{S_2}$).

Two Identical Three-State Subunits

If the symmetrical dimer of Fig. 7-20 requires a three-state cycle (as in Fig. 5-21), then the unperturbed diagram for a subunit is that in Fig. 7-26(a), and the diagram for a dimer is shown in Fig. 7-26(b). In both cases, inverse rate constants (with primes) are omitted for simplicity. In the dimer, states 12 and 21 are equivalent, etc. However, this symmetry does not allow linearization

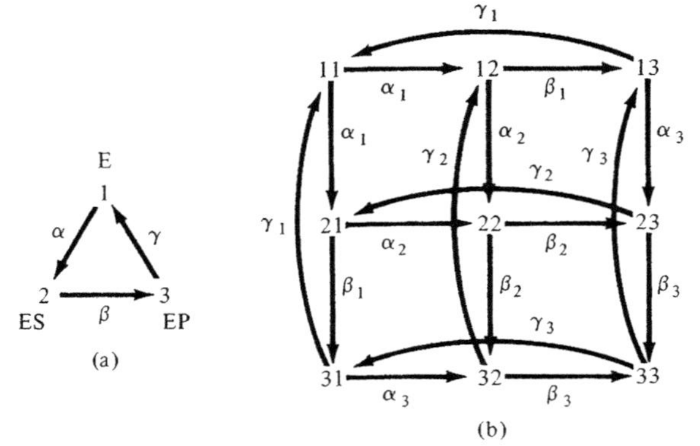

Fig. 7-26. (a) Three-state cycle for an unperturbed subunit of a symmetrical dimer. (b) Diagram for the symmetrical dimer. Subscripts refer to the state of the nonreacting neighbor. Inverse rate constants (omitted) have primes, in both (a) and (b).

of the diagram, as in Fig. 7-20(c). Hence a detailed balance type of solution of the steady-state rate equations does not exist in this case. The subscripts in Fig. 7-26(b) refer to the state of the nonreacting subunit. Following our usual system,

$$\alpha_1 = \alpha\left(\frac{y_{12}}{y_{11}}\right)^{f_\alpha}, \qquad \alpha_1' = \alpha'\left(\frac{y_{12}}{y_{11}}\right)^{f_\alpha - 1}, \tag{29.62}$$

etc.

With the specification of all of the rate constants in Fig. 7-26(b), as in Eq. (29.62), the steady-state properties of the dimer, including arbitrary interaction effects, are determined in principle. There are six (not nine) linear algebraic equations in the p_{ij} to solve (because of symmetry, $p_{12} = p_{21}$, etc.). The flux per dimer can then be found from the p_{ij} in several ways. For example,

$$J = 2[(\alpha_1 p_{11} - \alpha_1' p_{12}) + (\alpha_2 p_{21} - \alpha_2' p_{22}) + (\alpha_3 p_{31} - \alpha_3'{}_{32})]. \tag{29.63}$$

We give an explicit solution of the p_{ij} equations for one special case only: one-way transitions; $\beta = \gamma$; the only unusual intersubunit interaction is in state 33 (i.e., all y_{ij} are equal except y_{33}); and $f_\beta = f_\gamma = \frac{1}{2}$. The kinetic diagram for this special case appears in Fig. 7-27, where

$$Y \equiv (y_{33}/y_{ij})^{1/2} \tag{29.64}$$

and $ij \neq 33$. It is obvious from this diagram that the population of state 33 will be enhanced by $Y > 1$ and reduced by $Y < 1$. When $Y \to 0$, we expect $p_{33} \to 0$, as in Fig. 5-24(c). One finds, for arbitrary Y,

$$\begin{aligned} p_{11} &= (1 + x)(1 + Y + Yx)\Sigma^{-1}, \\ p_{12} = p_{21} &= x[(1 + Y + Yx) + x^2(2 + x)]\Sigma^{-1}, \\ p_{13} = p_{31} &= x(1 + x)(1 + Y + Yx)\Sigma^{-1}, \\ p_{23} = p_{32} &= x^2(1 + x)(2 + x)\Sigma^{-1}, \end{aligned} \tag{29.65}$$

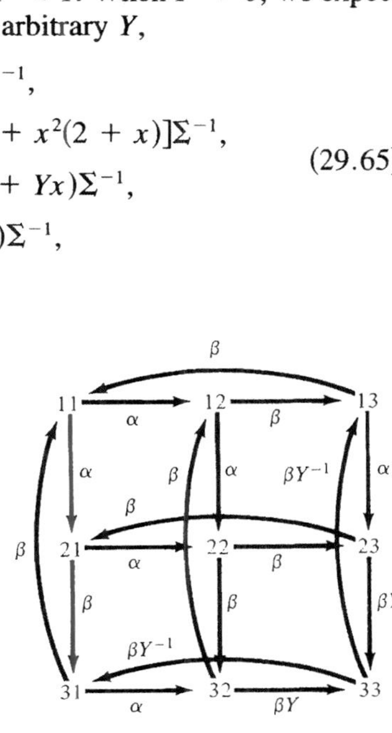

Fig. 7-27. Special case of Fig. 7-26(b) in which there are one-way transitions, $\beta = \gamma$, $f_\beta = f_\gamma = \frac{1}{2}$, and all y_{ij} are equal except y_{33}. See text for further details.

$$p_{22} = [x^2(1 + Y + Yx) + x^3(2 + x)]\Sigma^{-1},$$
$$p_{33} = Y^2x^2(1 + x)(2 + x)\Sigma^{-1}, \tag{29.65}$$

where

$$\Sigma = (1 + Y) + (5 + 6Y)x + (11 + 8Y + 2Y^2)x^2$$
$$+ (10 + 3Y + 3Y^2)x^3 + (3 + Y^2)x^4, \tag{29.66}$$
$$x \equiv \alpha/\beta.$$

Both α and x are proportional to c_S if Fig. 7-26(a) refers to S $\rightarrow$ P. The dimer flux J is then, from Eq. (29.63),

$$J/2\beta = x(p_{11} + p_{21} + p_{31})$$
$$= x[(1 + Y)(1 + x^3) + (3 + 4Y)(x + x^2)]\Sigma^{-1}. \tag{29.67}$$

When $Y = 1$, this reduces to the unperturbed flux per subunit,

$$J_o/\beta = x/(1 + 2x). \tag{29.68}$$

When $Y = 0$ and $x = 1$ (i.e., state 33 is excluded and $\alpha = \beta = \gamma$), $J/2\beta = 4/15$, in agreement with Fig. 5-24(c). In the limit of large x,

$$\frac{J}{2\beta} \rightarrow \frac{1 + Y}{3 + Y^2} \qquad (x \rightarrow \infty). \tag{29.69}$$

For large Y, $J/2\beta \rightarrow 1/Y \rightarrow 0$.

In Fig. 7-28, $J/2\beta$ is plotted against x for three values of Y. It is easy to

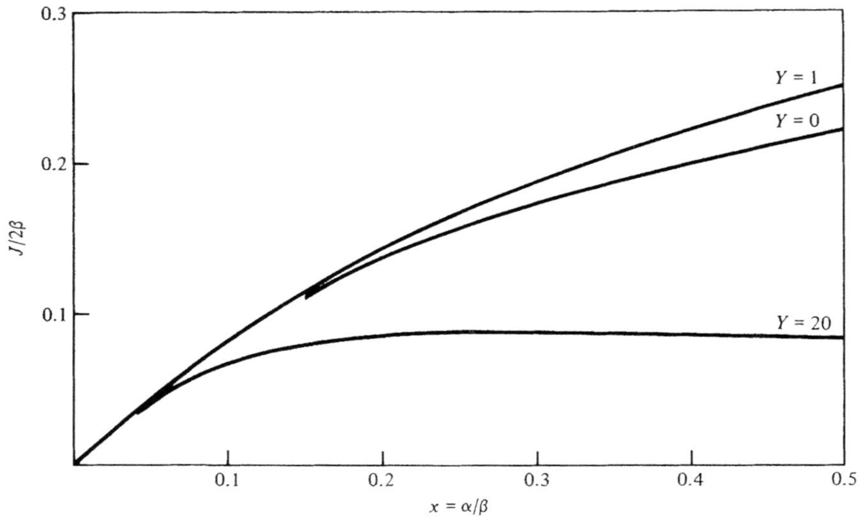

Fig. 7-28. Curves of $J/2\beta$ as a function of $x = \alpha/\beta$, from Eq. (29.67) and Fig. 7-27.

show analytically that the maximum flux as a function of Y occurs at $Y = 1$ for any value of x. That is, if state 33 is either unusually stable ($Y > 1$) or unusually unstable ($Y < 1$), the flux is reduced. In the former case, the dimer tends to become "frozen" in state 33; in the latter case, state 33 is less available than usual for participation in the enzymatic activity. When Y is large enough, $J(x)$ passes through a maximum at finite x. This occurs in the $Y = 20$ case. The maximum is more apparent when Y is still larger.

30. Interactions that Affect Rate Constants but Not Equilibrium Constants

Intersubunit interactions would ordinarily be expected to alter both equilibrium and rate constants, as in all of the examples of this chapter up to this point. In this section we digress briefly, by a consideration of a few rather explicit examples, to include another possible effect of interactions: rate constants are altered but not equilibrium constants.

Our discussion of adenylate cyclase in Section 18 and Fig. 5-38 could be made more general by using the methods of this section, but we do not pursue this topic here.

To introduce the subject, we return first to the arbitrary case discussed in Eqs. (5.1)–(5.8) and represented in Fig. 1-8. The unperturbed inverse rate constants for $1 \leftrightarrows 2$ are α_o, β_o, and the perturbed rate constants are α, β. Figure 7-29 shows the case of interest here. The perturbation, from neighbor interactions, say, does not alter the free energy levels G_1 and G_2, but it increases the free energy barrier between the two states by $w^\ddagger$. Consequently,

$$K = K_o = \frac{\alpha}{\beta} = \frac{\alpha_o}{\beta_o} \tag{30.1}$$

$$\alpha = \alpha_o e^{-\omega^\ddagger/kT}, \qquad \beta = \beta_o e^{-\omega^\ddagger/kT}. \tag{30.2}$$

Both rate constants are decreased by the same factor; the equilibrium constant is unchanged. In the examples of this section, we shall have in mind cases in

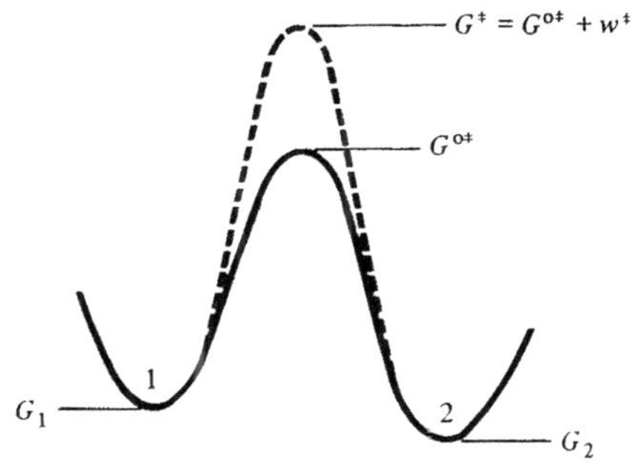

Fig. 7-29. Effect of a perturbation on the free energy along the reaction coordinate between states 1 and 2 in a case in which only the barrier height is altered.

which $w^{\ddagger} > 0$. However, $w^{\ddagger} < 0$ is also possible. In the latter case the perturbation acts as a catalyst: both rate constants are increased.

In a formal way, Fig. 7-29 is a special case of Fig. 1-8. In Eq. (5.4),

$$G = G^{o\ddagger} + w^{\ddagger},$$
$$w^{\ddagger} = (1 - f)w_1 + fw_2. \tag{30.3}$$

If w_1 and w_2 are both small and $|f|$ is large,

$$w^{\ddagger} = f(w_2 - w_1). \tag{30.4}$$

Thus it is possible to have a finite $w^{\ddagger}$, as in Fig. 7-29, even if $w_1 \to 0$, $w_2 \to 0$, provided that $|f| \to \infty$.

One-Conformation Two-State Heterologous Dimer

This first example is very simple. We consider an unsymmetrical dimer of identical enzyme molecules that catalyze S → P by a two-state cycle. The four states of the dimer are illustrated in Fig. 7-30(a); the two-state cycle, with

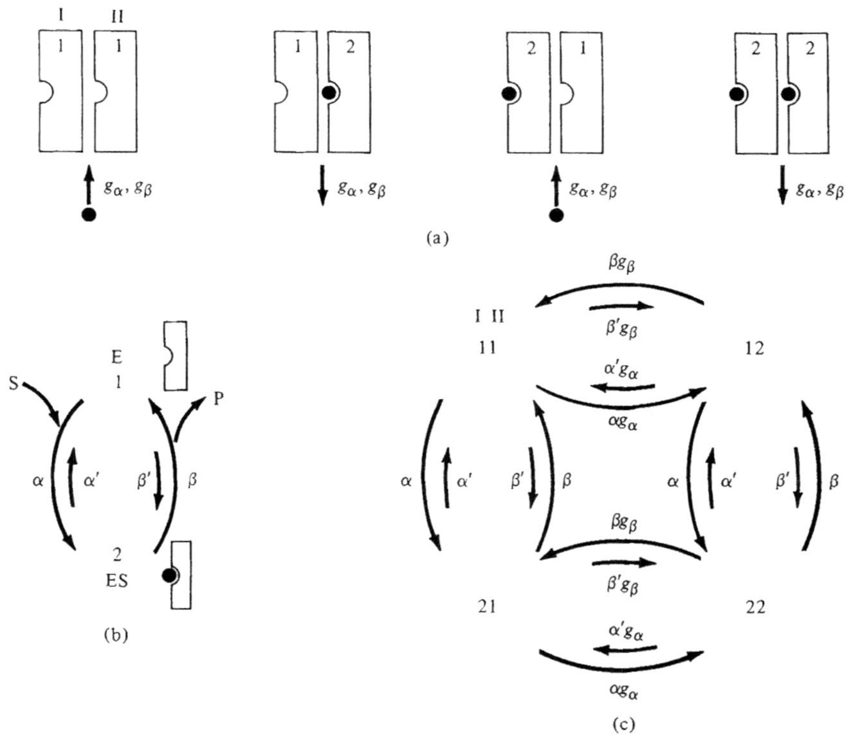

Fig. 7-30. (a) Four states of a dimer with two identical molecules; g_α and g_β are inhibition factors. (b) Kinetic cycle for a separated subunit. (c) Kinetic diagram for the dimer. See text for further details.

rate constants, for a separated (unperturbed) enzyme molecule is given in Fig. 7-30(b); and the kinetic diagram for the dimer is shown in Fig. 7-30(c).

The left subunit of the dimer is designated I; the right subunit is II. The subunit interaction free energies in the dimer are assumed to be the same: $w_{11} = w_{12} = w_{21} = w_{22}$. The two subunits have different properties, even though the molecules are identical, because of the location of the binding site for S and P [Fig. 7-30(a)]. The binding or release of S or P is uninhibited on subunit I of the dimer. Consequently the rate constants for the cycling of I in Fig. 7-30(c) are the same as in Fig. 7-30(b) (unperturbed). However, in this model, both S and P have difficulty entering or leaving the site on II because of the proximity of I. The effect may be different for S and P: α and α' are multiplied by a factor g_α and β and β' are multiplied by a factor g_β, where g_α and g_β are both examples of $e^{-w^\ddagger/kT}$ in Eq. (30.2). Because we are assuming steric hindrance, $0 \leqslant g_\alpha, g_\beta \leqslant 1$ (i.e., $w_\alpha^\ddagger$ and $w_\beta^\ddagger \geqslant 0$). These altered rate constants are included in Fig. 7-30(c) for the cycling of II. Note that we are assuming that the inhibitory effects of I on the rate constants of II are the same whether I is in state 1 (E) or 2 (ES). However, this need not be the case (i.e., we might have different factors g_α, g_β and h_α, h_β).

The S $\rightarrow$ P fluxes for I and II in the dimer are especially easy to determine in this example because each subunit has rate constants that are independent of the state of the other subunit. Hence the fluxes are independent:

$$J_{\mathrm{I}} = \frac{\alpha\beta - \alpha'\beta'}{\alpha + \alpha' + \beta + \beta'}, \tag{30.5}$$

$$J_{\mathrm{II}} = \frac{(\alpha\beta - \alpha'\beta')g_\alpha g_\beta}{(\alpha + \alpha')g_\alpha + (\beta + \beta')g_\beta}. \tag{30.6}$$

The flux J_{I} is also the unperturbed flux J_o [Fig. 7-30(b)]. The flux J_{II} is of course reduced by factors $g_\alpha, g_\beta < 1$.

Two-Conformation Two-State Heterologous Dimer

The second model is related to but more complicated than the first. Again we have an unsymmetrical dimer of identical two-state enzyme molecules that catalyze S $\rightarrow$ P. Here, however, the binding of S induces a conformational change, as indicated schematically in Fig. 7-31(a). The unperturbed rate constants are again those in Fig. 7-30(b). Because of the conformational change, there are two different "fittings" of adjacent I and II surfaces. Hence we assume $w_{11} = w_{12}$ and $w_{21} = w_{22}$ [Fig. 7-31(a)] and define

$$z \equiv \frac{y_{11}}{y_{21}} = \frac{y_{12}}{y_{22}}, \tag{30.7}$$

where $y_{ij} = e^{-w_{ij}/kT}$. For example, as drawn in Fig. 7-31(a), we might expect $z > 1$ (i.e., 11 and 12 have stronger attractive interactions than 21 and 22).

The access of S and P to the binding site is uninhibited in I but inhibited

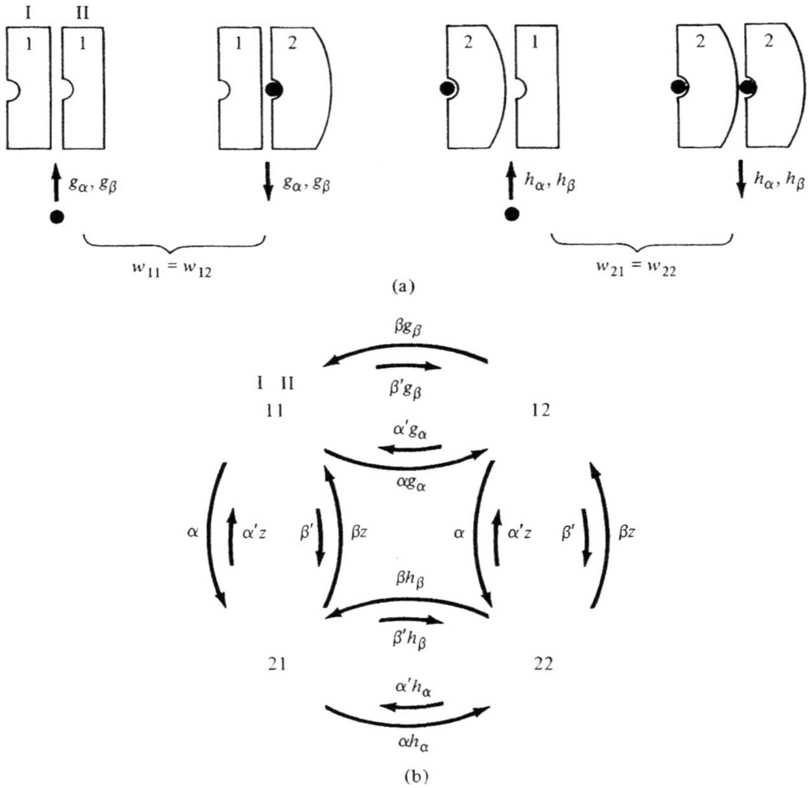

Fig. 7-31. (a) More complicated version of Fig. 7-30(a) in which binding of S induces a conformational (shape) change in E. (b) Corresponding kinetic diagram. See text for further details.

in II. Because of the conformational change, we might have inhibition factors g_α, g_β when I is in state 1 and different factors h_α, h_β when I is in state 2. Again, as drawn in Fig. 7-31(a), we might expect $h_\alpha > g_\alpha$ and $h_\beta > g_\beta$ (i.e., stronger inhibition of access in states 11 and 12). There is no other effect of interactions on the transitions of II (11, 12 and 21, 22) because $w_{11} = w_{12}$ and $w_{21} = w_{22}$.

The transitions between 11 and 21 and between 12 and 22 (cycling of I in both cases) involve the interaction parameter z [Eq. (30.7)] but no steric inhibition. In our usual fashion, we write, for the perturbed rate constants of the 11, 21 cycle,

$$\alpha\left(\frac{y_{21}}{y_{11}}\right)^{f_\alpha}, \qquad \alpha'\left(\frac{y_{21}}{y_{11}}\right)^{f_\alpha - 1},$$
$$\beta\left(\frac{y_{11}}{y_{21}}\right)^{f_\beta}, \qquad \beta'\left(\frac{y_{11}}{y_{21}}\right)^{f_\beta - 1}. \tag{30.8}$$

If we assume, to be explicit, diffusion-controlled binding of both S and P to I, then $f_\alpha = 0$ and $f_\beta = 1$. Thus these four rate constants become α, $\alpha' z$, βz, β'. The same four rate constants apply as well in the 12, 22 cycle because of Eq. (30.7). Hence the kinetic diagram for the dimer, with rate constants, appears as in Fig. 7-31(b).

Because the rate constants for I are independent of the state of II [Fig. 7-31(b)], we can write J_{I} immediately:

$$J_{\text{I}} = \frac{(\alpha\beta - \alpha'\beta')z}{\alpha + \beta' + (\alpha' + \beta)z}. \tag{30.9}$$

If $z > 1$, J_{I} is larger than the unperturbed flux [Eq. (30.5)].

To avoid lengthy algebra in deriving J_{II}, we consider three special cases only. First, if $g_\alpha = h_\alpha$ and $g_\beta = h_\beta$ (the conformational change does not alter the degree of inhibition), then the rate constants for II are the same for either state of I, as in Fig. 7-30(c). Hence, in this case, J_{II} is given by Eq. (30.6).

In the second special case, $g_\alpha = g_\beta \equiv g$ and $h_\alpha = h_\beta \equiv h$ (as would be expected if S and P are molecules with similar shapes). In this case there is a detailed balance solution of the steady-state rate equations:

$$\begin{gathered} p_{11} = 1/\xi, \qquad p_{12} = x/\xi, \qquad p_{21} = xz^{-1}/\xi, \qquad p_{22} = x^2z^{-1}/\xi, \\ p_2 = \tfrac{1}{2}(p_{12} + p_{21}) + p_{22}, \\ \xi \equiv (1 + x)(1 + xz^{-1}), \\ x \equiv (\alpha + \beta')/(\beta + \alpha'), \end{gathered} \tag{30.10}$$

where p_2 is the fractional binding of S. Note that g and h do not appear in the p_{ij} expressions. The flux for II is then

$$\begin{aligned} J_{\text{II}} &= (\alpha g p_{11} - \alpha' g p_{12}) + (\alpha h p_{21} - \alpha' h p_{22}) \\ &= \frac{(\alpha\beta - \alpha'\beta')(g + hxz^{-1})}{(\beta + \alpha')(1 + x)(1 + xz^{-1})}. \end{aligned} \tag{30.11}$$

For comparision, the unperturbed flux is

$$J_{\text{o}} = \frac{(\alpha\beta - \alpha'\beta')}{(\beta + \alpha')(1 + x)}. \tag{30.12}$$

Values of g and $h < 1$ reduce J_{II} compared to J_{o}. The interaction parameter z has little effect on J_{II}; it has no effect at all if $g = h$.

In the third special case, $\alpha' = 0$ and $\beta' = 0$ (both inverse transitions are negligible; S $\rightarrow$ P is far from equilibrium). If we compare Figs. 7-24(b) and 7-31(b), we see that the latter can be considered to be a special case of the former, insofar as the formal calculation of J_{II} is concerned [we already have J_{I} in Eq. (30.9)]. That is,

$$\begin{gathered} \alpha_1 = \alpha_2 = \alpha, \qquad \beta_1 = \beta_2 = \beta z, \\ \gamma_1 = \alpha g_\alpha, \qquad \delta_1 = \beta g_\beta, \qquad \gamma_2 = \alpha h_\alpha, \qquad \delta_2 = \beta h_\beta. \end{gathered} \tag{30.13}$$

If we make these substitutions in Eqs. (29.49), (29.51), and (29.52)(and use J_2 for J_{II}), we find

$$J_{\text{II}} = \frac{\alpha\beta A}{(\alpha + \beta z)B}, \tag{30.14}$$

where

$$\begin{aligned} A = {} & \alpha^2(g_\alpha + 1)h_\alpha h_\beta + \beta^2(h_\beta + z)g_\alpha g_\beta z \\ & + \alpha\beta[(g_\alpha z + h_\beta)g_\beta h_\alpha + (g_\alpha h_\beta + g_\beta h_\alpha)z], \end{aligned} \tag{30.15}$$

$$\begin{aligned} B = {} & \alpha^2(g_\alpha + 1)h_\alpha + \beta^2(h_\beta + z)g_\beta \\ & + \alpha\beta(g_\alpha z + h_\beta + g_\alpha h_\beta + g_\beta h_\alpha). \end{aligned} \tag{30.16}$$

The special case $g_\beta = 0$ resembles Fig. 5-37(a) and the special case $g_\alpha = g_\beta = 0$ resembles Fig. 5-37(b), where transitions (but not states) are omitted from the kinetic diagram.

Two-Conformation Three-State Heterologous Dimer

This model is similar to the one in Fig. 7-31, except that we use a three-state enzyme and we take all w_{ij} as equal. The dimer has nine possible states. In order to be able to present algebraic results, we consider only a very limited special case. A more general model would require numerical solution of nine linear algebraic equations by computer.

As shown in Fig. 7-32(a) (which refers to an unperturbed subunit), we limit ourselves to the case where all three forward rate constants are equal (α) and backward rate constants are negligible. Each subunit is assumed to have two conformations, one for E (state 1) and the other for ES and EP [Fig. 7-32(a)].

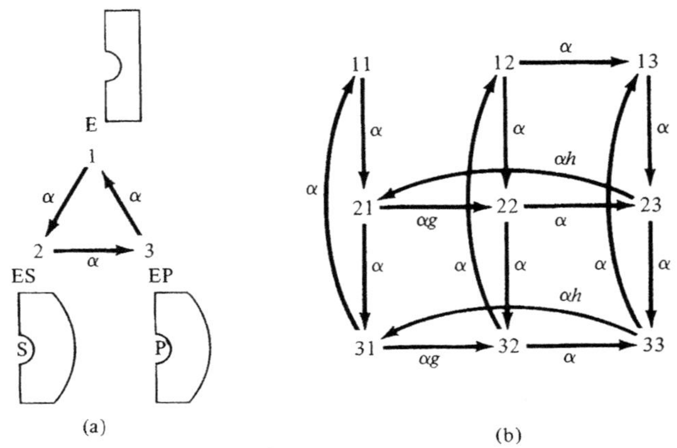

Fig. 7-32. (a) Cycle for three-state separated subunit. (b) Diagram for dimer. This is a three-state version of Fig. 7-31. See text for details.

In the dimer, when I is in state 1 (E), movement of S and P into and out of the binding site on II is assumed to be completely blocked because of the close fitting of I and II. Hence, in Fig. 7-32(b), the transitions $11 \to 12$ and $13 \to 11$ are omitted. When I is in state 2 (ES) or 3 (EP) (the conformation of I is the same in both states), S can reach the site on II with an inhibitory factor g and P can leave the site on II with an inhibitory factor h [note the rate constants αg and αh in Fig. 7-32(b)]. Finally, we assume that the actual catalytic step $S \to P$ (i.e., $2 \to 3$) on II is unaffected by the neighbor I. Also, as usual for this model, all transitions in I are assumed to be unaffected by II (because the site on I is not adjacent to II). Thus the remaining rate constants in Fig. 7-32(b) are all α.

The diagram in Fig. 7-32(b) is seen to be a generalization of Fig. 5-41(c) in that the inhibitory factors g and h for II are now included.

Because I has the same (unperturbed) rate constants for all states of II, J_{I} has the unperturbed value $J_o = \alpha/3$. To find J_{II}, we need first the p_{ij}. These are obtained by direct solution of nine steady-state rate equations (including normalization), for example,

$$\frac{dp_{21}}{dt} = 0 = \alpha p_{11} + \alpha h p_{23} - (\alpha + \alpha g)p_{21}. \tag{30.17}$$

The result is

$$\begin{aligned}
p_{11} = p_{31} &= h(14 + 7g + 7h + 2gh)\Sigma^{-1}, \\
p_{21} &= h(14 + 7g + 7h + 3gh)\Sigma^{-1}, \\
p_{12} &= gh(6 + 3g + 3h + gh)\Sigma^{-1}, \\
p_{22} &= gh(10 + 5g + 5h + 2gh)\Sigma^{-1}, \\
p_{32} &= gh(12 + 6g + 6h + 2gh)\Sigma^{-1}, \\
p_{13} &= g(14 + 7g + 12h + 5gh + 3h^2 + gh^2)\Sigma^{-1}, \\
p_{23} &= g(14 + 7g + 8h + 3gh)\Sigma^{-1}, \\
p_{33} &= g(14 + 7g + 6h + 2gh)\Sigma^{-1},
\end{aligned} \tag{30.18}$$

where Σ, the sum of the numerators, is

$$\Sigma = 3(14g + 14h + 7g^2 + 25gh + 7h^2 + 8g^2h + 8gh^2 + 2g^2h^2). \tag{30.19}$$

The two $S \to P$ fluxes are then

$$J_{\mathrm{I}} = \alpha(p_{21} + p_{22} + p_{23}) = \alpha/3 = J_o, \tag{30.20}$$

$$\begin{aligned}
J_{\mathrm{II}} &= \alpha(p_{12} + p_{22} + p_{32}) \\
&= \alpha gh(28 + 14g + 14h + 5gh)\Sigma^{-1}.
\end{aligned} \tag{30.21}$$

If $g = h = 1$, $J_{\mathrm{II}} = 61\alpha/255$, as in Fig. 5-41(c). This is less than J_o because of the two missing transitions. It should also be noticed that the numbers at the nine states in Fig. 5-41(c) agree with the numerators in Eqs. (30.18) if we put $g = h = 1$. If g and $h < 1$, then J_{II} is reduced further. If g and h are both very small ($\ll 1$),

$$\begin{aligned} p_{11} &= p_{31} = p_{21} = h/3(g + h), \\ p_{13} &= p_{23} = p_{33} = g/3(g + h), \end{aligned} \tag{30.22}$$

$$J_{\text{II}} = \frac{2\alpha gh}{3(g + h)}. \tag{30.23}$$

If $g = h$, $J_{\text{II}} = gJ_{\text{o}}$.

31. Dimers with Two-Conformation Subunits

If the subunits in the models of Sections 27 and 29 can exist in two conformations rather than only one, the models have many more possible states. For example, a dimer will have four times as many states. For this reason, complexes with two-conformation subunits will almost always have to be handled numerically, by computer. In this section we present two examples of such models (without numerical calculations) and then two special cases of the second model that can be treated analytically. In the first two models, as in many others in this book, the objective is to show the reader how to set up the formalism in order to be ready to carry out numerical calculations.

Membrane Transport by a Dimer with Two-Conformation Subunits

We begin with the notation for and properties of a single (unperturbed) subunit. We then consider an isologous dimer of two such subunits, with intersubunit interactions.

A ligand L is transported across a membrane by binding on a two-conformation subunit [see Fig. 3-5(a)]. The conformations are designated A(○) and B(□). As indicated in Fig. 7-33(a), conformation A can exchange ligand with side A (at c_A) only, whereas conformation B can exchange ligand with side B (at c_B) only. The subunit can exist in four states, numbered as in Fig. 7-33(a). A conformational change is possible either with L bound ($2 \rightleftarrows 3$) or not bound ($1 \rightleftarrows 4$). The cycles in Figs. 7-33(b) and 7-33(c) show the equilibrium constants and rate constants, respectively. These are related as follows:

$$\begin{aligned} &\text{bind L on } \bigcirc\text{:} \quad K_A = \alpha_A^*/\beta_A, \\ &\text{bind L on } \square\text{:} \quad K_B = \alpha_B^*/\beta_B, \\ &\bigcirc \to \square\text{:} \quad K = \kappa/\kappa', \\ &\text{Ⓛ} \to \boxed{\text{L}}\text{:} \quad K_L = KK_B/K_A = \kappa_L/\kappa_L'. \end{aligned} \tag{31.1}$$

The positive direction is chosen as counterclockwise; one cycle in the positive direction transports one L from side A to side B. At equilibrium, $c_A = c_B$ and

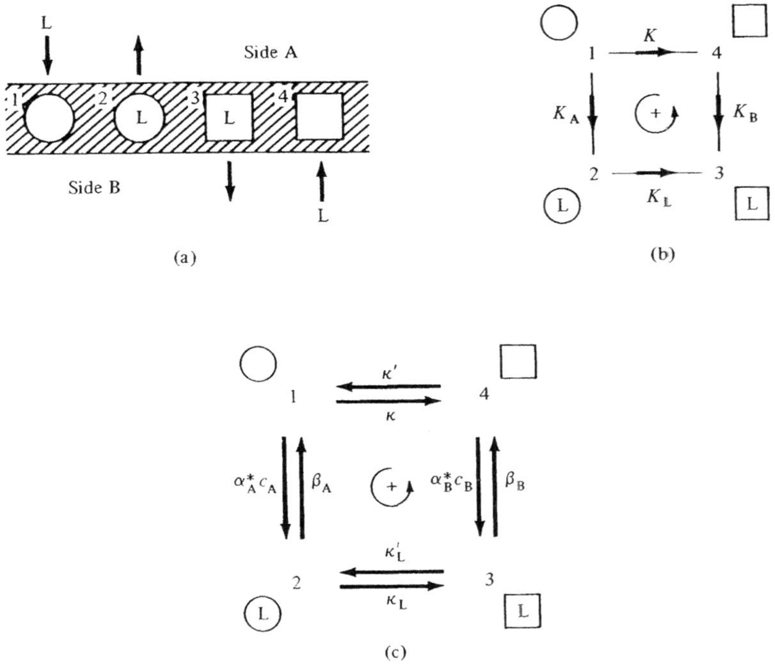

Fig. 7-33. (a) Possible states of a transporting subunit with a conformational change. (b) Four-state transport cycle with specification of equilibrium constant notation (arrows). (c) Rate constant notation for the same cycle.

$\Pi_+ = \Pi_-$ (Section 8). This leads to an interrelation among the rate constants, required for self-consistency:

$$\alpha_A^* \kappa_L \beta_B \kappa' = \alpha_B^* \kappa_L' \beta_A \kappa. \tag{31.2}$$

The thermodynamic force X (Section 8) is given by

$$e^{X/kT} = \frac{\Pi_+}{\Pi_-} = \frac{c_A}{c_B}. \tag{31.3}$$

The unperturbed flux for one subunit is

$$J_o = \frac{\Pi_+ - \Pi_-}{\Sigma} = \frac{\Pi_-(e^{X/kT} - 1)}{\Sigma}, \tag{31.4}$$

where Σ is a sum of 16 terms (there are four partial diagrams and thus 4×4 directional diagrams for the four states), which we omit.

We now consider an isologous (symmetrical) dimer of two of the subunits represented in Fig. 7-33. The new ingredient is that there are intersubunit interactions that perturb some of the equilibrium and rate constants. These interactions are assumed to depend only on the conformations A and B, and not on whether ligand is bound. The interaction parameters, defined as usual,

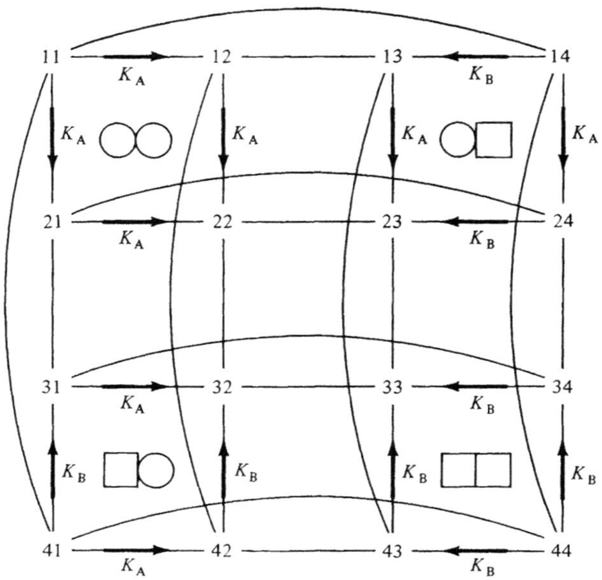

Fig. 7-34. Kinetic diagram for isologous dimer of subunits in Fig. 7-33. Binding equilibrium constants are indicated. See text for explanation.

are y_{AA}(○○), y_{AB}(○□, □○), and y_{BB}(□□). We now need a 16-state diagram for the dimer, as shown in Fig. 7-34. Each subunit is numbered as in Fig. 7-33.

All of the four dimer states in the upper left of Fig. 7-34 are in conformation state ○○ (subunit states 1 and 2 have conformation ○; subunit states 3 and 4 have □). All four states at the upper right have conformation state ○□, etc. The binding and release transitions in the diagram are marked K_A or K_B, as in Fig. 7-34. In all of these cases, the equilibrium and rate constants are the same (unperturbed) as in Figs. 7-33(b) and 7-33(c) because there are no conformational changes and therefore no interaction changes involved. These processes occur *within* the four groups (upper left, etc.) referred to above. Interactions perturb an equilibrium constant and the associated rate constants only if the interaction free energy *changes* in the process (Section 30 is the exception).

The unlabelled transitions in Fig. 7-34 all involve conformational changes ($2 \rightleftarrows 3$ or $1 \rightleftarrows 4$). Hence, interaction free energy factors appear in the equilibrium and rate constants for these transitions. Figure 7-35 gives the perturbed equilibrium constants for the K_L ($2 \rightleftarrows 3$) process and Fig. 7-36 does the same for the K ($1 \rightleftarrows 4$) process. To simplify these figures, we have introduced

$$Y \equiv y_{AB}/y_{AA}, \qquad Z \equiv y_{BB}/y_{AB}. \tag{31.5}$$

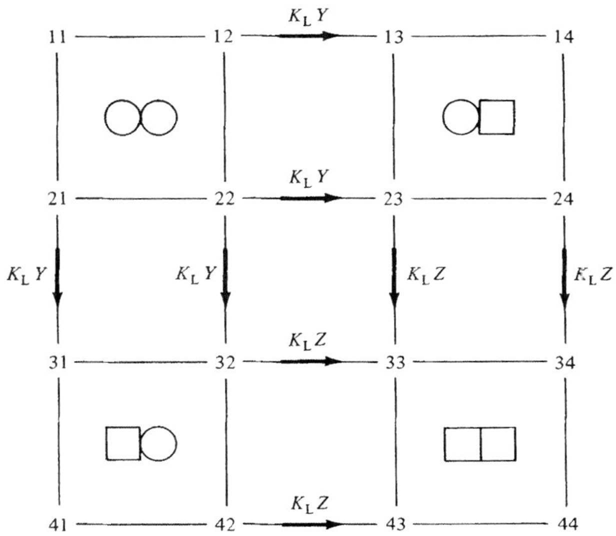

Fig. 7-35. Part of the diagram in Fig. 7-34. Perturbations of K_L are shown. See text for details.

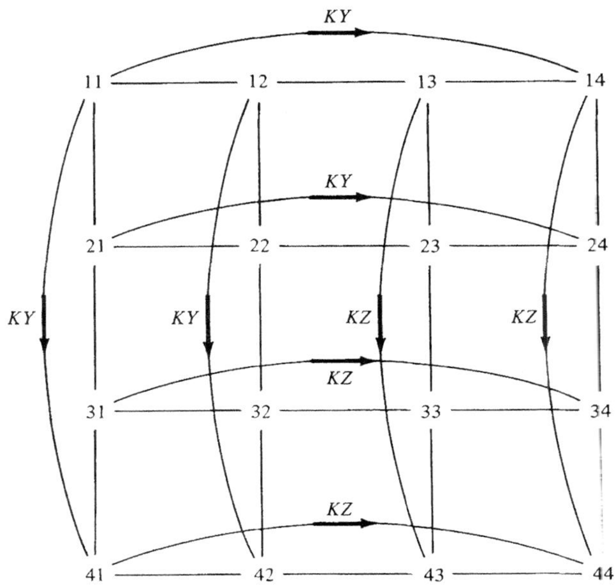

Fig. 7-36. Perturbations of K in the diagram of Fig. 7-34. See text for details.

The corresponding rate constants are not shown in the figures but are managed in our usual fashion. Because only one kind of process (○ ⇄ □) is involved, we use a single parameter f. Then the four perturbed equilibrium constants that appear in Figs. 7-35 and 7-36 are related to separate rate constants by

$$K_L \frac{y_{AB}}{y_{AA}} = \kappa_L \left(\frac{y_{AB}}{y_{AA}}\right)^f \Big/ \kappa_L' \left(\frac{y_{AB}}{y_{AA}}\right)^{f-1}, \qquad (31.6)$$

$$K_L \frac{y_{BB}}{y_{AB}} = \kappa_L \left(\frac{y_{BB}}{y_{AB}}\right)^f \Big/ \kappa_L' \left(\frac{y_{BB}}{y_{AB}}\right)^{f-1}. \qquad (31.7)$$

The two corresponding equations relating the perturbed K to κ and κ' are obtained simply by omitting the subscript L throughout Eqs. (31.6) and (31.7). Because the process here is a conformational change, a likely value for f is $f = \frac{1}{2}$.

All the rate constants in the dimer diagram have now been specified. The sixteen p_{ij} can be found numerically by solving 16 linear algebraic equations in the p_{ij} (including normalization). For example,

$$\frac{dp_{11}}{dt} = 0 = \beta_A(p_{12} + p_{21}) + \kappa' \left(\frac{y_{AB}}{y_{AA}}\right)^{f-1} (p_{14} + p_{41})$$
$$- 2\alpha_A^* c_A p_{11} - 2\kappa \left(\frac{y_{AB}}{y_{AA}}\right)^f p_{11}. \qquad (31.8)$$

As the symmetry of this equation and of the dimer diagram suggest, there are really only 10 independent p_{ij} ($p_{ij} = p_{ji}$ if $i \neq j$). Hence it is possible and advantageous to use 10 linear equations rather than 16.

Having found the steady-state p_{ij} for given c_A and c_B, the A → B flux per subunit of the dimer follows in several equivalent ways. For example,

$$J = \alpha_A^* c_A (p_{11} + p_{21} + p_{31} + p_{41}) - \beta_A (p_{12} + p_{22} + p_{32} + p_{42}). \qquad (31.9)$$

This is the net state 1 → state 2 flux for the right-hand subunit. Of course the left-hand subunit has the same flux, by symmetry.

The fraction of subunits with L bound and the fraction in a given conformation (○ or □) also follow directly from the steady-state p_{ij}.

Dimer of Two-Conformation Enzyme Molecules

We now consider a symmetrical (isologous) dimer comprised of subunits, each of which, by itself (unperturbed), would have the four-state diagram shown in Fig. 7-37(a). This is the same as Fig. 5-42(a). There are two possible conformations for the enzyme, designated A = ○ = E and B = □ = E*. The enzyme catalyzes S → P by two different two-state cycles (states 1, 2 and 3, 4). Figure 5-42(b) shows, rather schematically, the diagram for the dimer. There are 16 states in this diagram. The new feature that we introduce here are intersubunit interactions, with parameters y_{AA}, y_{AB},

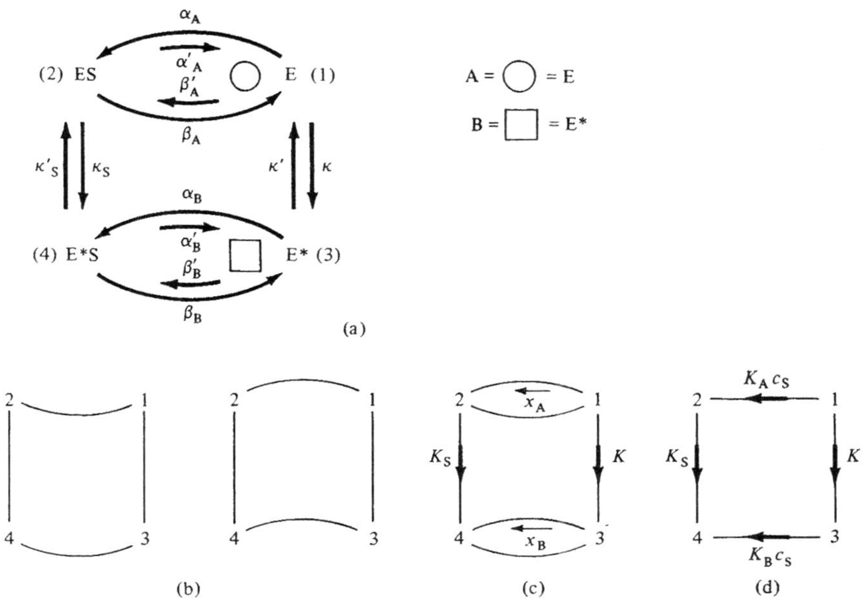

Fig. 7-37. (a) Diagram for an enzyme that catalyzes S → P by two different conformations, as in Fig. 5-42(a). (b) Two of the cycles of the diagram (these cycles have zero force), used in the text. (c), (d) Diagrams used in discussion of detailed balance in the text.

and y_{BB}, as in Figs. 7-35 and 7-36. We assume, as in the previous model, that the intersubunit interactions depend on conformations (○, □) but not on whether S is bound or not. Thus the interactions in this model do not affect the catalytic rate constants (αs and βs) themselves, but they can alter the flux by shifting conformations.

We return now to the single-subunit diagram, Fig. 7-37(a). The equilibrium constants for the conformational changes (○→ □) are

$$K \equiv \kappa/\kappa', \qquad K_S \equiv \kappa_S/\kappa'_S. \tag{31.10}$$

The thermodynamic force X for S → P is the same in the two two-state cycles:

$$e^{X/kT} = \frac{\alpha_A \beta_A}{\alpha'_A \beta'_A} = \frac{\alpha_B \beta_B}{\alpha'_B \beta'_B}. \tag{31.11}$$

The concentrations of S and P enter the kinetics via the relations

$$\alpha_A = \alpha^*_A c_S, \qquad \alpha_B = \alpha^*_B c_S, \qquad \beta'_A = \beta^*_A c_P, \qquad \beta'_B = \beta^*_B c_P, \tag{31.12}$$

where the rate constants with asterisks are second order. Not all of the rate constants in Fig. 7-37 are independent. Equation (31.11) gives one necessary relation among rate constants. Two other relations (only one is independent) follow from the cycles shown in Fig. 7-37(b). Because these cycles have zero force, $\Pi_+ = \Pi_-$ for each. Hence

$$\begin{aligned} \beta_A \kappa \beta'_B \kappa'_S &= \beta'_A \kappa' \beta_B \kappa_S, \\ \alpha'_A \kappa \alpha_B \kappa'_S &= \alpha_A \kappa' \alpha'_B \kappa_S. \end{aligned} \tag{31.13}$$

A very similar diagram has been discussed in Section 16.

In preparation for the two following subsections, we digress to remind the reader [see Eqs. (16.30) and (16.31)] why the relatively simple diagram in Fig. 7-37(a) does not have a detailed balance type of solution for the p_i at steady state. We define

$$x_A = \frac{\alpha_A + \beta'_A}{\beta_A + \alpha'_A}, \qquad x_B = \frac{\alpha_B + \beta'_B}{\beta_B + \alpha'_B}. \tag{31.14}$$

These quantities, for two-state cycles, are analogous to dimensionless equilibrium constants in an equilibrium system, as has been pointed out before. Figures 7-37(c) and 7-37(d) show these "equilibrium constants" for the steady-state system and for an analogous equilibrium case, respectively. In Fig. 7-37(d), we suppose that S simply binds on E(1) or E*(3) with binding constants K_A and K_B, respectively. There is no enzymatic activity. In this equilibrium case, where there is a detailed balance solution for the p_i, the four equilibrium constants are necessarily related by $K_A K_S = K K_B$. Correspondingly, in order for a "detailed balance" solution for the p_i to exist at steady state, we would need to have, in Fig. 7-37(c), $x_A K_S = K x_B$. However, despite the connection among rate constants in Eq. (31.11), there is no particular relation between x_A and x_B. These are kinetic, not thermodynamic, quantities. Thus, in general $x_A K_S \neq K x_B$. Consequently a true steady-state solution for the p_i must be found (see Section 16).

Figure 7-38 again shows the kinetic diagram for the dimer. The transitions in the upper left square of states all relate to the conformation O or A. The letter A in the small cycles means that the subscript A set of rate constants in Fig. 7-37(a) are to be used. There are no perturbations of these rate constants by interactions because the dimer conformation OO remains constant in all transitions. Similarly, subscript A and subscript B rate constants, from Fig. 7-37(a), are to be used in the designated small cycles of the other three squares of states in Fig. 7-38.

Perturbations, owing to intersubunit interactions, affect equilibrium and rate constants only for the conformational-change transitions. All of these perturbed equilibrium constants are included in Fig. 7-38, where Y and Z are defined in Eq. (31.5). The corresponding separate rate constants are [compare Eqs. (31.6) and (31.7)]

$$\begin{aligned} KY &= \kappa Y^f / \kappa' Y^{f-1}, \qquad & KZ &= \kappa Z^f / \kappa' Z^{f-1}, \\ K_S Y &= \kappa_S Y^f / \kappa'_S Y^{f-1}, \qquad & K_S Z &= \kappa_S Z^f / \kappa'_S Z^{f-1}. \end{aligned} \tag{31.15}$$

This completes the specification of all of the rate constants in Fig. 7-38.

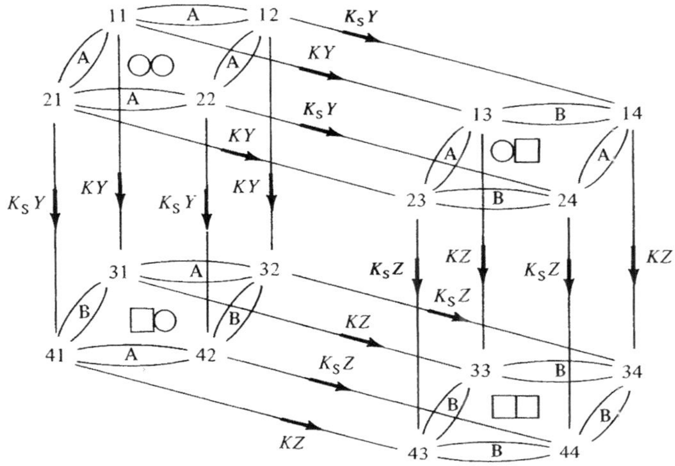

Fig. 7-38. Kinetic diagram for an isologous dimer of subunits as in Fig. 7-37. See also Fig. 5-42(b). Perturbed equilibrium constants are shown. See text for details.

Again, because of symmetry ($p_{ij} = p_{ji}$ for $i \neq j$), there are really only 10 independent states in Fig. 7-38. An example of a steady-state rate equation is

$$\frac{dp_{11}}{dt} = 0 = (\beta_A + \alpha'_A)(p_{12} + p_{21}) + \kappa' Y^{f-1}(p_{13} + p_{31})$$
$$-2(\alpha^*_A c_S + \beta^*_A c_P)p_{11} - 2\kappa Y^f p_{11}. \tag{31.16}$$

Solution of 10 linear algebraic equations, by computer, determines the p_{ij} as functions of c_S and c_P. The total S → P flux per subunit is then given, for example, by

$$J = \alpha^*_A c_S(p_{11} + p_{12} + p_{13} + p_{14}) - \alpha'_A(p_{21} + p_{22} + p_{23} + p_{24})$$
$$+ \alpha^*_B c_P(p_{31} + p_{32} + p_{33} + p_{34}) - \alpha'_B(p_{41} + p_{42} + p_{43} + p_{44}). \tag{31.17}$$

The α transitions are used here; a similar expression for J can be written using the β transitions. Knowledge of the p_{ij} also allows easy calculation of the fractional binding of S and the fraction of enzyme molecules in, say, conformation □.

We turn next to two special cases in which "detailed balance" solutions allow analytical expressions for J.

Special Case: Only E* Binds S or P

In this special case, state 2 (ES) in Fig. 7-37(a) does not exist: the conformation E* is an active enzyme for S → P but the conformation E is completely inactive; E does not bind S or P.

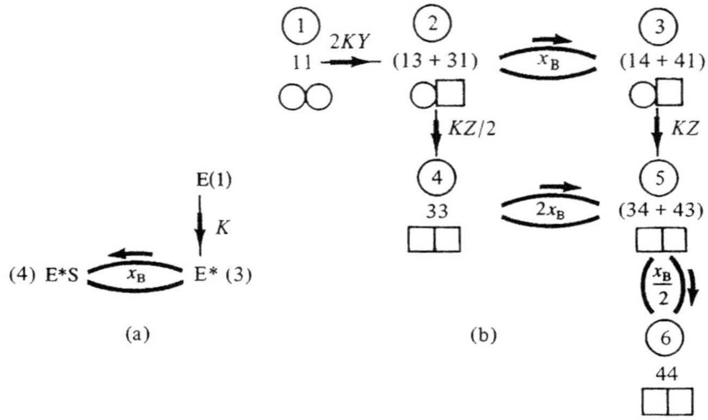

Fig. 7-39. (a) Special case of Fig. 7-37(a) in which state ES does not exist. (b) Same special case of Fig. 7-38, with combined and renumbered (numbers in circles) states. See text for details.

The diagram for a single (unperturbed) subunit is shown in Fig. 7-39(a). There is a detailed balance solution of the steady-state equations in this case, because of the linearity of the diagram:

$$p_1 = 1/\xi, \qquad p_3 = K/\xi, \qquad p_4 = Kx_B/\xi, \tag{31.18}$$
$$\xi = 1 + K(1 + x_B),$$

where K and x_B are given in Eqs. (31.10) and (31.14). The unperturbed flux per subunit is then

$$\begin{aligned} J_o &= \alpha_B p_3 - \alpha_B' p_4 \\ &= K(\alpha_B - \alpha_B' x_B)/\xi, \end{aligned} \tag{31.19}$$

where

$$\alpha_B - \alpha_B' x_B = \frac{\alpha_B \beta_B - \alpha_B' \beta_B'}{\beta_B + \alpha_B'} = \frac{\alpha_B^* c_S \beta_B - \alpha_B' \beta_B^* c_P}{\beta_B + \alpha_B'}. \tag{31.20}$$

We turn now to the dimer. Differential interactions between conformations can distort the distribution in Eqs. (31.18) and hence alter the flux. For example, if y_{AA} is relatively large (○○ is relatively stable), state 11 of the dimer would be heavily populated, thus reducing the flux by diversion of the enzyme into the inactive conformation. In Fig. 7-38, all dimer states with one or both subunits in state 2 are to be omitted. If we then combine (because of symmetry) ij states with $i \neq j$, we obtain the much simpler diagram for the dimer in Fig. 7-39(b), where Y and Z are defined in Eq. (31.5). To simplify notation, these states are renumberd 1 through 6. It will be noticed that there will be a "detailed balance" at steady state in the square of states 2345 because

$x_B \cdot KZ = (KZ/2) \cdot 2x_B$ (x_A is not involved, as in the previous subsection). The effective steady-state grand partition function for the dimer is ξ^2 from Eq. (31.18), with the inclusion, in addition, of interaction factors, just as in Chapter 6:

$$\begin{array}{ccccc} \xi = & y_{AA} & + 2y_{AB}K(1 + x_B) & + y_{BB}K^2(1 + x_B)^2. & (31.21) \\ & \bigcirc\bigcirc & \bigcirc\square & \square\square & \end{array}$$

The interaction factors will generally alter the relative weights of the three classes of terms. If we divide by y_{AA}, we obtain a new and more convenient ξ:

$$\begin{array}{ccccc} \xi = & 1 & + 2KY(1 + x_B) & + K^2YZ(1 + x_B)^2. & (31.22) \\ \text{States:} & 1 & 2,3 & 4,5,6 & \end{array}$$

The probabilities of the six states at steady state are

$$\begin{aligned} p_1 &= 1/\xi, & p_2 &= 2KY/\xi, & p_3 &= 2KYx_B/\xi, \\ p_4 &= K^2YZ/\xi, & p_5 &= 2K^2YZx_B, & p_6 &= K^2YZx_B^2/\xi. \end{aligned} \quad (31.23)$$

It should be noted that these probabilities are consistent with the "equilibrium" constants given in Fig. 7-39(b). The effect of the interaction parameters Y and Z on the p_i, in Eqs. (31.23), is very simple.

From the three small cycles in Fig. 7-39(b), the S → P flux per dimer is then

$$\begin{aligned} J &= (\alpha_B p_2 - \alpha_B' p_3) + (2\alpha_B p_4 - \alpha_B' p_5) + (\alpha_B p_5 - 2\alpha_B' p_6) \\ &= \frac{2KY(\alpha_B - \alpha_B' x_B)[1 + KZ(1 + x_B)]}{1 + 2KY(1 + x_B) + K^2YZ(1 + x_B)^2}, \end{aligned} \quad (31.24)$$

where $\alpha_B - \alpha_B' x_B$ can be rewritten as in Eq. (31.20). This simplifies to $2J_o$ [Eq. (31.19)], as expected, if $y_{AA} = y_{AB} = y_{BB}$. The flux is more sensitive to Y than to Z because $Y = y_{AB}/y_{AA}$ includes y_{AA}, which relates to the completely inactive state of the dimer (11 = ○○ = EE). The maximum flux will occur if states 4, 5, and 6 dominate. This requires large Z and YZ. In this case Eq. (31.24) simplifies to

$$J = \frac{2(\alpha_B \beta_B - \alpha_B' \beta_B')}{\alpha_B + \beta_B + \alpha_B' + \beta_B'}. \quad (31.25)$$

This is the flux for the lower right square of Fig. 7-38: the two subunits are independent and both have the B set of rate constants in Fig. 7-37(a) (for conformation □ = E*).

It should be noted that the parameter f [Eqs. (31.15)] does not appear at all in Eq. (31.24). This is because of the quasiequilibrium and the fact that (in this model) intersubunit interactions do not perturb the rate constants in the E*, E*S cycle.

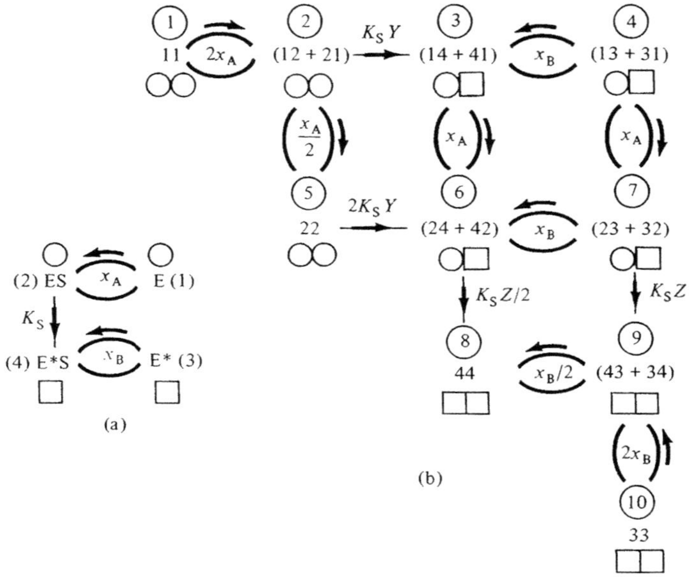

Fig. 7-40. As in Fig. 7-39, except that in this special case the conformational change E ⇄ E* occurs only when S is bound.

Special Case: Bound S Is Required for Conformational Change

In this example (again a special case of Figs. 7-37 and 7-38), we suppose that the conformational change E ⇄ E* cannot occur unless S is bound. Thus, for a single (unperturbed) subunit, the diagram appears as in Fig. 7-40(a). The rate constants are given in Fig. 7-37. Because the diagram in Fig. 7-40(a) is linear, there is a quasiequilibrium steady-state distribution and an effective grand partition function:

$$\begin{array}{rl} \xi & = 1 + x_A + K'(1 + x_B), \\ \text{States:} & \quad 1 \quad\; 2 \qquad\quad 3 \quad\; 4 \end{array} \tag{31.26}$$

where x_A and x_B are defined in Eq. (31.14) and $K' \equiv x_A K_S / x_B$. The physical significance of K' is that it is the effective steady-state "equilbrium constant" for the transitions 1 ⇄ 3, which in fact occur in this case at a negligible rate. (The real equilibrium constant for 1 ⇄ 3 is $K = \kappa/\kappa'$, where κ and κ' are very small.) Thus the steady-state probabilities for an isolated subunit are

$$p_1 = 1/\xi, \qquad p_2 = x_A/\xi, \qquad p_3 = K'/\xi, \qquad p_4 = K'x_B/\xi. \tag{31.27}$$

The unperturbed S → P flux for one subunit is then

$$J_o = J_{oA} + J_{oB}, \tag{31.28}$$

$$J_{oA} = \alpha_A p_1 - \alpha'_A p_2 = (\alpha_A - \alpha'_A x_A)/\xi, \tag{31.29}$$

$$J_{oB} = \alpha_B p_3 - \alpha'_B p_4 = K'(\alpha_B - \alpha'_B x_B)/\xi. \tag{31.30}$$

Here J_{oA} and J_{oB} are the separate contributions to J_o from the two conformations. Equation (31.20) gives alternate expressions for $\alpha_B - \alpha'_B x_B$. Similarly,

$$\alpha_A - \alpha'_A x_A = \frac{\alpha_A \beta_A - \alpha'_A \beta'_A}{\beta_A + \alpha'_A} = \frac{\alpha^*_A c_S \beta_A - \alpha'_A \beta^*_A c_P}{\beta_A + \alpha'_A}. \tag{31.31}$$

For the dimer, we begin by deleting all of the transitions in Fig. 7-38 that have a K label ($1 \rightleftarrows 3$). No states are eliminated. Then using the symmetry of the dimer, we can combine states to give the much simpler diagram in Fig. 7-40(b). The equilibrium constants (K_S), effective equilibrium constants (x_A, x_B), and interaction parameters (Y, Z) are included in the figure [compare Fig. 7-39(b)]. The states are renumbered for convenience. It will be observed that in each square of states in Fig. 7-40(b), the "equilibrium constants" satisfy a detailed balance [e.g., $K_S Y \cdot x_A = (x_A/2) \cdot 2K_S Y$]. Hence there is a quasiequilibrium solution of the steady-state rate equations. The effective grand partition function for the dimer is found by using ξ^2 from Eq. (31.26), and then introducing y_{AA}, y_{AB}, and y_{BB}, just as in Eq. (31.21). After division by y_{AA}, we have, for the dimer,

$$\xi = (1 + x_A)^2 + 2K'Y(1 + x_A)(1 + x_B) + K'^2 YZ(1 + x_B)^2. \tag{31.32}$$

States: 1, 2, 5   3, 4, 6, 7   8, 9, 10

The individual state probabilities are then

$$\begin{aligned} &p_1 = 1/\xi, \quad p_2 = 2x_A/\xi, \quad p_3 = 2K'Yx_B/\xi, \quad p_4 = 2K'Y/\xi, \\ &p_5 = x_A^2/\xi, \quad p_6 = 2K'Yx_A x_B/\xi, \quad p_7 = 2K'Yx_A/\xi, \\ &p_8 = K'^2 YZx_B^2/\xi, \quad p_9 = 2K'^2 YZx_B/\xi, \quad p_{10} = K'^2 YZ/\xi. \end{aligned} \tag{31.33}$$

The reader should verify that these are consistent with the "equilibrium constants" in Fig. 7-40(b). States 3, 4, 6, and 7 include a factor Y; states 8, 9, and 10 include a factor YZ. It is through these factors that intersubunit interactions perturb the state distribution and the flux.

The flux per dimer can now be obtained:

$$J = J_A + J_B, \tag{31.34}$$

$$\begin{aligned} J_A &= \alpha_A(2p_1 + p_2 + p_3 + p_4) - \alpha'_A(p_2 + 2p_5 + p_6 + p_7) \\ &= \frac{2(\alpha_A - \alpha'_A x_A)[1 + x_A + K'Y(1 + x_B)]}{\xi}, \end{aligned} \tag{31.35}$$

$$\begin{aligned} J_B &= \alpha_B(p_4 + p_7 + p_9 + 2p_{10}) - \alpha'_B(p_3 + p_6 + 2p_8 + p_9) \\ &= \frac{2K'Y(\alpha_B - \alpha'_B x_B)[1 + x_A + K'Z(1 + x_B)]}{\xi}, \end{aligned} \tag{31.36}$$

where ξ is given in Eq. (31.32). The parameter f does not appear in J_A and J_B. In the numerator of J_A, the terms $1 + x_A$ arise from the flux in states 1,

2, and 5 [upper left of Fig. 7-40(b)]; the terms in Y arise from states 3, 4, 6, and 7. Similarly, in the numerator of J_B, the terms in $Y(1 + x_A)$ originate from states 3, 4, 6, and 7, and the terms in YZ come from states 8, 9, and 10.

32. Two Simple Trimeric Enzyme Complexes

We have concentrated on dimers, above, to achieve relative simplicity. In this section we treat two examples of symmetrical trimeric (triangular) enzyme complexes: in the first, the individual enzyme subunits have two-state cycles; in the second, they have three-state cycles.

Three Identical Two-State Subunits

This subsection is a direct generalization of the first subsection of Section 29: we pass from a symmetrical dimer to a symmetrical triangular trimer. An individual unperturbed subunit has the two-state cycle and first-order rate constants already specified in Fig. 7-20(a). The flux per subunit is given in Eq. (29.2). If we use the S → P interpretation of the two-state cycle, $\alpha = \alpha^* c_S$ and $\beta' = \beta^* c_P$.

The trimer has eight states, with a diagram shown schematically in Fig. 7-11. The simple cycles in this figure are of the type shown in Fig. 7-20(a), but the rate constants of these cycles are perturbed by the presence of two neighbors, each of which can be either in state 1 or state 2. The integers on the cycles of Fig. 7-11 specify the number of neighbors of the cycling subunit that are in state 2. These integers are used as subscripts on the rate constants to distinguish the different perturbations encountered. Figure 7-13 is also applicable: because of symmetry in the trimer, the full diagram can be collapsed from Fig. 7-13(a) to Fig. 7-13(b). The passage from Fig. 7-20(b) to 7-20(c) is the analogue of this "collapse" for a dimer (incidentally, note that the subscript system used in Figs. 7-20(b) and 7-20(c) is not the same as that mentioned above and used below).

The detailed diagram is presented in Fig. 7-41. The subscript on the steady-state probability p_i is the r value (number of subunits in state 2). The interaction parameters included in Fig. 7-41 are the same as in Eqs.(27.43) and (27.44), but we have introduced $y_1 \equiv y_{12}/y_{11}$ and $y_2 \equiv y_{22}/y_{12}$, as in Eqs. (29.4). The equilibrium and rate constants in Fig. 7-41, including interaction perturbations, are then:

$$\frac{\alpha_0}{\alpha_0'} = \frac{\alpha}{\alpha'} y_1^2, \qquad \alpha_0 = \alpha y_1^{2f_\alpha}, \qquad \alpha_0' = \alpha' y_1^{2(f_\alpha - 1)},$$
$$\frac{\beta_0}{\beta_0'} = \frac{\beta}{\beta'} y_1^{-2}, \qquad \beta_0 = \beta y_1^{-2f_\beta}, \qquad \beta_0' = \beta' y_1^{2(1-f_\beta)}, \qquad (32.1)$$

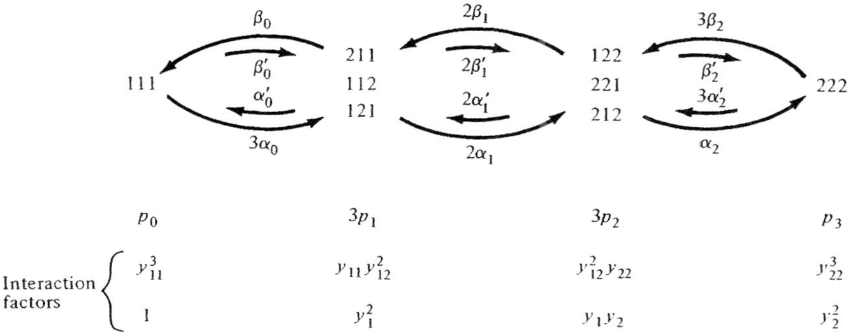

Fig. 7-41. Diagram for a symmetrical triangular trimer of two-state subunits. Subscripts specify the number of nonreacting neighbors that are in state 2. See text for further details.

$$\begin{aligned}
&\frac{\alpha_1}{\alpha_1'} = \frac{\alpha}{\alpha'} y_1 y_2, \qquad \alpha_1 = \alpha(y_1y_2)^{f_\alpha}, \qquad \alpha_1' = \alpha'(y_1y_2)^{f_\alpha - 1}, \\
&\frac{\beta_1}{\beta_1'} = \frac{\beta}{\beta'}(y_1y_2)^{-1}, \qquad \beta_1 = \beta(y_1y_2)^{-f_\beta}, \qquad \beta_1' = \beta'(y_1y_2)^{1-f_\beta}, \\
&\frac{\alpha_2}{\alpha_2'} = \frac{\alpha}{\alpha'} y_2^2, \qquad \alpha_2 = \alpha y_2^{2f_\alpha}, \qquad \alpha_2' = \alpha' y_2^{2(f_\alpha - 1)}, \\
&\frac{\beta_2}{\beta_2'} = \frac{\beta}{\beta'} y_2^{-2}, \qquad \beta_2 = \beta y_2^{-2f_\beta}, \qquad \beta_2' = \beta' y_2^{2(1-f_\beta)}.
\end{aligned} \tag{32.1}$$

These relations are the analogues of Eqs. (29.3) and (29.6); they follow from the interaction parameters given in Fig. 7-41, together with the introduction of f_α and f_β in our usual way.

We are now in a position to obtain the steady-state p_i. Because the collapsed diagram (Fig. 7-41) is linear, there is an effective detailed balance solution for the p_i:

$$\frac{3p_1}{p_0} = 3x_0, \qquad \frac{p_2}{p_1} = x_1, \qquad \frac{p_3}{3p_2} = \frac{x_2}{3}, \tag{32.2}$$

where

$$\begin{aligned}
x_0 &\equiv \frac{\alpha_0 + \beta_0'}{\beta_0 + \alpha_0'} = \frac{\alpha y_1^{2f_\alpha} + \beta' y_1^{2(1-f_\beta)}}{\beta y_1^{-2f_\beta} + \alpha' y_1^{2(f_\alpha - 1)}}, \\
x_1 &\equiv \frac{\alpha_1 + \beta_1'}{\beta_1 + \alpha_1'} = \frac{\alpha(y_1y_2)^{f_\alpha} + \beta'(y_1y_2)^{1-f_\beta}}{\beta(y_1y_2)^{-f_\beta} + \alpha'(y_1y_2)^{f_\alpha - 1}}, \\
x_2 &\equiv \frac{\alpha_2 + \beta_2'}{\beta_2 + \alpha_2'} = \frac{\alpha y_2^{2f_\alpha} + \beta' y_2^{2(1-f_\beta)}}{\beta y_2^{-2f_\beta} + \alpha' y_2^{2(f_\alpha - 1)}}.
\end{aligned} \tag{32.3}$$

Then, from Eqs. (32.2), the p_i are

$$p_1 = x_0/\xi, \qquad p_2 = x_0x_1/\xi, \qquad p_3 = x_0x_1x_2/\xi, \tag{32.4}$$

$$\theta = p_1 + 2p_2 + p_3,$$

$$\xi = 1 + 3x_0 + 3x_0x_1 + x_0x_1x_2, \tag{32.5}$$

where θ is the fraction of subunits with S bound.

The trimer steady-state flux is then (using α, α' transitions)

$$\begin{aligned} J &= 3[(\alpha_0p_0 - \alpha_0'p_1) + 2(\alpha_1p_1 - \alpha_1'p_2) + (\alpha_2p_2 - \alpha_2'p_3)] \\ &= \frac{3(\alpha\beta - \alpha'\beta')[y_1^{2(f_\alpha - f_\beta)} + 2(y_1y_2)^{f_\alpha - f_\beta}x_{01} + y_2^{2(f_\alpha - f_\beta)}x_1x_{02}]}{[\beta y_1^{-2f_\beta} + \alpha' y_1^{2(f_\alpha - 1)}](1 + 3x_0 + 3x_0x_1 + x_0x_1x_2)}, \end{aligned} \tag{32.6}$$

where x_{01} and x_{02} are the hybrid quantities [see Eqs. (32.3)]

$$\begin{aligned} x_{01} &= \frac{\alpha y_1^{2f_\alpha} + \beta' y_1^{2(1-f_\beta)}}{\beta(y_1y_2)^{-f_\beta} + \alpha'(y_1y_2)^{f_\alpha - 1}}, \\ x_{02} &= \frac{\alpha y_1^{2f_\alpha} + \beta' y_1^{2(1-f_\beta)}}{\beta y_2^{-2f_\beta} + \alpha' y_2^{2(f_\alpha - 1)}}. \end{aligned} \tag{32.7}$$

Equation (32.6), for a trimer, corresponds to Eq. (29.15) for a dimer. If $y_{11} = y_{12} = y_{22}$, Eq. (32.6) reduces to $J = 3J_o$ [Eq. (29.2)].

In the special case $f_\alpha + f_\beta = 1$, there is a quasiequilibrium solution:

$$x_0 = y_1^2x, \qquad x_1 = y_1y_2x, \qquad x_2 = y_2^2x, \tag{32.8}$$

$$\xi = 1 + 3y_1^2x + 3y_1^3y_2x^2 + (y_1y_2)^3x^3, \tag{32.9}$$

$$x \equiv (\alpha + \beta')/(\beta + \alpha'). \tag{32.10}$$

Equation (32.9) is formally equivalent to Eqs. (23.38) and (27.47). The trimer flux in this case becomes

$$J = \frac{3(\alpha\beta - \alpha'\beta')[y_1^{2f_\alpha} + 2y_1^2(y_1y_2)^{f_\alpha}x + y_1^3y_2^{2f_\alpha+1}x^2]}{(\beta + \alpha')[1 + 3y_1^2x + 3y_1^3y_2x^2 + (y_1y_2)^3x^3]}. \tag{32.11}$$

This result is equivalent to Eq. (27.50) [see Eqs. (29.22)]. A Michaelis–Menten enzyme has

$$\beta' = 0, \qquad \alpha = \alpha^* c_S, \qquad x = \frac{\alpha^* c_S}{\beta + \alpha'}. \tag{32.12}$$

In this case x is proportional to c_S. The asymptotic value of $J(c_S \to \infty)$ is (compare Fig. 7-41)

$$J = 3\beta_2 = 3\beta y_2^{2f_\alpha - 2}. \tag{32.13}$$

This is not always the maximum value of J (see Fig. 7-42).

In the special case $f_\alpha = 0$ ($f_\beta = 1$; diffusion-controlled binding of both S and P) and $y_{11} = y_{22}$ [as already used in Eqs. (27.37)], Eq. (32.11) for a Michaelis–Menten enzyme simplifies to:

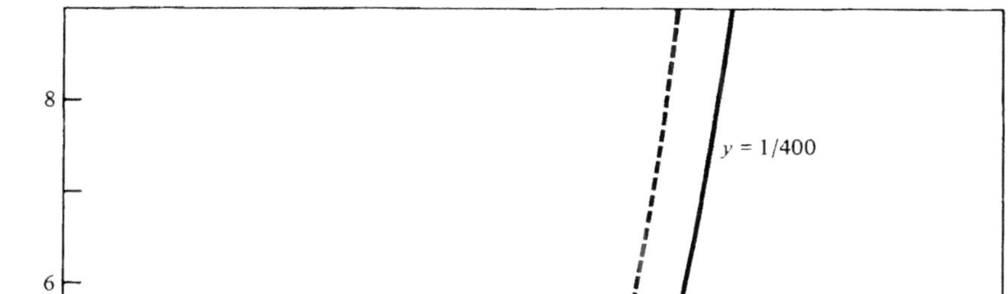

Fig. 7-42. Solid curves illustrate the special case, Eq. (32.14), where $y = y_2/y_1$. The dashed curves represent another special case of Eq. (32.11), with $\beta' = 0, f_\alpha = 0$, $y_1 = 1$. The solid $y = 1$ curve also applies to the second case with $y_2 = 1$.

$$J = \frac{3\beta x(1 + 2y^{-1}x + y^{-1}x^2)}{1 + 3y^{-1}x + 3y^{-1}x^2 + x^3}, \tag{32.14}$$

where x is related to c_S as in Eq. (32.12). This expression is similar to Eq. (27.51). When $y = 1$ (no cooperativity),

$$J = 3\beta x/(1 + x). \tag{32.15}$$

In this case, the three subunits are independent; each contributes $\beta x/(1 + x)$ to the trimer flux. The solid curves in Fig. 7-42 illustrate the effect of positive ($y = 20$) and negative ($y = \frac{1}{400}$) cooperativity on the flux. These curves are qualitatively similar to Fig. 7-10 (for a dimer), but the interaction effects are sharpened in Fig. 7-42 by the increased number of neighbors in the triangular trimer.

The dashed curves in Fig. 7-42 provide another example. In this case, in Eq. (32.11), we take $\beta' = 0$ (Michaelis–Menten), $f_\alpha = 0$ as before, and

$y_1 = 1$ (i.e., $y_{11} = y_{12}$). The remaining interaction parameter is $y_2 = y_{22}/y_{12}$. Thus 22 intersubunit interactions differ from 11 and 12 interactions, which are the same. The dashed curves are for $y_2 = \sqrt{20}$ (22 interactions relatively stable) and $y_2 = \frac{1}{20}$ (22 interactions unstable). These are the same contributions that y_2 makes to y [Eqs. (27.37)] for the solid curves in the figure.

The treatment in this subsection can be extended easily to a symmetrical tetrahedron of subunits, but we do not give the details. This is possible because all of the states with a given r value are equivalent, by symmetry. Hence a collapsed linear diagram, as in Fig. 7-41, is still possible. On the other hand, a square of subunits does not have this symmetry (there are two classes of states when $r = 2$), nor does a linear trimer (there are two classes of states for both $r = 1$ and $r = 2$). The schematic diagram for a linear trimer is shown in Fig. 7-43; this is the analogue of Fig. 7-11. In Fig. 7-43, the state 121 is not equivalent to states 211 and 112, and also state 212 is not equivalent to states 122 and 221. This is also reflected in the double indices i, j placed on the cycles. These indices specify the "environment" of the cycling subunit: i is the number of nearest neighbors in state 1, and j is the number of nearest neighbors in state 2. In Fig. 7-11, which has symmetry, the environment could be specified by means of a single index.

Three Identical Three-State Subunits

In this subsection we continue the discussion (which should be reviewed) of Figs. 5-32 and 5-33 in Section 17 by giving an analytical solution for the flux in one particular special case in which explicit intersubunit interactions are included.

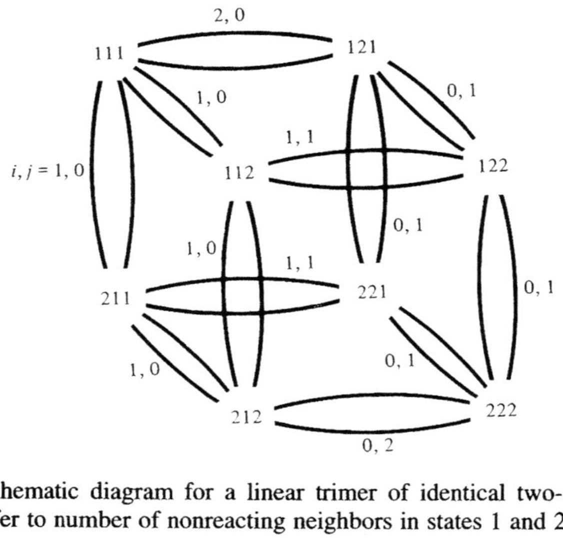

Fig. 7-43. Schematic diagram for a linear trimer of identical two-state subunits. Indices i, j refer to number of nonreacting neighbors in states 1 and 2, respectively.

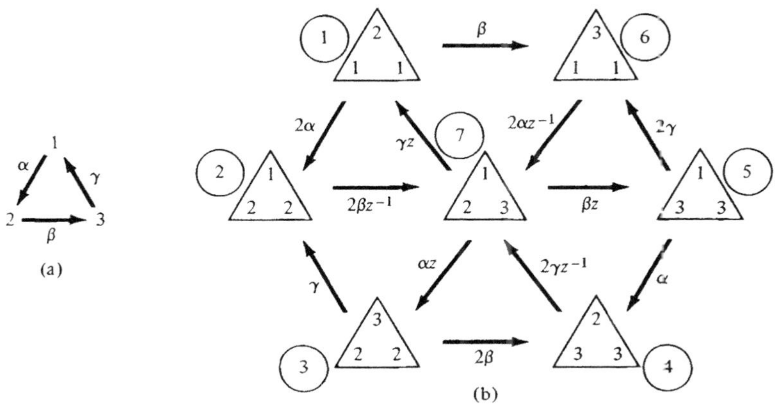

Fig. 7-44. (a) Diagram for a one-way three-state subunit. (b) Diagram for a symmetrical triangular trimer of these subunits in the special case explained in the text. See, also, Figs. 5-32 and 5-33.

Each isolated subunit of the trimer has an unperturbed one-way cycle, as shown in Fig. 7-44(a). The unperturbed flux per subunit is then

$$J_o = \frac{\alpha\beta\gamma}{\alpha\beta + \beta\gamma + \alpha\gamma}. \tag{32.16}$$

In the trimer we assume that subunit pairs in the same state have a relatively unfavorable intersubunit interaction, whereas pairs in different states have a relatively favorable interaction. To simplify the algebra, we assume further that all interaction parameters of the same type are equal:

$$\begin{aligned} y_{11} &= y_{22} = y_{33}, \\ y_{12} &= y_{13} = y_{23} \\ &= y_{21} = y_{31} = y_{32}. \end{aligned} \tag{32.17}$$

As mentioned above, $y_{ii} \ll y_{jk}$, where y_{ii} refers to any identical pair and y_{jk} to any nonidentical pair.

The three vertex states in Fig. 5-33(a) are assumed to be of negligible importance because each is destabilized by an interaction factor y_{ii}^3. The two central states in Fig. 5-33(a) are grouped together in Fig. 7-44(b); this is recognized by the factors of 2 that are included in the rate constants leading into the central state.

For a rate constant leading out of the central state, the associated equilibrium constant is perturbed by a factor

$$\frac{\text{outer}}{\text{central}} = \frac{y_{ik}^2 y_{ii}}{y_{ik}^3} = \frac{y_{ii}}{y_{ik}} \ll 1. \tag{32.18}$$

The reciprocal of this factor applies to a transition into the central state. If we assume further (again, to keep the algebra simple) that $f_\alpha = f_\beta = f_\gamma \equiv f$, and define

$$z \equiv \left(\frac{y_{ii}}{y_{ik}}\right)^f \ll 1, \tag{32.19}$$

then we have a perturbation factor z on rate constants out of the central state and a factor z^{-1} on rate constants into the central state. Each of the outer ring of states in Fig. 7-44 has one destabilizing ii pair; the central state has no such pair. Hence central → outer rate constants are reduced, whereas outer → central rate constants are increased by this interaction effect. Transitions within the outer ring of states are not influenced because all such states have one ii pair.

With all the rate constants in Fig. 7-44 specified, we can now find the p_i from the rate equations:

$$\begin{aligned} (2\alpha + \beta)p_1 &= \gamma z p_7, & 2\beta z^{-1}p_2 &= 2\alpha p_1 + \gamma p_3, \\ (2\beta + \gamma)p_3 &= \alpha z p_7, & 2\gamma z^{-1}p_4 &= 2\beta p_3 + \alpha p_5, \\ (2\gamma + \alpha)p_5 &= \beta z p_7, & 2\alpha z^{-1}p_6 &= 2\gamma p_5 + \beta p_1, \end{aligned} \tag{32.20}$$
$$p_1 + \cdots + p_7 = 1.$$

It is easy to express each of $p_1, \ldots, p_6$ in terms of p_7. We leave this to the reader. Then normalization of the p_i leads to

$$p_7 = \left[1 + \frac{\gamma z(2\alpha\beta + 2\alpha^2 z + \beta^2 z)}{2\alpha\beta(2\alpha + \beta)} + \frac{\alpha z(2\beta\gamma + 2\beta^2 z + \gamma^2 z)}{2\beta\gamma(2\beta + \gamma)} + \frac{\beta z(2\gamma\alpha + 2\gamma^2 z + \alpha^2 z)}{2\gamma\alpha(2\gamma + \alpha)}\right]^{-1}. \tag{32.21}$$

The quantity in brackets is like a grand partition function: the leading term, 1, relates to p_7; the terms in z arise from p_1, p_3, and p_5 [see Eqs. (32.20)]; and the terms in z^2 arise from p_2, p_4, and p_6. Thus, when z is small, state 7 dominates and states 2, 4, and 6 are very sparsely populated.

In the special case $\alpha = \beta = \gamma$,

$$p_1 = p_3 = p_5 = \frac{zp_7}{3}, \qquad p_2 = p_4 = p_6 = \frac{z^2 p_7}{2}, \tag{32.22}$$

$$p_7 = \frac{2}{2 + 2z + 3z^2}. \tag{32.23}$$

The total flux in the trimer is most simply written as

$$J = (\alpha + \beta + \gamma)zp_7 = \frac{(\alpha + \beta + \gamma)z}{[\,]}, \tag{32.24}$$

where [ ] is the expression in brackets in Eq. (32.21). When z is very small, $[\;] \to 1$ and $J \to (\alpha + \beta + \gamma)z$; state 7 is the only state with significant population. Equation (32.24) does not reduce to $3J_o$ [Eq. (32.16)] when $z = 1$ because the three vertex states have been omitted in Fig. 7-44 (z is presumed to be small). If $\alpha = \beta = \gamma$,

$$J = \frac{6\alpha z}{2 + 2z + 3z^2}. \tag{32.25}$$

This gives $J = 6\alpha/7$ when $z = 1$; with vertex states included, it would be $J = \alpha$ [Eq. (32.16)].

The conclusion we reach is that the mechanism discussed in relation to Eqs. (17.48) and (17.49) is physically plausible, but unfortunately it would be accompanied by a small flux, $J \cong (\alpha + \beta + \gamma)z$.

33. Calcium-ATPase as an Example

In this section we use a model of the Ca-ATPase of sarcoplasmic reticulum as a practical illustration of some of the general procedures introduced in this chapter and in Chapter 6. We limit ourselves here to the essential features of the system, model, and experimental results. Much more detailed explanations are available in the original paper.[5]

This membrane complex, *in vivo*, uses the free energy of hydrolysis of ATP to transport Ca^{2+} against its electrochemical potential gradient from "out" (cytoplasm) to "in" (sarcoplasmic reticulum). The general reaction scheme adopted is shown in Fig. 7-45. It will be noted that proton binding is much involved (see, also, Figs. 6-21 through 6-23). The three principal states of the enzyme, for present purposes, are designated ATP · E, *E, **E, or T, α, β, respectively. State T has high affinity binding sites for Ca^{2+}; states α and β have low affinity sites. States β and T have binding sites for Ca^{2+} and H^+ exposed to the outside; state α has binding sites exposed to the inside.

In the *in vitro* experiments of interest here (rate of ATP hydrolysis), the complex is in the membrane of "leaky" sarcoplasmic reticulum vesicles (there is no membrane potential); and Ca^{2+} and H^+ have the same concentrations (c_1 and c_2, respectively) inside and out. Hence the only thermodynamic force in

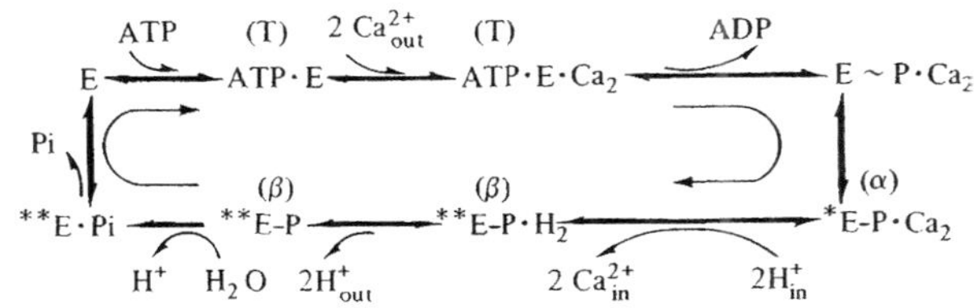

Fig. 7-45. Catalytic and transport cycle of sarcoplasmic reticulum ATPase.

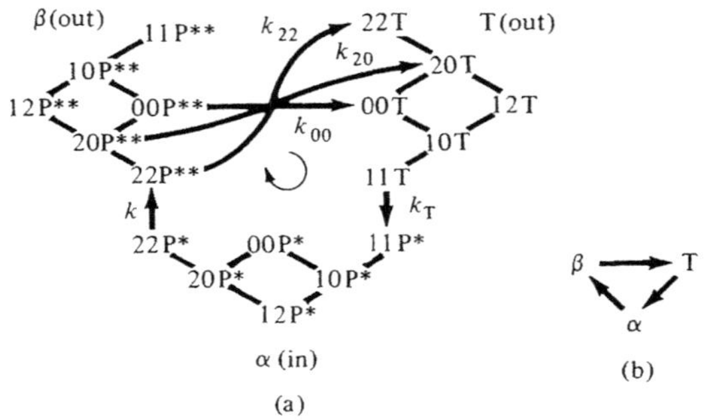

Fig. 7-46. (a) Kinetic diagram used to simulate steady-state ATPase activity. States α, β, and T are also indicated in Fig. 7-45. (b) Overall relationship between states α, β, and T.

the cycle of Fig. 7-45 arises from the hydrolysis of ATP. The cycle operates in the clockwise direction. In one cycle, one ATP is hydrolyzed, two Ca^{2+} are transported from out to in, and two H^+ are transported from in to out.

The actual reduced kinetic diagram that we work with in this analysis is shown in Fig. 7-46(a). As Fig. 7-46(b) indicates, this is a one-way cycle involving states T, α, and β only, but each includes substates that depend on the extent of binding of Ca^{2+}, H^+, or both. In our previous treatment of the equilibrium binding of Ca^{2+} and H^+ on Ca-ATPase (Section 25; Figs. 6-21 through 6-23), the only enzyme state in Fig. 7-45 considered (ATP was absent) was E; note that this state is not included in Fig. 7-46 (it is assumed to be a transient intermediate).

The enzyme represented in Figs. 7-45 and 7-46 contains one central site for the binding and hydrolysis of ATP and two equivalent sites for the *competitive* binding of Ca^{2+} and H^+. The latter two sites are on two equivalent domains or subunits that change conformation when either Ca^{2+} or H^+ is bound. For each of states T, α, β, the model is the same as in the first subsection of Section 25. These conformations, for each domain, are designated 0 (site empty), 1 (site occupied by Ca^{2+}), and 2 (site occupied by H^+). There is a set of interdomain interaction parameters y_{ij} $(i, j = 0, 1, 2)$, with $y_{ij} = y_{ji}$, for each of states T, α, β. None of these parameter sets should be expected to be the same as that used for state E of Figs. 6-21 through 6-23. Each of states T, α, β can exist in $3 \times 3 = 9$ substates, depending on binding. However, because of symmetry, these reduce to the six substates shown for each of T, α, β in Fig. 7-46(a). The notation 00 means both sites empty, 11 means both occupied by Ca^{2+}, etc. The short straight lines in Fig. 7-46(a) represent forward and backward transitions and rate constants for binding, as perturbed by interdomain interactions. The arrows, labelled with *k*s, refer to one-way interstate transitions. Interaction perturbations of these *k*s are considered to be embedded in the final *k* values selected; hence these perturbations need not be treated explicitly. Because of these one-way transitions in Fig. 7-46(a), there is tight coupling between ATP hydrolysis and Ca^{2+} (out → in) and H^+ (in → out) transport, with stoichiometry 1 ATP, 2 Ca^{2+}, and 2 H^+ in each cycle. We shall express the overall cyclic activity in terms of the steady-state rate of ATP hydrolysis, per enzyme unit, J.

The six substates of T are assumed to exist in a relatively fast equilibrium. This assumption effectively reduces the total number of states in Fig. 7-46(a) from 18 to 13.

In Figs. 6-21 through 6-23, the model used was a tetramer (i.e., four domains or subunits). Here, for simplicity, we work with a dimer, i.e., with the two domains already included in Fig. 7-46(a). Equilibrium binding on state E (Section 25) showed strong enough cooperativity so that a dimer was not adequate; at least a tetramer was needed (i.e., two Es). However, a dimer proves to be adequate, semiquantitatively, to fit the steady-state data. A tetramer (square), treated in the same way, would require 216 states in the diagram rather than only 13. We use a dimer, rather than the tetramer suggested by equilibrium binding, simply to avoid this complexity.

To summarize, the essential ion-binding aspects of the cycle kinetics are the following. State T has high affinity for Ca^{2+}. The binding of Ca^{2+} and H^+ in this state is maintained at equilibrium. The transition from T to α can occur only from 11T (both sites occupied by Ca^{2+}). State α has low affinity for Ca^{2+}. The transition $\alpha \rightarrow \beta$ can occur only from state 22P*, i.e., after both Ca^{2+} have been displaced by protons. State β also has low affinity for Ca^{2+}. The dominant transition from β to T is via k_{00} (i.e., after both protons have been released). In any case, both protons are released to the outside, from either β or T.

If the binding of Ca^{2+} and H^+ were to occur at equilibrium on any one of the three states T, α, β, the treatment in Eqs. (25.1)–(25.15) would apply. However, to conform with the notation used by Inesi and Hill,[5] we introduce the new interaction parameters [see Eqs. (25.4)]

$$Y^2 \equiv \left(\frac{y_{01}^2}{y_{00}y_{11}}\right)^{1/2} = y^{-1/2}, \tag{33.1}$$

$$Z^2 \equiv \left(\frac{y_{02}^2}{y_{00}y_{22}}\right)^{1/2} = z^{-1/2}, \tag{33.2}$$

$$U^2 \equiv \left(\frac{y_{12}^2}{y_{11}y_{22}}\right)^{1/2} = u^{-1/2}. \tag{33.3}$$

Positive cooperativity is associated with $y, z, u > 1$ and $Y, Z, U < 1$. Then the equilibrium ξ for any one of these three states becomes [see Eq. (25.8)]

$$\begin{aligned}\xi = \underset{00}{1} + \underset{01}{2Y^2\overline{K}_1c_1} + \underset{02}{2\,Z^2\overline{K}_2c_2} + \underset{12}{2U^2\overline{K}_1c_1\overline{K}_2c_2}& \\ + \underset{11}{(\overline{K}_1c_1)^2} + \underset{22}{(\overline{K}_2c_2)^2}.&\end{aligned} \tag{33.4}$$

The six binding constants in the three different states are then designated $\overline{K}_{1T}$, $\overline{K}_{2T}$, $\overline{K}_{1\alpha}$, etc., and the nine interaction parameters are designated Y_T, Z_T, U_T, Y_α, etc. The three corresponding grand partition functions [Eq. (33.4)] are then ξ_T, ξ_α, and ξ_β.

As already mentioned, we do in fact assume that state T is in an internal binding equilibrium. The probability P_{11T} that state T is in substate 11T is then, from Eq. (33.4),

$$P_{11T} = (\overline{K}_{1T}c_1)^2/\xi_T. \tag{33.5}$$

Having introduced the binding constants and interaction parameters, we can now obtain the perturbed rate constants among the substates of states α and β (this is not necessary for state T because of the assumed internal equilibrium). The discussion below applies to state α; for state β, we merely replace α everywhere by β. The equilibrium constant for $00 \rightleftarrows 10$ in Fig. 7-46(a), including interaction parameters, can be written in three alternative ways [Eqs. (25.3) and (25.6)]:

$$00 \rightleftarrows 10\text{:}\quad \frac{2y_{01\alpha}K_{1\alpha}^{\circ}}{y_{00\alpha}} = 2K_{1\alpha} = 2Y_\alpha^2\overline{K}_{1\alpha}. \tag{33.6}$$

In this system $K_{1\alpha}^{\circ}$ is not an operational equilibrium constant (the domains cannot be separated). On the other hand, $K_{1\alpha}$ and $\overline{K}_{1\alpha}$ are both operational. The choice of one or the other as the reference equilibrium constant, equal to the quotient of reference rate constants, is arbitrary. After a complete set of equilibrium constants, rate constants, and interaction parameters is found by matching the model with experiments, on the basis of one choice, the equilibrium and rate constants can all be converted to the other choice, if desired. See the next subsection for further consideration of this point. To be consistent with the earlier discussion of Figs. 6-21 through 6-23, we select $\overline{K}_{1\alpha}$ and $\overline{K}_{2\alpha}$ (rather than $K_{1\alpha}$ and $K_{2\alpha}$) as the reference equilibrium constants. With this choice, the required perturbed equilibrium constants (i.e., perturbed by Y_α, Z_α, U_α) all follow by taking quotients of pairs of terms in Eq. (33.4). In order to split the interaction parameters that appear in the perturbed equilibrium constants between forward and backward rate constants, we use $f = \frac{1}{2}$ as a plausible guess in every case because a conformational change is presumed to be involved (Section 25). Thus, for state α, we arrive at:

$$00 \rightleftarrows 10\text{:}\quad 2Y_\alpha^2\overline{K}_{1\alpha} = \frac{2\alpha_1 Y_\alpha}{\alpha_{-1}Y_\alpha^{-1}} \qquad \left(\overline{K}_{1\alpha} = \frac{\alpha_1}{\alpha_{-1}}\right), \tag{33.7}$$

$$00 \rightleftarrows 20\text{:}\quad 2Z_\alpha^2\overline{K}_{2\alpha} = \frac{2\alpha_2 Z_\alpha}{\alpha_{-2}Z_\alpha^{-1}} \qquad \left(\overline{K}_{2\alpha} = \frac{\alpha_2}{\alpha_{-2}}\right), \tag{33.8}$$

$$10 \rightleftarrows 11\text{:}\quad \frac{Y_\alpha^{-2}\overline{K}_{1\alpha}}{2} = \frac{\alpha_1 Y_\alpha^{-1}}{2\alpha_{-1}Y_\alpha}, \tag{33.9}$$

$$10 \rightleftarrows 12\text{:}\quad U_\alpha^2 Y_\alpha^{-2}\overline{K}_{2\alpha} = \frac{\alpha_2 U_\alpha Y_\alpha^{-1}}{\alpha_{-2}U_\alpha^{-1}Y_\alpha}, \tag{33.10}$$

$$20 \rightleftarrows 22: \ \frac{Z_\alpha^{-2}\overline{K}_{2\alpha}}{2} = \frac{\alpha_2 Z_\alpha^{-1}}{2\alpha_{-2} Z_\alpha}, \tag{33.11}$$

$$20 \rightleftarrows 12: \ U_\alpha^2 Z_\alpha^{-2}\overline{K}_{1\alpha} = \frac{\alpha_1 U_\alpha Z_\alpha^{-1}}{\alpha_{-1} U_\alpha^{-1} Z_\alpha}, \tag{33.12}$$

where α_1 and α_2 are second-order rate constants, whereas α_{-1} and α_{-2} are first order.

The numerators and denominators on the right-hand sides of Eqs. (33.7)–(33.12) are the perturbed α-state rate constants. These, together with the same set with β in place of α, complete the specification of rate constants in Fig. 7-46, except for two additional points: the rate constant k_{20} is taken to be equal to $(k_{00}k_{22})^{1/2}$, i.e., k_{20} is not considered to be independent; and, because T is treated as a single state, the effective rate constant from state T to 11P*, is $k_T P_{11T}$, where P_{11T} is given by Eq. (33.5).

There are 13 effective states in Fig. 7-46(a) (6 in each of α and β, plus state T). For any set of parameter choices, we then have to solve, by computer, 13 linear algebraic rate equations to find the 13 steady-state p_i. The steady-state flux (rate of hydrolysis of ATP)) is then found from

$$J = kp_{22P*} = k_T P_{11T} p_T. \tag{33.13}$$

In choosing parameters for states α and β, we consider the binding constants (e.g., $\overline{K}_{1\alpha}$) and "off" rate constants (e.g., α_{-1}) to be independent; the "on" rate constants are then not independent (e.g., $\alpha_1 = \alpha_{-1}\overline{K}_{1\alpha}$). The complete set of *independent* parameters for this model is then:

$$\begin{aligned} &\text{Binding constants: } \overline{K}_{1T}, \overline{K}_{2T}, \overline{K}_{1\alpha}, \overline{K}_{2\alpha}, \overline{K}_{1\beta}, \overline{K}_{2\beta}, \\ &\text{Rate constants: } \alpha_{-1}, \alpha_{-2}, \beta_{-1}, \beta_{-2}, k, k_T, k_{00}, k_{22}, \\ &\text{Cooperativity parameters: } Y_T, Z_T, U_T, Y_\alpha, Z_\alpha, U_\alpha, Y_\beta, Z_\beta, U_\beta. \end{aligned} \tag{33.14}$$

Figure 7-47 contains the experimental curves we are trying to fit with the present model. Each curve is a plot of ATP flux, J, as a function of pCa $= -\log c_1$, for a fixed pH $= -\log c_2$. The data points,[5] which exhibit some scatter, are omitted for clarity. The most interesting feature of this family of curves is that they rise with increasing $c_2(H^+)$ and then decrease again, with the highest curve at about pH = 7.2. In the model, J is small when c_2 is small (high pH) because the α state 00P* is highly populated. Also, J is small when c_2 is large (low pH), because the β state 22P** is highly populated (the rate constant k_{22} is much smaller than k_{00}).

Only a small fraction of the total parameter space [Eq. (33.14)] was sampled in trying to fit the curves in Fig. 7-47. The original paper[5] gives the procedure used in varying parameters. A rather satisfactory set was found to be:

$$\begin{aligned} \overline{K}_{1T} &= 0.2071 \times 10^6\ \text{M}^{-1}, & \overline{K}_{2\alpha} &= 90.34 \times 10^6\ \text{M}^{-1}, \\ \overline{K}_{2T} &= 1.585 \times 10^6\ \text{M}^{-1}, & \overline{K}_{1\beta} &= 0.15 \times 10^3\ \text{M}^{-1}, \end{aligned} \tag{33.15}$$

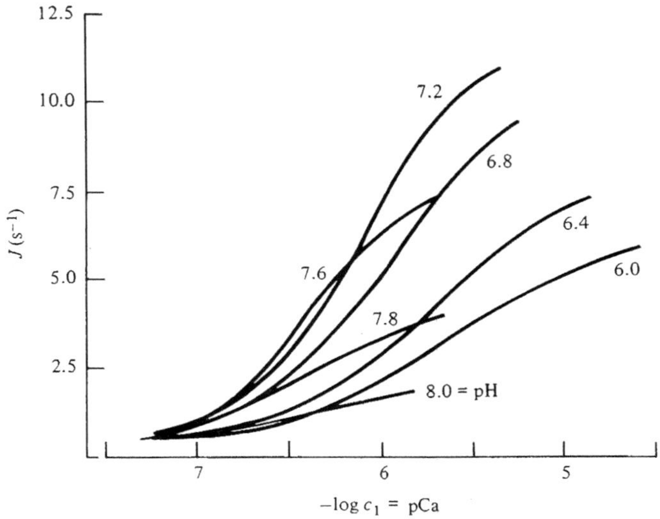

Fig. 7-47. Experimental steady-state ATP flux, as function of pCa, at different pH values.

$$
\begin{aligned}
\overline{K}_{1\alpha} &= 0.15 \times 10^3\ \mathrm{M}^{-1}, & \overline{K}_{2\beta} &= 57.06 \times 10^6\ \mathrm{M}^{-1},\\
\alpha_{-1} &= 163.2\ \mathrm{s}^{-1}, & k &= 261.1\ \mathrm{s}^{-1},\\
\alpha_{-2} &= 32.63\ \mathrm{s}^{-1}, & k_{\mathrm{T}} &= 978.9\ \mathrm{s}^{-1},\\
\beta_{-1} &= 130.5\ \mathrm{s}^{-1}, & k_{00} &= 326.3\ \mathrm{s}^{-1},\\
\beta_{-2} &= 326.3\ \mathrm{s}^{-1}, & k_{22} &= 9.463\ \mathrm{s}^{-1},\\
U_{\mathrm{T}} &= 6.0, & Z_\alpha &= 0.1.
\end{aligned}
\qquad (33.15)
$$

The other seven interaction parameters have the value 1.0. The values of U_{T} and Z_α correspond to negative cooperativity in state T and positive cooperativity in state α [see Eqs. (25.5) for details]. Note the relatively small binding constants for Ca^{2+} in states α and β. In view of Eqs. (25.6) and the interaction parameters above, $K_i = \overline{K}_i$ in all cases in Eqs. (33.15), except

$$K_{2\alpha} = z_\alpha^{-1/2}\overline{K}_{2\alpha} = Z_\alpha^2\overline{K}_{2\alpha} = 90.34 \times 10^4\ \mathrm{M}^{-1}. \qquad (33.16)$$

Figure 7-48 gives the family of curves, corresponding to the experimental curves in Fig. 7-47, that were calculated from the model and the above parameter set. The theoretical curves cover a much larger range in pCa. The agreement between theory and experiment is excellent except for pH 7.8 and 8.0, where the theoretical curves rise too quickly.

To see the significant roles played by $U_{\mathrm{T}} = 6.0$ and $Z_\alpha = 0.1$ in Fig. 7-48, we change U_{T} to 1.0 in Fig. 7-49 and Z_α to 1.0 in Fig. 7-50, with no other parameter changes. In Fig. 7-49, the separate curves become too compressed

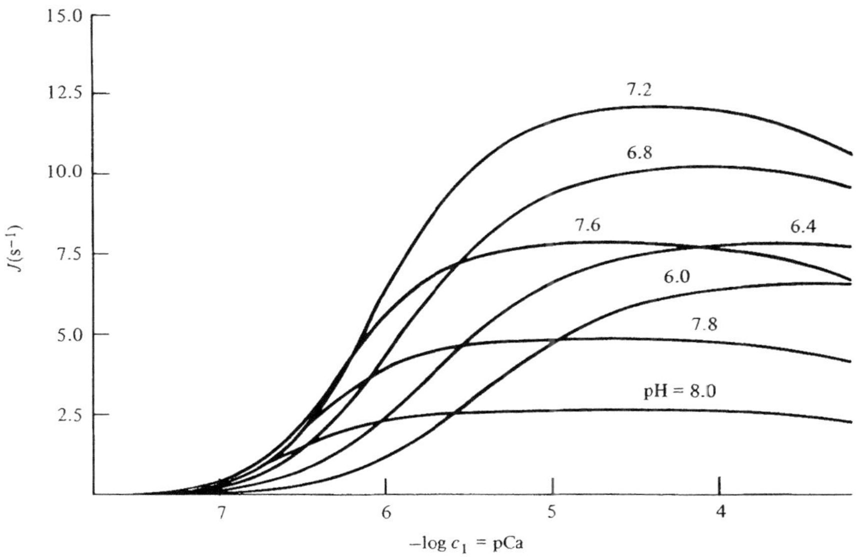

Fig. 7-48. Theoretical curves calculated from parameters in Eq. (33.15). The pCa range is larger here than in the experiments (Fig. 7-47).

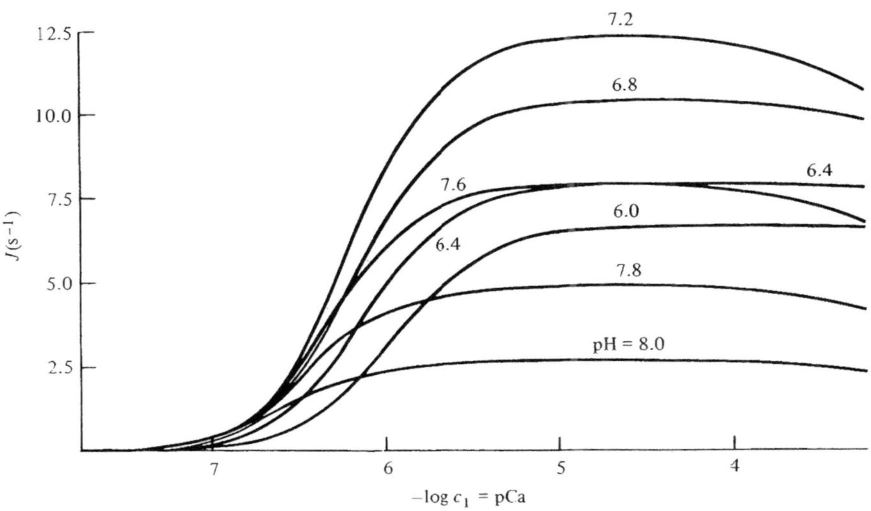

Fig. 7-49. The interaction parameters $U_T = 6.0$, $Z_\alpha = 0.1$ in Fig. 7-48 are changed here to $U_T = 1.0$, $Z_\alpha = 0.1$.

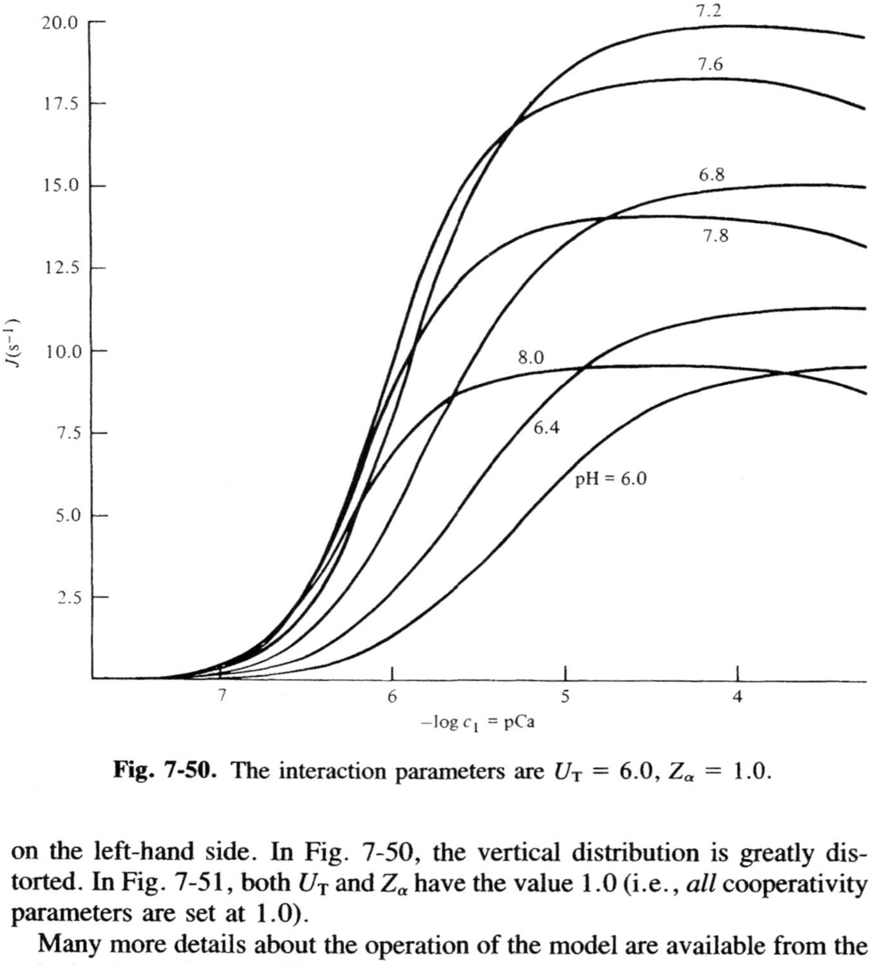

Fig. 7-50. The interaction parameters are $U_T = 6.0$, $Z_\alpha = 1.0$.

on the left-hand side. In Fig. 7-50, the vertical distribution is greatly distorted. In Fig. 7-51, both U_T and Z_α have the value 1.0 (i.e., *all* cooperativity parameters are set at 1.0).

Many more details about the operation of the model are available from the calculated p_i values, at different pCa and pH values. These are presented and discussed in the original paper.[5]

Alternative Choices of Reference Rate Constants

This is a short appendix illustrating, with a simpler model, a point that arose above in the discussion of Eq. (33.6).

Consider the symmetrical binding of a single ligand on a dimer with two identical subunits, as in Section 23. We use the on-off transitions of ligand to illustrate alternative choices of reference equilibrium constants and rate constants. The most elementary or fundamental way to write the grand partition function ξ is [Eq. (23.8)]

$$\xi = 1 + 2\frac{y_{12}}{y_{11}} K_o c + \frac{y_{22}}{y_{11}} K_o^2 c^2. \tag{33.17}$$

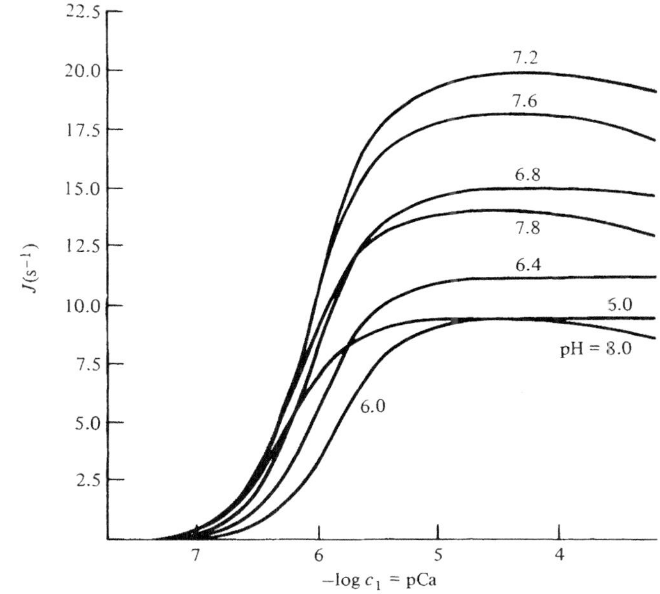

Fig. 7-51. The interaction parameters are $U_T = 1.0$, $Z_\alpha = 1.0$.

The three kinetic processes of interest here are (schematically):

$$\cdot \rightleftarrows \odot \qquad K_o = \frac{\alpha_o^*}{\beta_o}, \tag{33.18}$$

$$\cdot\cdot \rightleftarrows \odot\cdot \qquad \frac{y_{12}}{y_{11}} K_o = \frac{\alpha_o^* (y_{12}/y_{11})^f}{\beta_o (y_{12}/y_{11})^{f-1}}, \tag{33.19}$$

$$\odot\cdot \rightleftarrows \odot\odot \qquad \frac{y_{22}}{y_{12}} K_o = \frac{\alpha_o^* (y_{22}/y_{12})^f}{\beta_o (y_{22}/y_{12})^{f-1}}, \tag{33.20}$$

Equation (33.18) relates to the binding of L on a separated subunit. The unperturbed (or reference) equilibrium and rate constants are K_o, α_o^* (on), and β_o (off). Equation (33.19) relates to the first stage of binding ligand on the dimer. Because of intersubunit interactions, the equilibrium constant is perturbed to $y_{12}K_o/y_{11}$, and the corresponding perturbed rate constants are shown in numerator and denominator, on the right-hand side. Equation (33.20) gives the same information for the second stage of binding on the dimer (starting from the ratio of the last two terms in ξ).

The above treatment deals with the most elementary parameters. However, it may be more convenient to use the initial binding constant K in place of K_o and to rewrite ξ in the form [Eq. (23.10)]

$$\xi = 1 + 2Kc + yK^2c^2, \tag{33.21}$$

$$K \equiv \frac{y_{12}}{y_{11}} K_o, \qquad y \equiv \frac{y_{11}y_{22}}{y_{12}^2}. \tag{33.22}$$

Here K becomes the reference equilibrium constant, and we denote the corresponding reference rate constants by α^* and β. These are reference parameters, but they are *not* strictly unperturbed parameters because $K = y_{12}K_o/y_{11}$ and α^* and β are given by Eq. (33.19):

$$\alpha^* = \alpha_o^* (y_{12}/y_{11})^f, \qquad \beta = \beta_o (y_{12}/y_{11})^{f-1}. \tag{33.23}$$

All of these reference parameters (K, α^*, β) include interaction parameters. With this choice of reference parameters, and Eq. (33.21), we have, in place of Eqs. (33.19) and (33.20),

$$\cdot\cdot \rightleftarrows \odot\cdot \qquad K = \frac{\alpha^*}{\beta}, \tag{33.24}$$

$$\odot\cdot \rightleftarrows \odot\odot \qquad yK = \frac{\alpha^* y^f}{\beta y^{f-1}}. \tag{33.25}$$

In this case there are no perturbations of the reference parameters in the first stage of binding, but there are in the second stage. Substitution of the definitions of K, a^*, β, and y into Eq. (33.25) confirms the equivalence of Eqs. (33.25) and (33.20). The essential point is that, in Eqs. (33.24) and (33.25), interaction parameters are already embedded in the reference parameters so that the perturbations, *relative to the reference parameters,* are different than in Eqs. (33.19) and (33.20).

If $\overline{K}$ is selected as the convenient reference binding constant to work with, as in Eq. (23.20), then

$$\xi = 1 + 2y^{-1/2}\overline{K}c + \overline{K}^2c^2, \qquad \overline{K} \equiv \left(\frac{y_{22}}{y_{11}}\right)^{1/2} K_o. \tag{33.26}$$

If we denote the reference rate constants here by $\overline{\alpha}^*$ and $\overline{\beta}$, then

$$\overline{K} = \frac{\overline{\alpha}^*}{\overline{\beta}} = \frac{\alpha_o^* \, (y_{22}/y_{11})^{f/2}}{\beta_o \, (y_{22}/y_{11})^{(f-1)/2}}. \tag{33.27}$$

This shows how the new reference rate constants are related to the unperturbed rate constants α_o^* and β_o. The analogues of Eqs. (33.24) and (33.25) are (from ratios of the terms of ξ)

$$\cdot\cdot \rightleftarrows \odot\cdot \qquad y^{-1/2}\overline{K} = \frac{\overline{\alpha}^* y^{-f/2}}{\overline{\beta} y^{-(f-1)/2}}, \tag{33.28}$$

$$\odot\cdot \rightleftarrows \odot\odot \qquad y^{1/2}\overline{K} = \frac{\overline{\alpha}^* y^{f/2}}{\overline{\beta} y^{(f-1)/2}}. \tag{33.29}$$

These equations show the perturbations of the new reference parameters $\overline{K}$, $\overline{\alpha}^*$, and $\overline{\beta}$ in the two stages of binding. If the interaction parameters embedded in $\overline{K}$, $\overline{\alpha}^*$, and $\overline{\beta}$ are taken into account, Eqs. (33.28) and (33.29) are seen to be equivalent to Eqs. (33.19) and (33.20).

References

1. Hill, T. L. and Chay, T. (1979) Proc. Natl. Acad. Sci. USA **76**, 3203.
2. Hill, T. L. (1977) *Free Energy Transduction in Biology* (Academic, New York).
3. Hill, T. L. and Kedem, O. (1966) J. Theor. Biol. **10**, 399.
4. Hill, T. L. and Levitzki, A. (1980) Proc. Natl. Acad. Sci. USA **77**, 5741.
5. Inesi, G. and Hill, T. L. (1983) Biophys. J. **44**, 271.

General References

King, E. L. and Altman, C. (1956) J. Phys. Chem. **60**, 1375.
Levitzki, A. (1978) *Quantitative Aspects of Allosteric Mechanisms* (Springer-Verlag, New York).
Kurganov, B. I. (1982) *Allosteric Enzymes* (Wiley, New York).
Frieden, C. (1970) J. Biol. Chem. **245**, 5788.
Hill, T. L. (1978) J. Theor. Biol. **75**, 391.
Hill, T. L. (1978) Proc. Natl. Acad. Sci. USA **75**, 1101.

III

One- or Two-Dimensional Lattices of Units with Explicit Interactive Cooperativity

8

One-Dimensional Lattices of Interacting Units at Equilibrium

Chapters 8 (equilibrium) and 9 (steady state) extend the discussion in Chapters 6 (equilibrium) and 7 (steady state) to large one-dimensional lattices of units. Two-dimensional lattices, at equilibrium or steady state, are considered in Chapter 10. The primary question is: how do neighbor interactions affect the equilibrium and steady-state properties of units when they are organized into a large lattice? The same methods can be applied to three-dimensional lattices, and there are some biological examples of these,[1,2] but we omit this subject (like transients) for reasons of space.

Perhaps the most useful method for treating one-dimensional lattices, with interactions, at equilibrium is the matrix method.[3–6] This is introduced in Section 34 and used in the remainder of this chapter. The matrix method can also be applied to finite systems; this feature is included here, primarily for pedagogical purposes. This establishes a connection with some of the results in earlier chapters on finite linear or circular systems.

One of the main applications of the matrix method has been to helix-coil transitions in proteins and nucleic acids. We omit this topic here because it is somewhat outside the scope of this book and because of its excellent coverage in the books by Poland and Scheraga,[4] Poland,[5] and Davidson.[6] These books also give reviews of the elements of matrix algebra, another subject omitted here.

34. The Matrix Method in One-Dimensional Equilibrium Problems

This section serves two purposes. It introduces the matrix method, which is needed for the more complicated problems of Sections 35 and 36, and it also treats certain basic and important one-dimensional systems. The matrix method is applied in this section to linear arrays of *two-state* units. In Sections 35 and 36, we shall see that this method can be extended easily to linear arrays of units with more than two states.

Circular Finite Two-State System

We consider a circular (closed) ring of m sites, on each of which a ligand L at concentration c can bind with binding constant K. In addition, nearest-neighbor bound L molecules interact with a free energy w. We define $y = e^{-w/kT}$. Equations (22.22) and (22.38) give the grand partition function ξ for the cases $m = 3$ and $m = 4$. We shall introduce the matrix method using the $m = 3$ case, for which

$$\begin{array}{cccccccc} \xi = & 1 & + & 3x & + & 3yx^2 & + & y^3x^3, \\ & \cdot & & \odot & & \odot & & \odot \\ & \cdot\ \cdot & & \cdot\ \cdot & & \odot\ \cdot & & \odot\ \odot \end{array} \qquad (34.1)$$

where $x = Kc$. The matrix method is a new formal way of generating ξ. Obviously, it is not needed for this case, because ξ is already known, but the *method* can be extended to cases where ξ is not otherwise known.

Each of the eight terms in Eq. (34.1), e.g., yx^2, can be considered to be a product of a factor 1 for each empty site, a factor x for each occupied site, and an additional interaction factor y for each nearest-neighbor pair of occupied sites. Because y involves pairs of sites, we can take care of the various factors $(1, x, y)$ in every one of the eight terms in ξ by using a product of three new factors, one for each of the three *pairs* in the ring of three sites:

$$\begin{array}{lccc} \text{Site pair:} & 1,\ 2 & 2,\ 3 & 3,\ 1 \\ & A(\nu_1, \nu_2) & A(\nu_2, \nu_3) & A(\nu_3, \nu_1), \end{array} \qquad (34.2)$$

where ν_i is an index that specifies the state of the ith site: $\nu_i = 1$ (site empty) or $\nu_i = 2$ (site occupied by L). There are $2^3 = 8$ possible choices of the ν_i; with each choice we associate a product as in Eq. (34.2); ξ will then be the sum of the eight products.

Because each A above belongs to one pair of sites, A should include a factor y if both of these sites are occupied (both in state 2). However, there is some arbitrariness or possible redundancy about the assignment of 1s (empty sites) and xs (occupied sites) to A because each site index occurs twice in the product (34.2). There are three obvious possible ways to handle this: (a) assign 1 or x to the first site index in each A (i.e., 1 for state 1, x for state 2); (b) assign 1 or x to the second site index in each A; or (c) assign 1 or $x^{1/2}$ to both site indices in each A. These three different assignments are shown in Fig. 8-1 for all eight states of the trimer. The sites are numbered as in the upper right of the figure. Each pair of sites for each trimer state has a factor A indicated. The product (34.2) is the product of the three factors in any figure; the result is the same for (a), (b), and (c). These products are terms

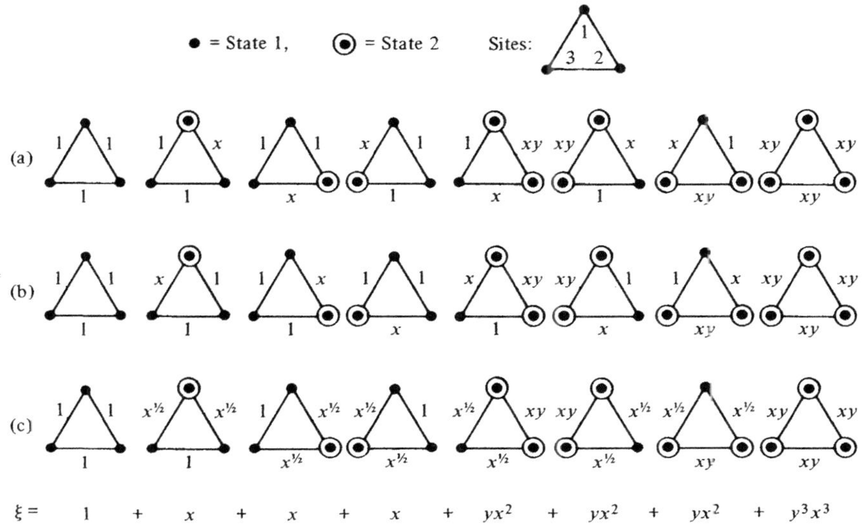

Fig. 8-1. The labels on sides of triangles are the factors A for each pair of sites, depending on the assignment choice (a), (b), or (c), as explained in the text. For any triangle, the product of the three As gives a term in ξ (below).

in the sum, ξ, shown at the bottom of the figure. This sum can be written in the following ways for (a), (b), or (c):

$$\xi = \sum_{\nu_1,\nu_2,\nu_3=1}^{2} A(\nu_1, \nu_2)A(\nu_2, \nu_3)A(\nu_3, \nu_1) \tag{34.3}$$

$$= \sum_{\nu_1,\nu_2} A(\nu_1, \nu_2) \sum_{\nu_3} A(\nu_2, \nu_3)A(\nu_3, \nu_1) \tag{34.4}$$

$$= \sum_{\nu_1,\nu_2} A(\nu_1, \nu_2)A^2(\nu_2, \nu_1) \tag{34.5}$$

$$= \sum_{\nu_1} A^3(\nu_1, \nu_1) = A^3(1, 1) + A^3(2, 2) \tag{34.6}$$

$$= \mathrm{Tr}\, \mathbf{A}^3. \tag{34.7}$$

The four quantities $A(\nu_1, \nu_2)$, for $\nu_1, \nu_2 = 1,2$, are considered here to be the elements of a 2×2 matrix, and similarly for $A(\nu_2, \nu_3)$ and $A(\nu_3, \nu_1)$. The sum over ν_3 in Eq. (34.4) follows the index rule of matrix multiplication, leading to Eq. (34.5), where $A^2(\nu_2, \nu_1)$ is the ν_2, ν_1 element of the matrix $\mathbf{A}^2$. Similarly, the sum over ν_2 in Eq. (34.5) gives the ν_1, ν_1 element of the matrix $\mathbf{A}^3$. The final sum over ν_1, in Eq. (34.6), is over the diagonal elements of $\mathbf{A}^3$. This is the so-called trace of $\mathbf{A}^3$, Tr $\mathbf{A}^3$.

The matrix **A** is

	$\nu_{k+1} =$ 1	2
$\nu_k =$ 1	$\cdot\ \cdot$ $A(1, 1)$	$\cdot\ \odot$ $A(1, 2)$
2	$\odot\ \cdot$ $A(2, 1)$	$\odot\ \odot$ $A(2, 2)$

(34.8)

All three pairs of sites are equivalent; the matrix elements are the same for any of the three pairs $k, k + 1 = 1, 2;\ 2, 3;\ 3, 1$. Because site 1 follows site 3 in the order 1, 2, 3, 1, we consider $k + 1$ to be 1 when $k = m = 3$. If we now use the three different ways to assign x factors to the elements of **A**, we have

$$\text{(a)}\ \mathbf{A} = \begin{pmatrix} 1 & 1 \\ x & yx \end{pmatrix}, \tag{34.9}$$

$$\text{(b)}\ \mathbf{A} = \begin{pmatrix} 1 & x \\ 1 & yx \end{pmatrix}, \tag{34.10}$$

$$\text{(c)}\ \mathbf{A} = \begin{pmatrix} 1 & x^{1/2} \\ x^{1/2} & yx \end{pmatrix}. \tag{34.11}$$

In all three cases, if we calculate $\mathbf{A}^3$ by direct matrix multiplication, we find the diagonal elements

$$\begin{aligned} A^3(1, 1) &= 1 + 2x + yx^2, \\ A^3(2, 2) &= x + 2yx^2 + y^3x^3. \end{aligned} \tag{34.12}$$

The sum of these, the trace of $\mathbf{A}^3$ [Eq. (34.7)], gives the same ξ as in Eq. (34.1).

There is an important physical significance to the separate terms in Eq. (34.6). Having summed the terms in ξ over ν_2 and ν_3, that is, over all possible states of sites 2 and 3, the final sum over ν_1 gives the relative weights or probabilities of site 1 being in state 1 or state 2. These relative weights are $A^3(1, 1)$ and $A^3(2, 2)$, respectively; but, of course, all three sites are equivalent, so these weights apply to any one site, having averaged over all possible states of the other two sites. Hence the probability that any site is in state 2 is

$$\theta = \frac{A^3(2, 2)}{A^3(1, 1) + A^3(2, 2)} = \frac{A^3(2, 2)}{\xi} \tag{34.13}$$

$$= (x + 2yx^2 + y^3x^3)\xi^{-1}. \tag{34.14}$$

This agrees with Eq. (22.24).

There is an important alternative to Eq. (34.7) as a means of finding ξ. This alternative is much simpler to use in complicated cases. For this purpose, we need the eigenvalues of **A**, which we denote by γ_1 and γ_2. These satisfy

$$\mathbf{A}\boldsymbol{\phi}_i = \gamma_i \boldsymbol{\phi}_i \qquad (i = 1, 2), \tag{34.15}$$

where the $\boldsymbol{\phi}_i$ are the eigenvectors (which are column vectors) belonging to the γ_i. The γ_i are found from the characteristic equation

$$\begin{vmatrix} A_{11} - \gamma & A_{12} \\ A_{21} & A_{22} - \gamma \end{vmatrix} = 0, \tag{34.16}$$

where we have simplified the notation in Eq. (34.8) in an obvious way. The three matrices in Eqs. (34.9)–(34.11) have the same eigenvalues because A_{11}, A_{22}, and $A_{12}A_{21}$ are the same in all three cases. From any one of Eqs. (34.9)–(34.11), we find

$$\gamma_1, \gamma_2 = \frac{1 + yx \pm [(1 - yx)^2 + 4x]^{1/2}}{2}. \tag{34.17}$$

We assign the + sign to γ_1, making it the larger eigenvalue.

If we multiply Eq. (34.15) from the left by **A**, and then again by **A**, we find

$$\mathbf{A}^3\boldsymbol{\phi}_i = \gamma_i^3 \boldsymbol{\phi}_i \qquad (i = 1, 2). \tag{34.18}$$

That is, the eigenvalues of $\mathbf{A}^3$ are simply γ_1^3 and γ_2^3. Another basic matrix theorem that we need (but do not prove) is that the trace of a square matrix is equal to the sum of its eigenvalues. This is actually a property of determinants. If we apply this theorem to $\mathbf{A}^3$, using Eq. (34.7),

$$\xi = \operatorname{Tr} \mathbf{A}^3 = \gamma_1^3 + \gamma_2^3. \tag{34.19}$$

If we now take γ_1 and γ_2 from Eq. (34.17), we again find the same ξ as in Eq. (34.1).

The above argument can be generalized without complication for any $m \geq 3$ (i.e., to an arbitrarily large ring of m sites with nearest-neighbor interactions between bound ligand molecules). Starting with Eq. (34.3), we have

$$\xi = \sum_{\nu_1,\ldots,\nu_m=1}^{2} A(\nu_1, \nu_2)A(\nu_2, \nu_3) \ldots A(\nu_m, \nu_1) \tag{34.20}$$

$$= A^m(1, 1) + A^m(2, 2) = \operatorname{Tr} \mathbf{A}^m \tag{34.21}$$

$$= \gamma_1^m + \gamma_2^m, \tag{34.22}$$

where the matrix **A** [Eqs. (34.9)–(34.11)] and its eigenvalues γ_1 and γ_2 [Eq. (34.17)] are unchanged, i.e., are independent of m. Also,

$$\theta = A^m(2, 2)/\xi. \tag{34.23}$$

The cases $m = 1$ and $m = 2$ require special comment. If we put $m = 1$ and $m = 2$ in Eq. (34.22), we find [see, also, Eq. (23.30)]

$$\begin{aligned} m &= 1: \ \xi = 1 + yx, \\ m &= 2: \ \xi = 1 + 2x + y^2x^2, \end{aligned} \tag{34.24}$$

whereas for a single site and for a pair of sites we expect

$$\begin{aligned} m &= 1: \ \xi = 1 + x, \\ m &= 2: \ \xi = 1 + 2x + yx^2. \end{aligned} \tag{34.25}$$

The reason for the discrepancy is that Eqs. (34.24) apply to an idealized ring of sites in which site m is a nearest-neighbor of site 1, even for $m = 1$ and $m = 2$ (Fig. 8-2).

In the limit of a very large ring of sites, $m \to \infty$, because $\gamma_1 > \gamma_2$, we can completely neglect γ_2^m compared to γ_1^m in Eq. (34.22). In this case

$$\ln \xi = m \ln \gamma_1 = m \ln\left\{\frac{1 + yx + [(1 - yx)^2 + 4x]^{1/2}}{2}\right\}. \tag{34.26}$$

We shall return to this case in the next subsection.

A second special case of interest is $y = 0$, with m arbitrary: there is a very strong repulsion between neighboring bound ligand molecules on the ring of m sites, with the consequence that any two neighboring sites are never both occupied at the same time. We have already encountered this problem in Eqs. (17.60)–(17.63) and (17.65). Here we can give an alternative general formula for ξ, with $m \geq 2$ arbitrary [$m = 1$ is excluded because of Eq. (34.24)]. From Eq. (34.17), putting $y = 0$,

$$\gamma_1 = \frac{1 + (1 + 4x)^{1/2}}{2}, \qquad \gamma_2 = \frac{1 - (1 + 4x)^{1/2}}{2}. \tag{34.27}$$

If we substitute these expressions into Eq. (34.22), we find, after cancellation of the odd powers of $(1 + 4x)^{1/2}$,

$$\xi = \frac{1}{2^{m-1}} \sum_{j=0}^{m'} \frac{m!\,(1 + 4x)^j}{(2j)!\,(m - 2j)!} \qquad (m \geq 2), \tag{34.28}$$

where

$$\begin{aligned} m' &= m/2 \qquad (m \text{ even}) \\ &= (m - 1)/2 \qquad (m \text{ odd}). \end{aligned}$$

Equation (34.28) is a polynomial in x; Eqs. (17.63) are recovered except that $\xi = 1$ for $m = 1$ [put $y = 0$ in the $m = 1$ Eq. (34. 24)].

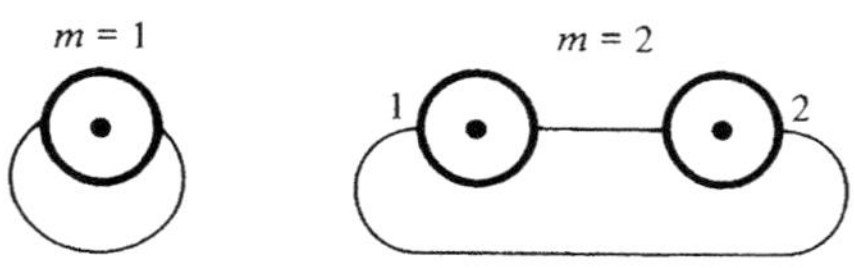

Fig. 8-2. In an idealized ring of sites, site m is a nearest neighbor of site 1 even if $m = 1$ or $m = 2$. The light lines indicate interactions [Eq. (34.24)].

Equations (17.64) and (17.66) give analogous expressions for $\xi_{\text{o}}^{(m)}$, the grand partition function of an *open* chain of m binding sites, again with $y = 0$. The analogue of Eq. (34.28), for an open chain, can be found by using the differential form of Eq. (17.72):

$$\frac{\theta\xi}{x} = \xi_{\text{o}}^{(m-3)} = \frac{1}{m}\frac{\partial\xi}{\partial x} \qquad (m \geq 4). \tag{34.29}$$

On differentiating Eq. (34.28), we obtain

$$\xi_{\text{o}}^{(m-3)} = \frac{1}{2^{m-3}}\sum_{j=1}^{m'}\frac{(m-1)!\,j(1+4x)^{j-1}}{(2j)!\,(m-2j)!} \qquad (m \geq 4). \tag{34.30}$$

Equations (17.64) are special cases of Eq. (34.30).

Very Long Chain of Two-State Subunits

In this subsection we consider the important special case $m \to \infty$, but we introduce the full set of interaction parameters, just as in Section 23. Because $m \to \infty$, the results for a very large ring are the same as for a very long open chain, because end effects are negligible in this limit. This is a second generalization. To be more explicit, the model we study here is an extremely large ring or open chain of identical subunits, each of which has a site for binding L, which is at concentration c. A subunit with an empty site is in state 1; a subunit with its site occupied by L is in state 2. The interaction parameters between nearest-neighbor subunits are the usual $y_{ij} = e^{-w_{ij}/kT}$ $(i, j = 1, 2)$. The unperturbed binding constant of L (i.e., on an isolated subunit) is K_{o}. This is the same model and notation introduced in Section 23, but here m is very large. We confine ourselves from the outset to the symmetrical case $y_{12} = y_{21}$.

To summarize the notation, we give here the grand partition function ξ for the $m = 3$ (ring) case:

$$\xi = y_{11}^3 + 3y_{11}y_{12}^2x_{\text{o}} + 3y_{12}^2y_{22}x_{\text{o}}^2 + y_{22}^3x_{\text{o}}^3, \tag{34.31}$$

where $x_{\text{o}} = K_{\text{o}}c$. This should be compared with Eqs. (23.37) and (34.1).

Because this is a macroscopic thermodynamic system, from this point we use M for the number of subunits and Ξ for the grand partition function. The matrix method introduced above is directly applicable, but the matrix **A** needs modification. Because we need certain other properties of **A** in addition to its eigenvalues, it is advantageous to use the symmetrical form of **A**, as in Eq. (34.11). Because each ij pair introduces a subunit interaction factor y_{ij} into a grand partition function term [see Eq. (34.31)], the required **A** is

$$\mathbf{A} = \begin{pmatrix} y_{11} & y_{12}x_{\text{o}}^{1/2} \\ y_{12}x_{\text{o}}^{1/2} & y_{22}x_{\text{o}} \end{pmatrix}, \tag{34.32}$$

which is symmetric. The eigenvalues of **A** are then [Eq. (34.16)]

$$\gamma_1, \gamma_2 = \frac{y_{11} + y_{22}x_o \pm \sqrt{\ }}{2}, \tag{34.33}$$

with

$$\sqrt{\ } = [(y_{11} - y_{22}x_o)^2 + 4y_{12}^2x_o]^{1/2}. \tag{34.34}$$

We associate γ_1 with the + sign; hence $\gamma_1 > \gamma_2$. The grand partition function is then

$$\Xi = \gamma_1^M = A^M(1, 1) + A^M(2, 2) \tag{34.35}$$

and

$$\theta = A^M(2, 2)/\Xi. \tag{34.36}$$

The term γ_2^M has been dropped because M is very large and $\gamma_1 > \gamma_2$. An explicit expression for θ, obtained from Eq. (34.36), will be given below.

We digress at this point to establish the connection with macroscopic thermodynamics. First we observe that, in writing Eq. (34.31), we have (by convention) omitted a factor q_P^3 from every term, where q_P is the partition function of an isolated subunit with no ligand bound. When the subunits are clustered into the trimer and ligand molecules may be bound, the y_{ij} take care of intersubunit interactions and K_o includes any intrinsic free energy change in a subunit induced by the binding of L. Similarly, a factor q_P^M has been omitted from every term in Ξ. As a result, the ligand molecules bound on the long chain of M subunits may be regarded formally as a one-component system, with basic thermodynamic equations

$$dA = -SdT - \Phi dM + \mu dN, \tag{34.37}$$

$$\mu\overline{N} = A + \Phi M, \tag{34.38}$$

$$d(\Phi M) = SdT + \Phi dM + \overline{N}d\mu, \tag{34.39}$$

where μ is the ligand chemical potential, $\overline{N}$ is the mean number of (bound) ligand molecules in the system, and Φ is the so-called spreading pressure (ΦM replaces the pV of ordinary thermodymanics). The connection with Ξ and γ_1 is

$$\Phi M = kT \ln \Xi = MkT \ln \gamma_1, \tag{34.40}$$

$$\Phi = kT \ln \gamma_1. \tag{34.41}$$

Here γ_1 is a function of μ and T; x_o in Eq. (34.33) for γ_1 is a disguised form of μ, through the relations

$$x_o = K_o c, \qquad \mu = \mu^o + kT \ln c \tag{34.42}$$

(i.e., the bound ligand is in equilibrium with ligand in solution at c, and hence both phases have the same μ). Equations (34.33) and (34.41) already provide us with an explicit expression for the thermodynamic function $\Phi(x_o)$, or $\Phi(c)$. Also, we can find $\theta(c)$ (the fractional binding of ligand) from [Eq. (34.39)]

$$\overline{N} = \left(\frac{\partial \Phi M}{\partial \mu}\right)_M = \lambda\left(\frac{\partial \ln \Xi}{\partial \lambda}\right)_M = x_o\left(\frac{\partial \ln \Xi}{\partial x_o}\right)_M, \qquad (34.43)$$

$$\theta = \frac{\overline{N}}{M} = \lambda\frac{\partial \ln \gamma_1}{\partial \lambda} = x_o\frac{\partial \ln \gamma_1}{\partial x_o}, \qquad (34.44)$$

where $\lambda = e^{\mu/kT}$. Actually, we shall obtain θ, below, from Eq. (34.36) rather than from Eq. (34.44).

It is worth another digression to note that the mathematical interpretation of the last form of Eq. (34.43) is that $\overline{N}$ results because the power of x_o in each term in Ξ is simply N: the power of x_o counts the number of bound ligand molecules in each term and the differential operation averages this count over all terms, using proper weights. This is a useful observation because the same kind of argument applies to any other factors, with exponents, in the terms of Ξ that are based on counting. Equation (34.31) provides a simple illustration: the powers of x_o count N and the powers of y_{ij} count N_{ij}, the number of nearest-neighbor pairs of type ij. Consequently, in the present problem,

$$\frac{\overline{N}_{ij}}{M} = y_{ij}\frac{\partial \ln \gamma_1}{\partial y_{ij}} \qquad (ij = 11,\ 12,\ 22). \qquad (34.45)$$

Because of symmetry ($y_{12} = y_{21}$) in the present model, the $ij = 12$ differentiation will give $\overline{N}_{12} + \overline{N}_{21} = 2\overline{N}_{12}$. An alternative calculation of the $\overline{N}_{ij}$ is used below.

We return now to the matrix treatment of our model. It is of some interest to find explicit expressions not only for θ and the $\overline{N}_{ij}$ (nearest-neighbor pairs), but also for the $\overline{N}_{ijk}$ (nearest-neighbor triplets). In fact the latter will be needed in Chapter 9. For this purpose we go back to [Eq. (34.20)]

$$\Xi = \sum_{\nu_1,\ldots,\nu_M=1}^{2} A(\nu_1, \nu_2)A(\nu_2, \nu_3) \ldots A(\nu_M, \nu_1). \qquad (34.46)$$

If we sum, in order, over $\nu_M, \ldots, \nu_3$, we have

$$\Xi = \sum_{\nu_1,\nu_2} A(\nu_1, \nu_2)A^{M-1}(\nu_2, \nu_1). \qquad (34.47)$$

Having summed over all possible states of sites 3, 4, . . . , M, the remaining sum over ν_1, ν_2 contains four terms that give the relative weights or probabilities of sites 1 and 2 being in the possible state combinations $\nu_1, \nu_2 = 11$, 12, 21, 22. However, all nearest-neighbor pairs of sites in a ring or very long chain are equivalent, so these results apply to any pair of sites. Thus we have

$$\frac{\overline{N}_{ij}}{M} = \frac{A(i, j)A^{M-1}(j, i)}{\Xi} \qquad (i, j = 1, 2). \qquad (34.48)$$

Division by M is appropriate because the total number of pairs in the chain is M; in fact, this is also the total number of triplets in the chain. Explicit equations for the $\overline{N}_{ij}$ will be given below.

We now repeat the argument, summing Eq. (34.46) over $\nu_M, \ldots, \nu_4$. This gives

$$\Xi = \sum_{\nu_1,\nu_2,\nu_3} A(\nu_1, \nu_2)A(\nu_2, \nu_3)A^{M-2}(\nu_3, \nu_1). \tag{34.49}$$

This sum over eight terms contains the relative weights of all possible states of nearest-neighbor triplets of sites. Consequently,

$$\frac{\overline{N}_{ijk}}{M} = \frac{A(i, j)A(j, k)A^{M-2}(k, i)}{\Xi} \qquad (i, j, k = 1, 2). \tag{34.50}$$

In order to use Eqs. (34.36), (34.48), and (34.50) for θ, $\overline{N}_{ij}$, and $\overline{N}_{ijk}$, we need an expression for $A^s(i, j)$, with s arbitrary and large. We omit the proof, but it can be shown after a little matrix algebra that, if **A** is real and symmetric and if the eigenvectors $\boldsymbol{\phi}_1$ and $\boldsymbol{\phi}_2$ of **A** are normalized, then

$$A^s(i, j) = \gamma_1^s\phi_1(i)\phi_1(j) + \gamma_2^s\phi_2(i)\phi_2(j) \qquad (i, j = 1, 2), \tag{34.51}$$

where $\phi_1(i)$ is the ith element of the column vector $\boldsymbol{\phi}_1$, etc. Because $\gamma_1 > \gamma_2$ and s is very large in our applications, the term in γ_2^s is negligible. Thus, recalling that $\Xi = \gamma_1^M$, we obtain

$$\theta = \phi_1^2(2), \tag{34.52}$$

$$\overline{N}_{ij}/M = A(i, j)\phi_1(i)\phi_1(j)/\gamma_1 \qquad (i, j = 1, 2), \tag{34.53}$$

$$\overline{N}_{ijk}/M = A(i, j)A(j, k)\phi_1(k)\phi_1(i)/\gamma_1^2 \qquad (i, j = 1, 2). \tag{34.54}$$

To apply Eqs. (34.52)–(34.54) we need the elements of $\boldsymbol{\phi}_1$ from [Eq. (34.15)]

$$\begin{pmatrix} y_{11} & y_{12}x_o^{1/2} \\ y_{12}x_o^{1/2} & y_{22}x_o \end{pmatrix}\begin{pmatrix} \phi_1(1) \\ \phi_1(2) \end{pmatrix} = \gamma_1\begin{pmatrix} \phi_1(1) \\ \phi_1(2) \end{pmatrix}. \tag{34.55}$$

This leads to

$$\begin{aligned} y_{11}\phi_1(1) + y_{12}x_o^{1/2}\phi_1(2) &= \gamma_1\phi_1(1), \\ y_{12}x_o^{1/2}\phi_1(1) + y_{22}x_o\phi_1(2) &= \gamma_1\phi_1(2). \end{aligned} \tag{34.56}$$

Arbitrarily working with the second of these equations, we have [Eq. (34.33)]

$$\frac{\phi_1(1)}{\phi_1(2)} = \frac{\gamma_1 - y_{22}x_o}{y_{12}x_o^{1/2}} = \frac{y_{11} - y_{22}x_o + \sqrt{\ }}{2y_{12}x_o^{1/2}}. \tag{34.57}$$

If we introduce a normalization constant C such that

$$\begin{aligned} \phi_1(1) &= C(y_{11} - y_{22}x_o + \sqrt{\ }), \\ \phi_1(2) &= C(2y_{12}x_o^{1/2}), \\ \phi_1(1)^2 &+ \phi_1(2)^2 = 1, \end{aligned} \tag{34.58}$$

we find

$$C^{-2} = 2\sqrt{\ }(y_{11} - y_{22}x_o + \sqrt{\ }). \tag{34.59}$$

Returning now to Eqs. (34-52)–(34.54), we obtain the final results:

$$\overline{N}/M = \theta = 2y_{12}^2x_o/\sqrt{\ }(y_{11} - y_{22}x_o + \sqrt{\ }), \tag{34.60}$$

$$\overline{N}_{11}/M = y_{11}(y_{11} - y_{22}x_o + \sqrt{\ })/\sqrt{\ }(\ \ ), \tag{34.61}$$

$$\overline{N}_{12}/M = \overline{N}_{21}/M = 2y_{12}^2x_o/\sqrt{\ }(\ \ ), \tag{34.62}$$

$$\overline{N}_{22}/M = y_{22}x_o(y_{22}x_o - y_{11} + \sqrt{\ })/\sqrt{\ }(\ \ ), \tag{34.63}$$

$$\overline{N}_{111}/M = y_{11}^2(y_{11} - y_{22}x_o + \sqrt{\ })/\sqrt{\ }[\ \ ], \tag{34.64}$$

$$\overline{N}_{211}/M = \overline{N}_{112}/M = 2y_{11}y_{12}^2x_o/\sqrt{\ }[\ \ ], \tag{34.65}$$

$$\overline{N}_{121}/M = y_{12}^2x_o(y_{11} - y_{22}x_o + \sqrt{\ })/\sqrt{\ }[\ \ ], \tag{34.66}$$

$$\overline{N}_{212}/M = y_{12}^2x_o(y_{22}x_o - y_{11} + \sqrt{\ })/\sqrt{\ }[\ \ ], \tag{34.67}$$

$$\overline{N}_{221}/M = \overline{N}_{122}/M = 2y_{22}y_{12}^2x_o^2/\sqrt{\ }[\ \ ], \tag{34.68}$$

$$\overline{N}_{222}/M = y_{22}^2x_o^2(y_{22}x_o - y_{11} + \sqrt{\ })/\sqrt{\ }[\ \ ], \tag{34.69}$$

where

$$(\ \ ) = y_{11} + y_{22}x_o + \sqrt{\ }, \tag{34.70}$$

$$[\ \ ] = y_{11}^2 + 2y_{12}^2x_o + y_{22}^2x_o^2 + (y_{11} + y_{22}x_o)\sqrt{\ }, \tag{34.71}$$

$$\sqrt{\ } = [(y_{11} - y_{22}x_o)^2 + 4y_{12}^2x_o]^{1/2}.$$

There are a number of normalization relations among $\overline{N}$, $\overline{N}_{ij}$, and $\overline{N}_{ijk}$, which we omit. There are also various intuitively obvious "quasichemical" relations that happen to be exact for one-dimensional cooperative systems. Examples are

$$\frac{\overline{N}_{11}\overline{N}_{22}}{\overline{N}_{12}\overline{N}_{21}} = \frac{y_{11}y_{22}}{y_{12}^2} \equiv y, \tag{34.72}$$

$$\frac{\overline{N}_{121}}{\overline{N}_{111}} = \frac{y_{12}^2x_o}{y_{11}^2}, \qquad \frac{\overline{N}_{221}}{\overline{N}_{211}} = \frac{y_{22}x_o}{y_{11}}, \tag{34.73}$$

$$\frac{\overline{N}_{222}}{\overline{N}_{212}} = \frac{y_{22}^2x_o}{y_{12}^2}, \qquad \frac{\overline{N}_{212}\overline{N}_{111}}{\overline{N}_{211}^2} = 1. \tag{34.74}$$

In Section 23, for a ring of $m = 3$ sites, we rewrote Eq. (34.31) in two other, more compact, ways (after factoring out y_{11}^3):

$$\xi = 1 + 3x + 3yx^2 + y^3x^3, \tag{34.75}$$

$$\xi = 1 + 3y^{-1}\overline{x} + 3y^{-1}\overline{x}^2 + \overline{x}^3, \tag{34.76}$$

$$x = Kc, \qquad \overline{x} = \overline{K}c, \qquad y = \frac{y_{11}y_{22}}{y_{12}^2}, \tag{34.77}$$

$$K = \frac{y_{12}^2}{y_{11}^2} K_o, \qquad \bar{K} = \frac{y_{22}}{y_{11}} K_o = yK. \tag{34.78}$$

The analogous set of three equations for ξ for a ring (square) of $m = 4$ sites was also discussed in Section 23. The same three choices are available for macroscopic M. In each term of Eq. (34.31) we have assigned a factor x_o for each bound L and a factor y_{ij} for each ij pair of sites. Comparison of Eq. (34.31) with Eqs. (34.75) and (34.76) shows that, in effect, in Eq. (34.75) we assign a factor x to each bound L and a factor y to each 22 pair of sites. Similarly, in Eq. (34.76), we assign a factor $\bar{x}$ to each bound L and a factor $y^{-1/2}$ to each 12 or 21 pair of sites. There is no loss of generality in the latter two assignments because of the concomitant redefinition of the thermodynamic activity from x_o to x or to $\bar{x}$ (x and $\bar{x}$ include some of the y_{ij}).

On the basis of the above three assignments, the matrices **A** that correspond to Eqs. (34.9)–(34.11) are

$$\boxed{x_o} \qquad\qquad \boxed{x} \qquad\qquad \boxed{\bar{x}}$$

$$\text{(a)} \begin{pmatrix} y_{11} & y_{12} \\ y_{12}x_o & y_{22}x_o \end{pmatrix} \begin{pmatrix} 1 & 1 \\ x & yx \end{pmatrix} \begin{pmatrix} 1 & y^{-1/2} \\ y^{-1/2}\bar{x} & \bar{x} \end{pmatrix}, \tag{34.79}$$

$$\text{(b)} \begin{pmatrix} y_{11} & y_{12}x_o \\ y_{12} & y_{22}x_o \end{pmatrix} \begin{pmatrix} 1 & x \\ 1 & yx \end{pmatrix} \begin{pmatrix} 1 & y^{-1/2}\bar{x} \\ y^{-1/2} & \bar{x} \end{pmatrix}, \tag{34.80}$$

$$\text{(c)} \begin{pmatrix} y_{11} & y_{12}x_o^{1/2} \\ y_{12}x_o^{1/2} & y_{22}x_o \end{pmatrix} \begin{pmatrix} 1 & x^{1/2} \\ x^{1/2} & yx \end{pmatrix} \begin{pmatrix} 1 & y^{-1/2}\bar{x}^{1/2} \\ y^{-1/2}\bar{x}^{1/2} & \bar{x} \end{pmatrix}. \tag{34.81}$$

The rows here relate to the same choices (a), (b), and (c) already discussed in connection with Eqs. (34.9)–(34.11). The columns refer, respectively, to employment of x_o, x, or $\bar{x}$ as the activity of the ligand. The current model can be treated using any one of these nine matrices. In our work above, we have used the lower left matrix, Eq. (34.32). The middle column of matrices is the same as Eqs. (34.9)–(34.11), but the model here is more general. The increase in generality is the same as that which occurs in passing from Section 22 to Section 23.

Corresponding to the alternative choices x_o, x, and $\bar{x}$ in Eqs. (34.79)–(34.81), the results in Eqs. (34.60)–(34.71) can all be rewritten in terms of x or $\bar{x}$, in place of x_o. We do this for θ only. Upon using Eqs. (34.77) and (34.78), we find

$$\theta = \frac{2x}{[(1 - yx)^2 + 4x]^{1/2}\{1 - yx + [(1 - yx)^2 + 4x]^{1/2}\}}, \tag{34.82}$$

$$\theta = \frac{2y^{-1}\bar{x}}{[(1 - \bar{x})^2 + 4y^{-1}\bar{x}]^{1/2}\{1 - \bar{x} + [(1 - \bar{x})^2 + 4y^{-1}\bar{x}]^{1/2}\}}. \tag{34.83}$$

A convenient feature of the choice of $\bar{x}$ as the ligand activity is that $\theta = \frac{1}{2}$ when $\bar{x} = 1$, for any y. In fact, $\theta(\ln \bar{x})$ is symmetrical about $\ln \bar{x} = 0$, for any y. This can be expressed by

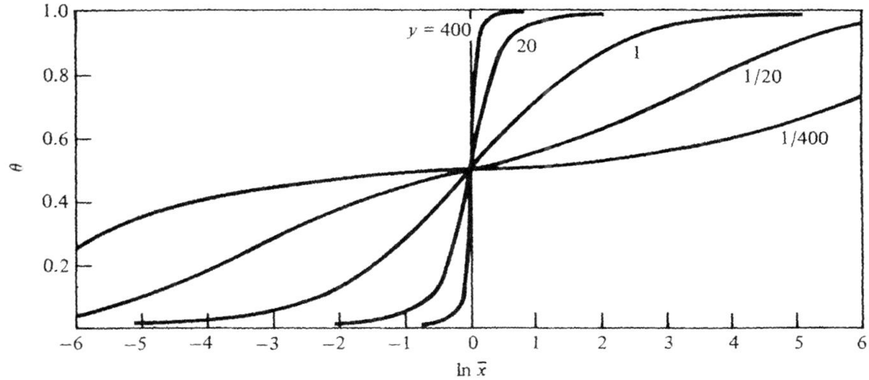

Fig. 8-3. Plots of θ as a function of $\ln \bar{x}$ from Eq. (34.83) for several values of y.

$$\theta(\bar{x}) + \theta(1/\bar{x}) = 1. \tag{34.84}$$

In Eq. (34.82), $\theta = \frac{1}{2}$ occurs at $x = 1/y$.

Figure 8-3 shows plots of $\theta(\ln \bar{x})$ from Eq. (34.83) for several values of y. When $y = 1$, $\theta = \bar{x}/(1 + \bar{x})$ (no cooperativity). These curves should be compared with Fig. 6-4 for a square ($m = 4$); here $M \to \infty$. In Fig. 6-4, θ is a function of $\ln x$, not $\ln \bar{x}$ (the difference corresponds simply to a shift of the curves along the horizontal axis). As expected, the cooperativity is stronger for the same y in the $M \to \infty$ case. However, there is no phase transition for any value of y. For this kind of system, a two- or three-dimensional lattice is required to produce a first-order phase transition (with large enough y).

The slope of $\theta(\ln \bar{x})$ at $\bar{x} = 1$ (or $\theta = \frac{1}{2}$) is found from Eq. (34.83) to be

$$\left(\frac{\partial \theta}{\partial \ln \bar{x}}\right)_{\theta=1/2} = \frac{\sigma_M^2}{M} = \frac{y^{1/2}}{4}. \tag{34.85}$$

The connection with the variance follows from Eq. (1.14). This slope approaches ∞ when $y \to \infty$. The Hill coefficient is $n = y^{1/2}$. In contrast, for $m = 4$ [Eq. (22.40)],

$$\left(\frac{\partial \theta}{\partial \ln \bar{x}}\right)_{\theta=1/2} = \frac{\sigma_4^2}{4} = \frac{y(1 + y)}{1 + 6y + y^2}. \tag{34.86}$$

This slope approaches 1, a finite value, when $y \to \infty$. The Hill coefficient is $n = \textit{slope} \times 4$. Both equations give a slope of $\frac{1}{4}$ when $y = 1$ (no cooperativity), and $n = 1$. When $y = 20$, as in Figs. 6-4 and 8-3, Eq. (34.85) gives a slope of 1.118, whereas Eq. (34.86) gives 0.806.

As a final topic in this subsection, we illustrate the use of the $\bar{x}$ matrices in Eqs. (34.79)–(34.81) by deriving the analogue of Eq. (34.86) for arbitrary m. From any one of the three $\bar{x}$ matrices, we have

$$\xi = \gamma_1^m + \gamma_2^m, \tag{34.87}$$

$$\gamma_1, \gamma_2 = \frac{1 + \bar{x} \pm [(1 - \bar{x})^2 + 4z^2\bar{x}]^{1/2}}{2}, \tag{34.88}$$

where we have written $z^2 = y^{-1}$ to simplify the equations below. Again $\gamma_1 \equiv \gamma_+$ and $\gamma_2 \equiv \gamma_-$. For example, for $m = 1$ and $m = 2$, Eq. (34.87) leads to

$$m = 1: \ \xi = 1 + \bar{x}, \tag{34.89}$$

$$m = 2: \ \xi = 1 + 2z^2\bar{x} + \bar{x}^2. \tag{34.90}$$

We can obtain θ from

$$\theta = \frac{\bar{x}}{m} \frac{\partial \ln \xi}{\partial \bar{x}}, \tag{34.91}$$

and then a second differentiation with respect to $\bar{x}$ gives the desired result. The expression for θ is complex, and we omit it. After rather lengthy algebra, we find,[7] at $\bar{x} = 1$, $\theta = \frac{1}{2}$,

$$\left(\frac{\partial \theta}{\partial \ln \bar{x}}\right)_{\theta=1/2} = \frac{\sigma_m^2}{m} = \frac{1}{4z} \frac{\{1 - [(1 - z)/(1 + z)]^m\}}{\{1 + [(1 - z)/(1 + z)]^m\}}. \tag{34.92}$$

The Hill coefficient, n, introduced in Section 11, is four times this quantity. Equation (34.92) agrees with Eq. (34.85) when $m \to \infty$ and with Eq. (34.86) when $m = 4$. When $y \gg 1$ (strong positive cooperativity),

$$\left(\frac{\partial \theta}{\partial \ln \bar{x}}\right)_{\theta=1/2} \cong \frac{\tanh mz}{4z}. \tag{34.93}$$

Incidentally, a single differentiation of ξ in Eq. (34.87) with respect to z gives (at $\bar{x} = 1$, $\theta = \frac{1}{2}$)

$$\bar{N}_{12} + \bar{N}_{21} = z\left(\frac{\partial \ln \xi}{\partial z}\right)_{\bar{x}=1} \tag{34.94}$$

$$= \frac{mz}{1 + z} \frac{\{1 - [(1 - z)/(1 + z)]^{m-1}\}}{\{1 + [(1 - z)/(1 + z)]^m\}} \tag{34.95}$$

$$\cong \frac{mz \tanh mz}{1 + z} \qquad (y \gg 1, \ m \gg 1). \tag{34.96}$$

Alternate Method for Certain Large Systems[8]

Here we provide an alternate way to obtain Eq. (34.41), without using the matrix method. The main value of this alternate method is that it can be used for much more complicated problems, e.g., problems that involve complex helix-coil transitions in polynucleotides. However, we shall not discuss this particular class of problems explicitly.

The following discussion applies to very large systems only. The system and notation is the same as in Eqs. (34.31)–(34.44). The chain of M subunits, in any configuration of states, is comprised of alternating sequences of state 1 (empty subunit) only and state 2 (subunit with bound ligand) only, with alternating "intersections" or boundaries of types 12 and 21. For example,

$$\ldots 111\underset{\uparrow}{}22\underset{\uparrow}{}1111\underset{\uparrow}{}2\underset{\uparrow}{}1\underset{\uparrow}{}2\underset{\uparrow}{}1111\underset{\uparrow}{}22222 \ldots$$

The arrows here indicate the boundaries between the two kinds of sequences. Let $N' = N_{12} + N_{21}$ be the total number of boundaries.

We now consider a hypothetical very large system of M subunits, with N ligand molecules bound, but restricted to exactly N' boundaries. In the real system, with N and M fixed, N' would fluctuate. The number of boundaries is a new extensive variable for this hypothetical system. Equation (34.37) becomes

$$dA = -SdT - \Phi dM + \mu dN + \mu' dN', \tag{34.97}$$

where $\mu' \equiv \partial A/\partial N'$. On integrating,

$$A = -\Phi M + \mu N + \mu' N'. \tag{34.98}$$

Then

$$d(\mu' N') = -SdT + \overline{M} d\Phi - \overline{N} d\mu + \mu' dN', \tag{34.99}$$

$$\mu' N' = -kT \ln \mathrm{T}(N'), \tag{34.100}$$

where $\mathrm{T}(N')$ is the appropriate partition function for the independent variables T, Φ, μ, and N'. With this choice of variables, there are fluctuations in N and M.

In a long chain with N' boundaries, there will be $N'/2$ sequences of type 1 (. . .111. . .) and $N'/2$ sequences of type 2 (. . .222. . .). Each type 1 sequence can, in principle, fluctuate in size from $n_1 = 1$ to $n_1 = \infty$ subunits, and similarly for type 2 sequences. Let ψ_1 and ψ_2 be the partition functions of single sequences of the two types, including the fluctuations in n_1 and n_2. Details about ψ_1 and ψ_2 will be given below. We then have

$$\mathrm{T}(N') = e^{-\mu' N'/kT} = \psi_1^{N'/2} \psi_2^{N'/2} y_{12}^{N'}, \tag{34.101}$$

where $y_{12} = e^{-w_{12}/kT}$ is the familiar interaction parameter for a 12 or 21 pair of subunits (each boundary contributes a factor y_{12} to T). From Eq. (34.101),

$$e^{-2\mu'/kT} = \psi_1 \psi_2 y_{12}^2. \tag{34.102}$$

This is a relation between the intensive variables of the system, T, Φ, μ, and μ'.

If we now remove the artificial restraint that fixes the number of boundaries in Eq. (34.97), N' will seek its equilibrium value such that A is minimized: $\partial A/\partial N' = 0$. However, $\mu' \equiv \partial A/\partial N'$; hence the equilibrium condition is $\mu' = 0$. Consequently, in the unrestrained system at equilibrium,

$$1 = \psi_1 \psi_2 y_{12}^2. \tag{34.103}$$

This is a relation between the intensive variables T, Φ, and μ [see Eq. (34.37)].

We return now to the details of the sequence partition functions ψ_1 and ψ_2. In the unrestrained system [Eqs. (34.37)–(34.40)],

$$e^{\Phi M/kT} = \Xi = \sum_N Q(N, M, T)\lambda^N \qquad (\lambda = e^{\mu/kT}), \tag{34.104}$$

$$\Upsilon = \sum_M \left(\sum_N Q\lambda^N\right) e^{-\Phi M/kT}, \tag{34.105}$$

where summation over N' in $\Upsilon(N')$ would also give Υ. This is the partition function of the "completely open" system, with fluctuations in N and M, and independent variables T, Φ, μ. It is clear from the form of Eq. (34.105) that each of the M subunits contributes a factor $\phi \equiv e^{-\Phi/kT}$ to each term of Υ, each bound ligand contributes a factor $q\lambda = K_o c = x_o$ (see Section 6), and each nearest-neighbor pair of type ij contributes a factor y_{ij}. As we have already seen above, the factors x_o and y_{ij} appear in the terms of Ξ; the new feature in Υ is ϕ. Using these various factors in the two kinds of individual sequences of the chain, we have

$$\psi_1 = \sum_{n_1=1}^{\infty} \phi^{n_1} y_{11}^{n_1-1} = \frac{\phi}{1 - y_{11}\phi}, \tag{34.106}$$

$$\psi_2 = \sum_{n_2=1}^{\infty} \phi^{n_2} x_o^{n_2} y_{22}^{n_2-1} = \frac{\phi x_o}{1 - y_{22}\phi x_o}. \tag{34.107}$$

The sum in Eq. (34.106) is appropriate because a sequence of n_1 subunits in state 1 has $n_1 - 1$ pairs of type 11, and similarly for ψ_2.

As in Eq. (34.45), the differentiation $\partial \ln \psi_i / \partial \ln \phi$ readily gives $\bar{n}_i$ ($i = 1, 2$). The $\bar{n}_i$ can be very useful, as is illustrated below in the first subsection of Section 35.

From Eqs. (34.103), (34.106), and (34.107), we then have

$$1 = \frac{y_{12}^2 \phi^2 x_o}{(1 - y_{11}\phi)(1 - y_{22}\phi x_o)}. \tag{34.108}$$

This is a quadratic equation in ϕ that is easily seen to agree with Eqs. (34.33) and (34.41), in view of the connection

$$\phi \equiv e^{-\Phi/kT} = 1/\gamma_1. \tag{34.109}$$

Equation (34.108) determines Φ as a function of μ and T; $\theta(\mu, T)$ then follows from Eq. (34.43). It is easy to show that

$$y_{11}\phi < 1 \qquad \text{and} \qquad y_{22}\phi x_o < 1, \tag{34.110}$$

and hence that the sums ψ_1 and ψ_2 do not diverge, provided that we use $\phi = 1/\gamma_1$ and not $\phi = 1/\gamma_2$ from Eq. (34.33). That is, the correct sign must be taken in the quadratic in ϕ, Eq. (34.108).

The above analysis merely recovers what we have already found by the matrix method, namely, Eqs. (34.33) and (34.41). However, the same general method can be applied to much more complicated linear systems provided that these systems are comprised of alternating sequences of two different types. The sequences themselves may have arbitrarily complicated internal structures, and the boundaries may also include new features. The essence of the method is: the macroscopic system is made up of repeating units; each unit consists of one completely open sequence of each type and two boundaries; hence the intensive properties of the macroscopic system can be deduced from an analysis of a single unit [Eqs. (34.103) and (34.108), in the example above].

Open Finite Two-State Systems

The matrix method above, for circular systems, has to be modified somewhat in order to treat finite open (linear) systems, because of end effects. We give an introductory account of this procedure, using three examples. Because the matrix method for circular systems (above) is simpler, it should be used for large systems, circular or linear.

We begin with the linear version of Eq. (34.31), where $m = 3$:

$$\begin{array}{ccccccccccc} \xi = & y_{11}^2 & + & 2y_{11}y_{12}x_o & + & y_{12}^2 x_o & + & 2y_{12}y_{22}x_o^2 & + & y_{12}^2 x_o^2 & + & y_{22}^2 x_o^3 . \\ & \cdot\,\cdot\,\cdot & & \odot\,\cdot\,\cdot & & \cdot\,\odot\,\cdot & & \odot\,\odot\,\cdot & & \odot\,\cdot\,\odot & & \odot\,\odot\,\odot \end{array} \tag{34.111}$$

Our object is to devise, for this system, the analogue of Eqs. (34.3)–(34.7). That is, we want to rederive Eq. (34.111) using matrix procedures. In this first example, we start with the x_o matrix in Eq. (34.80):

$$\mathbf{A} = \begin{pmatrix} y_{11} & y_{12}x_o \\ y_{12} & y_{22}x_o \end{pmatrix}. \tag{34.112}$$

The product $A(\nu_1, \nu_2)A(\nu_2, \nu_3)$ will take care of the interaction parameters y_{ij} for both pairs of sites, 12 and 23 (the sites are numbered from the left in the order 1, 2, 3). We should not include a factor $A(\nu_3, \nu_1)$ in this product, as in Eq. (34.3), because there is no interaction between sites 3 and 1 in an open $m = 3$ system. Because x_o is assigned to the second index in $\mathbf{A}$, the above product takes into account binding on sites 2 and 3 (note the second indices ν_2 and ν_3). However, binding on site 1 (with a factor x_o for state 2) is not included in the product; an additional factor must be inserted for this purpose. Another requirement is that our matrix manipulations must lead finally to a scalar result, as in Eq. (34.7), because ξ is a scalar. All of these features appear in the following expression:

$$\xi = \sum_{\nu_1,\nu_2,\nu_3=1}^{2} r(\nu_1)A(\nu_1, \nu_2)A(\nu_2, \nu_3)c(\nu_3), \tag{34.113}$$

where **r** and **c** are row and column vectors

$$\mathbf{r} = (1, x_o), \qquad \mathbf{c} = \begin{pmatrix} 1 \\ 1 \end{pmatrix}. \tag{34.114}$$

The row vector takes care of binding on site 1 and the column vector produces the desired scalar, upon summing over ν_3. If we sum over ν_2 in Eq. (34.113),

$$\xi = \sum_{\nu_1, \nu_3} r(\nu_1) A^2(\nu_1, \nu_3) c(\nu_3) \tag{34.115}$$

$$= (1, x_o)\mathbf{A}^2 \begin{pmatrix} 1 \\ 1 \end{pmatrix}. \tag{34.116}$$

If this product is calculated by direct multiplication, using Eq. (34.112) for **A**, Eq. (34.111) is recovered.

As a second example, we treat the same system ($m = 3$, open), starting with the x_o matrix in Eq. (34.79):

$$\mathbf{A} = \begin{pmatrix} y_{11} & y_{12} \\ y_{12}x_o & y_{22}x_o \end{pmatrix}. \tag{34.117}$$

Here x_o is assigned to the first index in **A**. Hence $A(\nu_1, \nu_2)A(\nu_2, \nu_3)$ takes into account the interaction factors y_{ij} and also binding on sites 1 and 2. Binding on site 3 must be represented by a separate factor. Thus, in this case, we start again with Eq. (34.113), but here **A** is given by Eq. (34.117) and

$$\mathbf{r} = (1, 1), \qquad \mathbf{c} = \begin{pmatrix} 1 \\ x_o \end{pmatrix}. \tag{34.118}$$

In this case the column vector introduces the binding on site 3. Consequently,

$$\xi = (1, 1)\mathbf{A}^2 \begin{pmatrix} 1 \\ x_o \end{pmatrix}, \tag{34.119}$$

with **A** given by Eq. (34.117). Direct calculation of this product again leads to Eq. (34.111). The roles of **r** and **c** in the two examples above are the same for large linear systems.

In the third example we treat binding on a linear (open) chain of m subunits, where m is arbitrary. Linear chains with arbitrary interactions do not have the symmetry about $\theta = \frac{1}{2}$ possessed by circular chains [see, for example, the discussion following Eq. (22.31)]. To simplify the algebra below, we restrict ourselves here to the special case $y_{11} = y_{22}$, which does lead to the above-mentioned kind of symmetry. For example, if we divide each term in Eq. (34.111) by y_{11}^2, and use

$$y_{11} = y_{22}, \qquad y = \frac{y_{11}y_{22}}{y_{12}^2} \equiv \frac{1}{z^2}, \qquad z = \frac{y_{12}}{y_{11}} = \frac{y_{12}}{y_{22}}, \tag{34.120}$$

then the new grand partition function is

$$\xi = 1 + (2z + z^2)x_o + (2z + z^2)x_o^2 + x_o^3. \tag{34.121}$$

This leads to a symmetrical binding curve [Eq. (15.19)], with $\theta = \frac{1}{2}$ at $x_o = 1$. There is no need to introduce $\bar{x}$ (i.e., $x_o = \bar{x}$). The symmetry arises, for any m, because of the fact that an exchange of state 1 and state 2 sites leaves the number of 12 and 21 boundaries (the power of z) unchanged.

We use the x_o matrix in Eq. (34.80) for $\mathbf{A}$, except that we divide each element by y_{11} to obtain

$$\mathbf{A} = \begin{pmatrix} 1 & zx_o \\ z & x_o \end{pmatrix}. \tag{34.122}$$

This corresponds to the notation in Eq. (34.121): each 12 or 21 interaction [see, also, Eq. (34.111)] is assigned a factor z and each bound ligand is assigned a factor x_o (second index). As in Eq. (34.113),

$$\xi = \sum_{\nu_1,\ldots,\nu_m=1}^{2} r(\nu_1)A(\nu_1, \nu_2) \cdots A(\nu_{m-1}, \nu_m)c(\nu_m), \tag{34.123}$$

where $\mathbf{r}$ and $\mathbf{c}$ are the same as in Eq. (34.114). Then, on summing over $\nu_{m-1}, \ldots, \nu_2$,

$$\begin{aligned} \xi &= \sum_{\nu_1,\nu_m} r(\nu_1)A^{m-1}(\nu_1, \nu_m)c(\nu_m) \\ &= (1, x_o)\mathbf{A}^{m-1}\begin{pmatrix} 1 \\ 1 \end{pmatrix}. \end{aligned} \tag{34.124}$$

This expression is inconvenient to use unless m is small; hence we introduce a modification of $\mathbf{A}^{m-1}$.

Let $\mathbf{T}$ be a matrix such that

$$\mathbf{T}^{-1}\mathbf{A}\mathbf{T} = \mathbf{\Lambda} \equiv \begin{pmatrix} \gamma_1 & 0 \\ 0 & \gamma_2 \end{pmatrix}, \tag{34.125}$$

where γ_1 and γ_2 are the eigenvalues of $\mathbf{A}$:

$$\gamma_1, \gamma_2 = \frac{1 + x_o \pm [(1 - x_o)^2 + 4z^2x_o]^{1/2}}{2}, \tag{34.126}$$

with $\gamma_1 \equiv \gamma_+$ and $\gamma_2 \equiv \gamma_-$. That is, this "similarity transformation" diagonalizes $\mathbf{A}$. The matrix $\mathbf{T}^{-1}$ is the inverse of $\mathbf{T}$:

$$\mathbf{T}\mathbf{T}^{-1} = \mathbf{T}^{-1}\mathbf{T} = \begin{pmatrix} 1 & 0 \\ 0 & 1 \end{pmatrix}. \tag{34.127}$$

If we multiply Eq. (34.125) by $\mathbf{T}$ from the left and by $\mathbf{T}^{-1}$ from the right,

$$\mathbf{A} = \mathbf{T}\mathbf{\Lambda}\mathbf{T}^{-1}. \tag{34.128}$$

Then repeated multiplication by $\mathbf{A}$ on the left-hand side of this equation and by $\mathbf{T}\mathbf{\Lambda}\mathbf{T}^{-1}$ on the right-hand side leads to the desired result:

$$\mathbf{A}^{m-1} = \mathbf{T}\mathbf{\Lambda}^{m-1}\mathbf{T}^{-1} = \mathbf{T}\begin{pmatrix} \gamma_1^{m-1} & 0 \\ 0 & \gamma_2^{m-1} \end{pmatrix}\mathbf{T}^{-1}. \tag{34.129}$$

Consequently, Eq. (34.124) becomes

$$\xi = (1, x_o)\mathbf{T}\begin{pmatrix} \gamma_1^{m-1} & 0 \\ 0 & \gamma_2^{m-1} \end{pmatrix}\mathbf{T}^{-1}\begin{pmatrix} 1 \\ 1 \end{pmatrix}. \tag{34.130}$$

This equation is easy to use once $\mathbf{T}$ and $\mathbf{T}^{-1}$ have been found.

To obtain $\mathbf{T}$, we need the eigenvectors $\boldsymbol{\phi}_i$ that belong to the eigenvalues γ_i of $\mathbf{A}$:

$$\begin{pmatrix} 1 & zx_o \\ z & x_o \end{pmatrix}\begin{pmatrix} \phi_i(1) \\ \phi_i(2) \end{pmatrix} = \gamma_i\begin{pmatrix} \phi_i(1) \\ \phi_i(2) \end{pmatrix}. \tag{34.131}$$

From the resulting equation [compare Eq. (34.56)],

$$\phi_i(1) + zx_o\phi_i(2) = \gamma_i\phi_i(1), \tag{34.132}$$

we have

$$\frac{\phi_i(1)}{\phi_i(2)} = \frac{zx_o}{\gamma_i - 1}. \tag{34.133}$$

This gives the ratio of the two elements of $\boldsymbol{\phi}_i$ (there is no need here to normalize $\boldsymbol{\phi}_i$). Then $\mathbf{T}$ is constructed from the eigenvector elements as follows:

$$\mathbf{T} = \begin{pmatrix} zx_o & zx_o \\ \gamma_1 - 1 & \gamma_2 - 1 \end{pmatrix}. \tag{34.134}$$

The cofactors of $\mathbf{T}$ are

$$\begin{pmatrix} \gamma_2 - 1 & 1 - \gamma_1 \\ -zx_o & zx_o \end{pmatrix}, \tag{34.135}$$

and the transpose of this matrix is

$$\begin{pmatrix} \gamma_2 - 1 & -zx_o \\ 1 - \gamma_1 & zx_o \end{pmatrix}. \tag{34.136}$$

If we divide the transpose by the determinant of $\mathbf{T}$, we obtain $\mathbf{T}^{-1}$:

$$\mathbf{T}^{-1} = \frac{1}{zx_o(\gamma_2 - \gamma_1)}\begin{pmatrix} \gamma_2 - 1 & -zx_o \\ 1 - \gamma_1 & zx_o \end{pmatrix}. \tag{34.137}$$

The reader who is unfamiliar with these recipes for finding $\mathbf{T}$ and $\mathbf{T}^{-1}$ should verify, by matrix multiplication, that Eqs. (34.125) and (34.127) are indeed satisfied.

If we now substitute Eqs. (34.134) and (34.137) into Eq. (34.130), and carry out the indicated operations, we arrive at

$$\xi = \frac{\gamma_1^{m-1}(1 - \gamma_2 + zx_o)(z - 1 + \gamma_1) + \gamma_2^{m-1}(\gamma_1 - 1 - zx_o)(z - 1 + \gamma_2)}{z(\gamma_1 - \gamma_2)}. \tag{34.138}$$

This is the analogue of Eq. (34.87) for a circle of m subunits, except that here we are limited to the special case $y_{11} = y_{22}$. Even with this limitation, ξ here is much more complicated because of end effects. A little algebra verifies that Eq. (34.138) gives the correct ξ for $m = 1, 2, 3$ (open chain).

Two lengthy differentiations[7] of Eq. (34.138) with respect to x_o give, first, θ [Eq. (34.91)], a complicated expression that we omit, and then, at $x_o = 1$, $\theta = \frac{1}{2}$,

$$\left(\frac{\partial \theta}{\partial \ln x_o}\right)_{\theta=1/2} = \frac{\sigma_m^2}{m} = \frac{1}{4z} - \frac{(1 - z^2)}{8z^2 m}\left[1 - \left(\frac{1 - z}{1 + z}\right)^m\right]. \tag{34.139}$$

The Hill coefficient, n, for any m, is four times this quantity. When $m = 1$, the right-hand side reduces to $\frac{1}{4}$, as it should. The results for $m = 2$ and $m = 3$ are also easy to check. Equation (34.139) corresponds to Eq. (34.92) for a circular chain. The two expressions agree, as expected, in the limit $m \to \infty$. For a large linear chain with strong positive cooperativity ($y \gg 1$),

$$\left(\frac{\partial \theta}{\partial \ln x_o}\right)_{\theta=1/2} \cong \frac{1}{4z}\left(1 - \frac{1}{2mz}\right) \qquad (z^2 \ll 1,\ mz \gg 1). \tag{34.140}$$

The analogue of Eq. (34.94) is relatively simple here:

$$\overline{N}_{12} + \overline{N}_{21} = z\left(\frac{\partial \ln \xi}{\partial z}\right)_{x_o=1} = \frac{(m - 1)z}{1 + z}. \tag{34.141}$$

This can be verified, for $m = 3$, using Eq. (34.121).

Thermodynamic Perturbation: Binding on a Linear Array of Sites

This discussion is a continuation of the last two subsections of Section 26, which should be reviewed for content and notation. Here we extend the argument to a long linear array of binding sites, with nearest-neighbor interactions between bound ligand molecules.

If we consider the on–off transitions of ligand from a particular site from a macroscopic point of view, then there is an average perturbation of these transitions owing to the presence of occasional bound ligand molecules on the nearest-neighbor sites. The perturbation of the average occupation of the site, θ, can be expressed in terms of the activity coefficient γ of ligand molecules in the linear system:

$$\frac{\theta}{1 - \theta} = \frac{\alpha^* c}{\beta} \cdot \frac{1}{\gamma}, \tag{34.142}$$

where γ (not an eigenvalue here!) is defined thermodynamically by Eq. (26.23) and α^* and β are unperturbed on and off rate constants. The ratio α^*/β is equal to the thermodynamic binding constant K. If we split γ between the on and off rate constants in the conventional way, using an average $\overline{f}$, the single average detailed balance relation (mean on rate = mean off rate) at equilibrium is

$$\alpha^* \gamma^{-\bar{f}} c(1 - \theta) = \beta \gamma^{1-\bar{f}} \theta. \tag{34.143}$$

Our objective in this subsection is to reexamine this on–off equilibrium from a more detailed point of view in order to discover the microscopic basis of the averaging in the macroscopic $\bar{f}$.

The special case of Eqs. (34.60)–(34.74) that we are interested in here is $y_{11} = 1, y_{12} = 1, y_{22} = y$ and $x_o = x = Kc$. In fact, Eq. (34.82) for θ applies to this special case; similar equations can easily be written for the quantities $\bar{N}_{ijk}/M$ referred to below, but these are omitted.

We need to consider an arbitrary ("central") site in the chain and its two nearest-neighbor sites. Thus, we are interested in triplets of sites. In Eqs. (34.64)–(34.69), we define $p_{ijk} \equiv \bar{N}_{ijk}/M$. The sum of the eight p_{ijk} is unity. Also,

$$\begin{aligned} \theta &= p_{121} + 2p_{221} + p_{222}, \\ 1 - \theta &= p_{111} + 2p_{211} + p_{212}. \end{aligned} \tag{34.144}$$

The probability that the two neighbors of the central site are both empty sites is $P_0 \equiv p_{111} + p_{121}$. If the two neighbors are both empty, the probability that the central site is occupied is

$$\theta_0 \equiv \frac{p_{121}}{p_{111} + p_{121}} = \frac{x}{1 + x}. \tag{34.145}$$

The probability that one neighbor site is occupied and one is empty is $P_1 \equiv 2(p_{211} + p_{221})$. In this case, the probability that the central site is occupied is

$$\theta_1 \equiv \frac{p_{221}}{p_{211} + p_{221}} = \frac{yx}{1 + yx}. \tag{34.146}$$

Finally, the probability that the two neighbors are both occupied is $P_2 \equiv p_{212} + p_{222}$. In this case, the probability that the central site is occupied is

$$\theta_2 \equiv \frac{p_{222}}{p_{212} + p_{222}} = \frac{y^2 x}{1 + y^2 x}. \tag{34.147}$$

The above relations will be needed below.

If the two neighbor sites are empty, the on–off equilibrium constant and rate constants for the central site are unperturbed:

$$\cdot\cdot\cdot \rightleftarrows \cdot \odot \cdot \qquad K = \alpha^*/\beta. \tag{34.148}$$

If one neighbor site is empty and one is occupied, the perturbed equilibrium and rate constants for the central site are

$$\odot \cdot\cdot \rightleftarrows \odot \odot \cdot \qquad \frac{\alpha^* y}{\beta} = \frac{\alpha^* y^f}{\beta y^{f-1}}, \tag{34.149}$$

where a microscopic f has been introduced in the usual way to split the perturbation y between forward and backward rate constants. If both neighbor sites are occupied, the corresponding quantities are:

$$\odot \cdot \odot \rightleftarrows \odot \odot \odot \qquad \frac{\alpha^* y^2}{\beta} = \frac{\alpha^* y^{2f}}{\beta y^{2(f-1)}}. \tag{34.150}$$

Equations (26.41) then give the associated three detailed balance relations for on–off transitions of the central site under the different conditions.

We now weight the three on and off rates in Eqs. (26.41) with P_0, P_1, and P_2, defined above, in order to calculate the mean on and off rates for the central site:

$$\begin{aligned} on &= \alpha^* c[P_0(1 - \theta_0) + P_1 y^f(1 - \theta_1) + P_2 y^{2f}(1 - \theta_2)] \\ &= \alpha^* c(p_{111} + 2y^f p_{211} + y^{2f} p_{212}), \end{aligned} \tag{34.151}$$

$$\begin{aligned} off &= \beta[P_0\theta_0 + P_1 y^{f-1}\theta_1 + P_2 y^{2(f-1)}\theta_2] \\ &= \beta[p_{121} + 2y^{f-1}p_{221} + y^{2(f-1)}p_{222}] \\ &= \beta x(p_{111} + 2y^f p_{211} + y^{2f} p_{212}). \end{aligned} \tag{34.152}$$

The mean on and off rates are seen to be equal, as expected, in view of the fact that $\alpha^* c = \beta x$ reduces to $\alpha^* = \beta K$.

To relate $\bar{f}$ to the microscopic analysis above, we equate the mean on rate in Eq. (34.151) to the left-hand side of Eq. (34.143), which is the on rate expressed in terms of $\bar{f}$. At the same time we use Eq. (34.142) to write

$$\frac{1}{\gamma} = \frac{\theta/x}{1 - \theta}. \tag{34.153}$$

Then

$$\left(\frac{\theta/x}{1 - \theta}\right)^{\bar{f}} = \frac{p_{111} + 2y^f p_{211} + y^{2f} p_{212}}{1 - \theta}. \tag{34.154}$$

Finally, we use Eqs. (34.144) for $1 - \theta$ and

$$\begin{aligned} \theta &= P_0\theta_0 + P_1\theta_1 + P_2\theta_2, \\ \theta/x &= P_0(1 - \theta_0) + P_1 y(1 - \theta_1) + P_2 y^2(1 - \theta_2) \\ &= p_{111} + 2yp_{211} + y^2 p_{212}. \end{aligned} \tag{34.155}$$

Thus

$$\left(\frac{p_{111} + 2yp_{211} + y^2 p_{212}}{p_{111} + 2p_{211} + p_{212}}\right)^{\bar{f}} = \frac{p_{111} + 2y^f p_{211} + y^{2f} p_{212}}{p_{111} + 2p_{211} + p_{212}}. \tag{34.156}$$

One can solve explicitly for $\bar{f}$ by taking logarithms. Equation (34.156) shows very clearly the relationship between the macroscopic $\bar{f}$ and the microscopic averaging that involves f. Equations (34.64)–(34.69) give explicit expressions for the p_{i1k}, if desired (put $y_{11} = y_{12} = 1$, $y_{22} = y$, $x_o = x$).

We repeat the main point here. If a macroscopic kind of on–off detailed balance for a site is used, as in Eq. (34.143), with an averaged $\bar{f}$ as the free energy "splitting" factor, the question arises as to the nature of the averaging implicit in $\bar{f}$. Because this particular model can also be handled exactly at the microscopic level, we are able to show that $\bar{f}$ has the meaning in Eq. (34.156).

Incidentally, an explicit formula for γ, from Eq. (34.82), is

$$\gamma = \frac{x(1-\theta)}{\theta} = \frac{(1-yx)^2 + 2x + (1-yx)\sqrt{}}{2}, \qquad (34.157)$$

where

$$\sqrt{} \equiv [(1-yx)^2 + 4x]^{1/2}.$$

Finally, we mention that it is very easy to generalize the above argument, keeping all the y_{ij} arbitrary, as in Eqs. (34.60)–(34.74). In this case we find for $\bar{f}$,

$$\left[\frac{(y_{12}/y_{11})^2 p_{111} + 2(y_{22}/y_{11})p_{211} + (y_{22}/y_{12})^2 p_{212}}{p_{111} + 2p_{211} + p_{212}}\right]^{\bar{f}} = \frac{(y_{12}/y_{11})^{2f} p_{111} + 2(y_{22}/y_{11})^{f} p_{211} + (y_{22}/y_{12})^{2f}}{p_{111} + 2p_{211} + p_{212}}. \qquad (34.158)$$

35. Several One-Dimensional Binding Problems

In Section 34, we introduced an alternate method that can be used in place of the matrix method for certain large one-dimensional systems. We employ this method here to provide a much simpler derivation than is otherwise available of the so-called McGhee–von Hippel binding isotherm.[9, 10] Following this, we return to the matrix method to treat three cooperative binding problems that require the use of units with more than two states (two states was the only case treated in Section 34). This will illustrate how the matrix method can be generalized to handle more complicated problems than those considered in Section 34.

The McGhee–von Hippel Binding Problem[9, 10]

The model is shown in Fig. 8-4. An r-mer (the ligand), at concentration c in solution, can bind on a long linear array of M sites. Each bound r-mer covers r sites. The r-mer itself is not an aggregate; we call it an r-mer only to

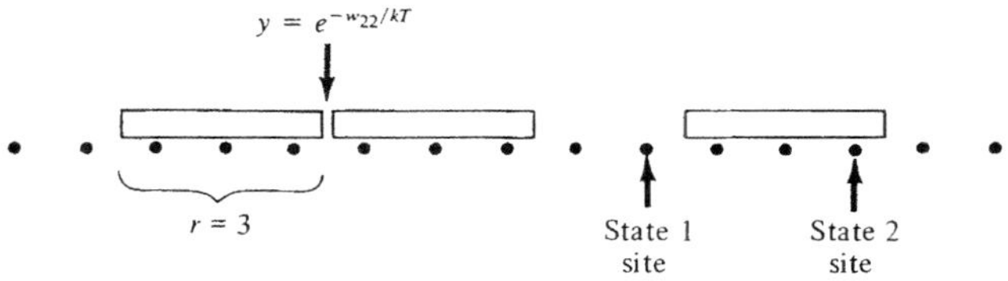

Fig. 8-4. Binding of an r-mer (here $r = 3$) on a long linear array of sites, with neighbor interactions (w_{22}).

designate its size. A free site is in state 1; a covered site is in state 2. If two r-mers occupy adjacent sites (as illustrated in Fig. 8-4), there is an associated interaction free energy w_{22}. The corresponding interaction parameter is $y = y_{22} = e^{-w_{22}/kT}$. There are no other interactions in the model ($y_{11} = 1$, $y_{12} = y_{21} = 1$). The initial binding constant (c small) of the ligand is K. Then

$$x = Kc = q\lambda, \qquad \lambda = e^{\mu/kT}, \tag{35.1}$$

as usual, where q is the partition function of a bound ligand with no neighbors, and μ is the chemical potential of the ligand molecules (in solution or bound).

The basic statistical thermodynamic equations are Eqs. (34.37)–(34.39), (34.104), and (34.105). N refers to the number of bound ligand molecules, each covering r sites. Also, we define ϕ and γ by

$$\Xi = e^{\Phi M/kT} = \gamma^M = (1/\phi)^M, \tag{35.2}$$

$$\gamma = e^{\Phi/kT}, \qquad \phi = e^{-\Phi/kT}, \qquad \phi = 1/\gamma. \tag{35.3}$$

Although we shall not use the matrix method, γ (*not* an activity coefficient here) plays the same thermodynamic role that the largest eigenvalue of an appropriate matrix would. The quantity ϕ has the same meaning and usefulness as in Eqs. (34.106) and (34.107). From Eq. (34.39), the intensive variables of the problem are related by

$$d\Phi = \theta d\mu \qquad (T \text{ constant}),$$
$$\theta = \left(\frac{\partial \Phi}{\partial \mu}\right)_T = x\left(\frac{\partial \ln \gamma}{\partial x}\right)_T, \tag{35.4}$$

where $\theta = \overline{N}/M$. The fraction of sites covered by ligand molecules is then $r\theta$. With the temperature constant, there is one independent intensive variable, which might be chosen as μ (or x), Φ (or γ), or even θ. Our primary objective below is to derive explicit formulas for $x(\gamma)$, $\theta(\gamma)$, $\gamma(\theta)$, and $x(\theta)$, in the order given. The last expression, $x(\theta)$, is usually attributed to McGhee and von Hippel,[10] but it was actually derived first by Zasedatelev *et al.*[9] Explicit formulas with x (or μ) as independent variable, $\gamma(x)$ and $\theta(x)$, are not available in the general case, but they are available in certain special cases (see below).

There are two important special cases of the above model that we shall return to below: $r = 1$ (this is the classical one-dimensional cooperative "lattice gas," or Ising problem); and $y = 1$ (this is the problem of binding "hard" or space-filling r-mers on a one-dimensional array of sites, where the only interaction is that no two bound r-mers can cover the same site).

We begin by modifying Eqs. (34.106) and (34.107) as needed for the present model. Here n_1 is the number of empty sites in a sequence, as before, but n_2 is the number of r-mers in a sequence of adjacent r-mers. There is again a factor ϕ in T [Eq. (34.105)] for each site, and also $y_{11} = 1$ and $y_{22} = y$. Because each r-mer covers r sites, ϕ^{n_2} in Eq. (34.107) becomes ϕ^{rn_2}. Thus the two sequence partition functions are

$$\psi_1 = \sum_{n_1=1}^{\infty} \phi^{n_1} = \frac{\phi}{1-\phi} = \frac{1}{\gamma - 1}, \tag{35.5}$$

$$\psi_2 = \sum_{n_2=1}^{\infty} \phi^{rn_2} x^{n_2} y^{n_2-1} = \frac{x\phi^r}{1 - yx\phi^r} = \frac{x}{\gamma^r - yx}. \tag{35.6}$$

It is very useful, below, to have $\bar{n}_1$ and $\bar{n}_2$ available (these are the mean sizes of the two sequences), so we digress to find these. In ψ_1, ϕ counts n_1 and, in ψ_2, x counts n_2. Thus

$$\bar{n}_1 = \phi \frac{\partial \ln \psi_1}{\partial \phi} = \frac{1}{1-\phi} = \frac{\gamma}{\gamma - 1}, \tag{35.7}$$

$$\bar{n}_2 = x \frac{\partial \ln \psi_2}{\partial x} = \frac{\gamma^r}{\gamma^r - yx}. \tag{35.8}$$

Because $y_{12} = 1$ in Eq. (34.103), we have

$$\psi_1 \psi_2 = 1 = \frac{x}{(\gamma - 1)(\gamma^r - yx)},$$

$$x = \frac{\gamma^r(\gamma - 1)}{y\gamma - y + 1}. \tag{35.9}$$

This is our first main result, the function $x(\gamma)$ mentioned above.

Next, we find $\theta(\gamma)$. This is obtained most simply by using

$$r\theta = \frac{r\bar{n}_2}{r\bar{n}_2 + \bar{n}_1}, \qquad \theta = \frac{(\bar{n}_2/\bar{n}_1)}{r(\bar{n}_2/\bar{n}_1) + 1}. \tag{35.10}$$

This is immediately obvious upon considering one average sequence of each type. From Eqs. (35.7) and (35.8), after eliminating x with Eq. (35.9),

$$\frac{\bar{n}_2}{\bar{n}_1} = \frac{(\gamma - 1)(y\gamma - y + 1)}{\gamma}. \tag{35.11}$$

If we substitute this in Eq. (35.10), we obtain

$$\theta = \frac{(\gamma - 1)(y\gamma - y + 1)}{r(\gamma - 1)(y\gamma - y + 1) + \gamma}. \tag{35.12}$$

This is the desired function $\theta(\gamma)$. A closely related expression is

$$\frac{\theta}{1 - r\theta} = \frac{\bar{n}_2}{\bar{n}_1} = \frac{(\gamma - 1)(y\gamma - y + 1)}{\gamma}. \tag{35.13}$$

Next, we can obtain $\gamma(\theta)$ by inverting Eq. (35.12), which is a quadratic in γ. This gives

$$\gamma = \frac{(2y - 1)(1 - r\theta) + \theta + R}{2y(1 - r\theta)}, \tag{35.14}$$

$$R \equiv \{[1 - (r + 1)\theta]^2 + 4y\theta(1 - r\theta)\}^{1/2}.$$

The positive root is selected by considering the case $y = 1$. An alternative form of Eq. (35.14), with R in the denominator, is easily found:

$$\gamma = \frac{2(y - 1)(1 - r\theta)}{(2y - 1)(1 - r\theta) + \theta - R}. \tag{35.15}$$

Finally, we deduce $x(\theta)$. We first use Eq. (35.13) to eliminate $\gamma - 1$ from Eq. (35.9). This leads to

$$x = \frac{\theta\gamma^{r-1}}{(1 - r\theta)}\left(\frac{\gamma}{y\gamma - y + 1}\right)^2. \tag{35.16}$$

Then, upon eliminating γ from Eq. (35.16), using Eq. (35.15), we get the required (McGhee–von Hippel) result:

$$x = \frac{\theta}{1 - r\theta}\left[\frac{2(y - 1)(1 - r\theta)}{(2y - 1)(1 - r\theta) + \theta - R}\right]^{r-1}\left[\frac{2(1 - r\theta)}{1 - (r + 1)\theta + R}\right]^2. \tag{35.17}$$

This can of course be used to find $\theta(x)$ numerically (see below).

Because $\bar{n}_1 \geq 1$ and $\bar{n}_2 \geq 1$, it is useful to rewrite $\bar{n}_1$ and $\bar{n}_2$ [Eqs. (35.7) and (35.8)] in the forms

$$\bar{n}_1 = \frac{1}{1 - A}, \qquad \bar{n}_2 = \frac{1}{1 - B}. \tag{35.18}$$

At the same time, we can express A and B as functions of θ. From Eqs. (35.7) and (35.15),

$$A = \frac{1}{\gamma} = \frac{(2y - 1)(1 - r\theta) + \theta - R}{2(y - 1)(1 - r\theta)}. \tag{35.19}$$

We can then find B from A using Eq. (35.13):

$$\frac{\theta}{1 - r\theta} = \frac{1 - A}{1 - B}. \tag{35.20}$$

This leads to

$$B = \frac{1 - (r - 2y + 1)\theta - R}{2\theta(y - 1)}. \tag{35.21}$$

Figure 8-5, which is the work of Wegner,[11] provides numerical examples. These are binding isotherms, at 37.1 °C, of tropomyosin (the ligand) on actin filaments at various concentrations of Mg^{2+}. In this case $r = 7$: one tropomyosin molecule covers seven actin subunits. The solid curves were calculated by Wegner from Eq. (35.17). Values of K and y are given in the figure caption. The steepest curve has $y = 1210$, which corresponds to $w_{22} = 7.10kT = 4.38$ kcal mol^{-1}.

We turn now to the special case $r = 1$: a ligand molecule occupies only one site and neighboring ligand molecules interact with each other, with interaction parameter $y_{22} = y = e^{-w_{22}/kT}$. This is the conventional one-dimensional Ising problem in the form of a "lattice gas." The matrix treatment in Eqs. (34.32)–(34.85) refers to the same problem if, in those equations, we put $y_{11} = y_{12} = 1$, $K_o = K$, $x_o = x$. Hence we are dealing merely with a self-consistency check here. Equations (35.9), (35.12), and (35.15) for $x(\gamma)$, $\theta(\gamma)$, and $\gamma(\theta)$ need not be rewritten; we simply put $r = 1$ in these equations and use

$$\begin{aligned} R &= [(1 - 2\theta)^2 + 4y\theta(1 - \theta)]^{1/2} \\ &= [1 - 4\theta(1 - \theta)(1 - y)]^{1/2}. \end{aligned} \tag{35.22}$$

However, it is now easy to invert $x(\gamma)$ to obtain $\gamma(x)$:

$$\gamma = \frac{1 + yx + [(1 - yx)^2 + 4x]^{1/2}}{2}. \tag{35.23}$$

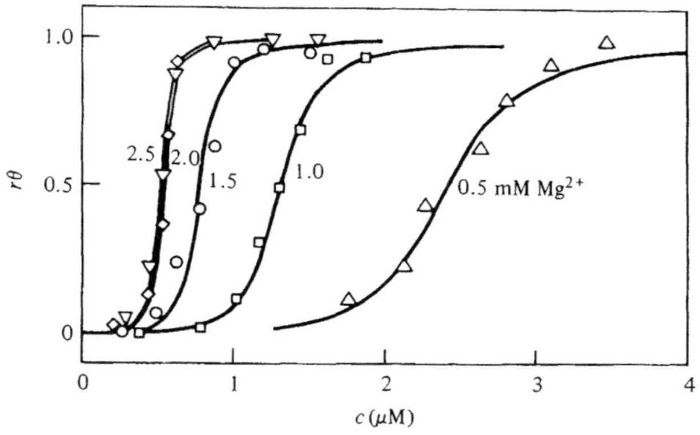

Fig. 8-5. Binding of tropomyosin on actin[11] at different concentrations of Mg^{2+}. Solid curves are from Eq. (35.17) with $r = 7$. The parameters K(M^{-1}) and y have the respective values, from right to left in the figure: 700, 590; 1020, 750; 1750, 730; 1560, 1190; and 1600, 1210.

This agrees with Eq. (34.33). Also, there is a simpler form of $\gamma(\theta)$ [Eq. (35.15)]:

$$\gamma = \frac{R+1}{R+1-2\theta}. \tag{35.24}$$

Equation (35.17) for $x(\theta)$ becomes

$$x = \frac{4\theta(1-\theta)}{(1-2\theta+R)^2}. \tag{35.25}$$

A simpler version of this is

$$x = \frac{R-1+2\theta}{y(R+1-2\theta)}. \tag{35.26}$$

Equations (35.24) and (35.26) are the same as Eqs. (14-16) and (14-13), respectively, in Ref. 12. Finally, after solving Eq. (35.26) for R and squaring, solution of the resulting quadratic equation in θ confirms $\theta(x)$ as given by Eq. (34.60). Thus, in the special case $r = 1$, explicit formulas can be written for $\gamma(x)$ and $\theta(x)$. That is, for this case, a complete set of six expressions based on x (or μ), γ(or Φ), and θ is available: $x(\gamma)$, $\theta(\gamma)$, $x(\theta)$, and their inverses.

Another important special case is $y = 1$. This model refers to the binding of a ligand molecule that occupies r sites on the one-dimensional lattice of sites, but there is no interaction between neighboring ligand molecules ($w_{22} = 0$ and $y = 1$ in Fig. 8-4). However, there remains the "parking" problem of distributing numerous r-mers along the array of sites. In this case ($y = 1$) Eqs. (35.9), (35.12), and (35.14) lead to

$$x = \gamma^{r-1}(\gamma - 1), \tag{35.27}$$

$$\theta = \frac{\gamma - 1}{r\gamma - r + 1}, \tag{35.28}$$

$$\gamma = \frac{1 - r\theta + \theta}{1 - r\theta}, \tag{35.29}$$

$$R = 1 - r\theta + \theta. \tag{35.30}$$

Substitution of Eq. (35.29) into Eq. (35.27) gives $x(\theta)$:

$$x = \frac{\theta}{1-r\theta}\left(\frac{1-r\theta+\theta}{1-r\theta}\right)^{r-1}. \tag{35.31}$$

This same result follows from Eq. (35.17) upon putting $y = 1 + \epsilon$ and letting $\epsilon \to 0$. The expressions for $\bar{n}_1$ and $\bar{n}_2$ are

$$\bar{n}_1 = \frac{\gamma}{\gamma - 1} = \frac{1 - r\theta + \theta}{\theta} = \frac{1}{1-A}, \tag{35.32}$$

$$\bar{n}_2 = \gamma = \frac{1 - r\theta + \theta}{1 - r\theta} = \frac{1}{1-B}, \tag{35.33}$$

$$A = \frac{1 - r\theta}{1 - r\theta + \theta}, \qquad B = \frac{\theta}{1 - r\theta + \theta}. \tag{35.34}$$

Within this special case, the dimer ($r = 2$) subcase is most important. For dimers,

$$x = \gamma(\gamma - 1) = \frac{\theta(1 - \theta)}{(1 - 2\theta)^2}, \tag{35.35}$$

$$\theta = \frac{\gamma - 1}{2\gamma - 1}, \qquad \gamma = \frac{1 - \theta}{1 - 2\theta}. \tag{35.36}$$

Upon solving Eq. (35.35) for $\theta(x)$, we find

$$\theta = \frac{2x}{(1 + 4x)^{1/2}[1 + (1 + 4x)^{1/2}]}. \tag{35.37}$$

This can be rearranged in several other ways. When x is small, $\theta \to x$. When x is large, $\theta \to \frac{1}{2}$. Figure 8-6 compares $\theta(x)$ from Eq. (35.37) (for a dimer) with $\theta = x/(1 + x)$ (for a noninteracting monomer). Both curves start with $\theta = x$.

Equation (35.37) is the same as Eq. (34.82) with $y = 0$, which, a little reflection will show, is what one should expect.

A More General Model with the Same Formal Properties

This is the same kind of generalization encountered in passing from Section 22 to Section 23. Figure 8-7(a) shows the model under consideration here. We have a linear array of subunits on which an r-mer (ligand) can bind ($r = 3$

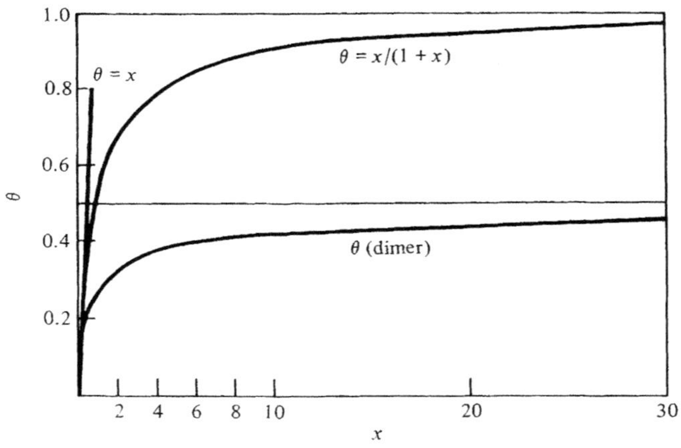

Fig. 8-6. $\theta(x)$ for a noninteracting dimer ($y = 1$), from Eq. (35.37). This is compared with two reference curves.

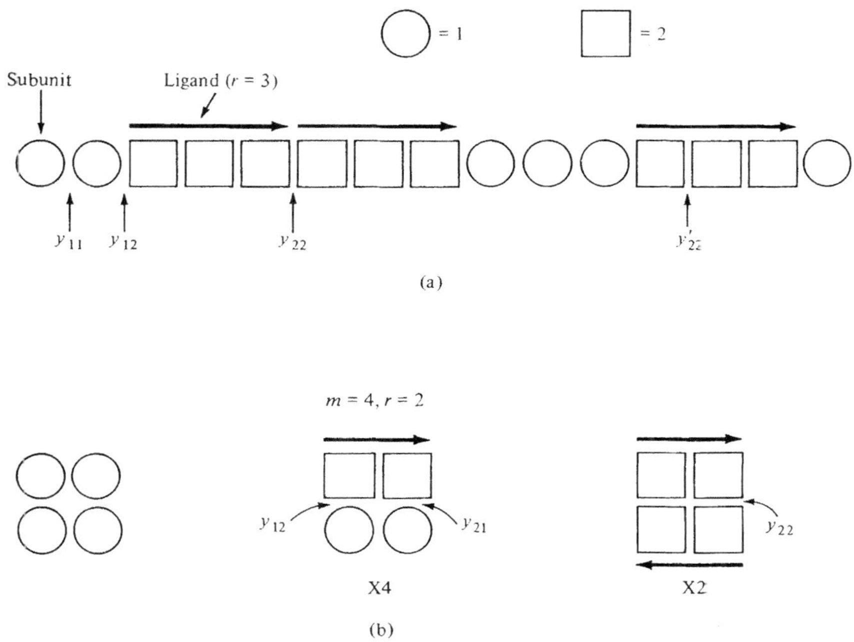

Fig. 8-7. (a) More general model of binding directional ligand ($r = 3$ here) on subunits with induced conformational change. See text for details. (b) Special case [see Eq. (35.38)] of small closed system with $m = 4$, $r = 2$.

in the figure). An uncovered subunit is always in state 1 (○); a covered subunit is always in state 2 (□). That is, binding of ligand induces the state (conformational) change. For generality, we assume that both the linear array and the ligand are directional (unsymmetrical). Hence $y_{12} \neq y_{21}$. Both y_{12} and y_{21} might include a *ligand–neighbor subunit* interaction. The interaction parameter y_{22} might include *subunit–subunit, ligand–ligand* and *ligand–neighbor subunit* contributions, whereas y'_{22} (in the interior of a covered group) refers to a subunit–subunit interaction only. Our object is to show that this model has the same formal properties as in the preceding subsection, despite the inclusion of all the y_{ij}.

To make the various contributions quite explicit, we begin with the special case of a square ($m = 4$) of subunits on which a dimer can be bound [Fig. 8-7(b)]. The complete grand partition function for this case is

$$\xi = q_1^4 y_{11}^4 + 4q_1^2 q_2^2 y_{11} y_{12} y_{21} y'_{22} q_b \lambda + 2q_2^4 y_{22}^2 y'^2_{22} q_b^2 \lambda^2, \qquad (35.38)$$

where q_1 and q_2 are partition functions for isolated subunits, the y_{ij} require no further explanation, q_b is the partition function of a dimer bound on a pair of state 2 subunits, and $\lambda = e^{\mu/kT}$, as usual (μ refers to the dimer). If we divide by q_1^4, to define a new ξ,

$$\xi = y_{11}^4 + 4(q_2/q_1)^2 y_{11} y_{12} y_{21} y'_{22} q_b \lambda + 2(q_2/q_1)^4 y_{22}^2 y'^2_{22} q_b^2 \lambda^2. \tag{35.39}$$

Next, we define the unperturbed binding constant of ligand, K_o, by

$$K_o c = (q_2/q_1)^2 y'_{22} q_b \lambda. \tag{35.40}$$

That is, K_o includes contributions from the induced conformational change in two subunits, the interior 22 interaction, and q_b. Then

$$\xi = y_{11}^4 + 4 y_{11} y_{12} y_{21} K_o c + 2 y_{22}^2 K_o^2 c^2. \tag{35.41}$$

The interaction parameters that appear here are all those outside of covered pairs of subunits (note that the number of these interaction parameters differs from term to term because y'_{22} is embedded in K_o). If we now divide ξ by y_{11}^4 to define another new ξ,

$$\xi = 1 + 4Kc + 2y^2 K^2 c^2, \tag{35.42}$$

$$K \equiv \frac{y_{12} y_{21}}{y_{11}^3} K_o, \qquad y \equiv \frac{y_{11} y_{22}}{y_{12} y_{21}}. \tag{35.43}$$

In these relations, K is the initial binding constant and y is the usual combination (but note that y includes y_{22}, not y'_{22}).

With this background, we return to the problem of binding r-mers on a long chain of subunits, as in Fig. 8-7(a). The definition of K_o in this more general case is

$$x_o = K_o c = (q_2/q_1)^r y'^{\,r-1}_{22} q_b \lambda. \tag{35.44}$$

The contributions that we include in the sequence partition functions ψ_1 and ψ_2 correspond to those in the terms in Eq. (35.41):

$$\psi_1 = \sum_{n_1=1}^{\infty} \phi^{n_1} y_{11}^{n_1-1} = \frac{\phi}{1 - y_{11}\phi} = \frac{1}{\gamma - y_{11}}, \tag{35.45}$$

$$\psi_2 = \sum_{n_2=1}^{\infty} \phi^{r n_2} x_o^{n_2} y_{22}^{n_2-1}$$

$$= \frac{x_o \phi^r}{1 - y_{22} x_o \phi^r} = \frac{x_o}{\gamma^r - y_{22} x_o}. \tag{35.46}$$

Then [Eq. (34.103)]

$$1 = \psi_1 \psi_2 y_{12} y_{21} = \frac{y_{12} y_{21} x_o}{(\gamma - y_{11})(\gamma^r - y_{22} x_o)}. \tag{35.47}$$

In passing from Ξ for M subunits in the $x_o = K_o c$ notation of Eq.(35.41) to a new Ξ in the $x = Kc$ notation of Eq. (35.42), we must divide the first Ξ by y_{11}^M. Because the first $\Xi = \gamma^M$ [Eq. (35.2)], this is equivalent to dividing γ (with x_o) by y_{11} to obtain a new γ (with x). Thus, in order to put Eq. (35.47) in x notation rather than x_o notation, we divide both numerator and denominator on the right by y_{11}^{r+1}, replace γ/y_{11} by the new γ, and replace [Eq. (35.43)]

$$\frac{y_{12}y_{21}x_o}{y_{11}^{r+1}} \quad \text{by } x. \tag{35.48}$$

This follows because, in the initial binding of a ligand molecule, 12 and 21 interactions (one each) appear and $r + 1$ 11 interactions disappear; also, $r - 1$ interior 22 interactions are included in K_o or x_o. With these changes in notation, Eq. (35.47) becomes

$$1 = \frac{x}{(\gamma - 1)(\gamma^r - yx)}. \tag{35.49}$$

Although the present model is more general, this is the same as Eq. (35.9). Also, to convert Eq. (35.45) into the form of Eq. (35.5) (with the new γ), we multiply the former equation by y_{11}. Similarly, multiplication of Eq. (35.46) by $y_{12}y_{21}/y_{11}$ converts it into Eq. (35.6) (with the new γ). The derived formal properties of the more general model are then all the same as for the McGhee–von Hippel model.

Three Illustrations of the Matrix Method

In the first of these examples, we treat the binding of a dimer, with neighbor interactions, by the matrix method. That is, we rederive some of the results in Eqs. (35.9)–(35.17) for the special case $r = 2$.

First, however, we require a straightforward generalization of Eqs. (34.3) and (34.35). If each site in a long chain of M sites can exist in ω different states, rather than in only two states, as in Eq. (34.3), then all of the ω^M terms in the complete grand partition function Ξ are generated if we sum each of $\nu_1, \ldots, \nu_M$ from 1 to ω. The factors $A(\nu_k, \nu_{k+1})$ are now elements in an $\omega \times \omega$ square matrix (rather than 2×2). We then have

$$\Xi = \sum_{\nu_1, \cdots, \nu_M = 1}^{\omega} A(\nu_1, \nu_2) \cdots A(\nu_{M-1}, \nu_M)\, A(\nu_M, \nu_1) \tag{35.50}$$

$$= \mathrm{Tr}\, \mathbf{A}^M = \gamma_1^M + \gamma_2^M + \cdots + \gamma_\omega^M \tag{35.51}$$

$$\cong \gamma_1^M. \tag{35.52}$$

The $\omega \times \omega$ matrix has ω eigenvalues but, if M is very large, as in the examples of the present subsection, only the largest eigenvalue γ_1 need be used in Eq. (35.52). The corresponding thermodynamic relations, for the binding of a single ligand, are Eqs. (34.37)–(34.44). Thus, if each site can exist in more than two states, the only change in our application of the matrix method is that we encounter a larger matrix with more eigenvalues; however, for a large system (M large), only the largest eigenvalue γ_1 is actually needed. In many examples an explicit formula for γ_1 cannot be found. In this case γ_1 must be differentiated numerically to find θ (see below).

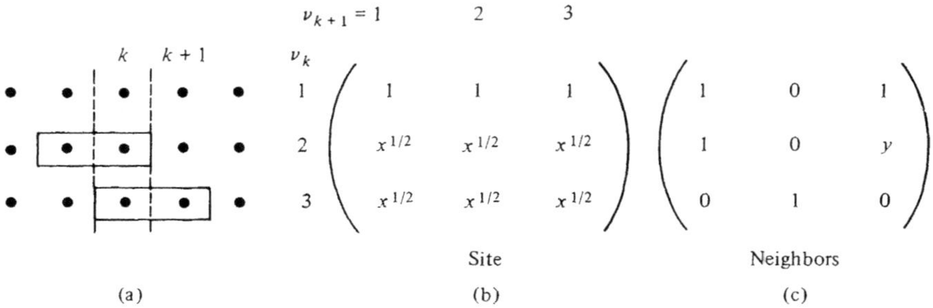

Fig. 8-8. (a) Three states of kth site in the binding of a dimer with interactions. (b) Site or binding factors that contribute to elements of the matrix **A**. (c) Interaction factors that contribute to **A**. See text for details.

We return now to the dimer binding problem. Figure 8-8(a) shows that the kth site can exist in three different states, $\nu_k = 1, 2, 3$. We shall assign binding factors, arbitrarily, to the first site index (ν_k) in $A(\nu_k, \nu_{k+1})$, as in Eq. (34.9). Figure 8-8(b) shows these site or binding factors only. This is not the complete **A.** An empty site ($\nu_k = 1$) has a factor 1; an occupied site ($\nu_k = 2$ or 3) has a factor $x^{1/2}$ because the bound dimer is shared by two sites. Thus, in the complete sum of products, Eq. (35.50), each bound dimer receives, correctly, a factor $x^{1/2}$ from both of the sites that it occupies. Figure 8-8(c) contains nearest-neighbor factors only. In the $\nu_k = 1$ row of Fig. 8-8(c), the factor 0 appears because it is not possible to have, simultaneously, site k in state 1 (empty) and site $k + 1$ in state 2. The factors 1 in the $\nu_k = 1$ row indicate simply that the state combinations $\nu_k, \nu_{k+1} = 11$ and 13 are consistent with each other. In the $\nu_k = 2$ row, the combination 22 is not possible (hence the factor 0) and the combination 23 produces a nearest-neighbor interaction (factor y). In the $\nu_k = 3$ row, the combinations 31 and 33 are not possible (factors 0), but 32 is a self-consistent combination. Each element of the required matrix **A** is the product of the corresponding elements in Figs. 8-8(b) and 8-8(c). Hence

$$\mathbf{A} = \begin{pmatrix} 1 & 0 & 1 \\ x^{1/2} & 0 & yx^{1/2} \\ 0 & x^{1/2} & 0 \end{pmatrix}. \tag{35.53}$$

The characteristic equation of **A** is

$$\begin{vmatrix} 1-\gamma & 0 & 1 \\ x^{1/2} & -\gamma & yx^{1/2} \\ 0 & x^{1/2} & -\gamma \end{vmatrix} = 0. \tag{35.54}$$

The eigenvalues of **A** are the roots of Eq. (35.54):

$$(1 - \gamma)(\gamma^2 - yx) + x = 0. \tag{35.55}$$

This is seen to be the same as Eq. (35.9), with $r = 2$. To obtain θ. in the matrix method, we differentiate Eq. (35.55) with respect to x to find $\partial\gamma/\partial x$. Then θ is given by $(x/\gamma)\partial\gamma/\partial x$ [see Eq. (34.44)]. If x is eliminated from this result, using Eq. (35.55), we recover Eq. (35.12), with $r = 2$. This suffices to show that the matrix method can easily reproduce our earlier main results for this model.

The second example [13] is somewhat related to the one just considered. The fragments HMM (heavy meromyosin) and S1 (subfragment 1 of HMM) of myosin apparently bind on an actin filament (considered here to be a linear chain of subunits, each with one binding site) without significant neighbor interactions. If both HMM (concentration c) and S1 (concentration c_S) are present, there is competitive binding on the actin sites (Fig. 8-9). Furthermore, HMM can bind to actin with one head (binding constant K_1) or two heads (constant K_2). The S1 binding constant is K_S. We define $x_1 = K_1c$, $x_2 = K_2c$, and $x_S = K_Sc_S$. Each actin site can exist in five different states, as shown in Fig. 8-9. States 1, 3, and 4 are equivalent to the three states in Fig. 8-8. The separate site and neighbor factors in **A** are:

$$\begin{array}{cc} & \nu_{k+1} = 1 \quad 2 \quad 3 \quad 4 \quad 5 \\ \begin{matrix} \nu_k = 1 \\ 2 \\ 3 \\ 4 \\ 5 \end{matrix} & \begin{pmatrix} 1 & 1 & 1 & 1 & 1 \\ x_1 & x_1 & x_1 & x_1 & x_1 \\ x_2^{1/2} & x_2^{1/2} & x_2^{1/2} & x_2^{1/2} & x_2^{1/2} \\ x_2^{1/2} & x_2^{1/2} & x_2^{1/2} & x_2^{1/2} & x_2^{1/2} \\ x_S & x_S & x_S & x_S & x_S \end{pmatrix} \\ & \text{Site} \end{array} \tag{35.56}$$

$$\begin{array}{cc} \begin{matrix} \nu_k = 1 \\ 2 \\ 3 \\ 4 \\ 5 \end{matrix} & \begin{pmatrix} 1 & 1 & 0 & 1 & 1 \\ 1 & 1 & 0 & 1 & 1 \\ 1 & 1 & 0 & 1 & 1 \\ 0 & 0 & 1 & 0 & 0 \\ 1 & 1 & 0 & 1 & 1 \end{pmatrix} \\ & \text{Neighbor} \end{array} \tag{35.57}$$

The explanation of the entries in Eqs. (35.56) and (35.57) is essentially the same as for Figs. 8-8(b) and 8-8(c). There are no interaction factors y_{ij}, by assumption. Upon multiplying corresponding elements, we obtain the complete **A**:

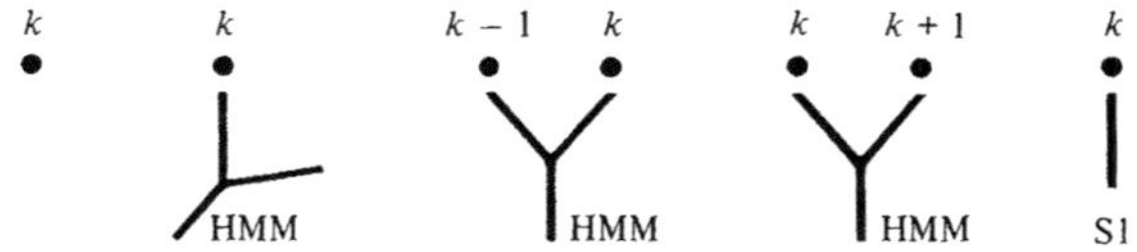

State	1	2	3	4	5
Binding Constant		K_1	K_2	K_2	K_S
Concentration		c	c	c	c_S

Fig. 8-9. Five possible states for site k in competitive binding of HMM and S1 on actin.

$$\mathbf{A} = \begin{pmatrix} 1 & 1 & 0 & 1 & 1 \\ x_1 & x_1 & 0 & x_1 & x_1 \\ x_2^{1/2} & x_2^{1/2} & 0 & x_2^{1/2} & x_2^{1/2} \\ 0 & 0 & x_2^{1/2} & 0 & 0 \\ x_S & x_S & 0 & x_S & x_S \end{pmatrix}. \tag{35.58}$$

The characteristic equation in this case has three roots $\gamma = 0$, and a remaining quadratic equation in γ with solution (largest eigenvalue)

$$\begin{aligned} \gamma_1 &= \tfrac{1}{2}(1 + x_1 + x_S + \sqrt{}), \\ \sqrt{} &= [(1 + x_1 + x_S)^2 + 4x_2]^{1/2}. \end{aligned} \tag{35.59}$$

Note that if $K_2 = 0$, $\gamma_1 = 1 + x_1 + x_S$. This is the correct grand partition function for a single site with competitive binding of HMM (K_1c) and S1 (with K_Sc_S).

To find the amount of binding of HMM and S1, we employ

$$\text{HMM:} \quad \theta = \frac{\bar{N}}{M} = c\,\frac{\partial \ln \gamma_1}{\partial c}, \tag{35.60}$$

$$\text{S1:} \quad \theta_S = \frac{\bar{N}_S}{M} = c_S\,\frac{\partial \ln \gamma_1}{\partial c_S}. \tag{35.61}$$

From these relations we find

$$\theta = \frac{x_1}{\sqrt{}} + \frac{2x_2}{\sqrt{}\,(1 + x_1 + x_S + \sqrt{})}, \tag{35.62}$$

$$\theta_S = \frac{x_S}{\sqrt{}}. \tag{35.63}$$

These give the expected results if $x_2 = 0$ or if $x_1 = 0$, $x_S = 0$ [see Eq. (35.57)].

Incidentally, it is easy to generalize these results to the case[3] of a mixture of several monovalent and several divalent species all competing for the same linear array of sites, but without neighbor interactions.

The third and final example requires a further straightforward generalization of the matrix method. Beginning with Eq. (34.3), we have used single sites as the basic repeating unit in the matrix method. This has been possible because, in all of our examples so far, the only interactions [or exclusion factors 0, as in Fig. 8-8(c)] have been between nearest-neighbor sites. However, if the interactions have a longer range than this, the repeating unit (a group of sites, say) must be chosen large enough so that all interactions are accounted for between *nearest-neighbor units*. The example we use to illustrate the above is the binding of a monomer ($r = 1$) with $x = Kc$ on a long linear array of M sites, but with both nearest-neighbor ($y = e^{-w/kT}$) and second-nearest-neighbor ($y_s = e^{-w_s/kT}$) interactions between bound monomers [see Eq. (22.33) for a simple model of the same type]. Examples of interaction factors are shown in Fig. 8-10(a). Because of the second-neighbor interaction, we need to choose a pair of sites as the repeating unit. The second-neighbor interaction y_s is then included in interactions between nearest-neighbor units k and $k + 1$ [Fig. 8-10(b)].

In the product in Eq. (35.50), there are M factors $A(\nu_k, \nu_{k+1})$, one for each site. For the present problem, we use a product of $M/2$ factors, one for each unit (a pair of sites). We then have

$$\Xi = \sum_{\nu_1, \ldots, \nu_{M/2}=1}^{\omega} A(\nu_1, \nu_2) \ldots A(\nu_{M/2}, \nu_1) \tag{35.64}$$

$$= \text{Tr}\, \mathbf{A}^{M/2} = \gamma_1^{M/2} + \gamma_2^{M/2} + \ldots + \gamma_\omega^{M/2} \tag{35.65}$$

$$\cong \gamma_1^{M/2}, \tag{35.66}$$

where each unit of two sites has ω possible states. If we needed to use a unit of three sites, $M/2$ would be replaced above by $M/3$, etc.

Figure 8-10(b) shows the four possible states for the kth unit. Again we

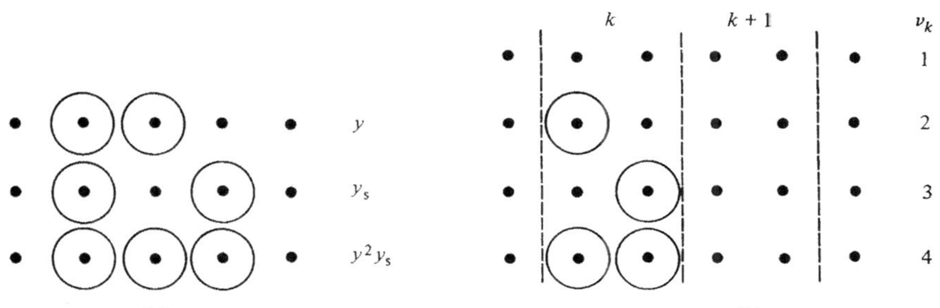

Fig. 8-10. (a) Examples of interaction factors in a model with both first (y) and second (y_s) neighbor interactions. (b) Four possible states of the kth unit of two sites.

subdivide the contributions to the elements of **A** into unit (two sites) and neighbor factors. Note that an interaction factor y should be included in the unit factor for state 4. The two sets of contributions to **A** are:

$$\begin{matrix} \nu_k = 1 \\ 2 \\ 3 \\ 4 \end{matrix} \begin{pmatrix} 1 & 1 & 1 & 1 \\ x & x & x & x \\ x & x & x & x \\ yx^2 & yx^2 & yx^2 & yx^2 \end{pmatrix} \tag{35.67}$$

Unit (two sites)

$$\begin{matrix} \nu_k = 1 \\ 2 \\ 3 \\ 4 \end{matrix} \begin{pmatrix} 1 & 1 & 1 & 1 \\ 1 & y_s & 1 & y_s \\ 1 & y & y_s & yy_s \\ 1 & yy_s & y_s & yy_s^2 \end{pmatrix} \tag{35.68}$$

Neighbor

The matrix **A** is then (from the products of elements)

$$\mathbf{A} = \begin{pmatrix} 1 & 1 & 1 & 1 \\ x & y_s x & x & y_s x \\ x & yx & y_s x & yy_s x \\ yx^2 & y^2 y_s x^2 & yy_s x^2 & y^2 y_s^2 x^2 \end{pmatrix}. \tag{35.69}$$

The characteristic equation of **A** is a quartic in γ. Without seeking algebraic shortcuts, the binding isotherm in such a problem can easily be calculated by numerical differentiation with respect to x of the largest root γ_1 of the quartic:

$$\Xi = \gamma_1^{M/2}, \qquad \frac{\partial \ln \Xi}{\partial \ln x} = \overline{N} = \frac{M}{2} \frac{\partial \ln \gamma_1}{\partial \ln x}, \tag{35.70}$$

$$\theta = \frac{\overline{N}}{M} = \frac{1}{2} \frac{\partial \ln \gamma_1}{\partial \ln x} \cong \frac{1}{2} \frac{\Delta \ln \gamma_1}{\Delta \ln x}. \tag{35.71}$$

This can be done even on an advanced hand calculator.

Incidentally, in the special case $y_s = 1$, this treatment is the two-site unit equivalent of the one-site unit approach in Eqs. (34.9) and (35.9) (with $r = 1$). From either of these equations we find ${}^{(1)}\gamma_1$, as given by Eq. (34.17), where the superscript refers to the one-site unit approach. When $y_s = 1$, the characteristic equation of **A** in Eq. (35.69) simplifies to a quadratic:

$$\gamma^2 - (1 + 2x + y^2x^2)\gamma + x^2(y - 1)^2 = 0. \tag{35.72}$$

The largest root is γ_1. Because

$$\Xi = {}^{(1)}\gamma_1^M = \gamma_1^{M/2}, \tag{35.73}$$

we expect ${}^{(1)}\gamma_1^2 = \gamma_1$, and this is easy to confirm. Thus, as one should anticipate, the same result is found from the matrix method in this special case ($y_s = 1$) whether the unit chosen is one site or two sites.

36. Cooperativity in the Tropomyosin–Actin–Myosin (S1) System

The power of the matrix method is illustrated well in this section, where the analysis of a relatively complicated model is made surprisingly simple by this method. At the same time the matrix method is generalized still further as we see that the elements of the matrix **A** can themselves be grand partition functions.

This model is believed to be a rather realistic one for the binding of the myosin fragment S1 (Section 35) and Ca^{2+} on regulated actin (i.e., actin saturated with tropomyosin-troponin; see Fig. 8-5). However, as usual, our emphasis will be on the basic theoretical method, though two examples of data fitting will be included at the end. We shall use the notation of the original paper to make it easier for the interested reader to obtain further details from that paper.[14]

We consider a very long actin filament saturated with troponin-tropomyosin, as indicated schematically in Fig. 8-11. We are interested in the equilibrium binding of S1 and of Ca^{2+} [Fig. 8-11(a)] as influenced by each other and by tropomyosin (the troponin is now always understood to be

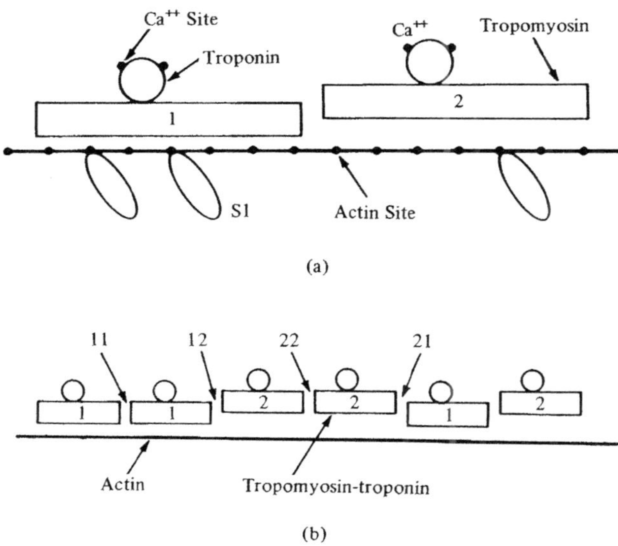

Fig. 8-11. (a) Binding of S1 and Ca^{2+} on actin saturated with tropomyosin-troponin. (b) Nearest-neighbor interactions in this model.

included). Each tropomyosin unit, including seven actin sites for S1 binding, can be in one of two states: state 1, with "weak" binding (constant K_1) of S1 on each of the seven actin sites; and state 2 with "strong" binding (constant K_2) of S1. The bound S1 molecules do not interact with each other. The intrinsic equilibrium constant for $2 \leftrightarrows 1$ in a hypothetical isolated tropomyosin unit, with no tropomyosin neighbors and with no bound S1 or Ca^{2+}, is L, which favors state 1 (weak) over state 2 (strong) ($L > 1$). In statistical mechanical language, $L = q_1/q_2$, where q_1 and q_2 are intrinsic unit partition functions in the two states. Each troponin has two equivalent binding sites for Ca^{2+} (the regulatory sites), with intrinsic binding constants K_a and K_b (in states 1 and 2, respectively). The interactions between nearest-neighbor tropomyosin molecules are of types 11, 12, 21, and 22 [Fig. 8-11(b)]. These boundary interactions are modulated by the extent of Ca^{2+} binding (0, 1, or 2) on the two Ca^{2+} sites of the *right-hand* member [Fig. 8-11(b)] of each nearest-neighbor pair (because of proximity of the right-hand troponin to the pair boundary).

State 1 is favored at low S1 concentrations ($L > 1$). However, state 2 dominates at high S1 concentrations because $K_2 \gg K_1$. The transition from state 1 to state 2 is cooperative for two reasons: because seven actin sites in a unit change state as a group; and, more importantly, because of the nearest-neighbor interactions between units (i.e., between the ends of tropomyosin molecules).

The above describes the model qualitatively; the details are filled in below. The beginning thermodynamic equations are

$$dA = -SdT + \mu_S dN_S + \mu_C dN_C + \mu dM, \tag{36.1}$$

$$A = \mu_S N_S + \mu_C N_C + \mu M, \tag{36.2}$$

where M is the number of tropomyosin units (the two possible states are in equilibrium and both have chemical potential μ), N_S and N_C are the numbers of S1 and Ca^{2+} bound on the M units, and μ_S and μ_C are the corresponding chemical potentials (on the one-dimensional array of units or in solution). Maximum binding is seven S1 and two Ca^{2+} per unit. Then

$$d(\mu M) = -SdT - \overline{N}_S d\mu_S - \overline{N}_C d\mu_C + \mu dM, \tag{36.3}$$

$$\mu M = -kT \ln \Xi, \tag{36.4}$$

where Ξ is the grand partition function of the linear system, open with respect to N_S and N_C but closed with respect to M. The chemical potential μ is the analogue of $-\Phi$ in Eqs. (34.37)–(34.41).

We shall apply the matrix method starting with Eq. (34.46), where M now refers to units, each of which can be in state 1 or state 2 (each of these states has many substates—see below). Interactions between nearest-neighbor units can be included as usual in $\mathbf{A}$; binding of Ca^{2+} will also appear in these nearest-neighbor contributions because, as mentioned above, the unit-unit interactions are assumed, in this model, to depend on the extent of Ca^{2+}

binding on the right-hand unit of a pair of units. The binding of S1 is assumed to depend on the state of the unit (see K_1 and K_2 above) but not on interactions between units. Hence S1 binding is included in unit factors, not neighbor factors (compare Fig. 8-8). We assume, also, that there are no interactions between neighboring bound S1 molecules within a unit, but if there were such interactions they could be included easily in the ξ_i below (as for the small complexes in Chapter 6). We can now complete the specification of the model by writing the matrix **A** explicitly:

$$\begin{array}{rl} & \nu_{k+1} = 1 \qquad\quad 2 \\ \mathbf{A} = & \begin{array}{c} \nu_k = 1 \\ 2 \end{array} \begin{pmatrix} \xi_1 Y_{11} & \xi_1 Y_{12} \\ \xi_2 Y_{21} & \xi_2 Y_{22} \end{pmatrix}, \end{array} \tag{36.5}$$

$$\begin{aligned} Y_{11} &\equiv x_{11} + 2K_a \rho y_{11} + K_a^2 \rho^2 z_{11}, \\ Y_{12} &\equiv x_{12} + 2K_b \rho y_{12} + K_b^2 \rho^2 z_{12}, \\ Y_{21} &\equiv x_{21} + 2K_a \rho y_{21} + K_a^2 \rho^2 z_{21}, \\ Y_{22} &\equiv x_{22} + 2K_b \rho y_{22} + K_b^2 \rho^2 z_{22}, \end{aligned} \tag{36.6}$$

$$\xi_1 \equiv L(1 + K_1 c)^7, \qquad \xi_2 \equiv (1 + K_2 c)^7, \tag{36.7}$$

where c is the concentration of free S1, ρ is the concentration of free Ca^{2+}, and x_{ij}, y_{ij}, z_{ij} are nearest-neighbor tropomyosin interaction parameters. For example, $x_{21} \equiv e^{-w_{21}/kT}$, where w_{21} is the 21 interaction free energy with no Ca^{2+} bound on the *right-hand* member (in state 1) of the pair, and y_{21}, z_{21} are corresponding parameters for a 21 pair with one and two Ca^{2+} bound, respectively. The implication is that the binding of Ca^{2+} induces changes in the structure of the nearby end of the tropomyosin. Incidentally, if two bound Ca^{2+} interact with each other, this effect can be included in the z_{ij}. All degrees of S1 binding are automatically taken care of by ξ_1 and ξ_2. All factors in Eq. (36.5) have the form of subsystem grand partition functions, the ξ_i referring to S1 binding and the Y_{ij} to Ca^{2+} binding plus unit–unit interactions. It will be noted, in Eq. (36.5), that the unit factors (ξ_1 and ξ_2) are assigned to the first unit index (ν_k) in $A(\nu_k, \nu_{k+1})$. This is arbitrary. However, Ca^{2+} binding must be associated with the second index in Y_{ij} because of our assumptions about the model ($\nu_{k+1} = 1 \rightarrow K_a$, $\nu_{k+1} = 2 \rightarrow K_b$).

In writing ξ_1 and ξ_2 in Eq. (36.7), we have used $q_2 = 1$ and $q_1 = L$ (in effect, ξ_1 and ξ_2 have been divided by q_2).

In summary, this rather complicated model is represented by a simple 2×2 matrix in which the elements themselves are products of grand partition functions that contain the details of the binding and of the neighbor interactions.

Because M is large [Eq. (34.35)],

$$\Xi = \gamma_1^M + \gamma_2^M \cong \gamma_1^M, \tag{36.8}$$

where γ_1 is the larger eigenvalue of **A**, found from the characteristic equation:

$$\gamma^2 - (\xi_1 Y_{11} + \xi_2 Y_{22})\gamma + \xi_1\xi_2(Y_{11}Y_{22} - Y_{12}Y_{21}) = 0, \tag{36.9}$$

$$\gamma_1 = \frac{a_1 + a_2 + [(a_1 - a_2)^2 + 4a_1a_2Y^{-1}]^{1/2}}{2}, \tag{36.10}$$

$$a_1 \equiv \xi_1 Y_{11}, \qquad a_2 \equiv \xi_2 Y_{22}, \qquad Y(\rho) = Y_{11}Y_{22}/Y_{12}Y_{21}. \tag{36.11}$$

It should be noticed especially that ξ_1 and ξ_2 are functions of c (S1), whereas Y_{ij} and Y are functions of ρ (Ca^{2+}). Thus a_1 and a_2 are functions of both c and ρ.

The physical significance of

$$a \equiv \frac{a_2}{a_1} = \frac{(1 + K_2c)^7 Y_{22}(\rho)}{L(1 + K_1c)^7 Y_{11}(\rho)} \tag{36.12}$$

is that this is the equilibrium constant, per tropomyosin unit, for the transition $1 \rightleftarrows 2$ between a filament with all units in state 1 and a filament with all units in state 2, at arbitrary concentrations of S1 (c) and Ca^{2+} (ρ). $L > 1$ favors state 1; large c favors state 2 ($K_2 \gg K_1$); ρ affects a by altering both the extent of Ca^{2+} binding and the effective nearest-neighbor interaction between 11 and 22 pairs in the two filaments [see Eqs. (36.6)]. Formally, a is the equivalent of $\bar{x}$ in Eq. (34.83).

We digress at this point to consider the alternate method introduced in Section 34. The analogue of Eq. (34.105) here is

$$\mathrm{T} = \sum_M \Xi(T, \mu_S, \mu_C, M)\lambda^M, \tag{36.13}$$

$$\lambda \equiv e^{\mu/kT}.$$

That is, in this completely open system, each of the M tropomyosin units accounts for a factor λ in Eq. (36.13). Then the sequence partition functions are

$$\psi_1 = \sum_{n_1=1}^{\infty} \lambda^{n_1}\xi_1^{n_1}Y_{11}^{n_1-1} = \frac{\lambda\xi_1}{1 - \lambda\xi_1Y_{11}}, \tag{36.14}$$

$$\psi_2 = \sum_{n_2=1}^{\infty} \lambda^{n_2}\xi_2^{n_2}Y_{22}^{n_2-1} = \frac{\lambda\xi_2}{1 - \lambda\xi_2Y_{22}}, \tag{36.15}$$

where n_1 and n_2 refer to strings of consecutive units all in state 1 or all in state 2. Then [Eq. (34.103)]

$$1 = \psi_1\psi_2Y_{12}Y_{21} = \frac{\lambda^2\xi_1\xi_2Y_{12}Y_{21}}{(1 - \lambda\xi_1Y_{11})(1 - \lambda\xi_2Y_{22})}, \tag{36.16}$$

From Eqs. (36.4) and (36.8),

$$\Xi = e^{-\mu M/kT} = (1/\lambda)^M = \gamma_1^M, \tag{36.17}$$

so that $\gamma_1 = 1/\lambda$. If we substitute $\lambda = 1/\gamma_1$ in Eq. (36.16), we arrive again at Eq. (36.9). Thus the two methods are equivalent.

In the special case of no binding and no neighbor interactions, Eq. (36.16) gives

$$\xi_1 = L, \qquad \xi_2 = 1, \qquad Y_{ij} = 1, \qquad \lambda = 1/(1 + L). \tag{36.18}$$

This is what we should expect because, from Eq. (36.2),

$$A = \mu M = -kT \ln Q = -kT \ln(q_1 + q_2)^M = -kT \ln(L + 1)^M, \tag{36.19}$$

where Q is the canonical partition function of M units, each of which has partition function $q_1 + q_2$ (because of the internal equilibrium between states 1 and 2).

We return now to Eq. (36.10) and the main argument. From Eqs. (36.3), (36.4), and (36.8),

$$\overline{N}_S = c \frac{\partial \ln \Xi}{\partial c} = Mc \frac{\partial \ln \gamma_1}{\partial c}, \tag{36.20}$$

$$\overline{N}_C = \rho \frac{\partial \ln \Xi}{\partial \rho} = M\rho \frac{\partial \ln \gamma_1}{\partial \rho}. \tag{36.21}$$

Then

$$\theta = \frac{\overline{N}_S}{7M} = \frac{1}{7} c \frac{\partial \ln \gamma_1}{\partial c}, \tag{36.22}$$

$$\sigma = \frac{\overline{N}_C}{M} = \rho \frac{\partial \ln \gamma_1}{\partial \rho}, \tag{36.23}$$

where θ is the fraction of the actin sites occupied by S1 ($0 \leq \theta \leq 1$) and σ is the mean number of Ca^{2+} bound per tropomyosin unit ($0 \leq \sigma \leq 2$). From Eqs. (36.10), (36.22), and (36.23), we can then derive, after considerable algebra,

$$\theta = p_1\theta_1 + p_2\theta_2, \tag{36.24}$$

$$\sigma = p_{11}\sigma_{11} + p_{12}\sigma_{12} + p_{21}\sigma_{21} + p_{22}\sigma_{22}, \tag{36.25}$$

where p_1 is the fraction of tropomyosin units in state 1, θ_1 is the fraction of state 1 actin sites occupied by S1, σ_{12} is the mean number of Ca^{2+} bound at a 12 boundary (right-hand tropomyosin, in state 2), p_{12} is the fraction of all nearest-neighbor pairs that are of type 12, etc. Conservation relationships are

$$\begin{gathered} p_1 + p_2 = 1, \qquad p_{12} = p_{21}, \qquad p_{11} + p_{22} + 2p_{12} = 1, \\ p_1 = p_{11} + p_{12}, \qquad p_2 = p_{22} + p_{12}. \end{gathered} \tag{36.26}$$

Explicit expressions for the above quantities are

$$\theta_i = K_i c/(1 + K_i c) \qquad (i = 1, 2), \tag{36.27}$$

$$\sigma_{ij} = \partial \ln Y_{ij}/\partial \ln \rho \qquad (i, j = 1, 2), \tag{36.28}$$

$$\sigma_{i1} = 2(K_a \rho y_{i1} + K_a^2 \rho^2 z_{i1}/Y_{i1} \qquad (i = 1, 2), \tag{36.29}$$

$$\sigma_{i2} = 2(K_b \rho y_{i2} + K_b^2 \rho^2 z_{i2})/Y_{i2} \qquad (i = 1, 2), \tag{36.30}$$

$$p_2 = \frac{2aY^{-1}}{\sqrt{\ }(1 - a + \sqrt{\ })}, \tag{36.31}$$

$$p_{11} = \frac{1 - a + \sqrt{\ }}{\sqrt{\ }(1 + a + \sqrt{\ }}, \tag{36.32}$$

$$p_{22} = \frac{a(a - 1 + \sqrt{\ })}{\sqrt{\ }(1 + a + \sqrt{\ })}, \tag{36.33}$$

$$p_{12} = p_{21} = \frac{2aY^{-1}}{\sqrt{\ }(1 + a + \sqrt{\ })}, \tag{36.34}$$

where

$$a \equiv a_2/a_1, \qquad \sqrt{\ } \equiv [(1 - a)^2 + 4aY^{-1}]^{1/2}. \tag{36.35}$$

Equations (36.31)–(36.34) are essentially the same as Eqs. (34.60)–(34.63) (only the notation is changed).

$Y > 1$ introduces positive cooperativity beyond the seven-actin-site effect [Eq. (36.7)]. The midpoint of the transition between states 1 and 2 ($p_2 = \frac{1}{2}$) occurs when $a = 1$ [Eq. (36.31)] for any Y. We cannot expect the usual symmetry of binding isotherm curves in simple Ising models (Section 34) because here $K_1 \neq K_2$ and $K_a \neq K_b$. In fact, this kind of symmetry is found here in $p_2(a)$, not in $\theta(c)$ or $\sigma(\rho)$. The function $p_2(a)$ is formally the same as $\theta(\bar{x})$ in Eqs. (34.83)–(34.85). The Hill coefficient should be defined here by [see Eq. (24.16)]

$$n = 4\left(\frac{\partial p_2}{\partial \ln a}\right)_{p_2=1/2} = Y^{1/2}. \tag{36.36}$$

Let us define θ' and c' as the values of θ and c, for any given ρ, at which $a = 1$, $p_2 = \frac{1}{2}$ (i.e., half of the units are in each state). Then θ' and c' are related by [Eqs. (36.24) and (36.27)]

$$2\theta' = \frac{K_1 c'}{1 + K_1 c'} + \frac{K_2 c'}{1 + K_2 c'}. \tag{36.37}$$

Also, from Eq. (36.12), we have

$$L'(\rho) \equiv \frac{LY_{11}(\rho)}{Y_{22}(\rho)} = \frac{(1 + K_2 c')^7}{(1 + K_1 c')^7}. \tag{36.38}$$

This equation defines a useful modification of L, $L'(\rho)$, and shows how L' can be found from c' or c' from L' (note that Y does not enter into the relation between L' and c'):

$$c' = \frac{L'^{1/7} - 1}{K_2 - L'^{1/7} K_1}. \tag{36.39}$$

Whereas $L = q_1/q_2$ is the equilibrium constant for $2 \leftrightarrows 1$ in isolated tropomyosin units, $L'(\rho)$ refers, per tropomyosin unit, to the process *all*-2 filament $\leftrightarrows$ *all*-1 filament, with Ca^{2+} at ρ, but in the absence of S1. This follows from the fact that, in Eq. (36.12), $L' = 1/a$ if $c = 0$. At low or high Ca^{2+}, we have the special cases

$$L'(0) = \frac{Lx_{11}}{x_{22}} \qquad (\rho \to 0),$$
$$L'(\infty) = \frac{LK_a^2 z_{11}}{K_b^2 z_{22}} \qquad (\rho \to \infty). \tag{36.40}$$

Having defined the new equilibrium constant $L'(\rho)$, Eq. (36.12) for a, which is another equilibrium constant, can be rewritten as

$$a(c, \rho) = \frac{a_2}{a_1} = \frac{(1 + K_2 c)^7}{L'(\rho)(1 + K_1 c)^7}. \tag{36.41}$$

In the practical calculation of $\theta(c)$ from Eq. (36.24), θ_1 and θ_2 depend on the parameters K_1 and K_2 [Eq. (36.27)], $p_1 = 1 - p_2$ and p_2 depend on a and Y [Eq. (36.31)], and a, in turn, depends on the parameters K_1, K_2, and L' [Eq. (36.41)]. Thus the complete set of parameters is K_1, K_2, Y, and L'.

The reader may have noticed the close resemblance of the present problem (aside from the special complications of the Ca^{2+} binding) to much of Section 24. Here we are discussing a very long chain of units; in Section 24 we considered small complexes. The notational connections are

$$\frac{q_1}{q_2} = L \leftrightarrow \frac{j_1}{j_2}, \qquad a(c = 0) = \frac{1}{L'} = \frac{Y_{22}}{LY_{11}} \leftrightarrow H_o,$$
$$a \leftrightarrow H, \qquad p_2 \leftrightarrow \delta, \qquad Y \leftrightarrow y. \tag{36.42}$$

A number of other symbols are the same in both places.

In the special case in which unit-unit interactions are absent, $x_{ij} = y_{ij} = z_{ij} = 1$. Note, in Eq. (36.6), that $Y = 1$ in this case but the separate $Y_{ij} \neq 1$ (unless $\rho = 0$). Then we find, as one might expect,

$$\theta = p_1\theta_1 + p_2\theta_2, \qquad \sigma = p_1\sigma_1 + p_2\sigma_2, \tag{36.43}$$

$$p_1 = 1/(1 + a), \qquad p_2 = a/(1 + a), \tag{36.44}$$

$$\sigma_1 \equiv \frac{2K_a\rho}{1 + K_a\rho}, \qquad \sigma_2 \equiv \frac{2K_b\rho}{1 + K_b\rho}, \tag{36.45}$$

where

$$a = \frac{(1 + K_2 c)^7 (1 + K_b\rho)^2}{L(1 + K_1 c)^7 (1 + K_a\rho)^2}, \tag{36.46}$$

and θ_1 and θ_2 are given by Eq. (36.27). Here the two ligands play similar roles, as in Eq. (24.92).

As a numerical example, we fit experimental data for $\theta(c)$ at, effectively, the limiting Ca^{2+} concentrations $\rho = 0$ and $\rho = \infty$ (actually, 1 mM EGTA and 0.5 mM $CaCl_2$). The S1 is in the form of S1 · ADP. The data are shown in Fig. 8-12. The four low $\rho = 0$ points near $c = 1\ \mu$M are presumed to represent metastable rather than equilibrium behavior.

We do not give details of the method of parameter selection; these are available in the original paper.[14] The values chosen were

$$K_1 = 3 \times 10^3\ \text{M}^{-1}, \qquad K_2 = 7.1 \times 10^5\ \text{M}^{-1},$$

$$\rho = 0:\ L'(0) = \frac{Lx_{11}}{x_{22}} = 63.7, \qquad Y(0) = \frac{x_{11}x_{22}}{x_{12}x_{21}} = 20.0, \tag{36.47}$$

$$\rho = \infty:\ L'(\infty) = \frac{LK_a^2 z_{11}}{K_b^2 z_{22}} = 7.16, \qquad Y(\infty) = \frac{z_{11}z_{22}}{z_{12}z_{21}} = 4.0.$$

These parameters lead to the smooth curves for $\theta(c)$ and $p_2(c)$ in Fig. 8-12. The corresponding values of c' and θ' are

$$\begin{aligned} \rho = 0:\ c' &= 1.15\ \mu\text{M}, \qquad \theta' = 0.226, \\ \rho = \infty:\ c' &= 0.46\ \mu\text{M}, \qquad \theta' = 0.124. \end{aligned} \tag{36.48}$$

The asymptotic θ curve at large c is $\theta \cong \theta_2(c)$ (i.e., $p_2 \cong 1$). The values of c' and θ' at which $a = 1, p_2 = \frac{1}{2}$ depend on L' but not on Y, but the steepness of the $\theta(c)$ and $p_2(c)$ curves at $p_2 = \frac{1}{2}$ is quite sensitive to Y. The larger value of Y at $\rho = 0$ produces the greater steepness in this case. The larger value of L' at $\rho = 0$ means that the relative stability of state 1 units at $c = 0$ is much

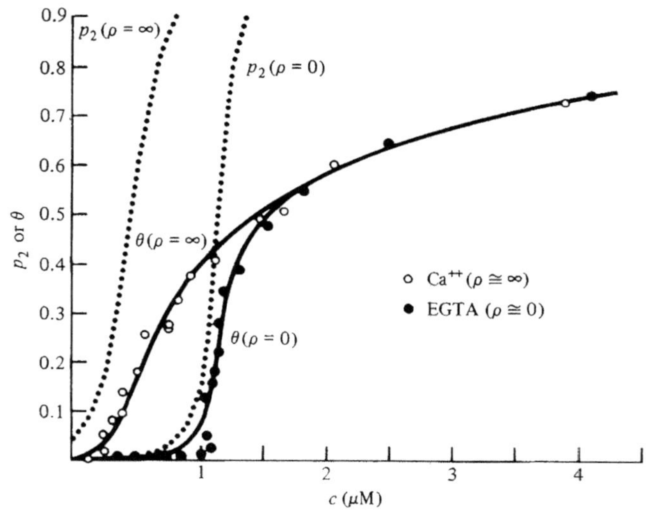

Fig. 8-12. Theoretical curves and experimental points for binding of S1 · ADP (c) on regulated actin at high and low Ca^{2+} (ρ). See text for details.

greater when $\rho = 0$ than when $\rho = \infty$. Therefore it requires a larger value of c' (S1 concentration) to bring the $1 \rightleftarrows 2$ equilibrium to $p_2 = \frac{1}{2}$ in this case. This is the reason the $\rho = 0$ curves are to the right of the $\rho = \infty$ curves in Fig. 8-12. The two $\theta(c)$ curves in Fig. 8-12 provide a glimpse of the possible origin of the regulation of muscle contraction by Ca^{2+}: around $c = 0.8$ μM, the attachment of S1 to actin is very sensitive to Ca^{2+}; mechanical force cannot be generated by a myosin molecule in a muscle fiber unless the S1 part of the molecule is attached to actin. We shall return to this subject in Section 39.

Modified Model for the Tropomyosin–Actin–S1 System

Different assumptions have been made in a modified model for the above system. As of this writing, it is not clear which model is better. Details of the theoretical treatment and application of the modified model are available in two papers.[15, 16] Here we merely describe the second model qualitatively. This will be useful because a simplified version of this is adopted in Section 39.

In the modified model, we assume that Tm-Tn (tropomyosin-troponin) in state 1 (Fig. 8-13) blocks all of the actin binding sites for S1. When no S1 is bound to a seven-site unit, the Tm-Tn is always in state 1. In order for even one S1 to be able to bind to a unit, as part of the binding process the S1 must push the entire Tm-Tn complex to a new position in relation to the actin sites. This is state 2 for Tm-Tn. The position of Tm-Tn in state 1, relative to the actin sites, is assumed to depend on the extent of Ca^{2+} binding on troponin, while the position of Tm-Tn in state 2 is assumed to depend on the nucleotide, if any, attached to S1, but not on Ca^{2+}. When Ca^{2+} is bound on Tm-Tn, the state 1 position or level, relative to the actin sites, is moved in the direction of the state 2 position.

After the first S1 is bound to a unit (with binding constant K_2/L), subsequent S1s can bind more easily, without doing any further work moving Tm-Tn relative to the actin sites. The binding constant in this case is K_2, with $K_2 \geq K_2/L$ (i.e., $L \geq 1$). Figure 8-14 shows, very schematically, the physical basis for the difference between K_2/L and K_2. Clearly, if ΔG is the (positive) free energy required to move Tm-Tn relative to the actin sites enough to allow binding, then $L \equiv e^{\Delta G/kT} \geq 1$. Note that K_1 in the previous

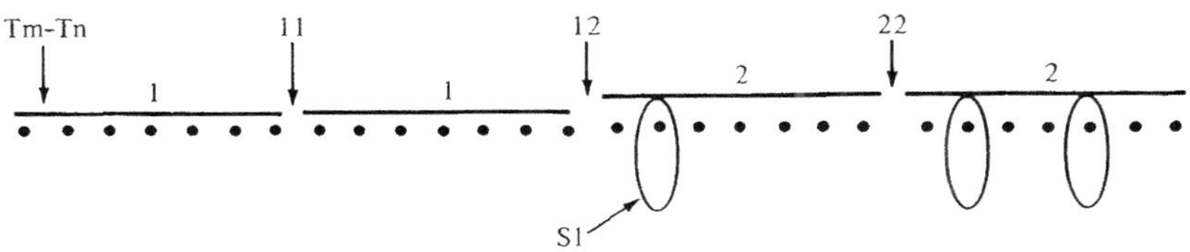

Fig. 8-13. Modified model for S1 binding on regulated actin, as explained in text.

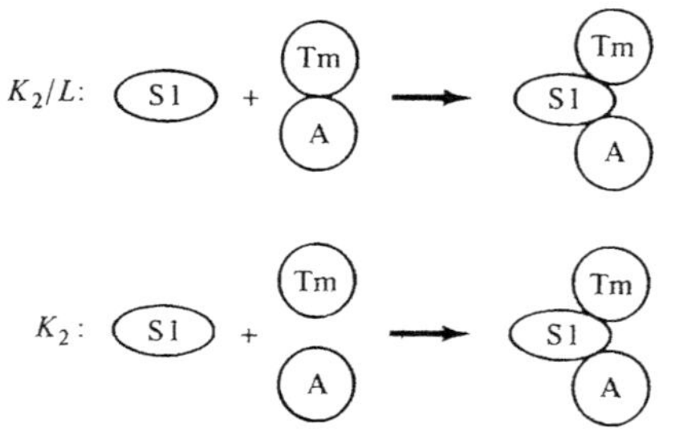

Fig. 8-14. Schematic representation of difference between binding constants K_2 and K_2/L.

model is eliminated here as an independent parameter.

In the present model there is *not* a dynamic equilibrium between empty (no S1 bound) Tm-Tn units in states 1 and 2. Correspondingly, L is not itself an isomeric equilibrium constant as it is in the previous model. Here Tm-Tn is in state 1 if and only if no S1 is bound to the unit; Tm-Tn is in state 2 if and only if one or more S1 are bound to the unit. Transitions occur between states 1 and 2 only when the first S1 is bound to a unit or the last S1 is released from a unit.

Again we assume that there are interactions between the ends of neighboring Tm-Tn units. This interaction is optimal (most attractive) when the two neighbors are at the same level, relative to the actin sites, and less attractive the greater the difference in level of the two neighbors.

References

1. Veenhuis, M., Harder, W., van Dijken, J., and Mayer, F. (1981) Mol. Cell. Biol. **1,** 949.
2. Kirsten, H., Gehring, H., and Christen, P. (1983) Proc. Natl. Acad. Sci. USA **80,** 1807.
3. Hill, T. L. (1957) J. Polymer Sci. **23,** 549.
4. Poland, D. and Scheraga, H. A. (1970) *Theory of Helix-Coil Transitions in Biopolymers* (Academic, New York).
5. Poland, D. (1978) *Cooperative Equilibria in Physical Biochemistry* (Clarendon, Oxford).
6. Davidson, N. (1962) *Statistical Mechanics* (McGraw-Hill, New York).
7. Hill, T. L. (1973) Biopolymers **12,** 257.
8. Hill, T. L. (1972) Proc. Natl. Acad. Sci. USA **69,** 1165.
9. Zasedatelev, A. S., Gurskii, G. V., and Volkenshtein, M. V. (1971) Molek. Biol. **5,** 245.
10. McGhee, J. D. and von Hippel, P. H. (1974) J. Mol. Biol. **86,** 469.
11. Wegner, A. (1979) J. Mol. Biol. **131,** 839.
12. Hill, T. L. (1960) *Introduction to Statistical Thermodynamics* (Addison-Wesley, Reading, MA).
13. Hill, T. L. (1978) Nature **274,** 825.

14. Hill, T. L., Eisenberg, E., and Greene, L. (1980) Proc. Natl. Acad. Sci. USA **77,** 3186.
15. Hill, T. L. (1981) Biophys. Chem. **14,** 31.
16. Hill, T. L., Eisenberg, E., and Greene, L. E. (1983) Proc. Natl. Acad. Sci. USA **80,** 60.

General References

Lifson, S. and Zimm, B. H. (1963) Biopolymers **1,** 15.
Tsuchiya, T. and Szabo, A. (1982) Biopolymers **21**, 979.

9

One-Dimensional Array of Interacting Two-State Units at Steady State

This chapter extends the subject dealt with in Chapter 8 to steady states. The treatment given here is merely an introduction to a field that is in its infancy. The present chapter is confined to analytical results; a few Monte Carlo results on one-dimensional lattices are postponed until Chapter 10 (where the Monte Carlo method is introduced for this type of problem, especially in two dimensions).

The principal one-dimensional biological example that is apparent at the time of writing is the cooperative Ca^{2+} regulation of the steady-state ATPase activity of myosin on regulated actin filaments (see Section 36), *in vitro* or *in vivo*. A preliminary and approximate theoretical treatment of the *in vitro* problem has been published,[1] but it will not be discussed here because it is too specialized. Experiments on this subject are in progress. The *in vivo* problem (Ca^{2+} regulation of isometric muscle contraction) will be introduced in an elementary form in Section 39.

From the point of view of statistical physics, the subject of this chapter adds a new "dimension" to the one-dimensional Ising problem: each of the interacting units in the one-dimensional lattice is driven in its cycle by a thermodynamic force X, which is zero at equilibrium but nonzero at steady state. The new dimension is the X axis. When $X = 0$, the equilibrium two-state Ising problem is easy to solve exactly (Section 34). When $X \neq 0$, the corresponding steady-state problem presumably requires Monte Carlo treatment for exact results, except in special cases. Two such special cases are the subjects of Sections 37 and 38: if $f_\alpha + f_\beta = 1$, the steady state is a quasiequilibrium state; and, near equilibrium ($X/kT \ll 1$), the exact L in the linear flux-force relation, $J = LX$, can be deduced. For the general one-dimensional steady-state problem, even with only two-state cycles, no systematic Monte Carlo study has been carried out.

Except for Section 39, the principal aim of Chapters 9–11 is to interest statistical physicists in this generalization of the classical equilibrium Ising problem. All of the fascinating questions having to do with cooperativity, critical phenomena, and phase transitions at equilibrium can be extended to steady-state systems. Steady states near equilibrium present the easiest special case, but the general problem relates to steady states arbitrarily far from equilibrium.

37. One-Dimensional Lattice in a Quasiequilibrium Steady State

We begin in a more general way than implied by the title of this section and turn to the quasiequilibrium special case[2] only when this is required in order to continue the analysis.

We consider a very long one-dimensional array of M identical two-state units, each of which is engaged in steady-state kinetic activity, as shown in Fig. 9-1. The rate constants in the figure are all first order and pertain to an unperturbed unit (no neighbor interactions). Usually α and β' are pseudo first order, and are proportional to ligand concentrations. The two-state cycle might refer to any enzymatic system whose kinetics can be reduced, effectively, to two states. There are nearest-neighbor interaction free energies w_{ij} ($i, j = 1, 2$) between units in the two possible states (Fig. 9-1). As usual, $y_{ij} \equiv e^{-w_{ij}/kT}$. We take $w_{12} = w_{21}$ for simplicity, but this is not necessary.

The thermodynamic force, X, driving the cycle in Fig. 9-1 in the positive direction, and the unperturbed steady-state flux per unit, J_o, are given by (Chapter 3)

$$e^{X/kT} = \frac{\alpha\beta}{\alpha'\beta'}, \qquad J_o = \frac{\alpha\beta - \alpha'\beta'}{\alpha + \alpha' + \beta + \beta'}. \tag{37.1}$$

The interactions between units perturb the four rate constants of the two-state cycle. There are three possible perturbations for any particular "central" unit: its two nearest neighbors might be in states 11, 12 (or 21), or 22. We use these indices as subscripts to designate the corresponding perturbed rate

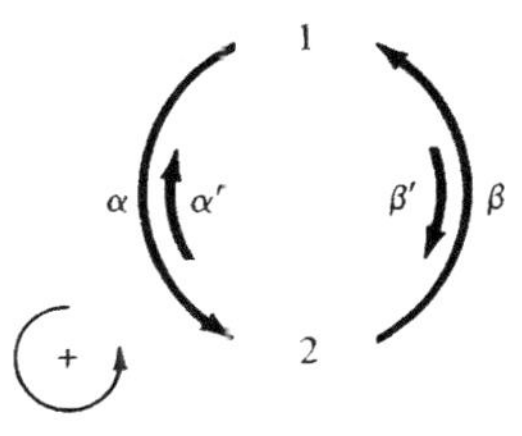

Fig. 9-1. Two-state cycle for an unperturbed unit.

constants of the cycle. Then, as in various examples in Chapter 7, the perturbed rate constants are related to unperturbed rate constants by:

$$111 \rightleftarrows 121 \qquad \alpha_{11} = \alpha\left(\frac{y_{12}}{y_{11}}\right)^{2f_\alpha}, \qquad \alpha'_{11} = \alpha'\left(\frac{y_{12}}{y_{11}}\right)^{2(f_\alpha - 1)}, \tag{37.2}$$

$$\beta_{11} = \beta\left(\frac{y_{11}}{y_{12}}\right)^{2f_\beta}, \qquad \beta'_{11} = \beta'\left(\frac{y_{11}}{y_{12}}\right)^{2(f_\beta - 1)}, \tag{37.3}$$

$$112 \rightleftarrows 122 \qquad \alpha_{12} = \alpha\left(\frac{y_{22}}{y_{11}}\right)^{f_\alpha}, \qquad \alpha'_{12} = \alpha'\left(\frac{y_{22}}{y_{11}}\right)^{f_\alpha - 1}, \tag{37.4}$$

or

$$211 \rightleftarrows 221 \qquad \beta_{12} = \beta\left(\frac{y_{11}}{y_{22}}\right)^{f_\beta}, \qquad \beta'_{12} = \beta'\left(\frac{y_{11}}{y_{22}}\right)^{f_\beta - 1}, \tag{37.5}$$

$$212 \rightleftarrows 222 \qquad \alpha_{22} = \alpha\left(\frac{y_{22}}{y_{12}}\right)^{2f_\alpha}, \qquad \alpha'_{22} = \alpha'\left(\frac{y_{22}}{y_{12}}\right)^{2(f_\alpha - 1)}, \tag{37.6}$$

$$\beta_{22} = \beta\left(\frac{y_{12}}{y_{22}}\right)^{2f_\beta}, \qquad \beta'_{22} = \beta'\left(\frac{y_{12}}{y_{22}}\right)^{2(f_\beta - 1)}. \tag{37.7}$$

The transitions are occurring here in the central units of the triplets. For example, in the α reaction, $1 \rightarrow 2$, in Eq. (37.2), two 12 interactions are produced, whereas two 11 interactions are lost. As usual, we assume that the same f_α and f_β apply to the different perturbations of the same reactions (α, α' or β, β').

It should be noted that, for any one of the three sets of four constants in Eqs. (37.2)–(37.7),

$$\frac{\alpha_{ij}\beta_{ij}}{\alpha'_{ij}\beta'_{ij}} = \frac{\alpha\beta}{\alpha'\beta'} = e^{X/kT} \qquad (ij = 11,\ 12,\ 22). \tag{37.8}$$

That is, the neighbor perturbations alter the rate constants (and hence the steady-state flux), but they have no effect on X. This is to be expected because X is the free-energy difference in the small-molecule reaction catalyzed by the units (enzymes); the value of X is *not* a property of the units themselves.

Let $\overline{N}_{ijk}$ be the mean number of triplets of type ijk at steady state. Because of symmetry ($w_{12} = w_{21}$), $\overline{N}_{112} = \overline{N}_{211}$ and $\overline{N}_{122} = \overline{N}_{221}$. The total number of triplets is M. Then the mean steady-state flux per unit, J, can be expressed in terms of either the α, α' reaction in the cycle or the β, β' reaction:

$$MJ = (\alpha_{11}\overline{N}_{111} - \alpha'_{11}\overline{N}_{121}) + 2(\alpha_{12}\overline{N}_{112} - \alpha'_{12}\overline{N}_{122}) + (\alpha_{22}\overline{N}_{212} - \alpha'_{22}\overline{N}_{222}) \tag{37.9}$$

$$= (\beta_{11}\overline{N}_{121} - \beta'_{11}\overline{N}_{111}) + 2(\beta_{12}\overline{N}_{122} - \beta'_{12}\overline{N}_{112}) + (\beta_{22}\overline{N}_{222} - \beta'_{22}\overline{N}_{212}). \tag{37.10}$$

These two expressions for J must be equal at an arbitrary steady state, but corresponding parenthetical expressions (e.g., α_{11}. . . and β_{11}. . .) need not be equal. Of course, at equilibrium ($X = 0$), *each* parenthetical expression would be equal to zero (detailed balance). Equations (37.9) and (37.10) cannot actually be used as they stand, without knowledge of the $\bar{N}_{ijk}$.

The above introduction is quite general. However, to make further progress analytically,[2] we limit the remainder of this subsection to the special case $f_\alpha + f_\beta = 1$. Without this limitation, we would have to resort to Monte Carlo calculations. As explained in detail in the discussion of Eqs. (27.52)–(27.57), when $f_\alpha + f_\beta = 1$, the steady-state distribution of states (θ, $\bar{N}_{ij}$, $\bar{N}_{ijk}$, etc.) is a quasiequilibrium distribution. In fact, the equilibrium Eqs. (34.60)–(34.74) apply here, except that we must replace x_o in those equations by

$$x = (\alpha + \beta')/(\beta + \alpha'). \tag{37.11}$$

This follows because the unperturbed ratio of state populations, p_2/p_1, is $x_o = K_o c$ in the equilibrium problem but is x [Eq. (37.11)] in the steady-state problem. This correspondence has been pointed out several times before [e.g., see Eq. (16.52)].

We return now to Eqs. (37.9) and (37.10) and eliminate $\bar{N}_{121}$, $\bar{N}_{122}$, and $\bar{N}_{222}$ using Eqs. (34.73) and (34.74):

$$\bar{N}_{121} = \frac{y_{12}^2 x \bar{N}_{111}}{y_{11}^2}, \qquad \bar{N}_{122} = \frac{y_{22} x \bar{N}_{112}}{y_{11}}, \qquad \bar{N}_{222} = \frac{y_{22}^2 x \bar{N}_{212}}{y_{12}^2}. \tag{37.12}$$

In Eqs. (37.9) and (37.10), we also use

$$\alpha - \alpha' x = \beta x - \beta' = \frac{\alpha\beta - \alpha'\beta'}{\beta + \alpha'} = (1 + x) J_o. \tag{37.13}$$

These are all equivalent expressions involving the unperturbed rate constants. Both Eqs. (37.9) and (37.10) then lead to

$$\begin{aligned} J &= \left(\frac{\alpha\beta - \alpha'\beta'}{\beta + \alpha'}\right)\left[\left(\frac{y_{12}}{y_{11}}\right)^{2f_\alpha}\frac{\bar{N}_{111}}{M} + 2\left(\frac{y_{22}}{y_{11}}\right)^{f_\alpha}\frac{\bar{N}_{112}}{M} + \left(\frac{y_{22}}{y_{12}}\right)^{2f_\alpha}\frac{\bar{N}_{212}}{M}\right] \\ &= \left(\frac{\alpha\beta - \alpha'\beta'}{\beta + \alpha'}\right)\frac{\{\ \}}{\sqrt{\ }\,[\ ]}, \end{aligned} \tag{37.14}$$

where, in the final form, we have introduced Eqs. (34.64), (34.65), and (34.67), $\sqrt{\ }$ and [ ] are given in Eq. (34.71) (with x in place of x_o), and

$$\begin{aligned} \{\ \} = &\left(\frac{y_{12}}{y_{11}}\right)^{2f_\alpha} y_{11}^2 (y_{11} - y_{22}x + \sqrt{\ }) + 4\left(\frac{y_{22}}{y_{11}}\right)^{f_\alpha} y_{11} y_{12}^2 x \\ &+ \left(\frac{y_{22}}{y_{12}}\right)^{2f_\alpha} y_{12}^2 x (y_{22}x - y_{11} + \sqrt{\ }). \end{aligned} \tag{37.15}$$

The corresponding expression for θ (fraction of units in state 2) is [Eq. (34.60)]

$$\theta = \frac{2y_{12}^2 x}{\sqrt{\ }\,(y_{11} - y_{22}x + \sqrt{\ })}. \tag{37.16}$$

This gives $\theta = \frac{1}{2}$ when x has the value y_{11}/y_{22}.

The reader will have noticed, in deriving the first line of Eq. (37.14) for this quasiequilibrium special case, that corresponding parenthetical expressions (e.g., α_{11}. . . and β_{11}. . .) are indeed equal. This occurs here because there is a *combined* detailed balance at steady state in 111 ⇄ 121 [Eq. (37.2)], etc., between $\alpha_{11} + \beta'_{11}$ transitions and $\beta_{11} + \alpha'_{11}$ transitions, etc. That is,

$$(\alpha_{11} + \beta'_{11})\overline{N}_{111} = (\beta_{11} + \alpha'_{11})\overline{N}_{121}, \tag{37.17}$$

etc. This relation would not hold in general if $f_\alpha + f_\beta \neq 1$.

The ratio J/J_o turns out to be relatively simple at $\theta = \frac{1}{2}$, i.e., at $x = y_{11}/y_{22}$. If we put this value of x in Eq. (37.14), we find, after simplification, that

$$\frac{J}{J_o} = \frac{1}{2}\left(1 + \frac{y_{11}}{y_{22}}\right)\left[\frac{\left(\frac{y_{12}}{y_{11}}\right)^{f_\alpha} + \left(\frac{y_{22}}{y_{12}}\right)^{f_\alpha} y^{-1/2}}{1 + y^{-1/2}}\right]^2, \tag{37.18}$$

where $y = y_{11}y_{22}/y_{12}^2$, as usual. Here J_o is not a constant; it also varies as x varies.

Equation (37.14), with Eq. (37.15), is the desired expression for the mean flux per unit in a long linear array of interacting units. The flux J is given as an explicit function of the unperturbed rate constants (in the packages x and J_o), the y_{ij}, and f_α. The results apply only to the special case $f_\alpha + f_\beta = 1$. Numerical illustrations will be given below. It should be mentioned that there are alternative ways to write Eq. (37.14) for J in which { }/$\sqrt{\ }$ [ ] is a function of θ instead of x [put $z = 2$ in Eqs. (20) and (21) of Ref. 3]. However, the independent variables in Eq. (37.14) are the most natural and convenient.

Because of the numerous parameters in Eq. (37.14) for J, there are many special cases. We mention a few in the following. If $y_{11} = y_{12} = y_{22}$ (i.e., state changes have no influence on interactions), then

$$J = J_o, \qquad \theta = x/(1 + x). \tag{37.19}$$

These are properties of unperturbed units.

If $y_{12}^2 = y_{11}y_{22}$ (i.e., $y = 1$), then

$$J = J_o\left(\frac{y_{22}}{y_{11}}\right)^{f_\alpha}\frac{y_{11}(1 + x)}{y_{11} + y_{22}x}, \tag{37.20}$$

$$\theta = \frac{y_{22}x}{y_{11} + y_{22}x}. \tag{37.21}$$

In this case w_{12} is the average of w_{11} and w_{22}. If $w_{11} \neq w_{22}$, this is not the same as Eq. (37.19).

If $y_{11} = y_{22}$ (but y_{12} is arbitrary), then

$$J = \left(\frac{\alpha\beta - \alpha'\beta'}{\beta + \alpha'}\right)\frac{y^{-1}[y^{1-f_\alpha}(1 - x + \sqrt{\ }) + 4x + y^{f_\alpha}x(x - 1 + \sqrt{\ })]}{\sqrt{\ }[1 + 2y^{-1}x + x^2 + (1 + x)\sqrt{\ }]}, \tag{37.22}$$

$$\theta = \frac{2y^{-1}x}{\sqrt{\ }(1 - x + \sqrt{\ })}, \tag{37.23}$$

where

$$y = \frac{y_{11}y_{22}}{y_{12}^2} = \left(\frac{y_{11}}{y_{12}}\right)^2 = \left(\frac{y_{22}}{y_{12}}\right)^2, \tag{37.24}$$

$$\sqrt{\ } = [(1 - x)^2 + 4y^{-1}x]^{1/2}. \tag{37.25}$$

In this case $\theta = \frac{1}{2}$ when $x = 1$.

In the final special case to be mentioned, we take $y_{11} = y_{12}$, but y_{22} is in general different. Then we find

$$J = \left(\frac{\alpha\beta - \alpha'\beta'}{\beta + \alpha'}\right)\frac{[1 - yx + \sqrt{\ } + 4y^{f_\alpha}x + y^{2f_\alpha}x(yx - 1 + \sqrt{\ })]}{\sqrt{\ }[1 + 2x + y^2x^2 + (1 + yx)\sqrt{\ }]}, \tag{37.26}$$

$$\theta = \frac{2x}{\sqrt{\ }(1 - yx + \sqrt{\ })}, \tag{37.27}$$

where

$$y = \frac{y_{11}y_{22}}{y_{12}^2} = \frac{y_{22}}{y_{12}} = \frac{y_{22}}{y_{11}}, \tag{37.28}$$

$$\sqrt{\ } = [(1 - yx)^2 + 4x]^{1/2}. \tag{37.29}$$

Here $\theta = 1/2$ when $x = 1/y$. Equation (37.27) is the same as Eq. (34.82).

The special case of Eqs. (37.26) and (37.27) in which $y = 0$ corresponds to $w_{22} \to \infty$. That is, there is a strong repulsion (e.g., a very bad fit) between nearest neighbors both in state 2. Hence 22 pairs do not occur; at most only one-half of the units can be in state 2 at any one time, and this limit can be achieved only if the states alternate, 1212. . . . Equation (17.59) relates to the same model for a finite ring of m units. Here the number of units, M, is very large. If we put $y = 0$ in Eqs. (37.26) and (37.27), we find

$$\theta = \frac{2x}{(1 + 4x)^{1/2}[1 + (1 + 4x)^{1/2}]}, \tag{37.30}$$

$$J = \left(\frac{\alpha\beta - \alpha'\beta'}{\beta + \alpha'}\right)\frac{\theta}{x}. \tag{37.31}$$

The equation for J agrees with Eq. (17.59) and θ is the same as in Eqs. (34.82) (with $y = 0$) and (35.37).

We conclude this subsection with a numerical illustration of Eqs. (37.26) and (37.27). We consider a Michaelis–Menten enzyme, E. In this case a unit (E) catalyzes the reaction S → P, state 1 is E, state 2 is ES, $\beta' = 0$, and $\alpha = \alpha^* c_S$ (α^* is second order). Then

$$x = \frac{\alpha}{\beta + \alpha'}, \qquad \frac{\alpha\beta - \alpha'\beta'}{\beta + \alpha'} = \frac{\alpha\beta}{\beta + \alpha'} = \beta x. \tag{37.32}$$

Thus, in Eq. (37.26), x is proportional to c_S and the leading parentheses $( ) = \beta x$. It is therefore convenient to plot J/β as a function of x. The solid curves in Fig. 9-2 show J/β as a function of x (proportional to c_S) for $y = 0$, 1, 10, and 40, all with $f_\alpha = \frac{1}{2}$. As usual, $y > 1$ corresponds to positive cooperativity. The limiting value ($x \to \infty$) of J/β when $y > 0$ is $y^{2(f_\alpha - 1)}$. Hence $J/\beta \to 1/y$ for the solid $y = 1, 10, 40$ curves (for which $f_\alpha = \frac{1}{2}$). In this limit ($x \to \infty$) we have $\theta \to 1$ (all state 2), and the dominant rate constant in the cycle is β_{22} in Eq. (37.7): the flux J is determined by the rate at which $2 \to 1$ in an all-2 array. For both $y = 0$ and $y = 1$, $J/\beta = \theta(x)$. For $y = 1$, $\theta = x/(1 + x)$. The broken curve in Fig. 9-2 illustrates the effect of f_α: this curve is for $y = 10$, $f_\alpha = 0$. Because $\theta(x)$ does not depend on f_α, J/β, for $y = 0$ and $y = 1$, is independent of f_α. Figure 9-3 shows $\theta(x)$ for $y = 0, 1,$ 10, and 40. The limiting value of θ ($x \to \infty$) when $y = 0$ is $\theta = 1/2$. In the other cases $\theta = 1/2$ is reached at $x = 1/y$. Conclusion: except at small x for positive cooperativity, the unit flux J is inhibited by either positive or negative cooperativity.

False Equilibrium for One-Way Cycle

There is also an equilibrium-like solution of the steady-state distribution problem in the special case of a one-way cycle ($\alpha' = 0$, $\beta' = 0$), even if

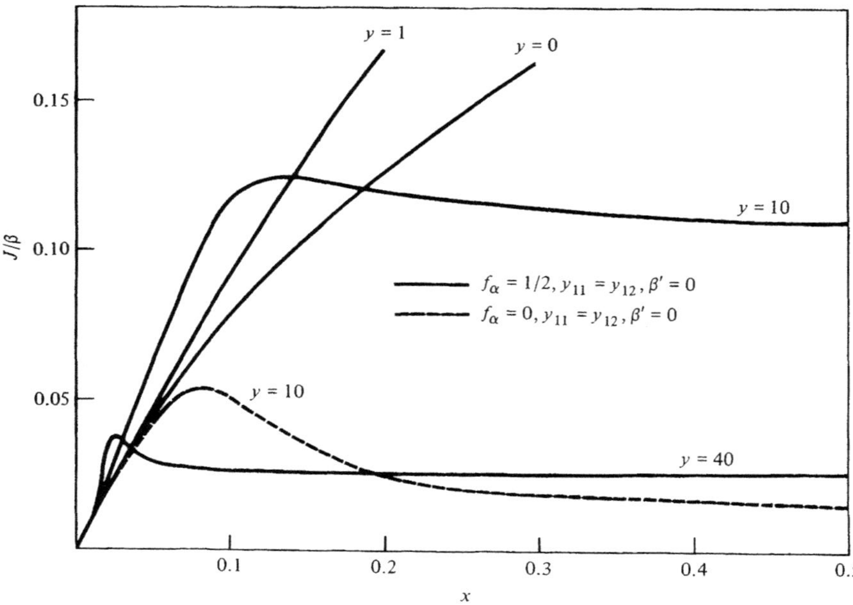

Fig. 9-2. Plot of J/β as a function of x (proportional to c_S) for a one-dimensional array of Michaelis–Menten enzyme molecules, from Eq. (37.26).

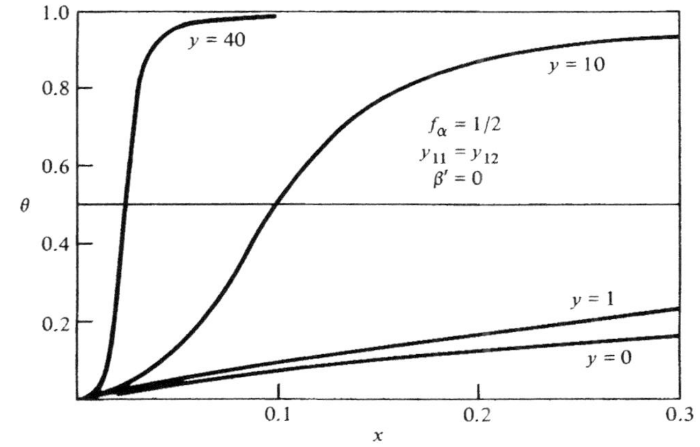

Fig. 9-3. Plot of θ against x [Eq. (37.27)] corresponding to the solid curves in Fig. 9-2.

$f_\alpha + f_\beta \neq 1$, that is, for any values of f_α and f_β. The argument was presented in connection with Eq. (27.59), and should be reviewed.

For a one-way cycle, $x = \alpha/\beta$ [Eq. (37.11)]. The steady-state distribution properties are given in this case, for arbitrary f_α and f_β, by Eqs. (34.60)–(34.74) provided that we replace x_o everywhere in these equations by $x = \alpha/\beta$ and replace each y_{ij} everywhere by $y_{ij}^{f_\alpha+f_\beta}$. The $\overline{N}_{ijk}$, modified in this way, then give the steady-state flux [compare Eqs. (37.9) and (37.10)]:

$$MJ = \alpha_{11}\overline{N}_{111} + 2\alpha_{12}\overline{N}_{112} + \alpha_{22}\overline{N}_{212} \tag{37.33}$$

$$= \alpha\left(\frac{y_{12}}{y_{11}}\right)^{2f_\alpha}\overline{N}_{111} + 2\alpha\left(\frac{y_{22}}{y_{11}}\right)^{f_\alpha}\overline{N}_{112} + \alpha\left(\frac{y_{22}}{y_{12}}\right)^{2f_\alpha}\overline{N}_{212} \tag{37.34}$$

$$= \beta_{11}\overline{N}_{121} + 2\beta_{12}\overline{N}_{122} + \beta_{22}\overline{N}_{222} \tag{37.35}$$

$$= \beta\left(\frac{y_{11}}{y_{12}}\right)^{2f_\beta}\overline{N}_{121} + 2\beta\left(\frac{y_{11}}{y_{22}}\right)^{f_\beta}\overline{N}_{122} + \beta\left(\frac{y_{12}}{y_{22}}\right)^{2f_\beta}\overline{N}_{222}. \tag{37.36}$$

If, in Eq. (37.36), we substitute $\beta = \alpha/x$,

$$\overline{N}_{121} = \left(\frac{y_{12}}{y_{11}}\right)^{2(f_\alpha+f_\beta)} x\overline{N}_{111}, \tag{37.37}$$

and similar expressions for $\overline{N}_{122}$ and $\overline{N}_{222}$ [from Eqs. (34.73) and (34.74)], we can verify that Eqs. (37.34) and (37.36) are indeed equivalent. Thus either of these equations may be used to calculate J, remembering that, in $\overline{N}_{ijk}$ [Eqs. (34.64)–(34.69)], $x_o \rightarrow x = \alpha/\beta$ and $y_{ij} \rightarrow y_{ij}^{f_\alpha+f_\beta}$. The same substitutions for x_o and y_{ij} are of course made in Eq. (34.60) to obtain θ.

It should be noted that Eqs. (37.34) and (37.36) are not only equivalent, they are equivalent term by term. That is, there is a "detailed balance" in this false equilibrium. For example,

$$\alpha_{11}\overline{N}_{111} = \beta_{11}\overline{N}_{121},$$
$$\frac{\overline{N}_{121}}{\overline{N}_{111}} = \frac{\alpha}{\beta}\left(\frac{y_{12}}{y_{11}}\right)^{2(f_\alpha + f_\beta)}. \tag{37.38}$$

If α and β were inverse rate constants, which they are not, in a true equilibrium, the perturbation factor multiplying α/β would be $(y_{12}/y_{11})^2$, as for α_{11}/α'_{11} in Eq. (37.2).

38. The Exact Linear Flux-Force Coefficient

Exact analytical results were presented in Section 37 for arbitrary steady states, but only for certain special parameter cases: $f_\alpha + f_\beta = 1$; or a one-way cycle ($\alpha' = 0$, $\beta' = 0$). In this section we obtain another exact analytical result. This is not limited to any special parameter case but is limited to steady states near equilibrium. With minor exceptions, we use the notation in the original paper on this subject.[4] However, the argument we give here is more straightforward.

The system or model is the same as in Section 37. Near equilibrium, J and X are both small and are related by

$$J = LX + \ldots, \tag{38.1}$$

where terms in $X^2, X^3, \ldots$ are negligible and L is a kinetic property of the equilibrium state itself. The last comment follows from the expansion

$$J = \left(\frac{\partial J}{\partial X}\right)_{X=0} X + \frac{1}{2}\left(\frac{\partial^2 J}{\partial X^2}\right)_{X=0} X^2 + \cdots. \tag{38.2}$$

Indeed, all of the coefficients in this expansion are kinetic properties of the equilibrium state, $X = 0$. However, we consider only the linear coefficient $L = (\partial J/\partial X)_{X=0}$. In fact, higher coefficients are not unique[5]: they depend on which rate constant (or constants) is varied in order to produce the departure from equilibrium ($X \neq 0$).

The unperturbed flux, J_o, in Eq. (37.1) can be written (see Section 9)

$$J_o = J_+ - J_-, \tag{38.3}$$

$$J_+ = \frac{\alpha\beta}{\alpha + \alpha' + \beta + \beta'}, \qquad J_- = \frac{\alpha'\beta'}{\alpha + \alpha' + \beta + \beta'}, \tag{38.4}$$

where J_+ and J_- are one-way cycle fluxes, i.e., the rates of cycle completions in + and − directions. (These fluxes are not to be confused with the flux in a one-way cycle, for which $\alpha' = 0$, $\beta' = 0$.) Then, from Eq. (37.1),

$$J_o = \frac{\alpha\beta(1 - e^{-X/kT})}{\alpha + \alpha' + \beta + \beta'} = J_+(1 - e^{-X/kT}). \tag{38.5}$$

Near equilibrium, upon expanding the exponential,

$$J_o = J_\pm^e\left(\frac{X}{kT}\right) = LX, \tag{38.6}$$

$$J_\pm^e \equiv J_+^e = J_-^e, \qquad L = J_\pm^e/kT. \tag{38.7}$$

At equilibrium, the two one-way cycle fluxes are equal; L is seen to be the equilbrium one-way cycle flux[6,7] (either direction), divided by kT.

At equilibrium, in Fig. 9-1,

$$X = 0, \qquad \alpha\beta = \alpha'\beta', \qquad \frac{\alpha}{\alpha'} = \frac{\beta'}{\beta} = K = \frac{p_2^e}{p_1^e}. \tag{38.8}$$

We use K here as a noncommittal equilibrium constant between states $1 \rightleftarrows 2$ at equilibrium (i.e., for generality, we avoid specification of the biochemical details implicit in Fig. 9-1). Then

$$J_\pm^e = \left(\frac{\alpha\beta}{\alpha + \alpha' + \beta + \beta'}\right)^e = \frac{\alpha\beta K}{(1 + K)(\alpha + \beta K)}, \tag{38.9}$$

where we have eliminated α' and β' using Eq. (38.8), and α and β are understood to be rate constants at equilibrium.

The above relates to unperturbed units. We turn now to the linear array of M interacting units, *at equilibrium*. There are three kinds of two-state cyclic activity in the units, at equilibrium, with rate constants as in Eqs. (37.2)–(37.7). These three cycle types are also represented in Fig. 9-4. Our procedure will be to calculate the equilibrium one-way cycle flux, $J_\pm^e$, for each of these three types of cycle, just as in Eq. (38.9), except that we must use the appropriate perturbed expressions for α, β, and K in each case. Equation (37.8) shows that the force (near equilibrium, here) is the same in all three cycles. Then, upon combining these three $J_\pm^e$, with weights according to the number of triplets of each class, we can deduce the total $J_\pm^e$ for the whole array of M units or the mean $J_\pm^e$ per unit. The total $J_\pm^e$ gives the desired

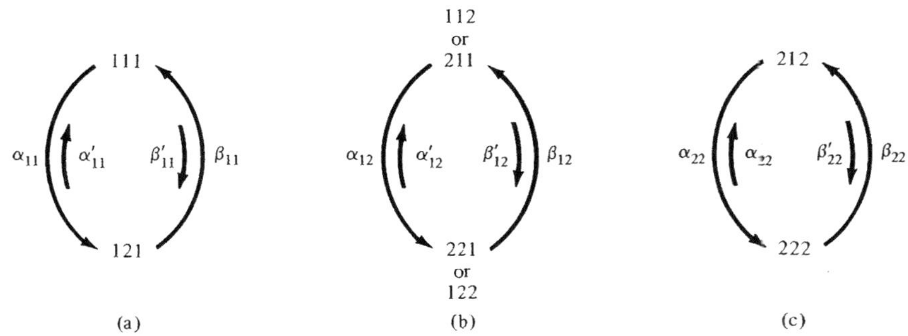

Fig. 9-4. Three kinds of cycles for the central molecule of triplets. The rate constants are given in Eqs. (37.2)–(37.7).

total flux near equilibrium because X is the same for all cycle types:

$$MJ = (\Sigma J^{e}_{\pm})\left(\frac{X}{kT}\right) = MLX, \tag{38.10}$$

where $\Sigma J^{e}_{\pm}$ represents the total $J^{e}_{\pm}$. L follows immediately from this relation.

Because we are considering cyclic activity *at equilibrium*, Eqs. (34.64)–(34.74) relating to the $\overline{N}_{ijk}$ apply here, except that $x_o = K_o c$ is replaced by the more general K in Eq. (38.8). Note that Eq. (37.11) becomes

$$x = \frac{\alpha + \beta'}{\beta + \alpha'} = K. \tag{38.11}$$

It is also convenient to introduce the following notation for interactions:

$$u = \frac{y_{11}}{y_{12}}, \qquad v = \frac{y_{22}}{y_{12}}, \qquad \frac{v}{u} = \frac{y_{22}}{y_{11}}, \qquad uv = y. \tag{38.12}$$

The mean number of triplets engaged in cyclic activity of the type in Fig. 9-4(a) is

$$\overline{N}_{111} + \overline{N}_{121} = \overline{N}_{111}(1 + Ku^{-2}), \tag{38.13}$$

having used Eq. (34.73). The perturbed equilibrium constant for these cycles (in place of K) is

$$\frac{\alpha_{11}}{\alpha'_{11}} = \frac{\beta'_{11}}{\beta_{11}} = \frac{\alpha}{\alpha'}\left(\frac{y_{12}}{y_{11}}\right)^2 = Ku^{-2}. \tag{38.14}$$

Then, following Eq. (38.9),

$$\begin{aligned}(\text{a}):\ J^{e}_{\pm} &= \frac{\alpha_{11}\beta_{11}Ku^{-2}}{(1 + Ku^{-2})(\alpha_{11} + \beta_{11}Ku^{-2})} \\ &= \frac{\alpha\beta u^{2(f_\beta - f_\alpha)}Ku^{-2}}{(1 + Ku^{-2})(\alpha u^{-2f_\alpha} + K\beta u^{2f_\beta - 2})}.\end{aligned} \tag{38.15}$$

For cycles as in Fig. 9-4(b), the corresponding relations are

$$2(\overline{N}_{112} + \overline{N}_{122}) = 2\overline{N}_{112}[1 + (v/u)K], \tag{38.16}$$

$$\frac{\alpha_{12}}{\alpha'_{12}} = \frac{\beta'_{12}}{\beta_{12}} = \frac{Kv}{u}, \tag{38.17}$$

$$\begin{aligned}(\text{b}):\ J^{e}_{\pm} &= \frac{\alpha_{12}\beta_{12}K(v/u)}{[1 + K(v/u)][\alpha_{12} + \beta_{12}K(v/u)]} \\ &= \frac{\alpha\beta(v/u)^{f_\alpha - f_\beta}K(v/u)}{[1 + K(v/u)][\alpha(v/u)^{f_\alpha} + K\beta(v/u)^{1-f_\beta}]}.\end{aligned} \tag{38.18}$$

Similarly, for Fig. 9-4(c),

$$\overline{N}_{212} + \overline{N}_{222} = \overline{N}_{212}(1 + Kv^2), \tag{38.19}$$

$$\frac{\alpha_{22}}{\alpha'_{22}} = \frac{\beta'_{22}}{\beta_{22}} = Kv^2, \tag{38.20}$$

$$\text{(c): } J^e_\pm = \frac{\alpha_{22}\beta_{22}Kv^2}{(1 + Kv^2)(\alpha_{22} + \beta_{22}Kv^2)}$$
$$= \frac{\alpha\beta v^{2(f_\alpha - f_\beta)}Kv^2}{(1 + Kv^2)(\alpha v^{2f_\alpha} + K\beta v^{2-2f_\beta})}. \tag{38.21}$$

The total equilibrium one-way cycle flux, as in Eq. (38.10), is then

$$\Sigma J^e_\pm = \overline{N}_{111}(1 + Ku^{-2})J^e_\pm(\text{a}) + 2\overline{N}_{112}[1 + (v/u)K]J^e_\pm(\text{b})$$
$$+\overline{N}_{212}(1 + Kv^2)J^e_\pm(\text{c}). \tag{38.22}$$

We rewrite Eqs. (34.64), (34.65), and (34.67) as

$$\frac{\overline{N}_{111}}{M} = \frac{u^2(u - vK + \sqrt{\ })}{\sqrt{\ }[\,]}, \tag{38.23}$$

$$\frac{\overline{N}_{112}}{M} = \frac{2uK}{\sqrt{\ }[\,]}, \tag{38.24}$$

$$\frac{\overline{N}_{212}}{M} = \frac{K(vK - u + \sqrt{\ })}{\sqrt{\ }[\,]}, \tag{38.25}$$

$$\sqrt{\ } = [(u - vK)^2 + 4K]^{1/2},$$

$$[\,] = u^2 + 2K + v^2K^2 + (u + vK)\sqrt{\ }.$$

If we now substitute Eqs. (38.15), (38.18), (38.21), and (38.23)–(38.25) into Eq. (38.22), we finally find, for L,

$$L = (\Sigma J^e_\pm)/MkT$$
$$= \frac{\alpha\beta K}{\sqrt{\ }[\,]kT}\left[\frac{u(u - vK + \sqrt{\ })u^{2(f_\beta - f_\alpha)}}{\alpha u^{1-2f_\alpha} + K\beta u^{2f_\beta - 1}}\right.$$
$$+ \frac{4K(uv)^{1/2}(v/u)^{f_\alpha - f_\beta}}{\alpha(v/u)^{f_\alpha - (1/2)} + K\beta(v/u)^{(1/2) - f_\beta}}$$
$$\left.+ \frac{vK(vK - u + \sqrt{\ })v^{2(f_\alpha - f_\beta)}}{\alpha v^{2f_\alpha - 1} + K\beta v^{1-2f_\beta}}\right]. \tag{38.26}$$

In the special case $u = v = 1$ (i.e., $y_{11} = y_{12} = y_{22}$), Eq. (38.26) reduces to the unperturbed L in Eqs. (38.7) and (38.9).

If the enzyme units catalyze S → P [as in Fig. 7-20(a)], then Eqs. (8.29)–(8.32) apply and

$$K = \frac{\alpha^* c^e_S}{\alpha'} = \frac{\beta^* c^e_P}{\beta},$$
$$\frac{c^e_P}{c^e_S} = K_{SP} = \frac{\alpha^*\beta}{\alpha'\beta^*}. \tag{38.27}$$

39. Simple Model for Regulation of Muscle Contraction by Calcium

The isometric mechanical force that can be generated by a muscle fiber rises very sharply with Ca^{2+} concentration, between 10^{-7}M and 10^{-5}M. The objective in this section is to use as simple a model as possible to sketch how this strong cooperative effect might arise. Even with this simple model we omit the final steps in the calculation because they would be of interest only to the specialist. The model is much simpler than the realistic one used in Section 36, but the source of cooperativity is the same: interaction of ends of tropomyosin molecules on actin filaments. In fact, here we shall use a simplified version of the last subsection of Section 36, which the reader should review.

The original paper[8] gives many more details and also a second, more complicated, model. We retain most of the notation of the original paper.

There are approximately two myosin molecules (from a myosin filament) per seven actin binding sites for myosin. To make our calculation simple, we shall assume here that there is only one myosin molecule per seven actin sites, and that only one S1 head of this molecule can attach to only one of these seven sites (the central one, as in Fig. 9-5). The corresponding model with two myosin molecules per seven actin sites is included in the original paper.[8]

One Tm-Tn (tropomyosin-troponin) unit includes one of each of these molecules (Tm, Tn) and seven actin subunits. A further simplification we make in the present model is that Tn has only one regulatory site for binding Ca^{2+} (instead of two). Either the binding of Ca^{2+} on the Tn of a unit or the attachment of myosin to the central actin of the unit will alter (we assume) the interaction of this unit with its two nearest-neighbor Tm-Tn units, as described below. We turn now to the details of this stripped-down model which, nevertheless, still retains the essential regulatory ingredients.

Equilibrium Attachment of Myosin

First we consider an equilibrium version of the model and then turn to the steady-state case. This procedure is useful because the steady-state treatment has a close formal resemblance to the equilibrium treatment.

Figure 9-6 illustrates the properties we assume for Tm-Tn when Ca^{2+} or S1, or both, are bound or attached. The properties are the following. Tm-Tn

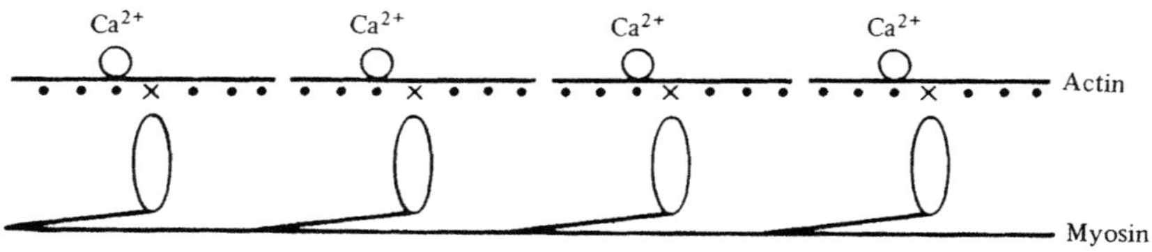

Fig. 9-5. Simplified actin-myosin model used in calculation of isometric force.

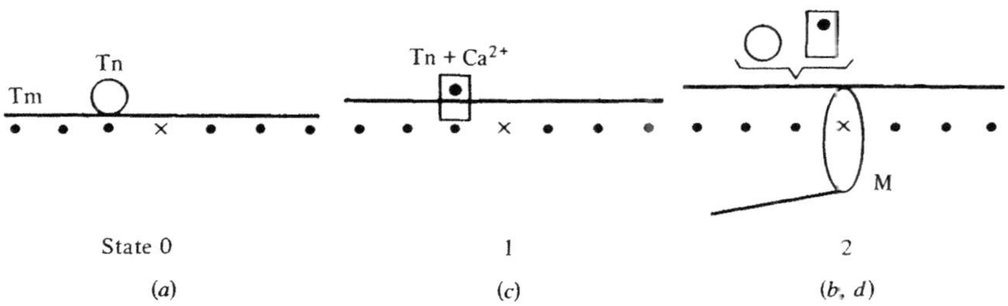

Fig. 9-6. Illustration of the properties of Tm-Tn when Ca^{2+} or M is bound. See text for details. The state labels a, b, c, d are from Fig. 9-7.

blocks the binding sites on the actin filament. When Ca^{2+} is bound, but not S1, there is a conformational change in Tn that moves Tm relative to the actin sites. This movement of Tm requires expenditure of free energy, which would be reflected in the binding constant of Ca^{2+}. Tm is moved even further, with a greater cost in free energy, when S1 is attached to a site of the unit, in the absence of Ca^{2+}. If the unit already has a Ca^{2+} bound, attachment of S1 requires less work against Tm: the attachment constant of S1 is correspondingly larger. Once S1 is attached to a unit, binding of Ca^{2+} requires no further work against Tm: this binding constant for Ca^{2+} is therefore maximal.

As will be seen below, the two essential ingredients in the regulation part of the problem, for the model adopted here, are: (a) Ca^{2+} binds more strongly to Tn if S1 is already attached to actin or, equivalently, S1 attaches more strongly to actin if Ca^{2+} is already bound to Tn; and (b) there is strong positive cooperativity in the system owing to nearest-neighbor Tm-Tm interactions. The latter property sharpens the response to the former.

The system of interest is the one-dimensional array of units in Fig. 9-5, with nearest-neighbor Tm-Tm interactions. An isolated individual unit (no neighbor interactions) can exist in four different states (a, b, c, d), as shown in Fig. 9-7. When Ca^{2+} is not bound to Tn (Fig. 9-7), the attachment equilibrium constant of M (myosin cross-bridge) to A (the specific actin site, x)

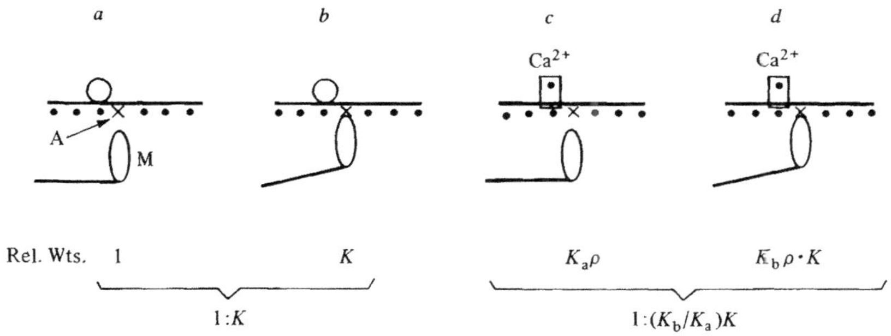

Fig. 9-7. Four different states of an individual unit in Fig. 9-5. Not all of the details in Fig. 9-6 are shown. The relative weights are terms in ξ, Eq. (39.1).

is K. This is a dimensionless constant (M is part of the myofilament structure and does not have a concentration) for the process $a \to b$. Included in K (i.e., reducing its value) is the work necessary to move Tm (Fig. 9-6). When M is not attached to A (Fig. 9-7), the binding constant of Ca^{2+} to Tn is K_a (process $a \to c$). The concentration of Ca^{2+} is ρ. When M is attached to A, the binding constant of Ca^{2+} to Tn is K_b (process $b \to d$). It is much easier to bind Ca^{2+} when M is attached (Fig. 9-6): $K_b \gg K_a$. We shall use the ratio $K_r \equiv K_b/K_a = 20$, as a realistic example.

The grand partition function for an isolated unit is then

$$\begin{array}{lllll} \xi = 1 + & K + & K_a\rho + & K_b\rho \cdot K, & \\ \text{State:} \quad a, 0 & b, 2 & c, 1 & d, 2 \end{array} \tag{39.1}$$

where the individual terms are proportional to the populations of the corresponding states (Fig. 9-7) at equilibrium. Thus, because $K_b \gg K_a$, M attaches to A much more readily when Ca^{2+} is bound to Tn (see, also, Fig. 9-6): when Ca^{2+} is *not* bound, the attachment constant of M is K; when Ca^{2+} *is* bound, the attachment constant of M is K_rK (Fig. 9-7).

At an arbitrary ρ, the probability θ that M is attached, in an isolated unit, follows from Eq. (39.1):

$$\theta = \frac{K(1 + K_b\rho)}{\xi}. \tag{39.2}$$

The two limiting cases of interest are:

$$\theta = \frac{K}{1 + K} \qquad (\rho = 0) \tag{39.3}$$

$$= \frac{K_rK}{1 + K_rK} \qquad (\rho = \infty). \tag{39.4}$$

For example, if $K = 0.2$ and $K_r = 20$, then $\theta = 0.167$ when $\rho = 0$ and $\theta = 0.8$ when $\rho = \infty$; the increase in Ca^{2+} concentration increases the attachment of M to A.

We now consider a one-dimensional array (Fig. 9-5) of these units, at equilibrium. At this point, Tm–Tm neighbor interactions have to be introduced explicitly. Figure 9-6 is pertinent here. In this figure the states are classified (0, 1, 2) according to the position of Tm relative to the actin sites. We assume an optimal Tm–Tm interaction if two neighboring units are in the same position or state (0, 1, or 2) and less favorable interactions the larger the position difference of two neighbors.[9] This assumption is consistent with equilibrium S1 binding data. If we let w_{ij} be the interaction free energy between two Tm neighbors in states i and j (or j and i), and define $y_{ij} = e^{-w_{ij}/kT}$, then we take (in accordance with the above discussion)

$$y_{00} = y_{11} = y_{22} \equiv y, \qquad y > y_{01}, \qquad y_{12} > y_{02}. \tag{39.5}$$

The exact equilibrium properties of the one-dimensional array can be deduced using the matrix method (Chapter 8). Because of the three states 0, 1, 2, we need a 3 × 3 matrix here. States *b* and *d* in Fig. 9-7 and Eq. (39.1) are both included in state 2. The single-unit grand partition function terms, namely, 1, $K_a\rho$, and $K(1 + K_b\rho)$, for states $k = 0, 1, 2$, are assigned to the rows of the matrix. These entries refer to the *k*th unit in the linear array. The columns relate to the $k + 1$th unit. The k, $k + 1$ neighbor interaction factor must be included in each matrix element. Thus the complete 3 × 3 matrix in this case is

$$\begin{pmatrix} y & y_{01} & y_{02} \\ K_a\rho y_{01} & K_a\rho y & K_a\rho y_{12} \\ K(1 + K_b\rho)y_{02} & K(1 + K_b\rho)y_{12} & K(1 + K_b\rho)y \end{pmatrix}. \tag{39.6}$$

If γ_1 is the largest eigenvalue of this matrix, then the grand partition function for a long array of M units is $\Xi = \gamma_1^M$. Because there is a factor K in the matrix for each attached M (myosin), the equilibrium fraction of units with M attached is given by [compare Eq. (34.45)]

$$\theta = K\frac{\partial \ln \gamma_1}{\partial K}. \tag{39.7}$$

This can be evaluated numerically for arbitrary ρ [see Eq. (35.71)].

In the limiting case $\rho \to 0$ (no Ca^{2+}), the matrix simplifies to (states *a* and *b*, Fig. 9-7)

$$\begin{pmatrix} y & y_{02} \\ Ky_{02} & Ky \end{pmatrix}. \tag{39.8}$$

Here we find in the usual way (Chapter 8),

$$\theta = \frac{2KY_o^{-1}}{R_o(1 - K + R_o)}, \tag{39.9}$$

$$R_o \equiv [(1 - K)^2 + 4KY_o^{-1}]^{1/2},$$

$$Y_o \equiv \frac{y^2}{y_{02}^2} > 1,$$

where Y_o is the effective cooperativity parameter in the absence of Ca^{2+}. Values of $Y_o > 1$ [see Eq. (39.5)] correspond to positive cooperativity. We shall take $Y_o = 20$ as typical (Section 36).

In the limiting case $\rho \to \infty$ (Tn saturated with Ca^{2+}), the matrix is (states *c* and *d*, Fig. 9-7)

$$\begin{pmatrix} K_a\rho y & K_a\rho y_{12} \\ KK_b\rho y_{12} & KK_b\rho y \end{pmatrix}. \tag{39.10}$$

From this we obtain

$$\theta = \frac{2K_rKY_\infty^{-1}}{R_\infty(1 - K_rK + R_\infty)}, \tag{39.11}$$

$$R_\infty \equiv [(1 - K_rK)^2 + 4K_rKY_\infty^{-1}]^{1/2},$$

$$Y_\infty = \frac{y^2}{y_{12}^2} > 1.$$

Note that Eq. (39.11) has the same functional form as Eq. (39.9), but K_rK replaces K and Y_∞ replaces Y_0. Although $Y_\infty > 1$ (positive cooperativity), $Y_\infty < Y_0$ [Eq. (39.5)]. We take $Y_\infty = 4$ (Section 36) as a representative value.

If each element in Eq. (39.6) is divided by y, we see that in this more general case (arbitrary ρ), a third effective interaction parameter appears: $Y_1 \equiv y^2/y_{01}^2 > 1$ [Eq. (39.5)]. A simple energetic argument in relation to Fig. 9-6 suggests that $Y_1 = Y_0/Y_\infty$, so we take $Y_1 = 5$, below, as typical.

The right-hand curve in Fig. 9-8 is a plot of θ (fraction of M attached) against $\log K$, using Eq. (39.9) (i.e., in the absence of Ca^{2+}) with $Y_0 = 20$. On this curve, $\theta = \frac{1}{2}$ at $K = 1$. The left-hand curve in Fig. 9-8 shows θ as a function of $\log K$, as calculated from Eq. (39.11) (i.e., with Ca^{2+} saturation), using $K_r = 20$ and $Y_\infty = 4$. This latter curve has $\theta = \frac{1}{2}$ at $K_rK = 1$, that is, at $K = 0.05$.

The two curves in Fig. 9-8 already contain an equilibrium version of muscle regulation by Ca^{2+}. If the attachment equilibrium constant for M, in the absence of Ca^{2+}, has a value significantly less than one, e.g., $K = 0.2$, while the attachment constant for M in the presence of Ca^{2+} has a value significantly greater than one, e.g., $K_rK = 4$, then the introduction of Ca^{2+} into the system can result in a dramatic increase in the probability θ of

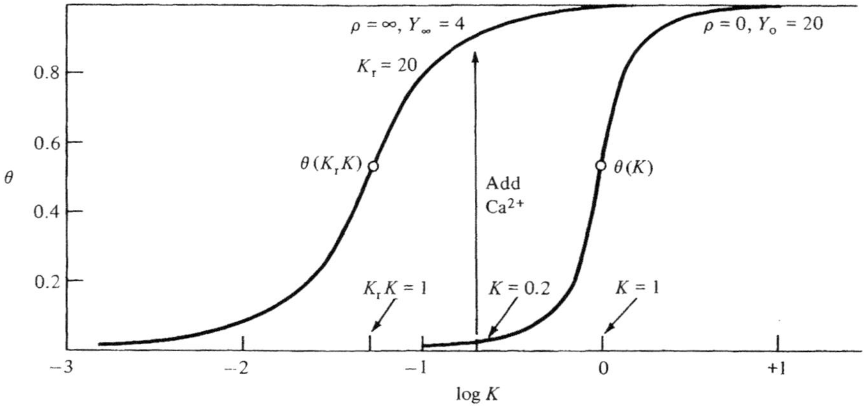

Fig. 9-8. Attachment of M as a function of log K, in the presence and absence of Ca^{2+}. See text for explanation.

attachment of M. This is illustrated by the long vertical arrow in Fig. 9-8 placed at $K = 0.2$ along the abscissa. Here θ increases from 0.015 to 0.916 when Ca^{2+} is added. Cooperativity plays a significant role in this result. In the absence of cooperativity ($Y_0 = Y_\infty = 1$), as already considered following Eq. (39.4), the corresponding increase in θ would be less impressive: $\theta = 0.167$ to 0.8. Note, also, that there would be significant attachment ($\theta = 0.167$) in the absence of Ca^{2+}.

The above effect of Ca^{2+} addition is shown in more detail in Fig. 9-9, where the solid curve is θ as a function of $\log K_a\rho$ *along* the vertical arrow in Fig. 9-8 (only the endpoints are given in Fig. 9-8). This curve (Fig. 9-9) has been calculated numerically from Eq. (39.7), using $Y_1 = 5$ as well as the other parameters already mentioned [see Eq. (35.71)].

Because a myosin molecule cannot exert force on an actin filament unless it is attached, the result in Fig. 9-9 for the equilibrium degree of attachment is what is needed qualitatively. Isometric force (*in vivo*) is activated when the Ca^{2+} concentration ρ is increased by a factor of about 100 (from 10^{-7} M to 10^{-5} M). This corresponds to 2 units on the abscissa of Fig. 9-9. The degree of cooperativity in this figure, that is, the sharpness in the rise in θ, is seen to be consistent with this observation. It should be emphasized that the values of the cooperativity parameters being used here are based on equilibrium S1 · ADP binding data (Section 36). Larger parameter values would lead to sharper curves. Indeed, the values of Y_0, Y_1, and Y_∞ are expected to depend on the conformation of the attached myosin and on the bound nucleotide in the attached state.

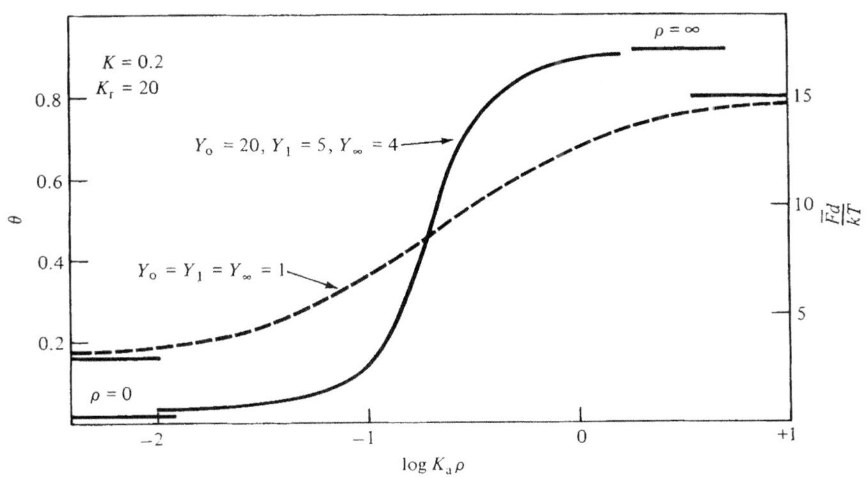

Fig. 9-9. Attachment of myosin as a function of ρ (Ca^{2+}) along the vertical arrow in Fig. 9-8. The broken curve applies in the absence of cooperativity. The right-hand scale is referred to later in the text.

The broken curve in Fig. 9-9 is what is found in the absence of cooperativity ($Y_0 = Y_1 = Y_\infty = 1$). This curve has been calculated from Eq. (39.2), using $K = 0.2$ and $K_r = 20$. It seems clearly inadequate as a basis for muscle regulation.

We turn now to the corresponding steady-state problem.

The Steady-State Isometric System

To retain maximum simplicity in the model, we adopt a two-state cycle, as shown in Fig. 9-10(a), to represent the ATPase activity of M, as well as attachment-detachment of M to actin. The schematic diagrams in Figs. 9-5 through 9-7 are still applicable, but attachment is now part of a two-state cycle and, in fact, can be achieved in principle by two different biochemical routes in Fig. 9-10(a) (release of P_i; and release of T ≡ ATP, binding of D ≡ ADP). Figures 9-10(b) and 9-10(c) show, with simpler notation, the same two-state cycle, without and with Ca^{2+} bound to Tn, and the unperturbed first-order rate constants that we adopt for these cycles for a single isolated Tm unit. These rate constants are perturbed by neighbor interactions in the array in Fig. 9-5, as discussed below.

Figures 9-11(a) and 9-11(b) present the corresponding free energy curves of the two states in the two cycles. In one complete cycle (either α, β or κ, λ), the free energy of Tm + Tn + M drops by an amount $X = \Delta\mu_{\text{ATP}}$, the free energy of hydrolysis of ATP (about 13 kcal mole^{-1}). This is the thermodynamic force that drives the cycles.

The free energy of the attached state is a function of z, a variable (usually called x) that locates the axial position of M relative to its specific actin site. The simplest choice for this function is a parabola. The minimum occurs where the attachment is optimal. $K_\alpha(z)$ is the attachment constant of M at z in Fig. 9-11(a) (no Ca^{2+}) via the α, α' transitions and $K_\beta(z)$ is the (very weak)

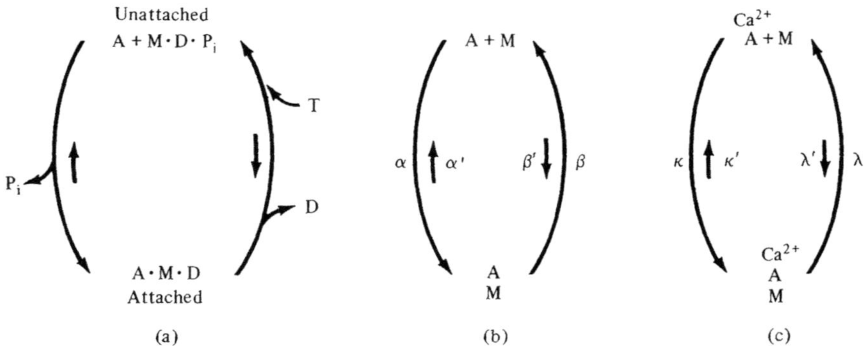

Fig. 9-10. (a) Two-state ATPase cycle, with attachment–detachment of M. D = ADP, T = ATP. Unperturbed rate constant notation for the cycle without (b) and with (c) Ca^{2+} bound.

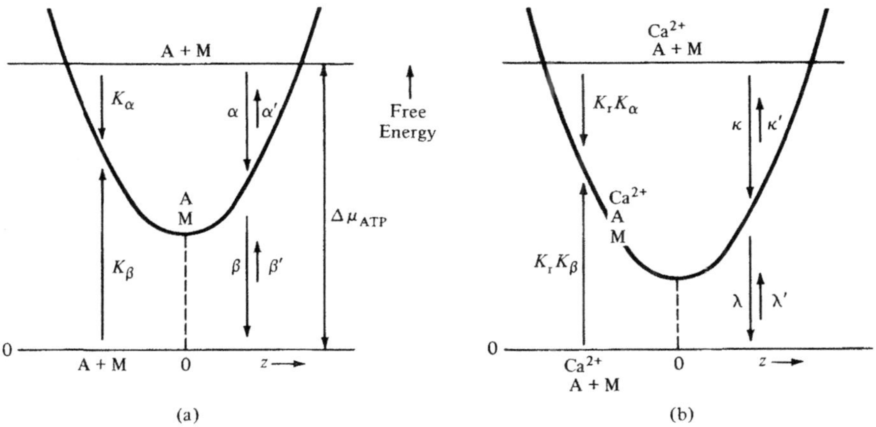

Fig. 9-11. Free-energy curves corresponding to the two cycles in Figs. 9-10(b) and 9-10(c). See text for explanation.

attachment constant via β', β. The corresponding (larger) attachment constants in Fig. 9-11(b) (Ca^{2+} bound) are $K_r K_\alpha(z)$ via the κ, κ' transitions and $K_r K_\beta(z)$ via λ', λ. $K_\alpha(z)$ and $K_\beta(z)$ are both analogues of K in Fig. 9-7. K_a is independent of z because it refers to binding of Ca^{2+} on Tn with M unattached (hence z cannot be involved). We take K_b to be a constant as well, for simplicity. With K_a and K_b chosen as constants, the two parabolas in Fig. 9-11 are the same except for vertical displacement: the left one (no Ca^{2+}) is higher by an amount $kT \ln K_r$. The deeper free-energy well in Fig. 9-11(b) corresponds to tighter attachment of M to A when Ca^{2+} is bound.

The eight rate constants in Fig. 9-11 are also, in general, all functions of z. We assume that Ca^{2+} binding and release are relatively fast processes, always at equilibrium. Hence corresponding rate constants need not be introduced.

In the isometric steady state each z value is independent, for computational purposes. This feature is what makes the steady isometric case relatively tractable.

Quasiequilibrium in the Steady State

If we now make several quite reasonable kinetic assumptions, the formal equations of the equilibrium treatment in the first subsection can also be used at steady state. This is possible because of our employment of two-state cycles.

Nearest-neighbor interactions (Fig. 9-5) will perturb the rate constants in Figs. 9-10(b) and 9-10(c). If we divide the perturbation free energy symmetrically between forward and backward transitions in all four cases, which is plausible because of the complexity of the condensed transitions in Fig.

9-10(a), this is expressed formally by taking the kinetic parameters $f_\alpha = f_\beta = f_\kappa = f_\lambda = \frac{1}{2}$.

The rate constants in Fig. 9-10(c) are also perturbations of those in Fig. 9-10(b), owing to the presence of Ca^{2+}. The free-energy effects are expressed by

$$\kappa/\kappa' = K_r(\alpha/\alpha') = K_r K_\alpha,$$
$$\lambda/\lambda' = K_r^{-1}(\beta/\beta') = 1/K_r K_\beta. \tag{39.12}$$

Physically, the presence of Ca^{2+} reduces the work M has to do against Tm in the attachment process (Fig. 9-6). The total effect should therefore appear in the rate constant for attachment. Formally, $f_\alpha^{Ca} = 1$ and $f_\beta^{Ca} = 0$, or

$$\kappa = \alpha K_r, \qquad \kappa' = \alpha',$$
$$\lambda = \beta, \qquad \lambda' = \beta' K_r, \tag{39.13}$$

for the separate rate constants.

With the relations

$$f_\alpha + f_\beta = 1, \qquad f_\kappa + f_\lambda = 1, \qquad f_\alpha^{Ca} + f_\beta^{Ca} = 1 \tag{39.14}$$

all satisfied, as above, the steady-state distribution of states in Fig. 9-5, including interactions, can be shown to have an equilibrium form [see the discussion of Eqs. (27.52)–(27.58)]. To be more specific, in Fig. 9-7 and in Eqs. (39.1)–(39.11), we merely replace K by

$$x_0(z) \equiv \frac{\alpha(z) + \beta'(z)}{\beta(z) + \alpha'(z)}, \tag{39.15}$$

just as in Eq. (37.11). No other parameter changes are needed. The analogue of the attachment constant $K_r K$ in Eqs. (39.4) and (39.11) is [using Eqs. (39.13)]

$$x_\infty(z) \equiv \frac{\kappa + \lambda'}{\lambda + \kappa'} = \frac{K_r(\alpha + \beta')}{\beta + \alpha'} = K_r x_0(z). \tag{39.16}$$

Thus K_r automatically plays the same role in the steady-state system as in the equilibrium system. As a consequence of these close analogies with the equilibrium attachment problem, many results in the first subsection carry-over to the present discussion.

Attachment at Steady State and at Equilibrium

In the equilibrium attachment of M in Fig. 9-7, the attachment constant K, with no Ca^{2+} bound, is increased to $K_r K$ when Ca^{2+} is bound. Consequently, the degree of attachment of M to A is increased by Ca^{2+}. At steady state (neglecting here, for simplicity, the reverse or primed rate constants), the quantities that determine the degree of attachment, i.e., the analogues of K

and K_rK at equilibrium, are rate constant ratios: $\alpha/\beta \cong x_o$, with no Ca^{2+} bound, and $\kappa/\lambda \cong x_\infty$ when Ca^{2+} is bound. The detachment steps, β and λ (Figs. 9-10 and 9-11), are not the inverses of the attachment steps, α and κ. Furthermore, several successive elementary steps are telescoped into both "attachment" and "detachment" in the simple two-state cycle [Fig. 9-10(a)] we are using. Thus, at steady state, the regulation of the degree of attachment of M · D to A [Fig. 9-10(a)] by Ca^{2+} is not a consequence of the effect of Ca^{2+} on the simple equilibrium binding of M · D to A (as in the first subsection) but rather of the effect of Ca^{2+} on the attachment and detachment rate constants α and β. According to Eq. (39.13), in our simple model the presence of Ca^{2+} does not affect β (i.e., $\lambda = \beta$), but α is increased by the factor K_r (i.e., $\kappa = \alpha K_r$). The rate constant α and the equilibrium constant K_α refer to the overall process: attachment of M · D · P_i to actin plus P_i release. The free-energy decrease in this process, $kT \ln K_\alpha$, is increased in magnitude by Ca^{2+} to $kT \ln K_rK_\alpha$. Here K_a refers to binding of Ca^{2+} on Tn with M · D · P_i unattached, whereas K_b refers to binding of Ca^{2+} on Tn with M · D attached to A. In the present steady-state model we need Ca^{2+} to be bound more strongly to Tn when M · D is attached to A ($K_b > K_a$). This has a kinetic effect at steady state, converting α to $\kappa = \alpha K_r$.

In the simple two-state cycle being used here, α and κ refer to attachment of M · D · P_i to actin plus P_i release. Recent experimental work[10] shows that P_i release is the Ca^{2+}-sensitive part of this overall process. A more detailed cycle would separate attachment from P_i release.

Calculation of the Isometric (Mechanical) Force

The argument so far shows that, with the assumptions made here, there is a very close parallel between the equilibrium and steady-state systems. To calculate the average mechanical force exerted by attached myosin molecules on the actin filament, still further details about the free-energy curves (Fig. 9-11) and about the z-dependence of the rate constants must be indulged in. These details would be of little interest to the general reader and are therefore omitted; they are available in the original paper[8] for specialists who care about them.

We discontinue the detailed discussion at this point and conclude here with a few general and qualitative comments, including the main result: the mechanical force exerted by an attached myosin on actin has to be averaged over z, the longitudinal position of the myosin relative to its actin site; at any z, the force exerted by an attached myosin is determined by the slope of the attached-state free-energy curve at that z; the probability of attachment, $\theta(z)$, is found by the matrix method from equilibrium-like equations [see Eqs. (39.7), (39.9), and (39.11)] because of the quasiequilibrium at steady state; the steady-state problem must be treated not only over a range in z but also over a range in ρ, the Ca^{2+} concentration [e.g., Eq. (39.6) with $x_o(z)$ in place of K]; after introducing still further simple but reasonable assumptions about the z-dependence of rate constants, the final approximate result is that the mean mechanical isometric force $\bar{F}$ depends on ρ in the same way that θ depends on ρ in Fig. 9-9 (i.e., $\bar{F}$ differs from θ, in this approximation, by a constant factor). In the right-hand scale in Fig. 9-9, d is the repeat distance 385Å. This is indeed the shape of curve observed experimentally[11] for $\bar{F}(\rho)$. In this simple treatment, steady-state attachment of myosin to actin is the essence of force generation. The experimental value of $\bar{F}d/kT$ at $\rho = \infty$ is about 30. The discrepancy of a factor of about 2 (Fig. 9-9) is presumably due to the assumption that only a single specific actin site (out of seven) is available for cross-bridge attachment.

References

1. Hill, T. L., Eisenberg, E., and Chalovich, J. M. (1981) Biophys. J. **35,** 99.
2. Hill, T. L. (1977) Proc. Natl. Acad. Sci. USA **74,** 4111.
3. Hill, T. L. (1977) Proc. Natl. Acad. Sci. USA **74,** 5227.
4. Hill, T. L. (1982) J. Chem. Phys. **76,** 1122.
5. Hill, T. L. (1979) Cell Biophys. **1,** 309.
6. Hill, T. L. (1982) Nature (London) **299,** 84.
7. Hill, T. L. (1983) Proc. Natl. Acad. Sci. USA **80,** 2589.
8. Hill, T. L. (1983) Biophys. J. **44,** 383.
9. Hill, T. L., Eisenberg, E., and Greene, L. E. (1983) Proc. Natl. Acad. Sci. USA **80,** 60.
10. Chalovich, J. M. and Eisenberg, E. (1982) J. Biol. Chem. **257,** 2432.
11. Hellam, D. C. and Podolsky, R. J. (1969) J. Physiol. (London) **200,** 807.

10

Monte Carlo Study of Equilibrium and Steady-State Two-Dimensional Lattices

This chapter follows closely published research papers on this subject. It is written primarily for statistical physicists rather than biochemists; the emphasis is on the critical point and on phase transitions. Hence a modest acquaintance with the two-dimensional equilibrium Ising problem is assumed. The Ising problem is dealt with in almost any statistical mechanics text for physicists. However, the number of exact analytical results is rather small. An easy introduction will be found in Ref. 1. Because of its limited applicability, at present, to biophysical and biochemical problems, it does not seem appropriate to repeat this material here. Our treatment of the equilibrium problem, using the Monte Carlo method, is intended only to provide the necessary reference material for the primary subject of this chapter: two-dimensional steady-state lattices of enzyme molecules. The Monte Carlo method can also be applied to one- and three-dimensional steady-state lattices; a few results for one-dimensional systems will be presented in Section 45. Almost all of the chapter will be concerned with lattices of two-state enzyme molecules, but one-way three-state enzyme molecules are considered briefly in Section 46.

Natural two-dimensional and three-dimensional[2,3] lattices of enzyme molecules are not unknown, but they are also not common, up to the present time. The author is not aware of any phase transition behavior in these systems, though this might have some biological usefulness. Bacteriorhodopsin is the best known example of a cycling molecule that forms a two-dimensional lattice *in vivo,* but it does not seem likely that kinetically significant interactions between these molecules in their two-dimensional hexagonal lattice could extend beyond triplets (i.e., $m = 3$, not M large, in our notation), if indeed there are even significant interactions within triplets. There have been speculations[4,5] about cooperativity in equilibrium lattices of membrane transport proteins, but only speculations.

A clean steady-state phase transition, associated with GTPase activity, does occur at the ends of a microtubule,[6] but this is a quite different kind of system.

The general plan of the chapter is the following: Section 40 provides an introduction; Sections 41–44 deal with square (two-dimensional) lattices of two-state molecules, Section 41 at equilibrium and Sections 42–44 at steady state; Section 45 considers a one-dimensional lattice of two-state molecules; and Section 46 treats a square lattice of three-state molecules.

40. The Model and the Monte Carlo Method

We shall start[7] with a general self-contained formulation of the two-state lattice problem, and then specialize for the actual numerical calculations. We consider a finite lattice of M identical two-state enzyme molecules. There are nearest-neighbor interactions between the molecules. These interactions perturb the rate constants and hence alter the kinetics of the system. The lattice may have one, two, or three dimensions; in any case, it has periodic boundary conditions (to make all M molecules equivalent). For example, in one dimension, the Mth molecule is a neighbor of the first molecule; the molecules are in a ring. Each molecule has z nearest neighbors in the lattice. An isolated (i.e., no neighbors) enzyme molecule has unperturbed first-order or pseudo-first-order rate constants as shown in Fig. 10-1. The thermodynamic force X, driving this cycle, is given by

$$e^{X/kT} = \alpha\beta/\alpha'\beta' . \tag{40.1}$$

This force is not affected by neighbor interactions. As usual, we define a variable x as

$$x \equiv (\alpha + \beta')/(\beta + \alpha') . \tag{40.2}$$

This variable governs the relative population of the two states; at equilibrium ($X = 0$), x is equal to K in Eq. (38.8) or to K_0c in a simple binding system (Chapter 6 or 8).

We now select β as a reference rate constant and set $\beta \equiv 1$. Thus, from here on, the other rate constants and the flux J (below) become dimensionless quantities that are measured in units of β, while the (dimensionless) time is measured in units of β^{-1}. Hence, we now have

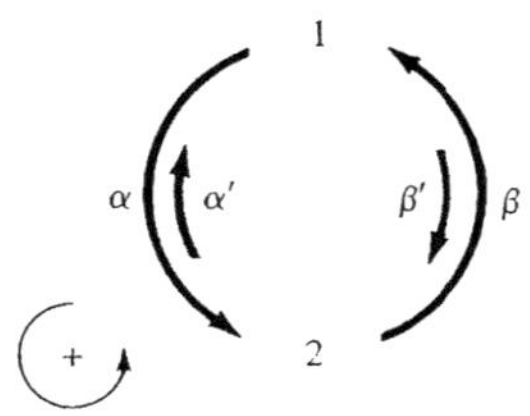

Fig. 10-1. Two-state cycle and first-order rate constants for an unperturbed enzyme molecule.

$$\begin{aligned} x &\equiv (\alpha + \beta')/(1 + \alpha'), \\ e^{X/kT} &= \alpha/\alpha'\beta' \equiv F. \end{aligned} \tag{40.3}$$

It is convenient, henceforth, to use F for the thermodynamic force variable instead of X.

The rate constants α and β' may be expressed in terms of x and F as follows:

$$\alpha = \frac{\alpha'(1 + \alpha')xF}{1 + \alpha'F}, \tag{40.4}$$

$$\beta' = \frac{(1 + \alpha')x}{1 + \alpha'F}. \tag{40.5}$$

The point of this is to replace the original rate constants of the model (Fig. 10-1) β, α', α, and β' by the new set of parameters $\beta \equiv 1$, α', x, and F. We shall be particularly interested, below, in θ and J as functions of x, with F held constant, where $\theta \equiv p_2$ is the steady-state probability of state 2 (Fig. 10-1).

We consider now the perturbed rate constants for a particular molecule (in either state) that has (instantaneously) a_2 nearest neighbors in state 2 and $a_1 = z - a_2$ nearest neighbors in state 1. The possible values of a_2 are 0, 1, . . . , z. We apply our usual approach [see, e.g., Eqs. (27.54) and (37.2)–(37.7)] to this more general problem. The perturbed rate constants are

$$\begin{aligned} \alpha_e &= \alpha \exp(f_\alpha(w_{1e} - w_{2e})/kT], \\ \alpha'_e &= \alpha' \exp[(1 - f_\alpha)(w_{2e} - w_{1e})/kT], \\ \beta_e &= \beta \exp[f_\beta(w_{2e} - w_{1e})/kT], \\ \beta'_e &= \beta' \exp[(1 - f_\beta)(w_{1e} - w_{2e})/kT], \end{aligned} \tag{40.6}$$

where e refers to the instantaneous environment of the given molecule (namely, a_2 molecules in state 2 and $z - a_2$ molecules in state 1), f_α and f_β are constants (usually but not necessarily between 0 and 1), and w_{ie} is the interaction free energy between the given molecule in state i and e. Hence,

$$\begin{aligned} w_{1e} - w_{2e} &= a_1(w_{11} - w_{12}) + a_2(w_{12} - w_{22}) \\ &= z(w_{11} - w_{12}) + a_2(2w_{12} - w_{11} - w_{22}), \end{aligned} \tag{40.7}$$

where w_{ij} is the interaction free energy between two nearest neighbors in states i and j (we take $w_{12} = w_{21}$). Equations (40.6) then become

$$\begin{aligned} \alpha_e &= \alpha(ry^{a_2})^{f_\alpha}, & \alpha'_e &= \alpha'(ry^{a_2})^{f_\alpha - 1}, \\ \beta_e &= \beta(ry^{a_2})^{-f_\beta}, & \beta'_e &= \beta'(ry^{a_2})^{1-f_\beta}, \end{aligned} \tag{40.8}$$

where

$$y_{ij} \equiv e^{-w_{ij}/kT}, \qquad r = (y_{12}/y_{11})^z, \qquad y = y_{11}y_{22}/y_{12}^2. \tag{40.9}$$

The final expressions for the perturbed rate constants, for arbitrary a_2, are obtained upon substituting $\beta = 1$ and Eqs. (40.4) and (40.5) into Eqs. (40.8).

In summary, the complete set of parameters of the model (aside from β and the type of lattice) are α', x, F, f_α, f_β, r, and y. The first three of these are intrinsic properties of the individual (isolated) enzyme molecules, while the last four relate to the nearest-neighbor interactions.

At equilibrium, $F = 1$. Furthermore, the choices of α', f_α, and f_β have no influence on equilibrium properties. For example, θ may be regarded as a function of rx and y. Equation (34.82) is a special case of this with x here $\leftrightarrow$ x_o there and rx here $\leftrightarrow$ x there.

For the reader not familiar with these matters, we mention here the qualitative phase-transition behavior of the infinite ($M = \infty$) two- or three-dimensional system at equilibrium ($F = 1$). Figure 10-2 shows θ as a function of ln rx for three values of y (Fig. 6-1, for example, is similar). The curves are all symmetrical about $\theta = \frac{1}{2}$. The $y = 1$ curve is the usual simple $\theta = rx/(1 + rx)$. For large enough y, $y = y_c > 1$, critical behavior is observed ($d\theta/d \ln rx = \infty$ at $\theta = \frac{1}{2}$). The value of y_c for a square lattice is given below [Eq. (41.1)]. For $y > y_c$, there is a first-order phase transition, a vertical jump in θ. In the two-phase equilibrium, one phase has $\theta = \theta' < \frac{1}{2}$ and the other phase has $\theta = 1 - \theta' > \frac{1}{2}$. In this infinite system, the positive cooperativity arising from intermolecular interactions leads to a drastic consequence: separation of the system into two phases. This cannot happen in such a clean-cut way in a finite system (e.g., $M = 100$, below) nor does it happen in a one-dimensional lattice even for $M = \infty$. The equilibrium calculations below (Section 41) are along the line $\theta = \frac{1}{2}$ in Fig. 10-2.

The Monte Carlo Method

We follow (via computer) the individual transitions of the system of M molecules. At an arbitrary point in a long sequence of transitions, each of the

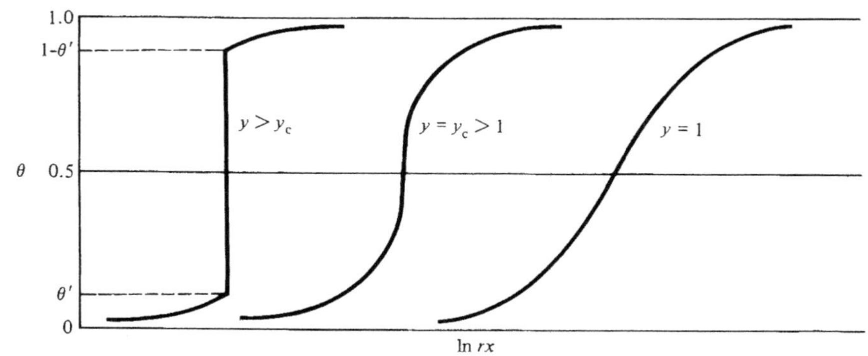

Fig. 10-2. Schematic curves illustrating phase-transition behavior in an infinite system. The right-hand curve has no cooperativity. The central curve is a critical curve. The left-hand curve shows a first-order phase transition.

M molecules is in a definite state (1 or 2) and has a definite value for a_2 $(0, 1, \ldots, z)$. This is referred to below as the "initial state" of the system. If a molecule is in state 1, the total transition probability for $1 \to 2$ is $\alpha_e + \beta_e'$ [from Eqs. (40.4), (40.5), and (40.8)]; if in state 2, the corresponding probability for $2 \to 1$ is $\beta_e + \alpha_e'$. Both quantities depend, of course, on the a_2 value for the particular molecule. In this way, every molecule in the system has an instantaneous total transition probability. A random number generator selects the one molecule (out of M) that makes the next transition, leading to the next state of the complete system; in this operation, the probability of a given molecule being chosen is proportional to its total transition probability. The mean time the system waits for this transition is the reciprocal of the sum of all M total transition probabilities. This mean time is the weight given to the initial state of the system upon taking time averages of various state properties over a long sequence of transitions. For steady-state purposes it is not necessary to refine this step by using a random-number-selected time for each transition, suitably distributed, rather than the mean time.

An example of a time-averaged state property is $\overline{M}_{a_2}^{(i)}$. This is the time average of $M_{a_2}^{(i)}$, which is the instantaneous number of molecules of the system that are in state i $(i = 1, 2)$ and that have a_2 $(a_2 = 0, 1, \ldots, z)$ nearest neighbors in state 2. Obviously, then,

$$\theta = M^{-1} \sum_{a_2=0}^{z} \overline{M}_{a_2}^{(2)}. \tag{40.10}$$

As another example, the computer can record the fraction $P(N_2)$ of the total time, during a long sequence of transitions, in which the system has exactly N_2 out of M molecules in state 2. An alternative expression for θ is then

$$\theta = \overline{N}_2/M = M^{-1} \sum_{N_2=0}^{M} N_2 P(N_2). \tag{40.11}$$

The variance in N_2, σ^2, is calculated in a similar way from $P(N_2)$.

At equilibrium $(F = 1)$, if we regard the system as open with respect to state 2,

$$P(N_2) = \frac{Q(N_2, M, T)\lambda^{N_2}}{\Xi(\mu, M, T)}, \tag{40.12}$$

where $\lambda = e^{\mu/kT}$ is the absolute activity (proportional to x). Q is the canonical partition function, and Ξ is the grand partition function.

The straightforward way (which has not been used because of the added expense) to measure the flux, in a long sequence of transitions, would be to record the total number of cycles completed by each of the M molecules, in either direction (+ = counterclockwise in Fig. 10-1). Then J, the mean net flux per molecule in the + direction, would be given by the total net number of cycles completed in the + direction by the entire system of M molecules, divided by M and by the total time used in the sequence of transitions.

Alternatively, either transition flux for each molecule could be recorded. The above procedure is essential if one is interested in fluctuations in the flux, as well as in the mean flux. It should be added that this procedure would require a modification of the above discussion: the random number generator must select among $2M$ possible events rather than among M, i.e., we would have to distinguish between α_e and β_e' transitions in $\alpha_e + \beta_e'$ and between β_e and α_e' transitions in $\beta_e + \alpha_e'$.

Instead, we use a more accurate method to calculate the mean flux J from a relatively modest number of transitions. We employ the $\overline{M}_{a_2}^{(i)}$, already referred to above, to calculate J_α (based on α transitions) and J_β (based on β transitions):

$$\begin{aligned} J_\alpha &= M^{-1} \sum_{a_2=0}^{z} [\alpha_e(a_2)\overline{M}_{a_2}^{(1)} - \alpha_e'(a_2)\overline{M}_{a_2}^{(2)}], \\ J_\beta &= M^{-1} \sum_{a_2=0}^{z} [\beta_e(a_2)\overline{M}_{a_2}^{(2)} - \beta_e'(a_2)\overline{M}_{a_2}^{(1)}], \end{aligned} \tag{40.13}$$

where $\alpha_e(a_2)$, etc., are given by Eqs. (40.8) with $\beta = 1$ and Eqs. (40.4) and (40.5) substituted. With a finite number of transitions, J_α and J_β will differ somewhat, owing to fluctuations in the $M_{a_2}^{(i)}$, so we take J as $(J_\alpha + J_\beta)/2$. Significant scatter in J will be noted in some of the figures. Longer simulation runs would reduce this scatter.

The unperturbed flux, incidentally, is

$$\begin{aligned} J_o &= \frac{\alpha\beta - \alpha'\beta'}{\alpha + \beta' + \beta + \alpha'} \\ &= \frac{\alpha' x(F-1)}{(1+\alpha' F)(1+x)}, \end{aligned} \tag{40.14}$$

where we have used $\beta = 1$ and Eqs. (40.4) and (40.5) in the second line. This has the familiar $x/(1+x)$ dependence on x. Similarly, the unperturbed expression for θ is

$$\theta_o = x/(1+x). \tag{40.15}$$

The nearest-neighbor correlation function

$$c_1 \equiv \langle s_k s_{k+1} \rangle \tag{40.16}$$

is of some interest and is used in the following sections. Here s_k is assigned the value -1 if molecule k is in state 1 and $+1$ if k is in state 2; $k+1$ is a nearest neighbor of k. For each state of the system, in a long sequence of states, the product $s_k s_{k+1}$ is summed for all molecules k in the lattice (M of these) and for all nearest neighbors of each molecule (z of these). The sum is divided by Mz and then a time average is taken over all states in the sequence. From this prescription, it follows that

$$c_1 = (Mz)^{-1} \sum_{a_2=0}^{z} [(z - 2a_2)\, \overline{M}_{a_2}^{(1)} + (-z + 2a_2)\overline{M}_{a_2}^{(2)}]. \qquad (40.17)$$

If a molecule always has neighbors in the same state as itself, $c_1 = 1$. If the neighbors are always in the other state, $c_1 = -1$. These are the extremes of positive and negative cooperativity, respectively.

The total interaction free energy W is closely related. This is the sum of the applicable w_{ij} for each of the $Mz/2$ nearest-neighbor pairs in a given state of the system, followed by time averaging over all states in the sequence. Consequently,

$$2W = \sum_{a_2=0}^{z} \{[(z - a_2)w_{11} + a_2 w_{12}]\, \overline{M}_{a_2}^{(1)} + [(z - a_2)w_{12} + a_2 w_{22}]\, \overline{M}_{a_2}^{(2)}\}. \qquad (40.18)$$

In the familiar special case and notation (used in the theory of ferro- or antiferromagnetism) $w_{11} = w_{22} = -J$ and $w_{12} = +J$, the connection between Eqs. (40.17) and (40.18) is obviously very close. In fact, we have, in this case,

$$(Mz/2)\, c_1 = W/(-J). \qquad (40.19)$$

This relation is needed in the next section. The quantity J here is *not* a flux (it is used, as above, by physicists).

41. Reference Monte Carlo Calculations at Equilibrium

Most of the Monte Carlo calculations in this chapter are on a 10×10 ($M = 100$) square lattice with periodic boundary conditions. Such boundary conditions for a 4×4 lattice are illustrated in Fig. 10-3. To establish the effects of the finite size, we begin by comparing 10×10 Monte Carlo results at equilibrium ($F = 1$) with some known exact properties of the infinite ($\infty \times \infty$) square lattice. Earlier Monte Carlo calculations of a similar nature[8] may be of interest to the reader. The methods and results in this section[7] will be needed for reference and guidance in the following sections on steady-state

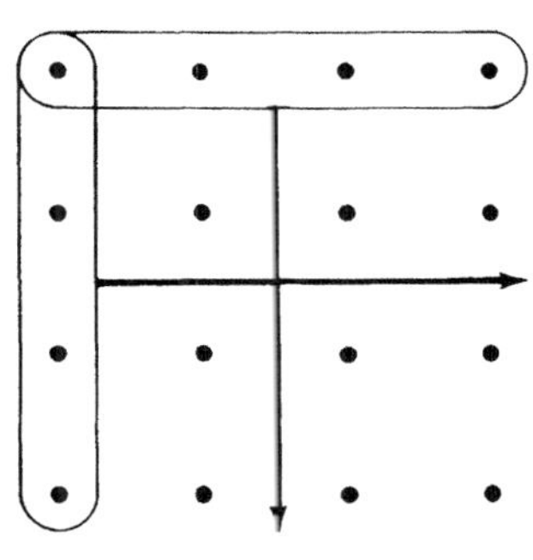

Fig. 10-3. Illustration of periodic boundary conditions. The top and bottom rows of sites are considered to be neighbors, as are the left and right columns of sites.

systems. They may also have some intrinsic interest of their own (for equilibrium statistical mechanics).

Unless otherwise noted, the Monte Carlo simulations in this section were done on a 10 × 10 square lattice ($M = 100$, $z = 4$), with periodic boundary conditions (i.e., the lattice is on a torus), and with parameter choices $F = 1$, $\alpha' = 1$, $r = 1$, $f_\alpha = 0$, $f_\beta = \frac{1}{2}$. This choice of f_α and f_β avoids quasi-equilibrium at steady state but is immaterial at equilibrium. Each run (except as noted) consisted of 102,000 transitions, usually starting with all molecules in state 2. However, the first 2000 transitions were discarded (approach to steady state) and all time averages (at steady state) were taken over the last 100,000 transitions. A larger number of transitions would be desirable, of course, but the expense would be greater.

For any y, the value of x that leads to $\theta = \frac{1}{2}$ (at equilibrium) is y^{-2} (square lattice, any M). This corresponds to zero magnetic field in the magnetism problem. At $\theta = \frac{1}{2}$, the correlation function c_1 is a known function of y for the infinite system. Green and Hurst[9] have, in effect, tabulated values of c_1 [$c_1 = -A/2H$ from their Table 3.1, in view of Eq. (40.19) above]. The lower solid curve in Fig. 10-4 is a plot of c_1 against T/T_c, from Green and Hurst, where T_c is the critical temperature for the infinite system. The critical y is

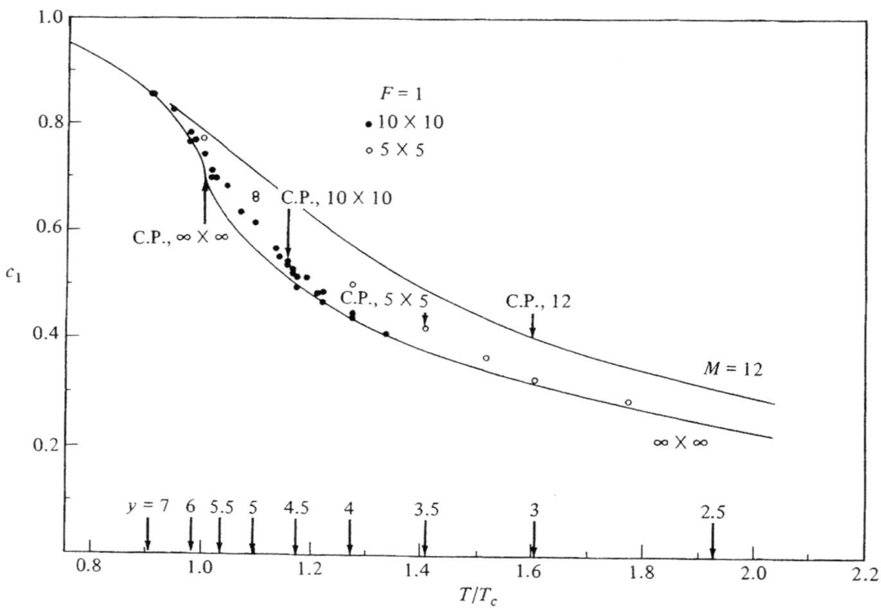

Fig. 10-4. Nearest-neighbor correlation function c_1 at equilibrium ($\theta = \frac{1}{2}$, $x = y^{-2}$), as a function of T/T_c or y. The $M = 12$ and the $\infty \times \infty$ curves are exact. The points are from Monte Carlo simulation runs. C. P. = critical point. See the text for details.

$$y_c = (\sqrt{2} - 1)^{-2} = 5.82843. \tag{41.1}$$

The relation between T/T_c and y is

$$T/T_c = \ln y_c/\ln y = 1.76275/\ln y. \tag{41.2}$$

This assumes that the w_{ij} in Eqs. (40.9) are independent of T. There is a first-order phase transition in this system for $T < T_c$ or $y > y_c$ (Fig. 10-2).

The filled circles in Fig. 10-4 have been obtained from 10×10 simulation runs, as described above, with various values of y but always with $x = y^{-2}$. The scatter in the points is due, of course, to the use of finite runs. The values of θ found fluctuate, as expected, around $\theta = \frac{1}{2}$, with large fluctuations at the larger values of y, owing to the phase transition (the nucleation time for a change from one phase, say, almost all state 1, to the other phase, almost all state 2, approaches the time of a run; but over an infinite time the system would spend equal amounts of time in the two phases). Correspondingly large fluctuations are not seen in c_1 because c_1 has the same value in the two phases.

As is to be expected, the singularity (infinite slope) at $T/T_c = 1$ in the c_1 curve for the infinite system is lost in the 10×10 data. In fact, as discussed below (including the definition of y_c for a finite system), the critical value of y for the 10×10 system is reduced to $y_c \cong 4.55$ (or $T/T_c \cong 1.163$). There is also no sign of a residual or smoothed-out "singularity" in the neighborhood of $y = 4.55$, though this could be due to the use of finite runs and the consequent scatter of the points.

The open circles in Fig. 10-4 have been found in exactly the same way, except that a 5×5 square lattice ($M = 25$) with periodic boundary conditions was used. The critical value of y (see below) for the 5×5 system is $y_c \cong 3.5$ ($T/T_c \cong 1.41$).

Figure 10-4 also includes a curve that gives $c_1(T/T_c)$ or $c_1(y)$ for a lattice of 12 molecules ($M = 12$) on the surface of a sphere (cubic close packing), again at $\theta = \frac{1}{2}$ ($x = y^{-2}$). Although $z = 4$ for this lattice, the connectivity of second and higher neighbors is not the same as for a square or rectangular lattice with periodic boundary conditions. However, we include the curve because it is relatively easy to calculate exactly (not by Monte Carlo) from previous work on this model.[10] Details are omitted. The critical value of y (see below) is $y_c \cong 3.0$ ($T/T_c \cong 1.60$).

In summary, we see in Fig. 10-4 that reduction in the size of systems with no boundaries from $M = \infty$ to 100, to 25, to 12 increases the correlation c_1 at a given y and decreases the value of y_c. The smaller systems are more cohesive: cooperativity is stronger because of the enhanced connectivity.

We turn now to the important question, bypassed above, of locating the approximate critical point (especially y_c) of a finite system, using finite simulation runs. There are no mathematical discontinuities in any case, because the system is finite, so conventional criteria, e.g.,

$$\partial \ln x/\partial\theta = \partial^2 \ln x/\partial\theta^2 = 0,$$

are inapplicable. We need a criterion that can be used for steady states as well as for equilibrium.

The criterion adopted here is based on the function $P(N_2)$, introduced above. At $\theta = \frac{1}{2}$, for $y > y_c$, $P(N_2)$ will have two peaks (near $N_2 = 0$ and $N_2 = M$), representing each of the two phases, with a minimum in between (at $N_2 = M/2$). The minimum is due to the extra interfacial free energy when both phases are present. (This is of order $kTM^{2/3}$ in three dimensions, $kTM^{1/2}$ in two dimensions, and $kTM^{0/1} = kT$ in one dimension.) For $y < y_c$, there will be a single peak (one phase) centered at $N_2 = M/2$. As y is gradually increased, starting from $y < y_c$, this single peak will broaden, become flat in the middle, and then divide into two peaks. The intermediate flat $P(N_2)$ curve is identified with y_c. If, for large M, $P(N_2)$ at $\theta = \frac{1}{2}$ were available as an analytical function, we could use

$$\partial P/\partial N_2 = \partial^2 P/\partial N_2^2 = 0$$

at $N_2 = M/2$ to define "flat." For small M, finite differences could be used in a similar way. However, with relatively short simulation runs on small systems, we evaluate flatness in the central region of $P(N_2)$ by eye and obtain an *estimate* of y_c. This procedure can obviously be applied to steady states as well as to equilibrium. However, when $F \neq 1$ we do not know in advance the relation between x and y at $\theta = \frac{1}{2}$; also, we can anticipate that $\theta_c \neq \frac{1}{2}$, in general.

Figure 10-5 shows $P(N_2)$ data for the $F = 1$, 10×10 system, at several

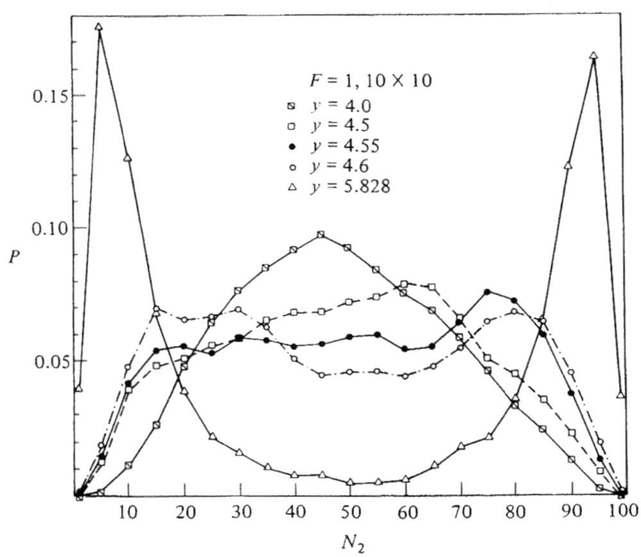

Fig. 10-5. Equilibrium $P(N_2)$ Monte Carlo (10×10) points at $x = y^{-2}$ in each case. The mean value of θ and the number of runs of 100,000 transitions each are $y = 4.0$, $\theta = 0.483$ (2); $y = 4.5$, $\theta = 0.489$ (2); $y = 4.55$, $\theta = 0.511$ (3); $y = 4.6$, $\theta = 0.493$ (2); and $y = 5.828$, $\theta = 0.490$ (1).

values of y, and with $x = y^{-2}$ in each case. The points in the figure are composite points (total probabilities for $N_2 = 3$ to 7, 8 to 12, etc., except for $N_2 = 0$ to 2 (first point) and 98 to 100 (last point). The number of runs used, of 100,000 transitions each, is indicated in parentheses in the figure caption. These same runs appear as individual points in Fig. 10-4. The $y = 4.55$ curve in Fig. 10-5 includes one additional run (not in Fig. 10-4) with $f_\alpha = 1, f_\beta = \frac{1}{2}$. This change in f_α should not and does not affect the $P(N_2)$ and c_1 results at equilibrium ($F = 1$) beyond the difference between any two repeated runs. (Thus, $c_1 = 0.522$ and 0.528 in the two runs with $f_\alpha = 0$, and $c_1 = 0.526$ in the one run with $f_\alpha = 1$.) Recall that $y = 5.828$ ((used in Fig. 10-5) is the critical value of y for the infinite system. The 10×10 system shows a clear phase separation at this y (more precisely, the open system is usually found in one phase or the other). It is clear from the figure why $y = 4.55$ is estimated as y_c for the 10×10 system.

Corresponding $P(N_2)$ curves for $F = 1$, 5×5 are given in Fig. 10-6, baased on 100,000 transitions each. Here the individual values of $N_2 = 0, 1, \ldots, 25$ are used in the plot. Clearly, $y_c \cong 3.5$, and again a rather complete phase separation occurs at $y = 5.828$.

Figure 10-7 gives the exact $P(N_2)$ (not Monte Carlo) for 12 molecules on a sphere, for $y = 1, 2, 3, 4$ and $x = y^{-2}$. The value $y = 3.0$ is *very* close to y_c, according to the criterion we are using here:

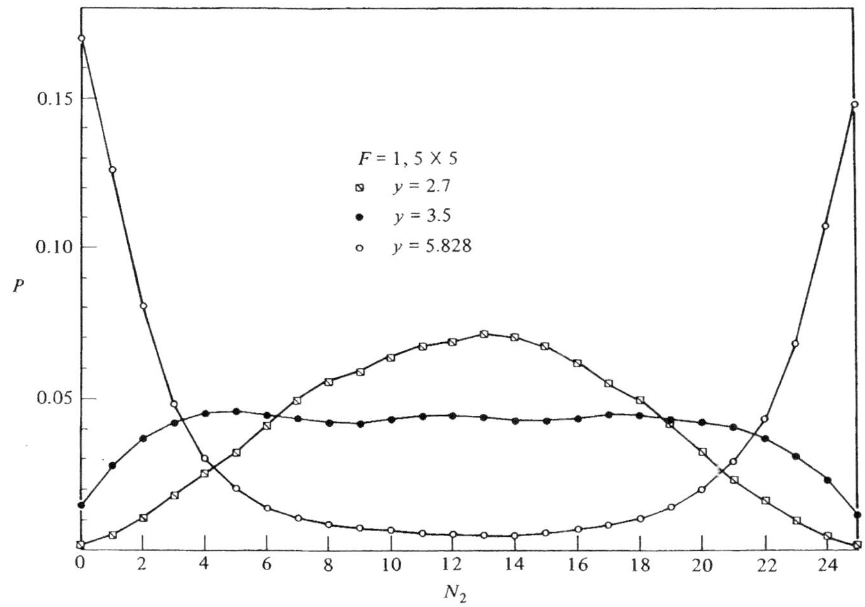

Fig. 10-6. Same as in Fig. 10-5 except 5×5: $y = 2.7$, $\theta = 0.499$ (1); $y = 3.5$, $\theta = 0.489$ (1); and $y = 5.828$, $\theta = 0.472$ (1).

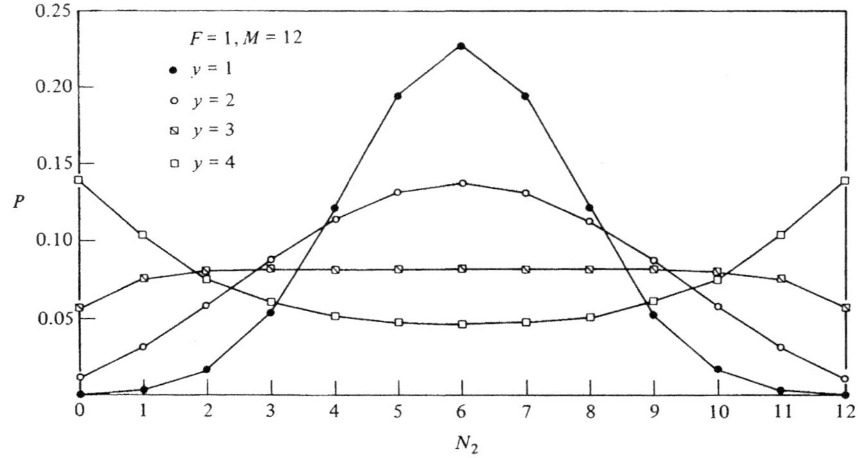

Fig. 10-7. Exact equilibrium $P(N_2)$ (not Monte Carlo) for 12 enzyme molecules on a sphere (cubic close packing), at $x = y^{-2}$ in each case.

$$P(6) = 0.08219; \qquad P(5) = P(7) = 0.08188.$$

We omit any examples of θ as a function of x for $F = 1$ because there is nothing novel about such curves at equilibrium [actually, the $\theta(x)$ points in Fig. 10-14, below, for the case $F = \infty$, $y = 4.55$ are, in effect, equilibrium points, as explained later].

42. Steady-State Monte Carlo Calculations at $F = 4$ and $F = \infty$

Here we give a sampling of Monte Carlo 10×10 steady-state results.[7] This will be supplemented by more examples in the following sections. In this section, all cases have parameter values (besides $\beta = 1$) $\alpha' = 1$, $r = 1$, $f_\alpha = 0$, and $f_\beta = \frac{1}{2}$ (as in Section 41 where $F = 1$). Because of the arbitrariness of the particular special cases studied, we do not attempt to find very precise critical constants.

The choice $F = \infty$ (used below) corresponds to Michaelis–Menten kinetics for the unperturbed enzyme molecules. From Eqs. (40.4) and (40.5), when $F = \infty$,

$$\alpha = (1 + \alpha')x, \qquad \beta' = 0. \tag{42.1}$$

We do not include cases with $F < 1$ because of a symmetry property. If, in Eqs. (40.8) [using $\beta = 1$ and Eqs. (40.4) and (40.5)], we leave r and y unchanged but replace F by $1/F$, f_α by $1 - f_\beta$, f_β by $1 - f_\alpha$, and α' by $1/\alpha'$, we find that the rate constants α_e and β'_e have, in effect, been exchanged in the new system, and also that β_e and α_e' have been exchanged. Furthermore, each of the new rate constants is multiplied by a factor $1/\alpha'$. Thus, in the new system, the cycle simply operates backwards, except at an overall rate altered by a factor α'. States 1 and 2 do not exchange roles. Therefore, the new and old steady-state functions $\theta(x)$ and $J(x)$, for given r and y, are related by

$$\begin{aligned} \theta(x; F^{-1}, 1 - f_\beta, 1 - f_\alpha, \alpha'^{-1}) &= \theta(x; F, f_\alpha, f_\beta, \alpha'), \\ J(x; F^{-1}, 1 - f_\beta, 1 - f_\alpha, \alpha'^{-1}) &= -\alpha'^{-1} J(x; F, f_\alpha, f_\beta, \alpha'), \end{aligned} \tag{42.2}$$

where the new functions are on the left and the old on the right. In the present section, we have the special case (α' is not involved)

$$\begin{aligned} \theta(x; F^{-1}, \tfrac{1}{2}, 1) &= \theta(x; F, 0, \tfrac{1}{2}), \\ J(x; F^{-1}, \tfrac{1}{2}, 1) &= -J(x; F, 0, \tfrac{1}{2}). \end{aligned} \tag{42.3}$$

We study two values of F ($F = 4$ and $F = \infty$). Figure 10-8 gives $\theta(x)$ and $J(x)$ points for $F = 4$ (i.e., $X = 1.386kT$) and $y = 5.0, 5.7$, and 6.5. Figure 10-9 indicates that (see Section 41) $y_c \cong 5.7$ and that there is a phase transition at $y = 6.5$. In Fig. 10-8, the points for $y = 5.7$ are from runs of 100,000

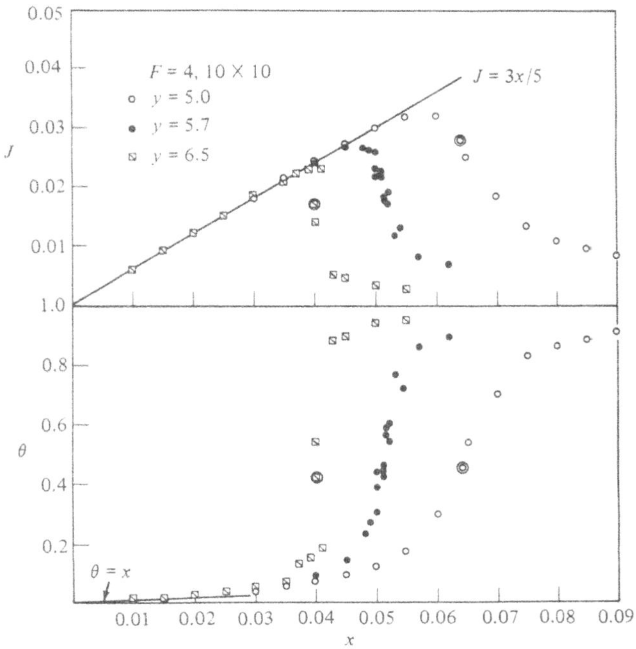

Fig. 10-8. $\theta(x)$ and $J(x)$ Monte Carlo points for the steady state $F = 4$, 10×10, $f_\alpha = 0$, $f_\beta = \frac{1}{2}$. The straight lines $J = 3x/5$ and $\theta = x$ represent asymptotic $x \to 0$ behavior. See the text for further details.

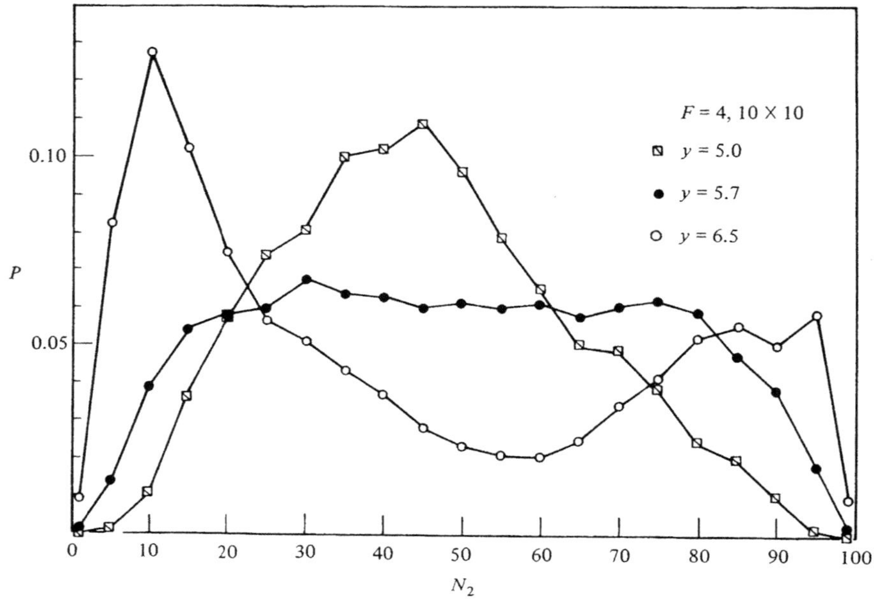

Fig. 10-9. $P(N_2)$ for the $F = 4$, 10×10 steady state: $y = 5.0$, $x = 0.064$, $\theta = 0.457$ (1); $y = 5.7$, $x = 0.0512$, $\theta = 0.499$ (5); and $y = 6.5$, $x = 0.040$, $\theta = 0.423$ (1). The curve and data for $y = 5.7$ are actually a composite of three runs at $x = 0.051$ and two runs at $x = 0.0515$.

transitions each, while those for $y = 5.0$ and 6.5 are based on 20,000 transitions each (except for the points with an extra circle, which have 100,000). The behavior of $\theta(x)$ in Fig. 10-8 is in fact typical of phase transitions *at equilibrium;* the same remark applies to $\theta(x)$ in Fig. 10-10 for $F = \infty$. In other words, there is no new qualitative feature in $\theta(x)$ at steady state as compared with equilibrium. A minor point should be mentioned, however: the well-known symmetry between θ and $1 - \theta$ in two-state equilibrium ($F = 1$) Ising systems is presumably lost at steady state ($F \neq 1$). The critical $\theta(x)$ curve in Fig. 10-8 ($y = 5.7$) does not have a vertical slope, of course, because the system is finite ($M = 100$).

The $\theta(x)$ data for $F = 4$, $y = 6.5$ in Fig. 10-8 serve as a good illustration of nucleation-time effects in a phase-transition region. As already mentioned, only 20,000 transtions were used for each of these points (except for the "extra circle" point). In the run at $x = 0.040$, the system happened to make one change of phase (near the middle of the run) and spent roughly equal amounts of time in the two phases. However, the system spent the entire time in phase 1 (i.e., the phase with state 1, primarily) in the runs at $x = 0.039$ and 0.041, and the entire time in phase 2 in the run at $x = 0.043$. In the extra circle run (100,000 transitions) at $x = 0.040$, the system made seven changes

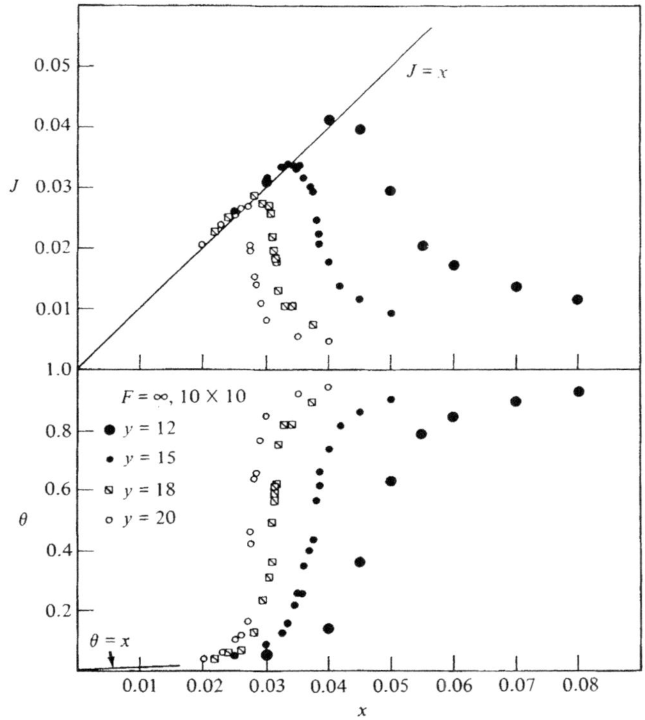

Fig. 10-10. $\theta(x)$ and $J(x)$ Monte Carlo points for the steady state $F = \infty$, 10×10, $f_\alpha = 0, f_\beta = \frac{1}{2}$.

of phase during the course of the run (judging from averages of θ taken every 2000 transitions). The point at $x = 0.041$ illustrates hysteresis: phase 2 is presumably the stable phase at this x, but not enough time was allowed (in 20,000 transitions) to nucleate the phase change from 1 to 2.

The top part of Fig. 10-8 gives $J(x)$ from the same runs as $\theta(x)$, calculated from Eqs. (40.13). The flux curves are, of course, a new property of steady states, not found at equilibrium (where $J = 0$, except for fluctuations). In this example, the two phases have different fluxes; in fact, phase 1 has a higher flux than phase 2. Phase 2 has a relatively low flux because favorable interactions between nearest-neighbor state 2 molecules tend to "freeze" molecules in state 2. With other choices of parameters (see below), the flux magnitudes of the two phases could be reversed or the two phases could have the same flux. In the present example ($F = 4$), we estimate $y_c \cong 5.7$, $\theta_c \cong 0.5$, $x_c \cong 0.0512$, and $J_c \cong 0.0205$. The values of θ_c and J_c are difficult to establish because θ and J change very rapidly with x near the critical point.

Because $y_{11} = y_{12}$ in this model (i.e., $r = 1$), interactions are not a factor unless there is an appreciable population of state 2. Consequently, the initial

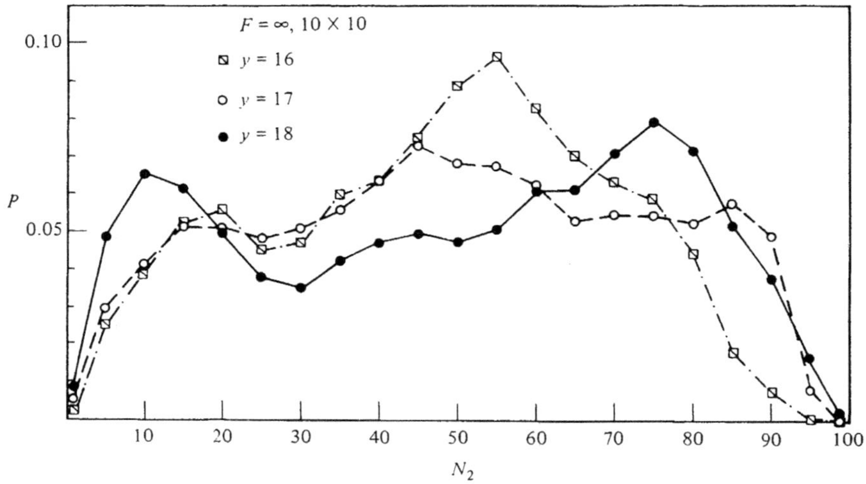

Fig. 10-11. $P(N_2)$ for the $F = \infty$, 10×10 steady state: $y = 16$, $x = 0.035$, $\theta = 0.471$ (1); $y = 17$, $x = 0.033$, $\theta = 0.495$ (2); and $y = 18$, $x = 0.0311$, $\theta = 0.497$ (1).

behavior (near $x = 0$) of $\theta(x)$ and $J(x)$ is the same as that of $\theta_o(x)$ and $J_o(x)$ in Eqs. (40.14) and (40.15), i.e., $\theta \to x$ and $J \to (F - 1)x/(F + 1)$ as $x \to 0$.

Figure 10-10 contains similar $\theta(x)$ and $J(x)$ results for $F = \infty$ and $y = 12$, 15, 18, and 20. The points for $y = 12$ are based on 20,000 transitions, and all others on 100,000 transitions. Figure 10-11 indicates that y_c is between 17 and 18, say, $y_c \cong 17.3$. Other approximate critical values are $\theta_c \cong 0.5$, $x_c \cong 0.0324$, and $J_c \cong 0.0230$.

Some critical constants are presented in Table 10-1. Values for θ_c are omitted because they cannot be estimated at all accurately from the Monte Carlo results. The values for J_c in the table are relatively uncertain. The equilibrium ($F = 1$) values for the $\infty \times \infty$ system are $y_c = 5.828$ and $x_c = y_c^{-2} = 0.02944$. It should be noted that the absolute J_c values in Table 10-1 are very small; by comparison, the maximum possible J_o values ($x \to \infty$) are $J_o = \frac{3}{5}$ for $F = 4$ and $J_o = 1$ for $F = \infty$. However, it should not be

Table 10-1. Monte Carlo Critical Constants ($f_\alpha = 0, f_\beta = \frac{1}{2}$).

F	y_c	x_c	J_c
1	4.55	0.0483	0
4	5.7	0.0512	0.0205
∞	17.3	0.0324	0.0230

Table 10-2. Approximate Properties at $\theta \cong \frac{1}{2}$, with $f_\alpha = 0, f_\beta = \frac{1}{2}$.

$F = 4$				$F = \infty$			
y	x	c_1	y^2x	y	x	c_1	yx
5.0	0.065	0.452	1.63	12	0.0475	0.457	0.57
5.7	0.0512	0.522	1.66	15	0.03775	0.506	0.57
6.5	0.040	0.634	1.69	16	0.035	0.511	0.56
				17	0.033	0.536	0.56
				18	0.0311	0.563	0.56
				20	0.0275	0.614	0.55

concluded that *all* models with phase transitions will have such small values of J_c.

If the three 10×10 y_c values in Table 10-1 all happened to increase in the same ratio 5.828/4.55 when we pass from a 10×10 to an $\infty \times \infty$ system, then we would find for the $\infty \times \infty$ system $y_c \cong 7.3$ for $F = 4$ and $y_c \cong 22.2$ for $F = \infty$. These should be compared with $y_c = 5.828$ for $F = 1$, $\infty \times \infty$.

The dominant qualitative effect in Table 10-1 is that stronger interactions are needed to induce a phase transition when $F > 1$ (steady state) than when $F = 1$ (the y_c ratio between $F = \infty$ and $F = 1$, for the 10×10 system, is 3.8). However, this is not always the case (see Section 43).

Table 10-2 contains additional approximate data on the $F = 4$ and $F = \infty$ 10×10 systems at $\theta \cong \frac{1}{2}$. The x values at $\theta \cong \frac{1}{2}$ are given, together with either y^2x or yx. It will be recalled that, when $F = 1$, $y^2x = 1$ at $\theta = \frac{1}{2}$. We note in Table 10-2 that y^2x is still fairly constant at $F = 4$ but that, at $F = \infty$, it is the product yx which is approximately constant.

Table 10-2 also shows that the correlation function c_1, at steady state, increases with y at $\theta \cong \frac{1}{2}$, as it does at equilibrium (10×10, Fig. 10-4). The approximate values of c_1 at the critical y_c are 0.525 ($F = 1$), 0.522 ($F = 4$), and 0.544 ($F = \infty$). There is very little difference among these numbers.

The variance in N_2, obtained from $P(N_2)$, is given in Table 10-3 in the form

Table 10-3. Approximate Values of σ^2/M at $\theta \cong \frac{1}{2}$, with $f_\alpha = 0, f_\beta = \frac{1}{2}$.

$F = 1$		$F = 4$		$F = \infty$	
y	σ^2/M	y	σ^2/M	y	σ^2/M
				12	2.01
4.0	3.55			16	4.60
4.5	4.97	5.0	3.36	17	5.99
4.55(c)[a]	5.95	5.7(c)	5.44	17.3(c)	6.38[b]
4.6	6.46	6.5	9.77	18	7.26
				20	8.95

[a](c) = critical
[b]Interpolated.

σ^2/M for the $\theta \cong \frac{1}{2}$ simulation runs in Figs. 10-5, 10-9, and 10-11, and Table 10-2. For reference, σ^2/M in a noninteracting two-state Ising system at equilibrium is $\theta(1 - \theta)$, or 0.25 at $\theta = \frac{1}{2}$. The critical values are all in the neighborhood of $\sigma^2/M = 6$. This number is significant, of course, only for the special case $M = 100$.

43. Steady-State Calculations at $F = \infty$ with Other Choices of f_α and f_β

With the choice $f_\alpha = 0$, $f_\beta = \frac{1}{2}$ in the previous section, stronger nearest-neighbor interactions are required to produce critical behavior at the $F = \infty$ steady state than at equilibrium (Table 10-1). The main point of this brief section[7] is to show that the opposite is true for the choice $f_\alpha = 1, f_\beta = \frac{1}{2}$. That is, there is enhancement of cooperativity at steady state relative to equilibrium.

We also include in this section a few 10×10 Monte Carlo calculations with $F = \infty$, $f_\alpha = \frac{1}{2}$, $f_\beta = \frac{1}{2}$. Because $f_\alpha + f_\beta = 1$, this should be a quasi-equilibrium case, i.e., we should find $y_c \cong 4.55$, $x_c = y_c^{-2}$, etc. (as at $F = 1$), although there will be a nonzero flux.

The Case $F = \infty, f_\alpha = 1, f_\beta = \frac{1}{2}$

The other parameters (besides $\beta = 1$) are the same as in Sections 41 and 42: $\alpha' = 1$ and $r = 1$.

We found above that, for a 10×10 system, $y_c \cong 4.55$ at $F = 1$, whereas $y_c \cong 17.3$ at $F = \infty$ with $f_\alpha = 0, f_\beta = \frac{1}{2}$. In the present case, with $F = \infty$, $f_\alpha = 1, f_\beta = \frac{1}{2}$, we find $y_c < 4.55$. Figure 10-12 presents Monte Carlo results (θ and J as functions of x) for $y = 3.51$, 3.8, and 4.1. The extra circle points are from runs with 100,000 transitions; the other points are based on 20,000 transitions. Note in Fig. 10-12 that, in contrast with Fig. 10-10, phase 2 has a higher flux than phase 1.

Figure 10-13 gives $P(N_2)$ at $\theta \cong \frac{1}{2}$ for the above three values of y. The $y = 4.1$ curve has two peaks, and the $y = 3.51$ curve only one. The $y = 3.8$ curve is intermediate. A rough guess is $y_c \cong 3.9$. In any case, $4.1 > y_c > 3.51$, which is less than 4.55. This confirms that weaker interactions at $F = \infty$ than at $F = 1$ suffice for critical behavior when $f_\alpha = 1$, $f_\beta = \frac{1}{2}$.

Another way to see the enhancement of interaction effects at steady state is to compare the $y = 4.1$ curve for $P(N_2)$ at $F = 1$ (equilibrium), $f_\alpha = 1$, $f_\beta = \frac{1}{2}$, $x = y^{-2}$ with the $y = 4.1$, $\theta \cong \frac{1}{2}$ curve at $F = \infty$ (see Fig. 10-13). The $F = 1$, $y = 4.1$ curve is subcritical (it is, in fact, part of the family of curves in Fig. 10-5 because f_α and f_β values are irrelevant at $F = 1$), while the $F = \infty$, $y = 4.1$ curve is supercritical (two peaks).

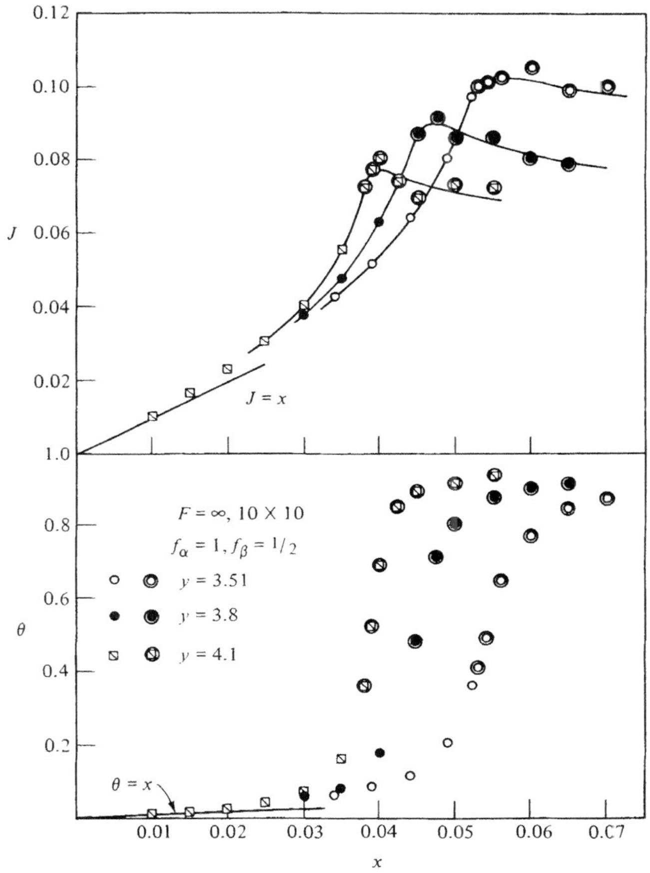

Fig. 10-12. $\theta(x)$ and $J(x)$ Monte Carlo points (and curves) for the steady state $F = \infty$, 10×10, $f_\alpha = 1$, $f_\beta = \frac{1}{2}$. The points with extra circles are from runs with 100,000 transitions, the others from 20,000 transitions.

Table 10-4 collects some information about the $\theta \cong \frac{1}{2}$ simulation runs in Fig. 10-13. If we asume $\sigma^2/M \cong 6$ at the critical point (see Table 10-3), Table 10-4 confirms that y_c is between 3.8 and 4.1. It also confirms (4.20 vs 7.35) the subcritical and supercritical assignments made in the preceding paragraph. If we take $y_c \cong 3.9$, the critical c_1 is about 0.51 for $F = \infty$. This is similar to the critical range 0.52 to 0.54 mentioned near the end of Section 42.

It is interesting that y^2x is approximately constant in Table 10-4 for $F = \infty$ ($f_\alpha = 1, f_\beta = \frac{1}{2}$), whereas $yx \cong$ constant for $F = \infty$ ($f_\alpha = 0, f_\beta = \frac{1}{2}$) in Table 10-2.

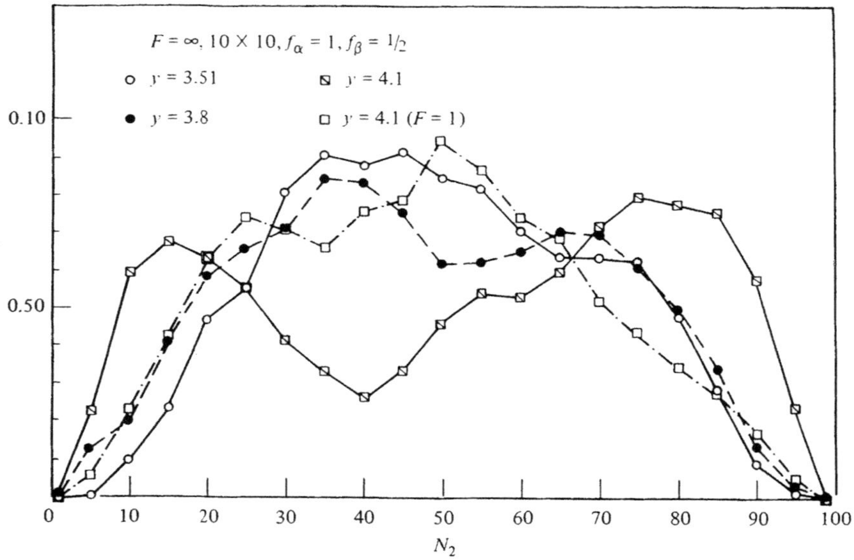

Fig. 10-13. $P(N_2)$ for the steady-state case in Fig. 10-12: $y = 3.51$, $x = 0.054$, $\theta = 0.495$ (1); $y = 3.8$, $x = 0.045$, $\theta = 0.483$ (1); and $y = 4.1$, $x = 0.039$, $\theta = 0.527$ (1). Also included, for comparison, is the $F = 1$, $y = 4.1$, $x = y^{-2}$, $\theta = 0.472$ (1) curve.

The Case $F = \infty, f_\alpha = \frac{1}{2}, f_\beta = \frac{1}{2}$

Here the only change in parameters is $f_\alpha = \frac{1}{2}$. With the choice $f_\alpha = \frac{1}{2}, f_\beta = \frac{1}{2}$, $\theta(x)$ is an equilibrium function at any F. Also, from approximate treatments,[11] we expect that the flux is the same in the two phases and hence that the critical flux curve $J(x)$ has a cusp at the critical point. The former property (θ) holds, in fact, for any combination $f_\alpha + f_\beta = 1$, but the latter (J) requires $f_\alpha = \frac{1}{2}, f_\beta = \frac{1}{2}$. Symmetry properties should be the same in the exact system as in the approximations. This is confirmed below for the particular case studied.

In Section 41 we have already found that $y_c \cong 4.55$ at $F = 1$ (this is independent of the choice of f_α, f_β). With the present parameters $F = \infty$, $f_\alpha = \frac{1}{2}, f_\beta = \frac{1}{2}$, as just explained, we ought to find $y_c \cong 4.55$ again. Figure

Table 10-4. Approximate Properties at $\theta \cong \frac{1}{2}$, with $f_\alpha = 1, f_\beta = \frac{1}{2}$.

F	y	x	c_1	y^2x	σ^2/M
∞	3.51	0.054	0.452	0.67	3.68
∞	3.8	0.045	0.490	0.65	4.59
∞	4.1	0.039	0.560	0.66	7.35
1	4.1	0.05949	0.463	1	4.20

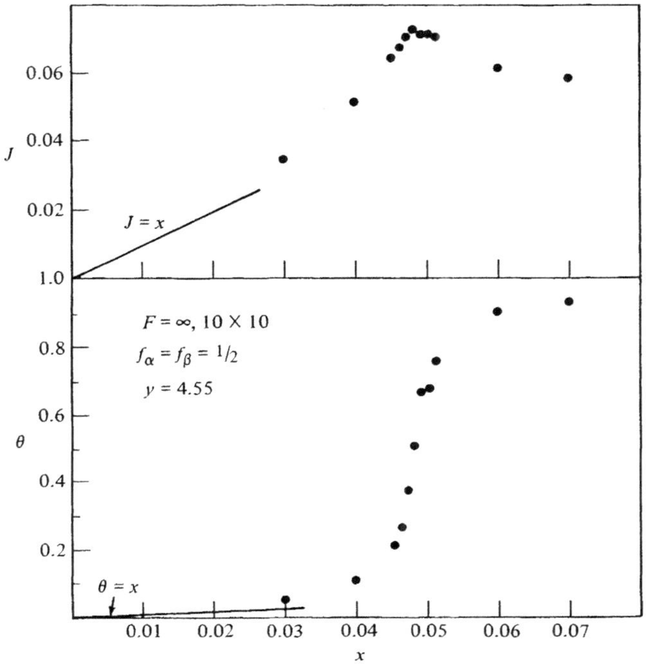

Fig. 10-14. $\theta(x)$ and $J(x)$ Monte Carlo points for the steady state $F = \infty$, 10×10, $f_\alpha = \frac{1}{2}$, $f_\beta = \frac{1}{2}$. See the text for further details.

10-14 contains $\theta(x)$ and $J(x)$ Monte Carlo points (100,000 transitions each) for $F = \infty, f_\alpha = \frac{1}{2}, f_\beta = \frac{1}{2}, y = 4.55$. As anticipated, at $x = y^{-2} = 0.04830$, θ is close to $\frac{1}{2}$ (0.5059), and for this x the $P(N_2)$ curve (not shown) is practically the same as the (critical) $y = 4.55$ curve in Fig. 10-5 (based on a total of 300,000 transitions). In fact, the agreement is so close that it has to be considered accidental. Also, in the present example, we find $\sigma^2/M = 5.950$ and $c_1 = 0.519$; from the three runs in Fig. 10-5 for $y = 4.55$, the averages are $\sigma^2/M = 5.947$ and $c_1 = 0.525$. The flux also behaves as expected, with a maximum at $x = y^{-2}$. Because the system is finite (10×10), this maximum is not very sharp (not a true cusp).

In summary, the results here confirm the expected quasiequilibrium behavior and, as an incidental bonus, provide the equilibrium critical $\theta(x)$ curve for the 10×10 system.

44. Steady-State Monte Carlo Calculations with $f_\alpha = f_\beta = -\frac{1}{2}$

As is clear from Fig. 1-8, if the kinetic parameter f has a value between 0 and $+1$, then the free energy of the transition state or "activated complex" is affected by the perturbation in a way intermediate between the effects on the initial and final states of the transition. However, the transition state could possibly have special characteristics in its structure such that its free energy is altered in a direction opposite to the perturbation effect on initial and final states. In this section[12] we study such an example, with $f_\alpha = f_\beta = -\frac{1}{2}$.

The main case we consider allows us to cover the complete range between equilibrium and a one-way cycle. At the end of the section, another kind of case (still with $f_\alpha = f_\beta = -\frac{1}{2}$) is studied briefly.

Explicitly, we begin by examining the case

$$\beta \equiv 1, \qquad f_\alpha = f_\beta = -\tfrac{1}{2}, \qquad r = 1, \qquad \alpha' = F^{-1/2}. \tag{44.1}$$

Thus we continue to use β as a reference rate constant and hence set $\beta \equiv 1$. We shall choose F between $F = 1$ and $F = \infty$ and adjust α' in each case according to the value of F, as in Eq. (44.1). With this choice for α', Eqs. (40.4) and (40.5) lead to

$$\alpha = x, \qquad \beta' = \alpha' x = F^{-1/2} x. \tag{44.2}$$

Having assigned f_α, f_β, and r, the parameters that we can vary are x, F, and y. Aside from $\beta \equiv 1$, the rate constants α, α', and β' are fixed by x and F. When $F = 1$,

$$\alpha = x, \qquad \alpha' = 1, \qquad \beta' = x. \tag{44.3}$$

When $F = \infty$,

$$\alpha = x, \qquad \alpha' = 0, \qquad \beta' = 0. \tag{44.4}$$

Thus the $F = \infty$ case corresponds to a one-way cycle.

In Fig. 10-15 we show $\theta(\ln x)$ for several values of F, all with $y = 0.220$. For clarity, in all the figures of this section we have drawn smooth curves through the "experimental" (Monte Carlo) points. The typical amount of scatter of the points can be seen from the examples in Sections 41–43. We

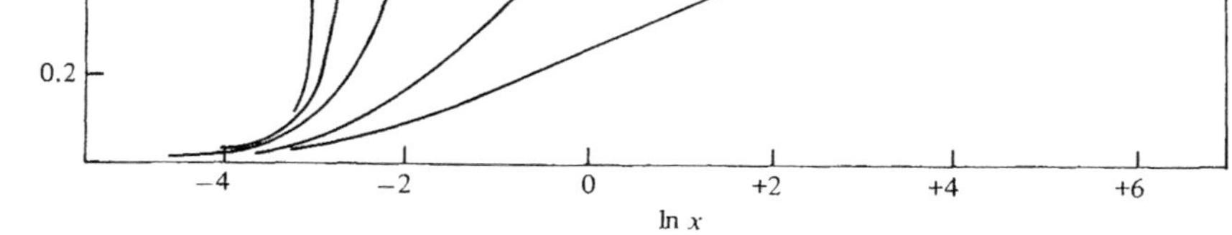

Fig. 10-15. θ as a function of $\ln x$, from equilibrium ($F = 1$) to a one-way cycle ($F = \infty$).

have generally discarded the first 5000 transitions (transient) in a Monte Carlo simulation run and used an additional 20,000 to 100,000 transitions (depending on the magnitude of the fluctuations) at steady state to obtain time averages. In a few cases, we have used up to 500,000 steady-state transitions. Returning to Fig. 10-15, we have chosen $y = 0.220$ because this is the reciprocal of $y = 4.55$. In Section 41 we found that $y \cong 4.55$ is the 10×10 critical value of y at equilibrium ($F = 1$) or at any value of F if $f_\alpha + f_\beta = 1$ (quasiequilibrium). Also, it is easy to see from Eqs. (40.6) that the transformation $w_{ij}, f_\alpha, f_\beta \rightarrow -w_{ij}, -f_\alpha, -f_\beta$ in a one-way-cycle system will leave $\theta(x)$ unchanged. Hence, even though $y = 0.220$ corresponds to relatively unfavorable interactions between neighboring molecules in the same state (11 or 22), the $\theta(\ln x)$ curve for $F = \infty$ in Fig. 10-15 is a critical curve. In fact, it is the same curve shown in Fig. 10-14. For $y < 0.22$ and $F = \infty$, there will be a phase transition. This odd steady-state property is a consequence of the negative f_α and f_β.

At the other extreme, the $F = 1$ case in Fig. 10-15 is the equilibrium Ising curve for a 10×10 system with repulsive interactions (antiferromagnet; negative cooperativity). The switch from $F = 1$ to $F = \infty$ in this *steady-state* example corresponds to changing the sign but not the magnitude of the intermolecular forces in an *equilibrium* system ($y = 0.220 \rightarrow y = 4.55$; antiferromagnet $\rightarrow$ ferromagnet: negative $\rightarrow$ positive cooperativity). The other curves in Fig. 10-15 provide the expected transition between the two extreme cases. The exact 10×10 $F = 1$ and $F = \infty$ curves would be symmetrical about $\theta = \frac{1}{2}$, as the figure suggests.

Figure 10-16 shows the nearest-neighbor correlation function c_1 as a func-

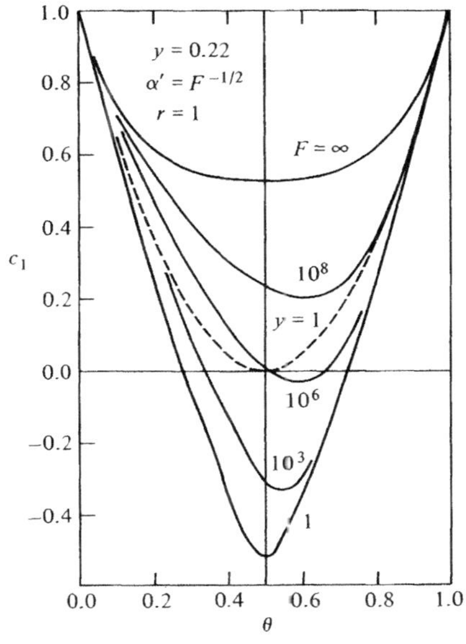

Fig. 10-16. Nearest-neighbor correlation function $c_1(\theta)$ for the same example as in Fig. 10-15.

tion of θ for the same example. Also included (broken curve) for comparison is $c_1(\theta) = (1 - 2\theta)^2$ for a (random) system with no neighbor interactions. The exact curves for $F = 1$ and $F = \infty$ (10 × 10 system) would again be symmetrical about $\theta = \frac{1}{2}$ (as in the $y = 1$ curve). The $F = \infty$ (critical) curve exhibits positive correlation (cooperativity) compared with $y = 1$, while the $F = 1$ curve shows the opposite effect. The mean value $c_1 = 0.525$ at $\theta = \frac{1}{2}$, in the $F = \infty$ case, was already reported in Section 41. The $F = 10^6$ curve in Fig. 10-16 shows positive cooperativity up to about $\theta = \frac{1}{2}$ and negative cooperativity for larger θ.

Very similar comments can be made about Fig. 10-17, which gives M/σ^2 as a function of θ, where σ^2 is the variance in N_2. In the $y = 1$ (random) case, $M/\sigma^2 = 1/\theta(1 - \theta)$. Fluctuations are seen to be relatively small in the $F = 1$ (antiferromagnetic) case. We have already found, in Table 10-3 for the $F = \infty$ case at $\theta = \frac{1}{2}$, $M/\sigma^2 = 1/5.95 = 0.168$.

Figure 10-18 contains flux curves and corresponds to Fig. 10-15. The $F = \infty$ curve in Fig. 10-18 is the same as $J(x)$ in Fig. 10-14. It is not hard to prove that this should be so [Eqs. (40.6)]. This critical curve would have

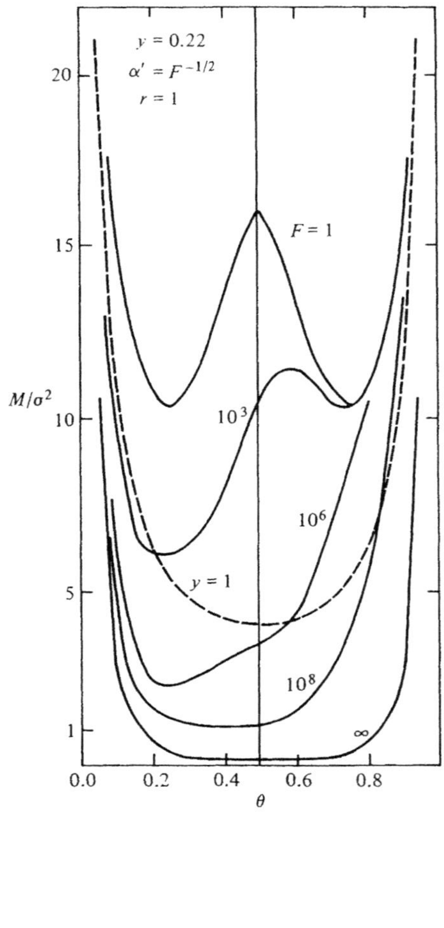

Fig. 10-17. Reciprocal of variance σ^2 in N_2 (number of molecules in state 2) as a function of θ, for the same example as in Fig. 10-15.

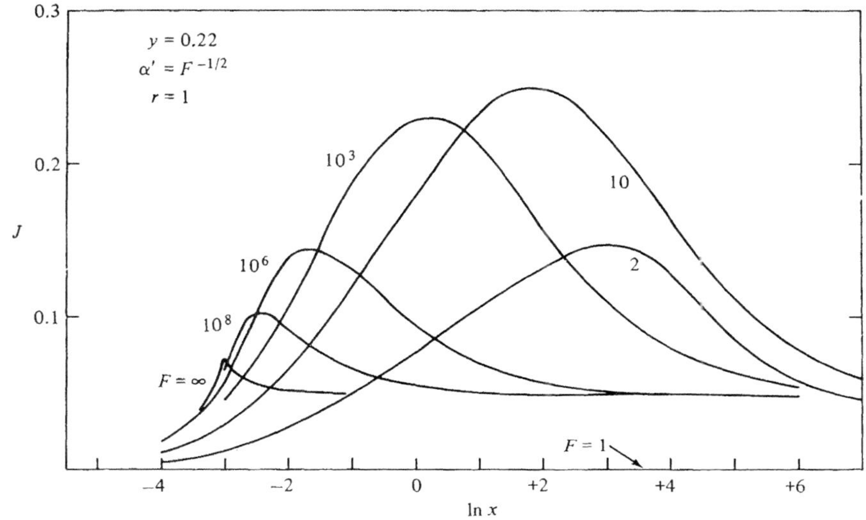

Fig. 10-18. Flux $J(\ln x)$ corresponding to Fig. 10-15.

a cusp if M were infinite. At the other extreme, the $F = 1$ flux curve is the abscissa (i.e., there is no net mean flux at equilibrium, except for fluctuations). We have included $J(\ln x)$ for $F = 2$ and 10 in order to see the nature of the transition between the $F = 1$ and $F = \infty$ cases. It is easy to show that if $y < 1$ and $F \gg 1$, then $J \to y^2$ as $x \to \infty$. This accounts for the asymptotic value $J \to 0.0484$ seen in Fig. 10-18.

In Figs. 10-19 and 10-20, which should be examined together, we keep F constant at $F = 10^6$ and vary y. When y is decreased from 1 to 0.5, positive cooperativity is observed for all θ; but this effect is reversed for $\theta > 0.22$

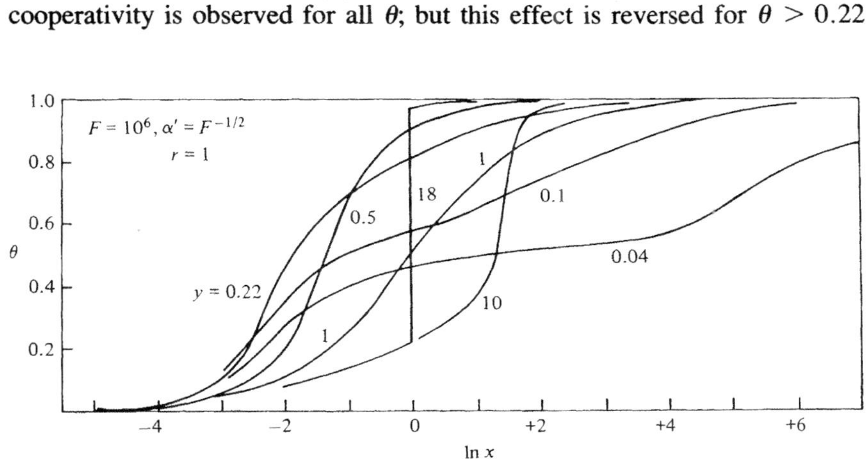

Fig. 10-19. $\theta(\ln x)$ for $F = 10^6$ and various values of y.

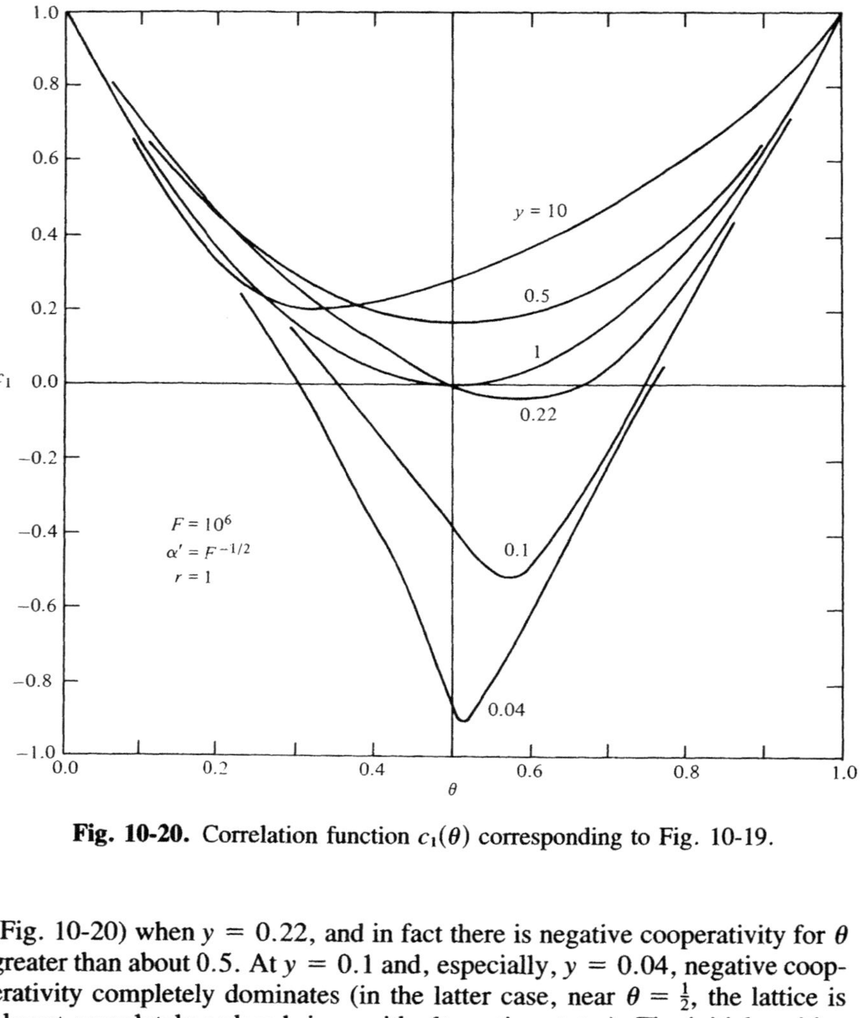

Fig. 10-20. Correlation function $c_1(\theta)$ corresponding to Fig. 10-19.

(Fig. 10-20) when $y = 0.22$, and in fact there is negative cooperativity for θ greater than about 0.5. At $y = 0.1$ and, especially, $y = 0.04$, negative cooperativity completely dominates (in the latter case, near $\theta = \frac{1}{2}$, the lattice is almost completely ordered, i.e., with alternating states). The initial positive cooperativity (as y is decreased below 1) does not persist long enough to produce critical behavior. The same is true (not shown) at $F = 10^8$ but is *not* true at $F = \infty$, as already mentioned in connection with Fig. 10-15: $y = 0.22$ is the 10×10 critical value of y, and a phase transition occurs for $y < 0.22$.

For $y > 1$, there is a phase transition for large enough y. Also, if we compare $y = 1$ and $y = 10$ in Fig. 10-20, we see that, in the $y = 10$ case, there is slight negative cooperativity for $\theta < 0.26$ and strong positive cooperativity for large θ.

The above paragraph refers to $F = 10^6$ and $y > 1$. At $F = \infty$, the situation is quite different. The 10×10 system exhibits only negative cooperativity when $y > 1$. No first-order phase transition is possible. Actually, the curve

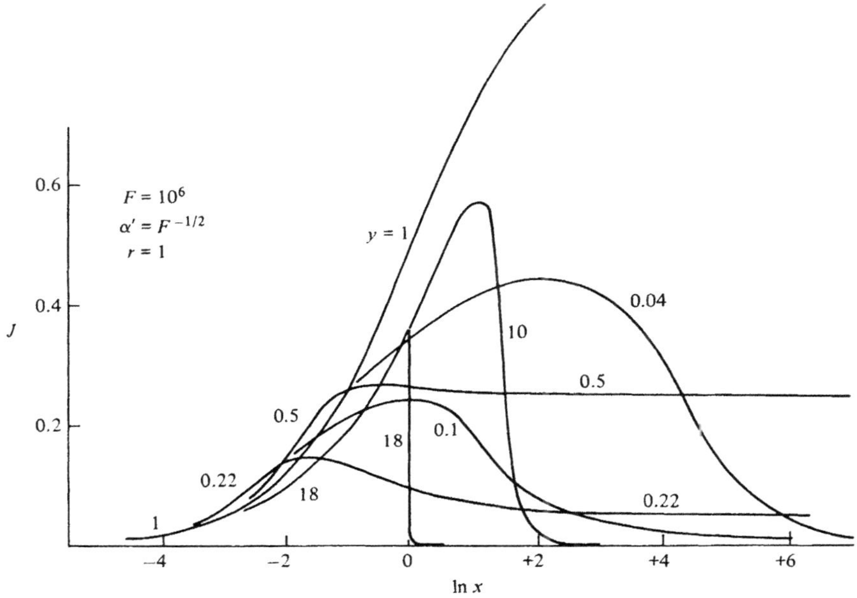

Fig. 10-21. Flux $J(\ln x)$ corresponding to Fig. 10-19.

labeled $F = 1$ in Fig. 10-15 is relevant here: it is also the $F = \infty$, $y = 4.55$ curve (the same is true in Figs. 10-16 and 10-17). Analogously, as already mentioned, the $F = \infty$ curve in Fig. 10-15 is also the $F = 1$, $y = 4.55$ curve.

The flux curves corresponding to Fig. 10-20 are given in Fig. 10-21. The peak fluxes in Fig. 10-21, when $y \leq 0.5$, coincide approximately with the minima in c_1 in Fig. 10-20. As already noted, when $y < 1$, $J \to y^2$ as $x \to \infty$.

Supplementary cases with $f_\alpha = f_\beta = -\frac{1}{2}$

Finally, in this subsection, we provide a few cases with $f_\alpha = f_\beta = -\frac{1}{2}$, $\alpha' = 1$, and $r = 0.05$. In Fig. 10-22, $F = 20$ is held constant and y is varied. In Fig. 10-23, $y = 15$ is held constant and F is varied. The case $F = 20$, $y = 15$ is common to both figures.

45. Steady-State One-Dimensional System with $F = \infty$, $f_\alpha = 1, f_\beta = \frac{1}{2}$

In Section 43 we found that there is enhancement of cooperativity at steady state ($F = \infty$) as compared to equilibrium in the case $f_\alpha = 1, f_\beta = \frac{1}{2}$. In this section[7] we find a similar but smaller effect in a one-dimensional (100×1) lattice.

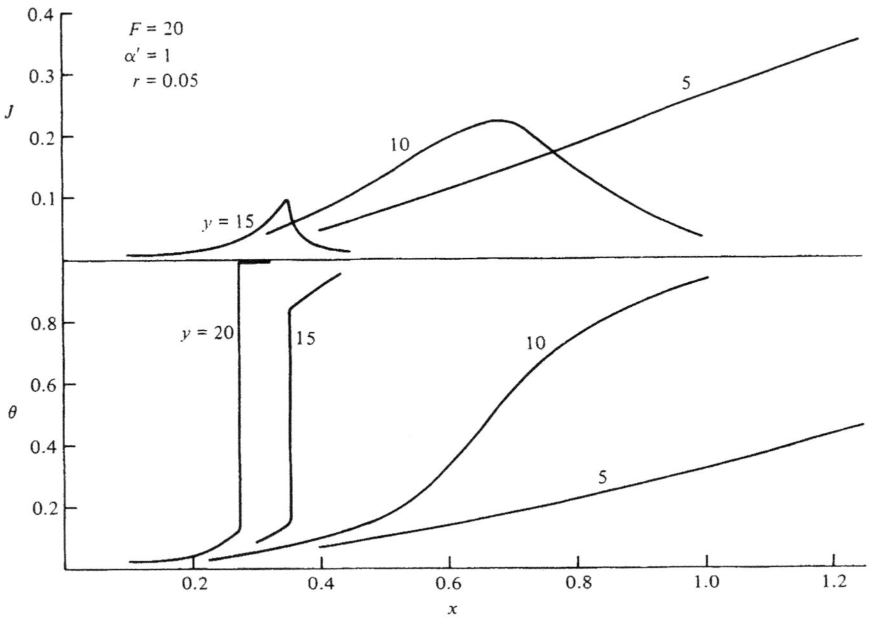

Fig. 10-22. $\theta(x)$ and $J(x)$ at $F = 20$ in the case $f_\alpha = f_\beta = -\frac{1}{2}$, $\alpha' = 1$, $r = 0.05$.

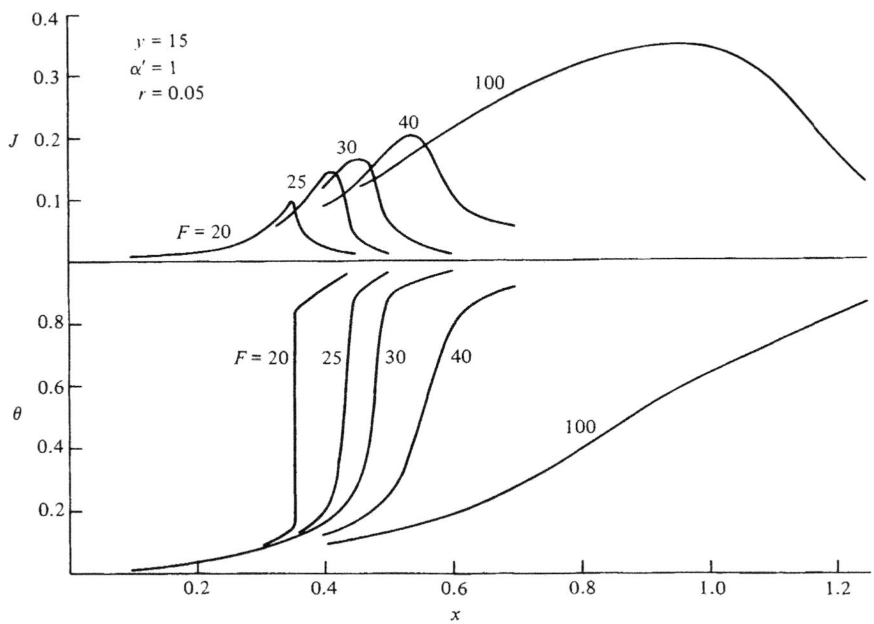

Fig. 10-23. Same as Fig. 10-22 but for $y = 15$.

The model is a closed ring of $M = 100$ two-state enzyme molecules, with $F = \infty$ (or $F = 1$), $r = 1$, $\alpha' = 1$, $f_\alpha = 1$, $f_\beta = \frac{1}{2}$. The computer program can be tested for this system by comparing the $F = 1$ $\theta(x)$ points at, say, $y = 10$ with the exact equilibrium $\theta(x)$ for $y = 10$, $M = \infty$ (Section 34). Agreement is found to be excellent; thus, the closed $M = 100$ system is, in effect, infinite, if $y = 10$. However, with the much larger values of y (e.g., $y = 400$, 600, or more) needed to approach critical behavior, the finite size ($M = 100$) of the system becomes a significant and perturbing factor. With the present choice of parameters (r, α', f_α, f_β), significantly larger values of M would be needed in order to avoid this problem.

Most of the calculations to be reported here are for $y = 400$ (i.e., $w_{12} - w_{22} = 5.99kT$). Longer simulation runs than in previous sections are needed in order to obtain reliable averages (propagation of a new phase, for example, takes much longer because it can proceed along only one dimension). We start the runs with $\theta = \frac{1}{2}$ and discard the first 15,000 transitions (transient). Figure 10-24 contains a few Monte Carlo $\theta(x)$ points for $F = \infty$, $y = 400$. All points are based on 300,000 transitions except the circled point at $x = 0.00135$, which is for 2.5×10^6 transitions. The $J(x)$ behavior resembles Fig. 10-12 as might be expected, but the points scatter badly when $\theta > \frac{1}{2}$. The value of J at $x = 0.00135$ is 0.00420.

Figure 10-25 gives the $P(N_2)$ curve corresponding to the circled point in Fig. 10-24: $x = 0.00135$, $\theta = 0.511$. The other $P(N_2)$ curve in Fig. 10-25 is for $F = 1$, $y = 400$, $x = y^{-1} = 0.0025$, $\theta = 0.483$ (10^6 transitions). With $r = 1$, we expect $\theta = \frac{1}{2}$ at $x = y^{-1}$ [Eq. (34.82)]. The $F = \infty$ curve is somewhat flatter than the $F = 1$ curve, verifying the expected cooperativity enhancement effect. However, the $F = \infty$ curve is subcritical.

Table 10-5 confirms the enhancement effect (see σ^2/M, c_1, $2\overline{N}_{12}$). The quantity $2\overline{N}_{12}$ is the mean number of phase boundaries (i.e., between a sequence of 1s and of 2s) in the $M = 100$ system. Necessarily, $\overline{N}_{12} = \overline{N}_{21}$. The $F = 1$, $M = \infty$ theoretical value of $2\overline{N}_{12}$ at $x \doteq y^{-1}$ is easily seen to be $M/(1 + y^{1/2})$ [Eq. (34.62)]. The Monte Carlo values are found from $\overline{M}_1^{(1)} + 2\overline{M}_2^{(1)}$ or from $\overline{M}_1^{(2)} + 2\overline{M}_0^{(2)}$ (the two quantities are equal). Incidentally, c_1 and $2\overline{N}_{12}$ are not independent: $c_1 = 1 - (4\overline{N}_{12}/M)$.

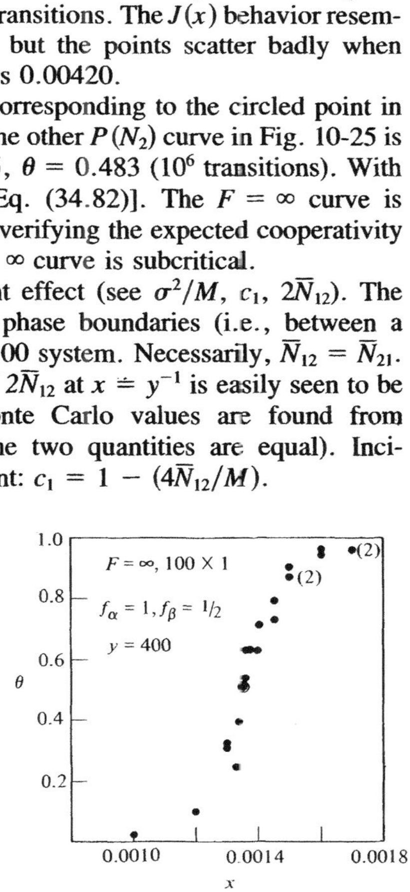

Fig. 10-24. $\theta(x)$ Monte Carlo points for the $F = \infty$, 100×1, $y = 400$ steady state. Note that x begins at 0.0008, not $x = 0$. The line $\theta = x$ (omitted) is practically on the abscissa. See the text for further details.

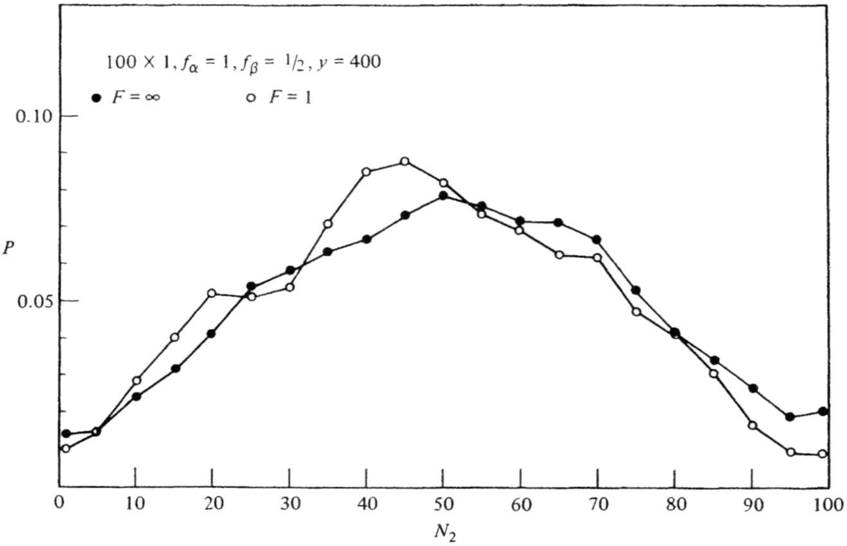

Fig. 10-25. $P(N_2)$ curves for 100×1 system with $y = 400$: $F = 1$, $x = y^{-1}$, $\theta = 0.483$ (10); $F = \infty$, $x = 0.00135$, $\theta = 0.511$ (25).

The theoretical σ^2/M in Table 10-5 for $F = 1, x = y^{-1}$ is calculated on the basis of a Gaussian distribution for $P(N_2)$, centered about $\theta = \frac{1}{2}$. One can then show, from Eq. (40.12) and the known Helmholtz free energy of the $M = \infty$ system, that $\sigma^2/M = y^{1/2}/4$. A more direct but less informative derivation is given in Eq. (34.85). This is valid for y not too large. To be more explicit about this limitation, we note that a Gaussian distribution has room to "fit" in the finite range $0 < N_2 < M$ only if we have, approximately, $3\sigma < M/2$, or $\sigma < M/6$. If we substitute this inequality in the σ^2/M expression above, we deduce the approximate condition for the validity of the Gaussian assumption, $y < M^2/81$ ($\cong 123$, if $M = 100$). To confirm this, in a 100,000 transition run at $y = 10$, $x = y^{-1}$, we obtained $\sigma^2/M = 0.790$ compared to

Table 10-5. One-Dimensional System, $y = 400, f_\alpha = 1, f_\beta = \frac{1}{2}$.

	$x = 0.0025$		$x = 0.00135$
	$F = 1$ MC[a]	$F = 1$ theor.	$F = \infty$ MC[b]
θ	0.483	0.500	0.511
σ^2/M	4.54	5.00	5.42
c_1	0.905	0.905	0.917
$2\overline{N}_{12}$	4.75	4.76	4.14

[a]10^6 transitions. MC = Monte Carlo.
[b]2.5×10^6 transitions.

Table 10-6. One-dimensional System, $y = 600, f_\alpha = 1, f_\beta = \frac{1}{2}$.

	$x = 0.001667$		$x = 0.000875$
	$F = 1$	$F = 1$	$F = \infty$
	MC[a]	theor.	MC[a]
θ	0.514	0.500	0.416
σ^2/M	5.46	6.12	6.36
c_1	0.919	0.922	0.934
$2\overline{N}_{12}$	4.06	3.92	3.31

[a]600,000 transitions. MC = Monte Carlo.

0.791 (theoretical); similarly, for $y = 100$, we found (100,000 transitions) $\sigma^2/M = 2.47$ compared to 2.50 (theoretical); but the agreement is not very good at $y = 400$, in Table 10-5, as expected. The Monte Carlo value of σ^2/M is smaller, presumably because the Gaussian tails of $P(N_2)$ are "piled up" at and near $N_2 = 0$ and $N_2 = 100$.

Table 10-6 contains similar results for $y = 600$ (i.e., $w_{12} - w_{22} = 6.40kT$), based on 600,000 transitions for both $F = 1$ and $F = \infty$. Cooperativity is again enhanced in the latter case. The Monte Carlo entries here are less reliable than in Table 10-5. Incidentally, 395,000 transitions lead to $\theta = 0.604$ at $x = 0.00090$ in the $F = \infty$ system, compared with $\theta = 0.416$ at $x = 0.000875$ (Table 10-6): $\theta(x)$ is quite steep when $y = 600$.

46. Monte Carlo Treatment of a One-Way Three-State Enzyme Lattice

In this section[13] we give a rather compressed discussion of a two-dimensional lattice of interacting one-way three-state enzyme molecules. The parameter choices are quite specialized. The objective is simply to illustrate how the Monte Carlo method can be extended to lattices of more complicated enzyme molecules that interact with each other.

The general procedure and notation here is the same as in Sections 40–45. We consider a square 10×10 lattice ($z = 4$) of enzyme molecules, with periodic boundary conditions. Each molecule can be in one of the three states in Fig. 10-26. We have to keep track of the states of each of the $M = 100$

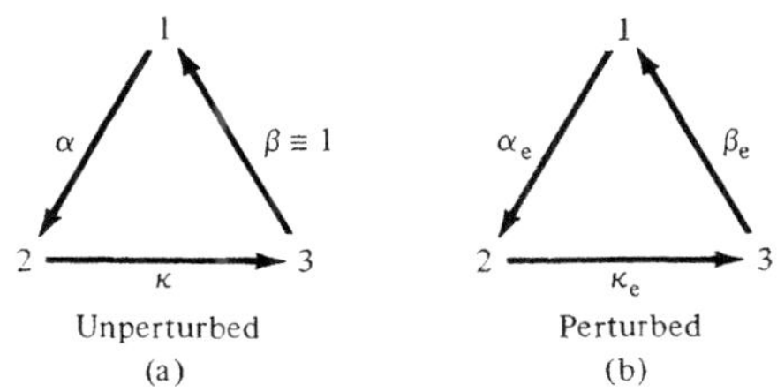

Fig. 10-26. (a) Unperturbed and (b) perturbed rate constants for a three-state one-way enzymatic cycle.

molecules in the lattice as transitions are made in the one-way cycles of various molecules. If a particular molecule at a particular time in a Monte Carlo simulation has a_i nearest neighbors in state i ($i = 1, 2, 3$), where $\Sigma_i a_i = z = 4$, then the operative (perturbed) rate constants for this molecule are, following our usual system,

$$\alpha_e = \alpha\left(\frac{y_{21}^{a_1} y_{22}^{a_2} y_{23}^{a_3}}{y_{11}^{a_1} y_{12}^{a_2} y_{13}^{a_3}}\right)^{f_\alpha} \qquad (1 \to 2), \tag{46.1}$$

$$\kappa_e = \kappa\left(\frac{y_{31}^{a_1} y_{32}^{a_2} y_{33}^{a_3}}{y_{21}^{a_1} y_{22}^{a_2} y_{23}^{a_3}}\right)^{f_\kappa} \qquad (2 \to 3), \tag{46.2}$$

$$\beta_e = \left(\frac{y_{11}^{a_1} y_{12}^{a_2} y_{13}^{a_3}}{y_{31}^{a_1} y_{32}^{a_2} y_{33}^{a_3}}\right)^{f_\beta} \qquad (3 \to 1). \tag{46.3}$$

Equation (46.1) applies if the molecule is in state 1, etc. Here we use $\beta \equiv 1$ as a reference rate constant, just as in the preceding sections. Also, $y_{ij} = y_{ji}$ ($i \neq j$). The a_i are stochastically fluctuating integers with values between 0 and $z = 4$.

We turn now to a particular special case:

$$f_\alpha = f_\kappa = f_\beta = \tfrac{1}{2}, \tag{46.4}$$

$$y_{12} = y_{21} = y_{13} = y_{31} = y_{23} = y_{32} \equiv y_{ij}. \tag{46.5}$$

That is, all unlike pairs of molecules are assumed to have the same interaction free energy; hence y_{ij} can be used as a single reference parameter. The only variable interactions are between like pairs of neighbors, 11, 22, or 33. We therefore define

$$s_k = (y_{kk}/y_{ij})^z \qquad (k = 1, 2, 3). \tag{46.6}$$

The s_k are the new interaction parameters we use below. In view of Eqs. (46.4)–(46.6), Eqs. (46.1)–(46.3) simplify to

$$\alpha_e = \alpha s_1^{-a_1/2z} s_2^{a_2/2z}, \tag{46.7}$$

$$\kappa_e = \kappa s_2^{-a_2/2z} s_3^{a_3/2z}, \tag{46.8}$$

$$\beta_e = s_3^{-a_3/2z} s_1^{a_1/2z}. \tag{46.9}$$

When the complete lattice reaches some particular state (there are 3^M different lattice states) after a transition in one of its M molecules, each molecule has a definite transition probability, given by Eqs. (46.7)–(46.9), that depends on its own state and on the states of its z nearest neighbors. In calculating time averages, the time assigned to this lattice state is the mean time until the next transition in the lattice; this time is equal to the reciprocal of the sum of all M transition probabilities. A random number generator decides which of the M molecules actually makes the next transition, the relative probability for each molecule being proportional to its own transition probability [Eqs. (46.7)–(47.9)].

In a long simulation at steady state involving thousands of consecutive transitions, time averages of various quantities of interest may be calculated. For example, $P_i(N_i)$ is the fraction of time in the simulation during which the lattice has exactly N_i molecules (out of M) in state i (N_i ranges from 0 to 100). Other averages of interest are $\overline{N}_i$, $\overline{N_i^2}$, $\overline{N_iN_j}$, $\overline{N}_{ii}$, and $\overline{M}_{a_ia_j}^{(i)}$. The first three of these are self-evident, the fourth is the mean number of nearest-neighbor pairs (out of a total of $Mz/2$) of type ii, and the last is the mean number of molecules (out of M) that are in state i and that have a_i nearest neighbors in state i and a_j nearest neighbors in state j (needed particularly for the flux). The six quantities referred to above are interrelated in several ways that we need not specify. Of course, many other properties as well can be calculated from the simulation record.

Derived quantities of special interest are then

$$p_i = \overline{N}_i/M, \ \sigma_{ij}^2 = \overline{N_iN_j} - \overline{N}_i\overline{N}_j \qquad (i, j = 1, 2, 3), \tag{46.10}$$

$$c_1 = \frac{2(\overline{N}_{11} + \overline{N}_{22} + \overline{N}_{33}) - (Mz/2)}{Mz/2}, \tag{46.11}$$

$$J_\alpha = M^{-1}\alpha \sum_{a_1}\sum_{a_2} s_1^{-a_1/2z} s_2^{a_2/2z} \overline{M}_{a_1a_2}^{(1)}, \tag{46.12}$$

$$J_\kappa = M^{-1}\kappa \sum_{a_2}\sum_{a_3} s_2^{-a_2/2z} s_3^{a_3/2z} \overline{M}_{a_2a_3}^{(2)}, \tag{46.13}$$

$$J_\beta = M^{-1} \sum_{a_3}\sum_{a_1} s_3^{-a_3/2z} s_1^{a_1/2z} \overline{M}_{a_3a_1}^{(3)}. \tag{46.14}$$

The last three quantities are the transition fluxes.

The six variances σ_{ij}^2 are interrelated (three are independent). The nearest-neighbor correlation function c_1 is defined as the number of like neighbor pairs (11, 22, 33) less the number of unlike pairs (12, 13, 23), divided by the total, $Mz/2$. The steady-state flux can be calculated in three ways; the results will differ somewhat in a finite simulation run, owing to fluctuations. For J, we use the mean of J_α, J_κ, and J_β.

Numerical Examples

We use the parameters $\kappa = 0.5$, $s_1 = 1$, and $s_2 = s_3 = 1000$, 1600, 3000, 6000, with variable α, which would usually be proportional to a concentration. There are favorable 22 and 33 interactions in these examples. Incidentally, with $z = 4$, $s_2 = s_3 = 6000$ corresponds to only $w_{22} - w_{ij} = w_{33} - w_{ij} = -2.17kT$, a not very large interaction free energy.

Figure 10-27 shows $p_2(\alpha)$ for the above four values of $s_2 = s_3$. Smooth curves have been drawn through the "experimental" Monte Carlo points. Each point has been obtained from a simulation consisting of 104,000 transitions, with the first 4000 transitions being discarded in taking time averages (to allow for the transient to steady state).

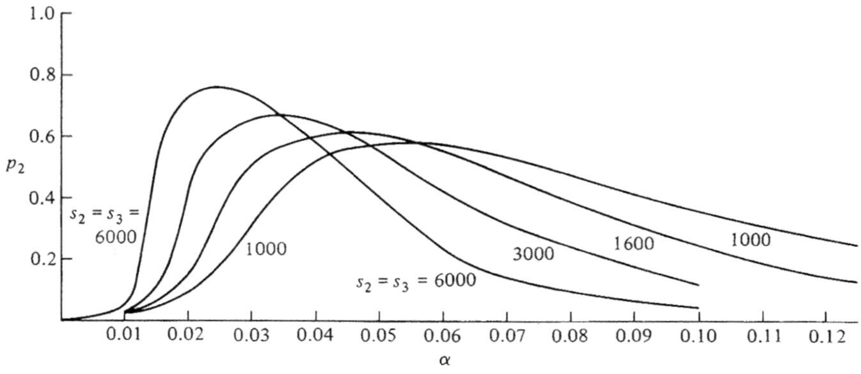

Fig. 10-27. Monte Carlo (10 × 10 square lattice) $p_2(\alpha)$ curves for $\kappa = 0.5$, $s_1 = 1$, and several values of $s_2 = s_3$.

Figure 10-28 gives p_1, p_2, p_3, and J as functions of α for the case $s_2 = s_3 = 6000$. J_o is $\alpha/(1 + 3\alpha)$. The crossing of p_1 and p_2 at low α and the crossing of p_2 and p_3 at higher α suggest that there may be two phase transitions taking place.

Figure 10-29, again for $s_2 = s_3 = 6000$, confirms that there are indeed two phase transitions. Here the evidence is in the form of two peaks in both the σ_{22}^2/M and the $c_1 - c_1^o$ curves. We consider these features in turn.

The variance σ_{22}^2/M in Fig. 10-29 is quite large in the range $\alpha = 0.0125$ to 0.06. For comparison, $\sigma_{22}^2/M = p_2 - p_2^2$, with a maximum of 0.25 at $p_2 = 0.50$, if there are no differential interactions (all $s_i = 1$); $\sigma_{22}^2/M = 8.3$ if $P_2(N_2)$ is a completely flat (constant) function ($P_2 = 0.01$, $p_2 = 0.50$), as

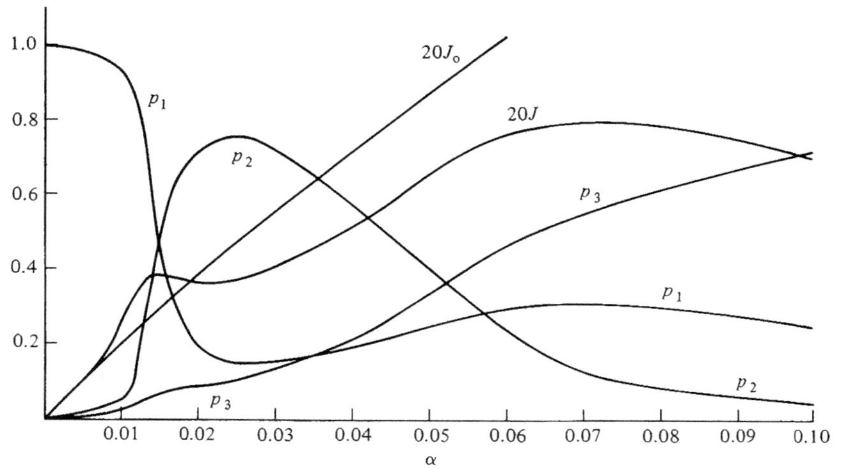

Fig. 10-28. Monte Carlo properties for the case $\kappa = 0.5$, $s_1 = 1$, $s_2 = s_3 = 6000$. The unperturbed flux, $J_o = \alpha/(1 + 3\alpha)$, is included for comparison.

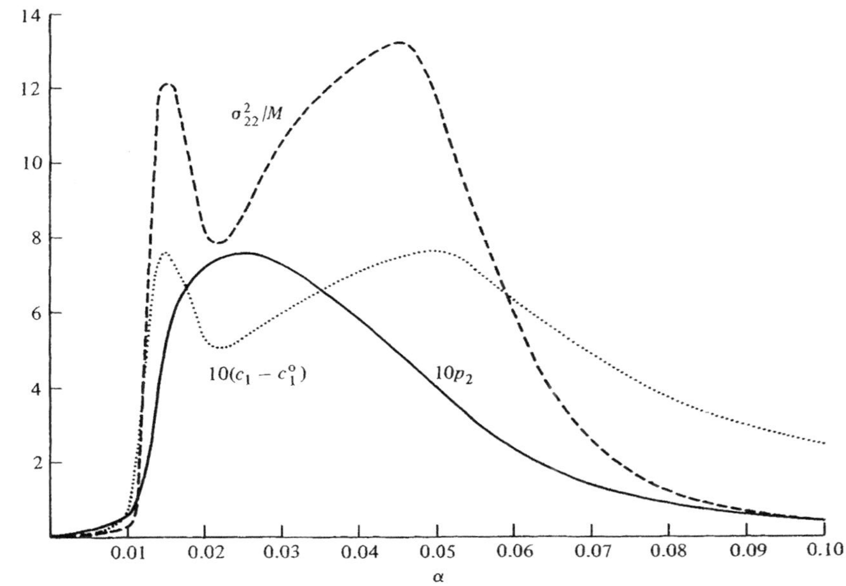

Fig. 10-29. Variance and correlation curves for the same case as in Fig. 10-28. See the text for definitions and significance of these functions. The $p_2(\alpha)$ curve, for comparison, is the same as in Fig. 10-27 ($s_2 = s_3 = 6000$) and Fig. 10-28.

at an idealized critical point; and $\sigma_{22}^2/M = 25$, the maximum value possible, if there are two equal peaks at $N_2 = 0$ and at $N_2 = 100$, $P_2(0) = P_2(100) = 0.50$, and $p_2 = 0.50$, as in an extreme phase transition. It is seen in Fig. 10-29 that the two peaks in σ_{22}^2/M are well above $\sigma_{22}^2/M = 8$ and occur near the inflections in the $p_2(\alpha)$ curve. The second transition is less sharp (broader σ_{22}^2/M peak) than the first. Corresponding $P_2(N_2)$ curves, for $\alpha = 0.015$ ($p_2 = 0.481$) and $\alpha = 0.045$ ($p_2 = 0.560$), are shown in Fig. 10-30. These curves are critical-like (fairly flat and nonzero) in the middle range in N_2, with minor peaks at the ends.

In Fig. 10-29, the reference correlation function c_1^0 is the value that would be found from the observed (Monte Carlo) p_1, p_2, and p_3, if molecules in the three enzyme states were distributed in the lattice at random. It is easy to show that

$$c_1^0 = 2(p_1^2 + p_2^2 + p_3^2) - 1. \tag{46.15}$$

The difference $c_1 - c_1^0$ is then the intermolecular interaction contribution to the correlation function. It is seen in Fig. 10-29 that $c_1 - c_1^0$ behaves qualitatively like σ_{22}^2/M. The two peaks in $c_1 - c_1^0$ reflect increased clustering of like molecules in the lattice, as expected in a phase transition.

Naturally, because of the finite size of the system ($M = 100$), we do not see the sharp discontinuities characteristic of a phase transition in an infinite system.

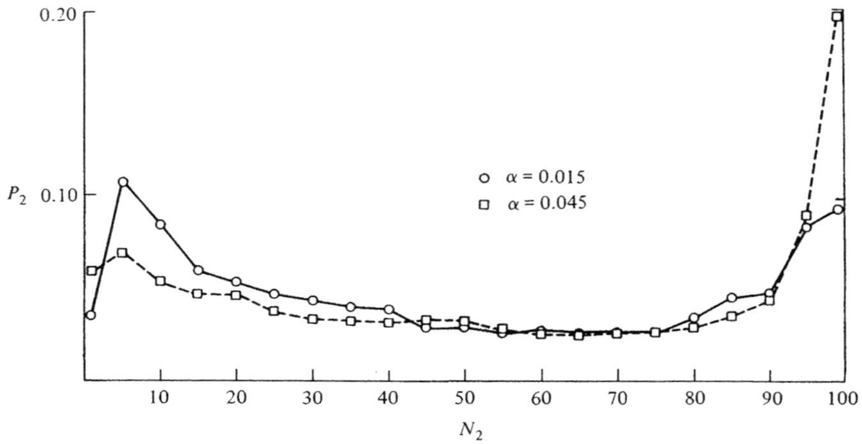

Fig. 10-30. The $P_2(N_2)$ curves at the two maxima in the variance in Fig. 10-29 [α = 0.015 (○) and 0.045 (□)]. The individual points in the figure are total probabilities for the intervals N_2 = 0–2, 3–7, 8–12, . . . , 93–97, 98–100, each plotted at the center of the interval.

References

1. Hill, T. L. (1960) *Statistical Thermodynamics* (Addison-Wesley, Reading, MA).
2. Veenhuis, M., Harder, W., van Dijken, J. P., and Mayer, F. (1981) Molec. Cell. Biol. **1,** 949.
3. Kirsten, H., Gehring, H., and Christen, P. (1983) Proc. Natl. Acad. Sci. USA **80,** 1807.
4. Changeux, J.-P., Thiery, J., Tung, Y., and Kittel, C. (1967) Proc. Natl. Acad. Sci. USA **57,** 335.
5. Hill, T. L. (1967) Proc. Natl. Acad. Sci. USA **58,** 111.
6. Hill. T. L. and Chen, Y. (1984) Proc. Natl. Acad. Sci. USA **81,** 5772.
7. Hill, T. L. and Chen, Y. (1978) J. Chem. Phys. **69,** 1126.
8. Chen, Y. and Hill, T. L. (1973) Proc. Natl. Acad. Sci. USA **70,** 62.
9. Green, H. S. and Hurst, C. A. (1964) *Order-Disorder Phenomena* (Interscience, New York).
10. Hill, T. L. (1964) *Thermodynamics of Small Systems, Part II* (Benjamin, New York), pp. 62–64 and 115–117.
11. Hill, T. L. and Stein, L. (1978) J. Chem. Phys. **69,** 1139.
12. Hill, T. L. and Chen, Y. (1978) Proc. Natl. Acad. Sci. USA **75,** 5260.
13. Hill, T. L. and Chen, Y. (1981) Proc. Natl. Acad. Sci. USA **78,** 4.

11

The Bragg–Williams or Mean-Field Approximation in Steady-State Systems

The Bragg–Williams (BW) approximation, also called the mean-field approximation, is well known for equilibrium lattice problems in statistical physics. The approximation is especially useful for a qualitative preliminary study of critical or phase-transition behavior (i.e., strong positive cooperativity). It has roughly the same status as the van der Waals equation for a fluid, which is useful for a first approach to the gas-liquid phase transition. However, the BW approximation, as used here, is not satisfactory for systems with negative cooperativity, for a reason to be mentioned below.

In this chapter we extend the BW method to steady-state systems of the type we have been considering. The treatment will be rather brief because we do not want to place too much emphasis on approximations. Many more details and examples have been published (see the list of 15 references in the Preface).

After the basic treatment in Section 47, Sections 48 and 49 examine the steady-state equivalent of the famous Maxwell equal-area theorem for equilibrium systems: when an approximate phase-transition theory provides a loop (e.g., the p-V loop from the van der Waals equation), where does one cut across the loop to locate the transition point? The final section, Section 50, uses the BW approximation to examine a higher-order steady-state transition: at the transition, not only is there a sudden switch in population among *states* of the kinetic diagram, but there is also a switch in importance (cycle flux) among *cycles* of the diagram. This could conceivably be a useful biological control mechanism.

47. Introduction to the Steady-State BW Approximation

The Equilibrium BW Approximation[1]

We introduce this method for an equilibrium lattice of M two-state units. Each unit can be in state 1 or state 2. The unperturbed equilibrium constant for $1 \rightleftarrows 2$ is K (in Section 34, K is $x_o = K_o c$). Each unit has z nearest neighbors in the infinite one-, two-, or three-dimensional lattice ($M \to \infty$). The nearest-neighbor interaction parameters are the usual y_{ij} ($i, j = 1, 2$), with $y_{12} = y_{21}$. At equilibrium, the probability that a unit is in state 2 is $p_2 \equiv \theta$; for state 1, $p_1 = 1 - \theta$. The approximation is to assume that the z neighbors of a given unit always have the *mean* population distribution, rather than the actual fluctuating distribution (as in the Monte Carlo treatment). On the average, then, a given unit has $z\theta$ neighbors in state 2 and $z(1 - \theta)$ neighbors in state 1.

When the given unit is in state 2, its interaction with its environment (neighbors) is characterized by $y_{22}^{z\theta}\, y_{12}^{z(1-\theta)}$. When it is in state 1, the corresponding expression is $y_{12}^{z\theta}\, y_{11}^{z(1-\theta)}$. Consequently, the perturbed equilibrium constant for $1 \rightleftarrows 2$, with average environment e, is

$$K_e = \frac{\theta}{1-\theta} = K\frac{y_{22}^{z\theta}y_{12}^{z(1-\theta)}}{y_{12}^{z\theta}y_{11}^{z(1-\theta)}}. \tag{47.1}$$

We can rewrite this as

$$\frac{\theta}{1-\theta} = KY = Kry^{z\theta}, \tag{47.2}$$

where

$$Y \equiv ry^{z\theta}, \qquad r \equiv \left(\frac{y_{12}}{y_{11}}\right)^z, \qquad y \equiv \frac{y_{11}y_{22}}{y_{12}^2}. \tag{47.3}$$

As already mentioned, K in Section 34 (where $z = 2$) is $x_o = K_o c$. Note that r and y are the same as in Eq. (40.9) (Monte Carlo). Equation (47.2) gives Kr as a function of θ or θ as an implicit function of Kr. Equation (47.2) has the familiar symmetry properties, but we leave this topic to the reader.

Figure 11-1 shows a few curves of θ as a function of Kr for three values of $y^{z/2}$. The central curve is the critical curve. The critical value of y is easily found to be $y_c = e^{4/z}$; the critical value of $y^{z/2}$ is therefore e^2. For $y > y_c$, there is a phase transition that appears here as a loop, as in the $y^{z/2} = 20$ curve in Fig. 11-1. Because of symmetry, the vertical cut crosses the loop at $\theta = \frac{1}{2}$. The phase transition is between points A and B. However, Maxwell's equal-area theorem applies only if θ is plotted against $\ln Kr$ (not Kr). It then obviously holds (for the $\theta = \frac{1}{2}$ cut) because of symmetry. Kr is the equivalent of x in Eqs. (34.77), (34.78), and (34.82).

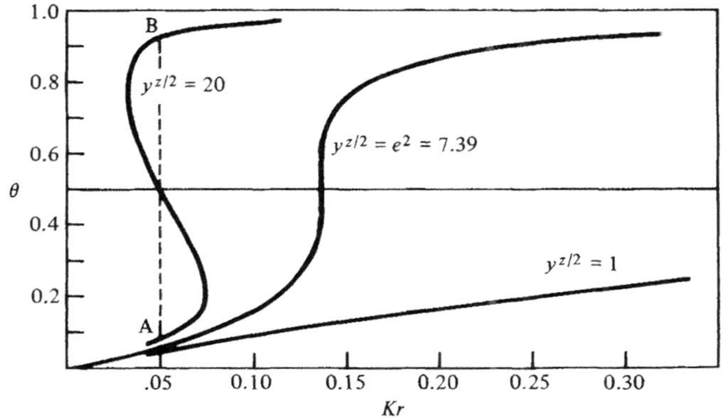

Fig. 11-1. θ as a function of Kr in the equilibrium BW approximation. The left-hand curve shows a phase transition; the central curve is the critical curve.

The BW approximation exaggerates the positive cooperativity. For example, for a square lattice ($z = 4$), BW gives $y_c = e = 2.72$, whereas the exact value [Eq. (41.1)] is $y_c = 5.83$. That is, stronger interactions are needed to achieve critical behavior than is suggested by BW. The quasichemical approximation[1] is an improvement over BW. It leads to

$$y_c = \left(\frac{z}{z-2}\right)^2 = 4 \quad \text{if} \quad z = 4. \tag{47.4}$$

This shortcoming in BW is also responsible for the incorrect prediction of a phase transition in the one-dimensional lattice, $z = 2$ (e.g., $y = 20$ in Fig. 11-1). The quasichemical approximation happens to be exact for a one-dimensional lattice; it predicts, correctly, $y_c = \infty$ in this case [Eq. (47.4)].

BW is unsuitable for systems with negative cooperativity ($y < 1$) because if y is very small the neighbors of a given unit in either state 1 or state 2 will tend to be in the opposite state. However, this is not allowed in the BW model: θ has the same value for both the central unit and its neighbors.

The BW Approximation at Steady-State

We start in a general way[2] and specialize to a two-state system in the next subsection. One advantage of the BW treatment is that it can be applied readily to a lattice of interacting enzyme molecules with an arbitrarily complicated kinetic diagram (see Section 50 for a four-state three-cycle diagram).

Consider an arbitrary transition pair $i \rightleftarrows j$ in the diagram of a particular ("central") molecule of the lattice. The unperturbed rate constants are α_{ij} and α_{ji}. The neighbors present an environment e. The interaction of the central molecule with e is w_{ie} when in state i and w_{je} when in state j. The perturbed rate constants are then [Eqs. (5.6) and (5.8)]

$$\alpha_{ije} = \alpha_{ij} e^{f_{ij}(w_{ie}-w_{je})/k_B T}, \tag{47.5}$$

$$\alpha_{jie} = \alpha_{ji} e^{(1-f_{ij})(w_{je}-w_{ie})/k_B T}, \tag{47.6}$$

where f_{ij} is the kinetic parameter (like f_α, f_β, etc.) for the particular transition pair $i \rightleftarrows j$.

We now introduce the BW approximation, taking the same point of view about the neighbors of a given molecule as in the equilibrium case above. Let p_i (an unknown, to begin with) be the probability of any state i for the "central" molecule. Then, in the BW approximation, the same p_i applies to the nearest neighbors. The interaction free energy between the central molecule in state i and its z nearest neighbors is then $w_{ie} = z\sum_k p_k w_{ik}$, in which the sum is over all possible states k. Similarly, if the central molecule is in state j, $w_{je} = z\sum_k p_k w_{jk}$. Thus, the perturbed rate constants for the arbitrary transition pair ij of the kinetic diagram are, from Eqs. (47.5) and (47.6),

$$\begin{aligned} \alpha_{ije} &= \alpha_{ij} \exp[z f_{ij} \sum_k p_k (w_{ik} - w_{jk})/k_B T], \\ \alpha_{jie} &= \alpha_{ji} \exp[z(1 - f_{ij}) \sum_k p_k (w_{jk} - w_{ik})/k_B T]. \end{aligned} \tag{47.7}$$

The steady-state probabilities p_i of the central molecule can now be expressed, in the same way as for an independent molecule, in terms of the rate constants α_{ije} of the kinetic diagram. However, the α_{ije} values in the present problem themselves depend on the p_i [Eqs. (47.7)]. Thus, we have a set of implicit equations in the p_i, together with $\sum_i p_i = 1$, from which a self-consistent set of p_i must be found (numerically). The flux is then calculated from the α_{ije} and the p_i as for independent molecules.

Molecules with a Two-State Cycle[2,3]

As an explicit example, consider the two-state cycle in Fig. 11-2. The perturbed rate constants, from Eqs. (47.7), are

$$\begin{aligned} \alpha_e &= \alpha Y^{f_\alpha}, \qquad & \alpha'_e &= \alpha' Y^{f_\alpha - 1}, \\ \beta_e &= \beta Y^{-f_\beta}, \qquad & \beta'_e &= \beta' Y^{1-f_\beta}, \end{aligned} \tag{47.8}$$

where Y is defined in Eq. (47.3) and $\theta = p_2 = 1 - p_1$. That is, Y depends on θ. Also, $p_2 = \theta$ as a function of rate constants is

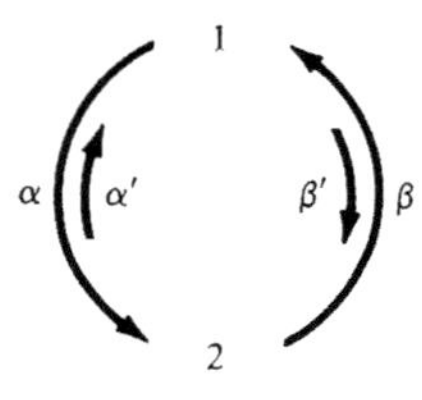

Fig. 11-2. Two-state cycle with unperturbed first-order rate constants.

$$\theta = \frac{\alpha_e + \beta_e'}{\alpha_e + \beta_e' + \beta_e + \alpha_e'}. \tag{47.9}$$

Equation (47.9) is the one implicit equation, here, in the one unknown θ. Having calculated θ from Eq. (47.9), the flux per molecule is given by

$$J = \frac{\alpha_e \beta_e - \alpha_e' \beta_e'}{\alpha_e + \beta_e' + \beta_e + \alpha_e'}. \tag{47.10}$$

Equations (47.9) and (47.10) are the familiar expressions for independent molecules. Here, each "independent" molecule is operating in the mean field of its neighbors.

We define, as before,

$$x \equiv \frac{\alpha + \beta'}{\beta + \alpha'}. \tag{47.11}$$

In the special case $f_\alpha + f_\beta = 1$, Eq. (47.9) takes the quasiequilibrium form

$$\theta = \frac{Yx}{1 + Yx}, \qquad Yx = \frac{\theta}{1 - \theta}. \tag{47.12}$$

Here x plays the role of K in Eq. (47.2). Thus a steady-state phase transition can occur. Also, in this special case, Eq. (47.10) becomes

$$J = \frac{J_o Y^{f_\alpha}(1 + x)}{1 + Yx}, \tag{47.13}$$

where the unperturbed flux is

$$J_o = \frac{\alpha\beta - \alpha'\beta'}{\alpha + \beta' + \beta + \alpha'}. \tag{47.14}$$

For numerical calculations, it is convenient to put Eqs. (47.9) and (47.10) in a form equivalent to that used for Monte Carlo calculations in Chapter 10. We again adopt $\beta \equiv 1$ as a reference rate constant and use Eqs. (40.3)–(40.5). The parameters β, α', α, and β' are thus replaced by $\beta \equiv 1$, α', x, and F. The unperturbed flux then becomes

$$J_o = \frac{\alpha' x(F - 1)}{(1 + \alpha' F)(1 + x)}. \tag{47.15}$$

In Michaelis–Menten kinetics,

$$F \to \infty, \qquad \beta' \to 0, \qquad \alpha \to (1 + \alpha')x, \qquad J_o \to \frac{x}{1 + x}. \tag{47.16}$$

If we replace β, α, and β' in Eqs. (47.8) by $\beta = 1$ and Eqs. (40.4) and (40.5), Eqs. (47.9) and (47.10) for θ and J become

$$x = \frac{(1 + \alpha'F)\theta(1 + \alpha'Y^{f-1})}{(1 + \alpha')(1 - \theta)Y(1 + \alpha'FY^{f-1})}, \tag{47.17}$$

$$J = \frac{\alpha'(1 + \alpha')xY^{f_\alpha}(F - 1)}{(1 + \alpha')xY(1 + \alpha'FY^{f-1}) + (1 + \alpha'F)(1 + \alpha'Y^{f-1})}, \tag{47.18}$$

where $f \equiv f_\alpha + f_\beta$. Equation (47.17) gives x as a function of θ; this can be used in Eq. (47.18) to calculate J. Note that f_α and f_β appear in Eq. (47.17) only as the sum f (in fact, only as $f - 1$). The principal interest is in the functions $\theta(x)$ and $J(x)$ for various assignments of the parameters α', F, r, y, f_α, and f_β.

The equilibrium case is $F = 1$. Then $J = 0$ and

$$x = \frac{\theta}{(1 - \theta)Y} \tag{47.19}$$

as in Eq. (47.2), with x in place of K. The kinetic parameters α', f_α, and f_β drop out of the problem in this case. As $x \to 0$, $\theta \to rx$. Critical properties, from Eq. (47.19), are $y_c = e^{4/z}$, $\theta_c = 1/2$, and $x_c = 1/re^2$.

It suffices, in numerical examples, to examine values of F in the range $F \geq 1$. This follows because values in the interval $0 \leq F < 1$ (i.e., X is negative) are related to $F > 1$ by a symmetry property. If, in Eqs. (47.17) and (47.18) without changing r or y, we replace F by F^{-1}, α' by α'^{-1}, f_α by $1 - f_\beta$, and f_β by $1 - f_\alpha$ (thus, f by $2 - f$), we find that $\theta(x)$ is unchanged, while $J(x)$ changes sign and is multiplied by a factor α'^{-1}. The enzyme molecules cycle backward with an altered flux. This same property holds in the exact system [Eqs. (42.2)].

Both θ and J are linear in x near $x = 0$:

$$\theta \to \frac{(1 + \alpha')(1 + \alpha'Fr^{f-1})rx}{(1 + \alpha'F)(1 + \alpha'r^{f-1})}, \tag{47.20}$$

$$J \to \frac{\alpha'(1 + \alpha')r^{f_\alpha}(F - 1)x}{(1 + \alpha'F)(1 + \alpha'r^{f-1})}. \tag{47.21}$$

When $r = 1$, these simplify to

$$\theta \to x, \qquad J \to \frac{\alpha'(F - 1)x}{(1 + \alpha'F)}, \tag{47.22}$$

i.e., near $x = 0$, θ and J behave like θ_o and J_o (when $r = 1$). This follows because the first state 2 molecules to appear in the lattice (as x is increased starting from $x = 0$), as a result of $1 \to 2$ transitions, have not altered their interactions with their environment (which consists of all state 1 molecules), i.e., $y_{12} = y_{11}$ when $r = 1$.

In the limit $x \to \infty$, we have $\theta \to 1$ and

$$J \to \frac{\alpha'(ry^z)^{f_\alpha - 1}(F - 1)}{1 + \alpha' F (ry^z)^{f-1}}. \tag{47.23}$$

The corresponding limit for J_o is

$$J_o \to \frac{\alpha'(F - 1)}{1 + \alpha' F}. \tag{47.24}$$

In the Michaelis–Menten case, $F \to \infty$, these become

$$J \to (ry^z)^{-f_\beta}, \qquad J_o \to 1. \tag{47.25}$$

In the quasiequilibrium, $f = 1$, case, we find Eq. (47.19) for $\theta(x)$, i.e., equilibrium behavior at any F, but a nonequilibrium flux:

$$J = \frac{\alpha' x Y^{f_\alpha}(F - 1)}{(1 + \alpha' F)(1 + xY)}. \tag{47.26}$$

In the important limit $F \to \infty$ (Michaelis–Menten), Eqs. (47.17) and (47.18) become, respectively,

$$x = \frac{\theta(1 + \alpha' Y^{f-1})}{(1 + \alpha')(1 - \theta) Y^f}, \tag{47.27}$$

$$J = \frac{(1 + \alpha') x Y^{f_\alpha}}{(1 + \alpha') x Y^f + (1 + \alpha' Y^{f-1})}. \tag{47.28}$$

The one-way cycle case is $\alpha' = 0$, $\beta' = 0$. Using $\beta = 1$ as usual, we have the simplified relations

$$x = \alpha, \qquad \alpha_e = xY^{f_\alpha}, \qquad \beta_e = Y^{-f_\beta}, \tag{47.29}$$

$$\theta = \alpha_e/(\alpha_e + \beta_e) = xY^f/(1 + xY^f), \tag{47.30}$$

$$x = \theta/(1 - \theta)Y^f, \tag{47.31}$$

$$J = \alpha_e \beta_e/(\alpha_e + \beta_e) = \beta_e \theta = xY^{f_\alpha}/(1 + xY^f). \tag{47.32}$$

Equations (47.31) and (47.32) also follow from Eqs. (47.27) and (47.28) ($F = \infty$, $\beta' = 0$) on putting $\alpha' = 0$.

The unusual feature of Eqs. (47.30) and (47.31) is that Y appears only in the form Y^f, which is to say that the three w_{ij} occur in $\theta(x)$ only as the products fw_{ij}. Consequently, if there is a $\theta(x)$ phase transition in this one-way steady-state system for some particular choice of the w_{ij} and f, the *same* $\theta(x)$ phase transition will occur in another case with parameters $-w_{ij}$ and $-f$. If, in fact, we change the sign of *both* f_α and f_β (rather than just the sign of the sum $f_\alpha + f_\beta$), the behavior of $J(x)$ will also be the same in the case $-w_{ij}$, $-f_\alpha$, $-f_\beta$ as in the case $w_{ij}, f_\alpha, f_\beta$. It is therefore possible in principle to have a first-order phase transition in a one-way steady-state system with $y < 1$.

As pointed out (for a two-state cycle) in the discussion of Fig. 10-15, the above conclusions about sign changes apply as well to an *exact* treatment of a lattice of interacting enzyme molecules, each with a one-way cycle with an arbitrary number of states ($\alpha_{ji} = 0$; f_{ij}, w_{ie}, $w_{je} \rightarrow -f_{ij}$, $-w_{ie}$, $-w_{je}$, where e is the "environment" of neighbors).

It should also be noted that Eq. (47.31) has the same form as the equilibrium equation (47.19), but with fw_{ij} replacing w_{ij}. This is a "false equilibrium" with an apparent "detailed balance."

Numerical Examples (Two-State Cycle)

An extensive survey of BW steady-state examples has been published.[3] In many cases there is rather exotic behavior predicted (some of which can be attributed to the approximation itself). We present only one rather conventional family of cases here.

We select the special case $F = 20$, $f = 0.5$, $\alpha' = 1$, $r = 1$. Figure 11-3

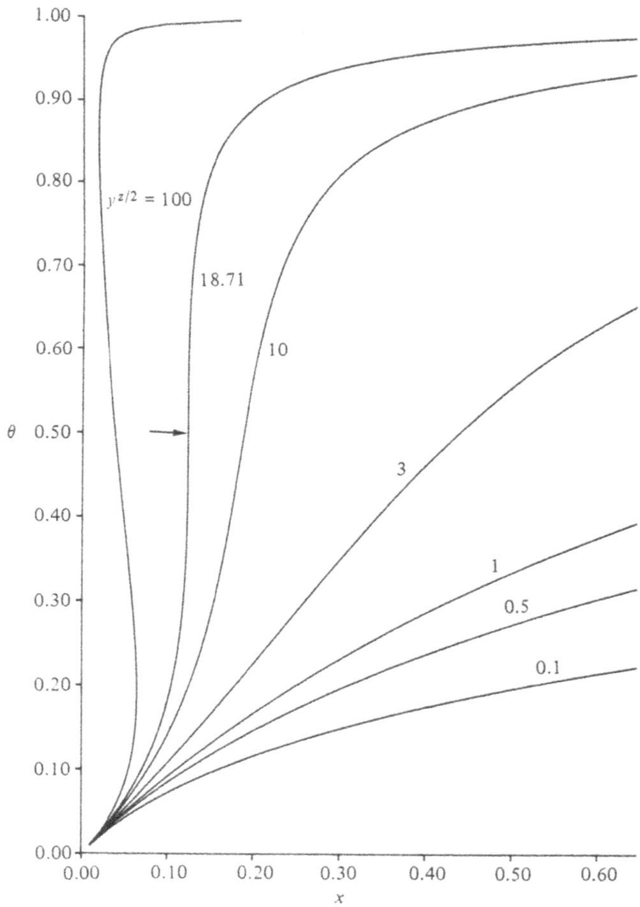

Fig. 11-3. Family of $\theta(x)$ curves for $F = 20$, $\alpha' = 1$, $r = 1$, $f = 0.5$. The critical point is marked by an arrow.

shows $\theta(x)$ curves for several values of $y^{z/2}$, calculated from Eq. (47.17). The behavior is qualitatively similar to Fig. 11-1, for an equilibrium system. However, the critical value of y is larger in Fig. 11-3: $y_c^{z/2} = 18.71$. Other critical properties are $\theta_c = 0.4980$ and $x_c = 0.1229$. Note that $\theta_c \neq \frac{1}{2}$; hence the symmetry of the equilibrium system has been lost (this is much more apparent in other cases). There is a loop and phase transition, at steady state, for $y > y_c$. Where is the vertical cut (see AB in Fig. 11-1) across the loop located? This is the subject of Section 48.

The $\theta(x)$ curves in Fig. 11-3 depend on f but not on f_α and f_β separately. Thus the critical curve holds for any combination $f_\alpha + f_\beta = 0.5$. This is not true for the flux curves, $J(x)$. Figure 11-4 shows a family of critical $J(x)$ curves [Eq. (47.18)], all with $f = \frac{1}{2}$ and $y_c^{z/2} = 18.71$. However, the choices of f_α and f_β are different. There is a cusp at the critical point when $f_\alpha = f_\beta$. When $f_\alpha > f_\beta$, the flux is larger in phase 2 (predominantly state 2) than in phase 1. The opposite is true when $f_\alpha < f_\beta$. The $f_\alpha = 1.0, f_\beta = -0.5$ curve goes off scale: $J_c = 1.67$ and $J(\infty) = 9.18$. The unperturbed flux [Eq. (47.15)] at $x = \infty$ is $J_o = 0.905$.

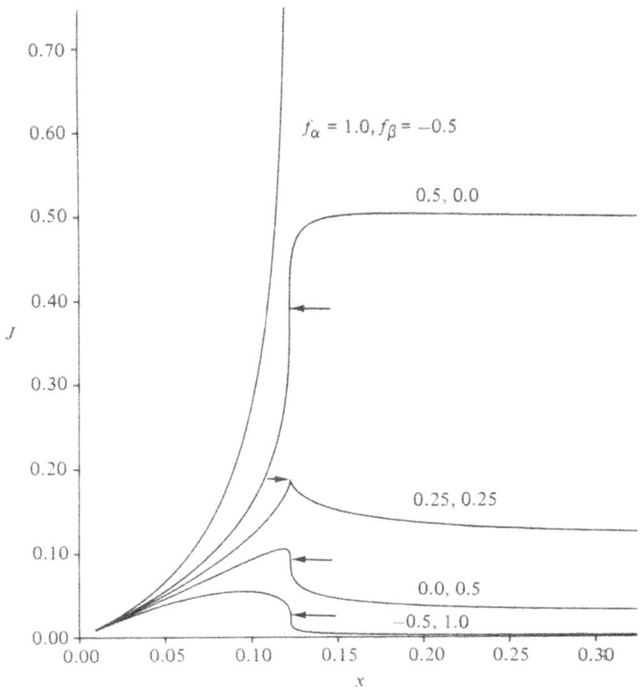

Fig. 11-4. Family of $J(x)$ curves (different f_α, f_β), all with $y_c^{z/2} = 18.711$, for $F = 20$, $\alpha' = 1$, $r = 1$, $f = 0.5$. Critical points are marked by arrows. For $f_\alpha = 1$, $f_\beta = -\frac{1}{2}$: $J_c = 1.67$ (off scale).

48. Maxwell Equal-Area Theorem for Two-State Cycles

If, for a lattice of interacting units at steady state, any approximate theory is used that produces a phase-transition loop in $\theta(x)$, where should the cut across the loop be placed? We discuss this subject in this section for two-state cycles, where the answer is simple, and in the next section for three-state cycles, where it is not.

Review for a One-Component Equilibrium System[4]

Certain features of this case provide the general approach to the steady-state problem as well. Hence a very brief summary is called for. Consider a lattice gas of M sites, N of which are occupied by molecules. There are attractive interactions between the molecules, leading to a phase transition. If an approximation of the BW or quasichemical type is used on the closed system (independent variables N, M, T), a loop is obtained, for example, in a plot of $\theta \equiv N/M$ against $\lambda = e^{\mu/kT}$. However, if the open version of the same system, to the same approximation, is treated (fluctuations in N are allowed; independent variables μ, M, T), the loop in θ plotted against λ is now missing and a vertical (or almost vertical, if M is large but finite) line takes its place. The "open" vertical line cuts the "closed" loop at just that λ prescribed by Maxwell's equal-area theorem (applied to a plot of θ against $\ln \lambda$).

The probability that the open system contains N molecules is proportional to $R(N) \equiv Q(N, M, T)\lambda^N$, where Q is the canonical partition function. Let $P(N)$ be $R(N)$ normalized. At an "ordinary" point, $P(N)$ has a single peak centered at $N = \theta(\lambda)M$ for a given λ. However, there are two peaks in $P(N)$ at or near a phase transition, at $N = \theta_1 M$ and $\theta_2 M$, one peak corresponding to each phase (with θ_1 or θ_2). At the center of the phase transition (with λ located by the vertical line mentioned above), $\theta = (\theta_1 + \theta_2)/2$, and the two peaks in $P(N)$ necessarily have the same area or weight.

Thus, although it is not at all necessary at equilibrium, the cut across the "closed" loop *could* be located on the λ axis by passing to the open system and noting where the vertical step in the plot of θ against λ occurs. Alternatively, and completely equivalently, one could note where the two peaks in $P(N)$ have the same area or weight. Exactly the same procedure (based on passing from the closed to the open system) can be followed at steady state. This procedure provides a general method, at least in principle, for steady-state systems. This is noteworthy because, as we shall see, thermodynamic-like methods cannot be applied to an arbitrary closed steady-state system—in fact, such methods can be applied, in general, only to a steady-state system comprised of *two-state* molecules.

Steady-State System of Two-State Molecules[4]

In this subsection we show that the treatment of any steady-state system of two-state enzyme molecules, using an approximation of the BW type, can be arranged to parallel the equilibrium treatment of the same kind of system. Hence we find a Maxwell theorem, etc.

We have a lattice system of M interacting enzyme molecules, where M is very large. Each molecule has an unperturbed (i.e., no interactions) kinetic diagram as shown in Fig. 11-5(a). Each molecule may be in state 1 or state 2; the system is therefore "open" with respect to the number of molecules N in state 2. In the general interaction problem, the open system has 2^M separate states that must be considered. The kinetic diagram for the entire system is therefore extremely complicated. However, in the BW and other approximations all of the $M!/N!(M - N)!$ states with exactly N molecules in state 2 are equivalent and hence can be grouped together. Therefore the kinetic diagram for the complete system simplifies tremendously: the diagram is *linear*, as shown in Fig. 11-5(b), with system states characterized by the value of the single variable N. The circumstance that, in the BW approximation, the kinetic diagram for the system is linear is the key point: the equilibrium-like treatment below is practically an automatic consequence of this feature.

In Fig. 11-5, α_{N-1} is the rate constant for the α transition $1 \to 2$ of any one of the $M - N + 1$ molecules in state 1 (when $N - 1$ molecules are in state 2). The value of α_{N-1} is α multiplied by a factor that takes nearest neighbors into account, according to the BW or other approximation. Similar remarks apply to the other three rate constants in Fig. 11-5.

If there are no interactions, $\alpha_{N-1} = \alpha$, etc., in Fig. 11-5. In this case Fig. 11-5(b) does not represent an approximation. If we define R_N as proportional to the steady-state probability of system-state N, with $R_0 \equiv 1$, then the R_N can be found *seriatim* (because of the linear diagram):

$$R_0 = 1, \qquad R_1 = MxR_0, \qquad R_2 = (M - 1)xR_1/2, \text{ etc.}, \tag{48.1}$$

where $x = (\alpha + \beta')/(\beta + \alpha')$. Thus we find

$$R_N = M!x^N/N!(M - N)! \qquad (0 \le N \le M). \tag{48.2}$$

In effect, because of the linear diagram, there is "detailed balance" between successive states of the diagram. At equilibrium, in lattice gas notation, $x = q\lambda$, where q ($\equiv q_2$) is the partition function of a single molecule in state 2 (with $q_1 \equiv 1$). That is,

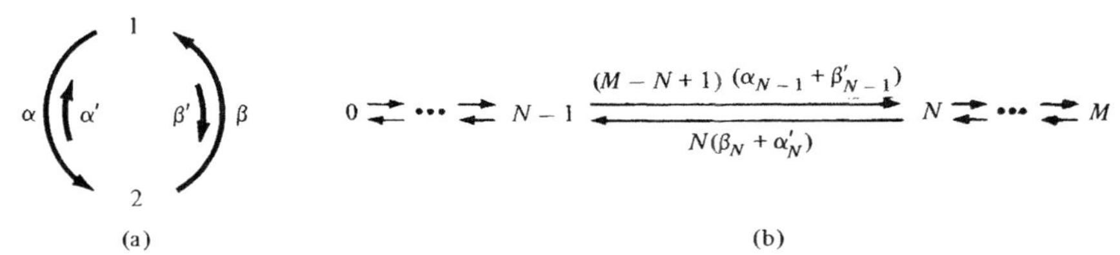

Fig. 11-5. (a) First-order rate constants and diagram for two-state enzyme molecule in the absence of interactions. (b) Kinetic diagram for a system of M two-state molecules, according to the BW (mean field) approximation. N is the number of molecules in state 2.

$$R_N = Q_N\lambda^N, \qquad Q_N = M!q^N/N!(M - N)!, \tag{48.3}$$

where Q_N is the canonical partition function. Equation (6.22) is equivalent.

With interactions included, R_N can still be written as $Q_N\lambda^N$ at equilibrium, either in an exact or in an approximate treatment. However, at steady state, with interactions, R_N cannot, *in an exact treatment,* be put in the form $\Gamma_N x^N$, as in Eq. (48.2). This follows because there is no detailed balance, real or simulated, in the exact system diagram at an arbitrary steady state, even with two-state enzyme molecules.

In the BW or other approximation, however, with interactions included, the linear diagram in Fig. 11-5(b) does allow steady-state R_N of the form $\Gamma_N x^N$. To arrange this, we make the same change of independent kinetic variables as in previous sections: instead of $\beta, \alpha', \alpha, \beta'$, we use $\beta \equiv 1, \alpha', x, F$ [Eqs. (40.3)–(40.5)]. Both α and β' are proportional to x. Consequently, when we write R_1/R_0, R_2/R_1, etc., essentially as in Eq. (48.1), a factor x appears at each stage (each α_{N-1} and β'_{N-1} have factors α and β', respectively, which in turn have factors x). Thus, by virtue of a simulated "detailed balance," we again obtain $R_N = \Gamma_N x^N$, but here Γ_N includes not only $M!/N!(M - N)!$, as in Eq. (48.2), but also a factor involving the intermolecular interactions of the model as taken care of by the approximation.

The final ingredient required, before proceeding to the main argument, is the observation that, when M is very large, the steady-state system we are considering here will have extensive and intensive properties, just as an equilibrium system does. Thus, for example, if we double the size of the system (M), holding all kinetic and interaction parameters constant (these are intensive variables), then all other extensive variables will also double; for example, the total enzyme flux, $\ln \Gamma_N$ [see Eq. (48.2), for example], N, etc.

The method to be outlined below is critically dependent on the definitions of (a) an intensive variable x that appears at each stage R_1/R_0, R_2/R_1, etc., and of (b) an extensive variable $\ln \Gamma_N$ (N large) such that the unnormalized probability ($R_0 \equiv 1$) R_N has the form $\Gamma_N x^N$.

The normalized probability $P(N)$ (we discontinue the use of subscript N in the remainder of this subsection) that the open BW system has N molecules in state 2 is

$$P(N; x, M) = R(N; x, M)/S(x, M), \tag{48.4}$$

$$R(N; x, M) = \Gamma(N, M)x^N, \quad S(x, M) = \sum_N R(N; x, M). \tag{48.5}$$

Here, S is analogous to a grand partition function and Γ to a canonical partition function (we are omitting the temperature T from the notation). It follows from Eqs. (48.4) and (48.5) that

$$\overline{N} = x(\partial \ln S/\partial x)_M, \tag{48.6}$$

$$\sigma^2/M \equiv (\overline{N^2} - \overline{N}^2)/M = x\partial\theta/\partial x, \tag{48.7}$$

where $\theta = \overline{N}/M$. The variance σ^2 is of order M; the peak in $P(N)$, for an "ordinary" point, occurring at, say, $N = N^*$ will be extremely sharp if M is very large. Therefore, we can replace $\ln S$ by the logarithm of the maximum term in S.

To do this, we use

$$\ln R = \ln \Gamma + N \ln x, \tag{48.8}$$

$$\partial \ln R/\partial N = 0 = (\partial \ln \Gamma/\partial N)_M + \ln x. \tag{48.9}$$

The value of N that satisfies Eq. (48.9) is $N = N^*$. In the equations below it is to be understood that N represents N^*. In effect, we thereby treat the open system as virtually closed, because of the small fluctuations in N. In Eq. (48.8) we replace $\ln R$ by $\ln S$ and use Eq. (48.9) to eliminate $\ln x$:

$$\ln S = \ln \Gamma - N(\partial \ln \Gamma/\partial N)_M. \tag{48.10}$$

We also have

$$d \ln \Gamma = (\partial \ln \Gamma/\partial N)_M \, dN + (\partial \ln \Gamma/\partial M)_N dM, \tag{48.11}$$

$$\ln \Gamma = N(\partial \ln \Gamma/\partial N)_M + M(\partial \ln \Gamma/\partial M)_N, \tag{48.12}$$

where Eq. (48.12) follows (Euler's theorem) because $\ln \Gamma$, N, and M are extensive properties, while the two derivatives are intensive properties of the system. Upon comparing Eqs. (48.10) and (48.12), we have

$$\ln S = M\Phi, \qquad \Phi \equiv (\partial \ln \Gamma/\partial M)_N. \tag{48.13}$$

Here Φ is the analogue of an equilibrium pressure or spreading pressure (in dimensionless units).

Equations (48.10) and (48.11) can be rewritten as

$$\ln S = M\Phi = \ln \Gamma + N \ln x, \tag{48.14}$$

$$d \ln \Gamma = -\ln x \, dN + \Phi dM. \tag{48.15}$$

Thus we find

$$d \ln S = d(M\Phi) = \Phi dM + N d \ln x \tag{48.16}$$

and also

$$d\Phi = \theta \, d \ln x, \qquad \Phi(x) = \int_0^x \theta(x') d \ln x', \tag{48.17}$$

where $\theta = N/M$. Equation (48.17) shows how to calculate Φ if $\theta(x)$ is known (e.g., from the BW treatment of the closed system).

The above equations all have well-known thermodynamic analogues, but it should be noted particularly that thermodynamics *per se* has not been used in the derivation.

We turn now to the situation at a phase transition. In the open steady-state system, just as at equilibrium, $P(N)$ will have two sharp peaks (phases) centered at $N_1 = \theta_1(x)M$ and $N_2 = \theta_2(x)M$. Suppose the two peaks have equal areas (total probabilities) at $x = x_t$. Then, at this x,

$$\theta = \bar{N}/M = (\theta_1 + \theta_2)/2. \tag{48.18}$$

This is where the vertical step in $\theta(x)$ occurs. For x slightly different from x_t, one or the other peak will be very small (a metastable state, possibly followed in a hysteresis loop). For an x at or near x_t, let $R_1(x)$ and $\sigma_1^2(x)$ be the largest $R(N)$ and the variance of the first peak, respectively, and let $R_2(x)$ and $\sigma_2^2(x)$ have the same meaning for the second peak. The condition that the two peaks have equal areas (to locate x_t) is equivalent to, for Gaussian peaks,

$$R_1(x_t)\sigma_1(x_t) = R_2(x_t)\sigma_2(x_t) \tag{48.19}$$

or, to terms of macroscopic order of magnitude,

$$\ln S_1(x_t) = \ln S_2(x_t) \qquad \text{or} \qquad \ln R_1(x_t) = \ln R_2(x_t). \tag{48.20}$$

Here, S_1 represents the sum in Eq. (48.5) over the first peak, etc., and the second of Eqs. (48.20) follows because $\ln R_i$ is of order M, while $\ln \sigma_i$ is of order $\ln M$. This is, of course, also the justification for the maximum term procedure in Eqs. (48.8)–(48.10).

Finally, we note that the equal peak area condition, $\ln S_1 = \ln S_2$, is the same as the condition $\Phi_1 = \Phi_2$, in view of Eq. (48.13). Thus, if we have derived $\theta(x)$ for a closed BW system, we can find $\Phi(x)$ from Eq. (48.17) and then locate the phase transition at $x = x_t$ by the condition $\Phi_1 = \Phi_2$. Although we have used the mathematically more complicated open system to deduce this procedure, we need not actually consider the open system in the *application* of the method. It suffices to work with the simpler closed system exclusively.

Maxwell's equal-area theorem holds here if applied to a plot of θ against $\ln x$. This follows from Eq. (48.17):

$$\Phi_2 - \Phi_1 = 0 = \int_1^2 \theta(x')d \ln x', \tag{48.21}$$

where the integral is over the loop from θ_1, x_t to θ_2, x_t. Maxwell's theorem is obviously equivalent to $\Phi_1 = \Phi_2$.

The above results exactly parallel those for an equilibrium system. Unfortunately, the formal resemblance is limited to two-state enzyme molecules.

The BW Model as an Explicit Example[4]

Equation (47.17) gives $x(\theta)$ for a closed steady-state BW system. If we use this equation and Eq. (48.17), after integrating by parts, we find for Φ, expressed as a function of θ,

$$\Phi = -\ln(1-\theta) - \left(\frac{z\theta^2}{2}\right)\ln y + \theta \ln\left[\frac{\alpha' + r^{1-f}\eta(\theta)}{\alpha' F + r^{1-f}\eta(\theta)}\right] + \left[\frac{H(\theta)}{z(1-f)\ln y}\right], \tag{48.22}$$

where

$$H(\theta) = \int_1^{\eta(\theta)} \ln\left(\frac{\alpha' F + r^{1-f}\eta'}{\alpha' + r^{1-f}\eta'}\right)\frac{d\eta'}{\eta'},$$
$$\eta(\theta) = y^{z(1-f)\theta}. \tag{48.23}$$

Usually the integral H must be evaluated numerically. Using Φ from Eq. (48.22), and Eq. (48.14),

$$\ln S = M\Phi, \qquad (1/M)\ln\Gamma = \Phi - \theta \ln x. \tag{48.24}$$

A number of examples of phase transitions (in the form of loops), based on Eq. (47.17), have been published.[3] In several of these cases, as a check, we have located x_t by: (a) use of Eq. (48.22) and $\Phi_1 = \Phi_2$; (b) numerical integration of Eq. (48.21) (Maxwell's theorem); and (c) for the open system, explicit step-by-step calculation of all the R_N and P_N (normalized) based on Fig. 11-5(b), for $M = 1000$, 5000, or 10,000. Equality of the two Gaussian peaks in P_N was used to find x_t. The three methods agree, as expected.

49. Location of a Phase Transition for BW Systems with Three-State Cycles

As a simple example of a more complicated system,[4] suppose we have M three-state enzyme molecules in a lattice. The diagram for a single unperturbed molecule is shown in Fig. 11-6. In the absence of interactions, the steady-state probabilities p_i for each of three states can easily be expressed in terms of the six rate constants α_{ij} belonging to Fig. 11-6. If we then define $x_2 \equiv p_2/p_1$ and $x_3 \equiv p_3/p_1$, the generalization of Eq. (48.2), with the same physical significance, is

$$R_{N_2N_3} = \frac{M!x_2^{N_2}x_3^{N_3}}{N_2!N_3!(M - N_2 - N_3)!}. \tag{49.1}$$

However, it is easy to see, from simple examples (M small), that for the BW approximation of the system *with interactions*, it is *not* possible in general to put $R_{N_2N_3}$ in the form

$$R_{N_2N_3} = \Gamma_{N_2N_3}x_2^{N_2}x_3^{N_3}, \tag{49.2}$$

with $\Gamma_{N_2N_3}$ independent of x_2 and x_3. This is a consequence of the absence of any sort of "detailed balance" in this case.

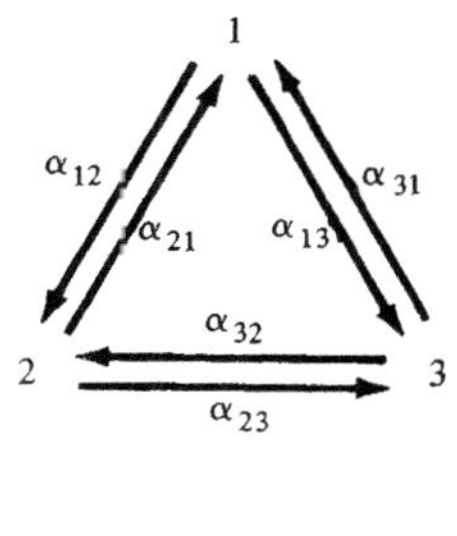

Fig. 11-6. First-order rate constants and cycle for a three-state enzyme molecule in the absence of interactions.

At equilibrium, for a two-component system (*not* based on Fig. 11-6), the right-hand side of Eq. (49.2) is $Q_{N_2N_3}\lambda_2^{N_2}\lambda_3^{N_3}$, in conventional notation, whether there are interactions or not and whether or not any approximation is used in handling the interactions. Because the three-state steady-state BW system does not follow this same pattern, equilibrium-like methods of locating a phase transition are inapplicable in this case (unlike the two-state steady-state situation above). This will generally be true for arbitrary steady-state BW systems.

To illustrate the approach that must be resorted to, we now consider an explicit BW three-state example.

BW Approximation for a One-Way Three-State Cycle[5]

We return to the model introduced in Fig. 10-26 and Eqs. (46.4)–(46.9). In the closed BW approximation, a_i/z in these equations is replaced by the averaged quantity p_i. Then Eqs. (46.7)–(46.9) become

$$\alpha_e = \alpha(s_2^{p_2}/s_1^{p_1})^{1/2}, \qquad \kappa_e = \kappa(s_3^{p_3}/s_2^{p_2})^{1/2}, \qquad \beta_e = (s_1^{p_1}/s_3^{p_3})^{1/2}, \tag{49.3}$$

where e refers to the average environment. Each molecule, in this approximation, is an "independent" molecule with an average environment. The kinetic diagram is that shown in Fig. 10-26(b). The steady-state probabilities for the closed BW system are then [from Fig. 10-26(b)]

$$p_1 = \frac{\kappa_e\beta_e}{\alpha_e\kappa_e + \kappa_e\beta_e + \beta_e\alpha_e}, \tag{49.4}$$

$$p_2 = \frac{\beta_e\alpha_e}{\alpha_e\kappa_e + \kappa_e\beta_e + \beta_e\alpha_e}, \tag{49.5}$$

$$p_3 = \frac{\alpha_e\kappa_e}{\alpha_e\kappa_e + \kappa_e\beta_e + \beta_e\alpha_e}. \tag{49.6}$$

Substitution of Eqs. (49.3) in Eqs. (49.4)–(49.6) provides implicit relations from which the p_i can be found numerically. Then the flux is obtained from

$$J = \alpha_e p_1 = \kappa_e p_2 = \beta_e p_3. \tag{49.7}$$

The above system is referred to as "closed" because the number of molecules in the steady-state lattice that are in state i, $N_i = p_i M$, in which M is the total number of molecules in the lattice ($M \to \infty$), has a definite nonfluctuating value in this treatment.

In a real enzyme lattice, on the other hand, the three N_is would of course fluctuate about their respective mean values $\overline{N}_i$ (just as in the Monte Carlo treatment in Section 46). This is what is meant by an "open" system. That is, the system is open with respect to the N_i (but not with respect to M).

In this section, we take over the basic BW approach, but apply it to a system with finite M and fluctuating N_i. Thus, suppose there are M equivalent molecules in the lattice (there are no edge or boundary effects even though M is finite) and let N_1, N_2, and N_3 have particular integral values, subject only to the conditions $N_1 + N_2 + N_3 = M$ and $N_i \geq 0$. Then, for this set of N_i, the rate constants of the kinetic cycle for the molecules in the lattice are assumed to be [compare Eqs. (49.3)]

$$\alpha_e = \alpha b_2^{N_2}/b_1^{N_1}, \qquad \kappa_e = \kappa b_3^{N_3}/b_2^{N_2}, \qquad \beta_e = b_1^{N_1}/b_3^{N_3}, \tag{49.8}$$

where $b_i = s_i^{1/2M}$. The rate constant α_e would apply to the N_1 molecules in state 1, κ_e to the N_2 molecules in state 2, and β_e to the N_3 molecules in state 3. In an exact Monte Carlo treatment (Section 46), these averaged rate constants and homogeneous groups (N_1, N_2, N_3) of molecules cannot be used. Instead, one must follow the stochastic history of each *individual* molecule in the lattice (with particular attention to the states of its nearest neighbors).

Say only N_2 and N_3 are independent variables; N_1 is given by $M - N_2 - N_3$. In the open BW approximation, detailed microstates of the lattice are grouped together according to N_2, N_3 values only. Let $P_{N_2N_3}$ be the probability that an open BW system of M molecules has particular values of N_2 and N_3. Then, from Eqs. (49.8), the kinetic master equation in $P_{N_2N_3}$ is

$$\begin{aligned} dP_{N_2N_3}/dt = {} & -P_{N_2N_3}[N_3(b_1^{M-N_2-N_3}/b_3^{N_3}) + \kappa N_2(b_3^{N_3}/b_2^{N_2}) \\ & + \alpha(M - N_2 - N_3)(b_2^{N_2}/b_1^{M-N_2-N_3})] \\ & + P_{N_2+1,N_3-1}(N_2 + 1)\kappa(b_3^{N_3-1}/b_2^{N_2+1}) \\ & + P_{N_2N_3+1}(N_3 + 1)(b_1^{M-N_2-N_3-1}/b_3^{N_3+1}) \\ & + P_{N_2-1,N_3}(M - N_2 - N_3 + 1)\alpha \\ & \times (b_2^{N_2-1}/b_1^{M-N_2-N_3+1}). \end{aligned} \tag{49.9}$$

As indicated schematically in the complete kinetic diagram for the whole lattice, Fig. 11-7, there are $(M + 1)(M + 2)/2$ substates N_2, N_3 and the same

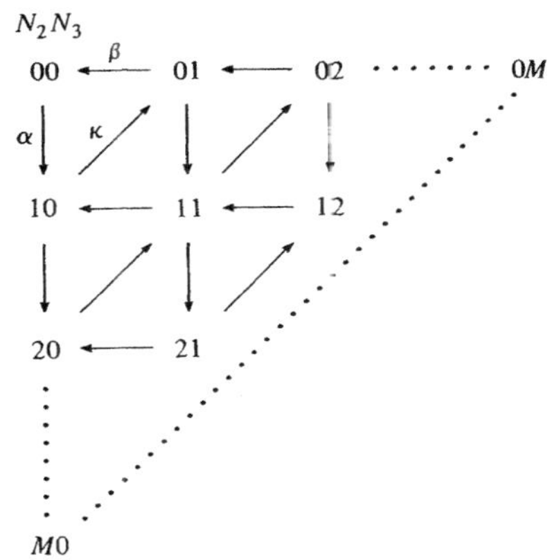

Fig. 11-7. Schematic kinetic diagram for an open system of M interacting enzyme molecules, according to the BW (mean-field) approximation. N_2 and N_3 are the numbers of molecules (out of M) in states 2 and 3.

number of equations [Eq. (49.9)]. However, along any one of the three edges in Fig. 11-7, two terms out of the six in Eq. (49.9) have to be dropped, and at any one of the three corners, four terms have to be dropped. Also, there is the normalization relation $\Sigma P_{N_2N_3} = 1$. At steady state, the left-hand side of Eq. (49.9) is zero. We then have to solve $(M + 1)(M + 2)/2$ linear algebraic equations (including normalization) in the same number of unknowns in order to find the steady-state $P_{N_2N_3}$. For example, if $M = 10$ or 20, the number of unknowns and equations is 66 or 231, respectively. Obviously, to go to very large values of M, a continuous version (differential equation) of Eq. (49.9) would be needed. However, we do not include this subject here.

If the set of steady-state $P_{N_2N_3}$ has been found from the algebraic equations, we can then calculate the three mean state probabilities $p_i(\alpha)$ and the flux $J(\alpha)$ (per molecule of the lattice) from

$$p_i(\alpha) = \overline{N}_i/\overline{M}, \qquad \overline{N}_i = \sum_{N_2N_3} N_i P_{N_2N_3}, \tag{49.10}$$

$$\begin{aligned} MJ &= \sum_{N_2N_3} P_{N_2N_3}\alpha(M - N_2 - N_3)b_2^{N_2}/b_1^{M-N_2-N_3} \\ &= \sum P_{N_2N_3}\kappa N_2 b_3^{N_3}/b_2^{N_2} = \sum P_{N_2N_3} N_3 b_1^{M-N_2-N_3}/b_3^{N_3}. \end{aligned} \tag{49.11}$$

The equality of the α, κ, and β fluxes at steady state can be used as a check. The variable α is usually proportional to a concentration.

Numerical Example

As a particular example, we choose the interesting bifurcation case in Ref. 6, in which there are two phase transitions: $s_1 = 1$, $s_2 = s_3 = 1000$, and $\kappa = 0.5$. Thus there are attractive 22 and 33 neighbor interactions. This is also the same system as in Figs. 10-28 through 10-30, except that $s_2 = s_3 = 6000$ in the Monte Carlo calculations (BW exaggerates the positive cooperativity). It is not possible to tell in the *closed* BW system where the two phase transitions occur (that is, where along the α axis the loops are to be cut vertically). However, loops cannot occur in the *open* BW system (the unstable part of a loop in the closed system corresponds to a minimum in $P_{N_2N_3}$ in the open system); the transition point can be located directly from the calculated (open) $p_i(\alpha)$ curves or, equivalently, from $P_{N_2N_3}(\alpha)$, which has two peaks of equal integrated magnitude at the transition point.

Figures 11-8 through 11-11 show p_1, p_2, p_3, and J, as functions of α, calculated for the *closed* BW system from Eqs. (49.3)–(49.7). The exact value $\kappa = 0.5$ produces the odd bifurcation behavior seen in these figures. Many other details are available in Ref. 6. The first transition (as α increases) occurs at about $\alpha = 0.026$ (see below): A → E via C (unstable). The second transition occurs at $\alpha = 0.29$ (see below), which is off scale. This transition is from branch P → N via Q (unstable). The same qualitative behavior (minus

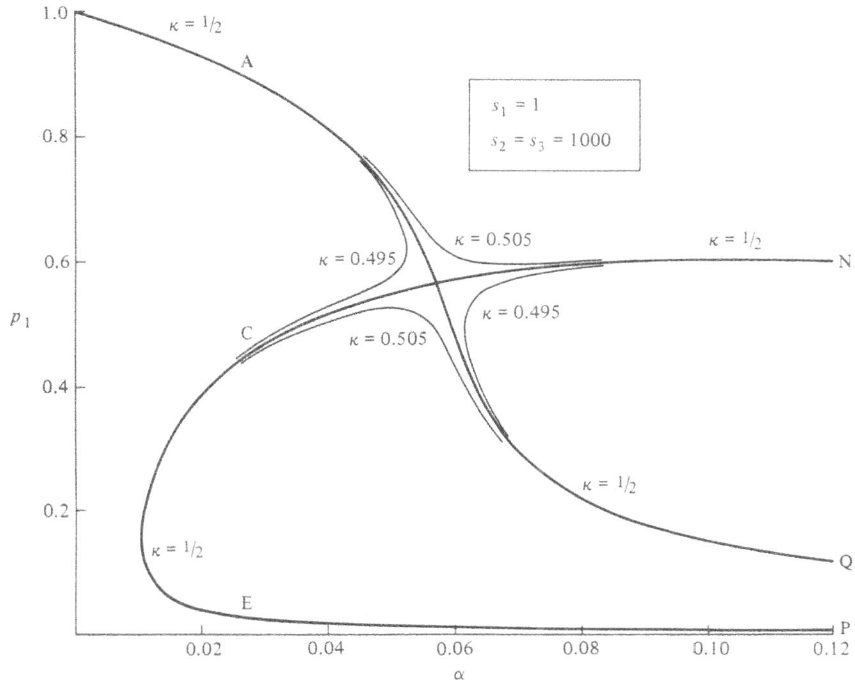

Fig. 11-8. Plot of state probability p_1 as a function of α for $s_1 = 1$, $s_2 = s_3 = 1000$ and $\kappa = 0.495$, 0.500, and 0.505.

the loops, of course) can be seen in the exact 10×10 Monte Carlo system, Fig. 10-28.

We turn now to the *open* BW calculation. The complete calculation for $M = 10$ and 20 has been made using the standard matrix inversion method to find $P_{N_2N_3}$. As a supplement, fragmentary calculations were carried out for $M = 30$ and $M = 50$, using an iteration procedure that fails, however, to provide practical convergence at and near the two transition points. With $M = 50$, there are 1326 unknowns ($P_{N_2N_3}$). Figure 11-12 shows the complete (up to $\alpha = 0.45$) $p_i(\alpha)$ curves for $M = 10$ and a partial $p_2(\alpha)$ curve for $M = 20$ (note the change in abscissa scale at $\alpha = 0.10$). Transitions occur where p_1 and p_2 cross ($\alpha \cong 0.028$ for $M = 10$) and where p_2 and p_3 cross ($\alpha \cong 0.29$). The second transition is much less precipitous than the first. Figures 11-8 through 11-10 give the corresponding closed BW ($M = \infty$) $p_i(\alpha)$ curves (with loops). The two transitions are also clearly of different types in the closed BW system. The p_1, p_2 and p_2, p_3 intersections, referred to above, are shown in more detail in Fig. 11-13, including several values of M. The location (on the α axis) of the p_1, p_2 intersection seems to be slightly M dependent, but the location of the p_2, p_3 intersection is not noticeably dependent on M.

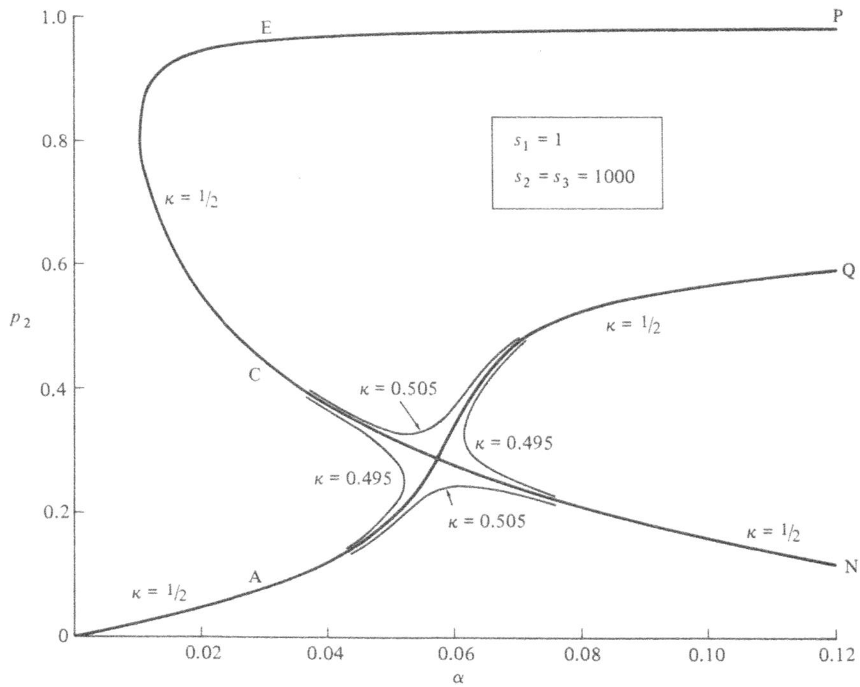

Fig. 11-9. Plot of $p_2(\alpha)$, as in Fig. 11-8 for p_1.

At $\alpha = 0.027$ for $M = 20$, $p_1 = 0.471$ and $p_2 = 0.482$ are almost equal. As expected, the corresponding set of 231 $P_{N_2N_3}$ values, arranged as in Fig. 11-7, has modest peaks of essentially equal integrated magnitude at the upper-left-hand (primarily state 1) and lower-left-hand (primarily state 2) corners.

At $\alpha = 0.29$ for $M = 20$, $p_2 = 0.461$ and $p_3 = 0.457$. The equal peaks in

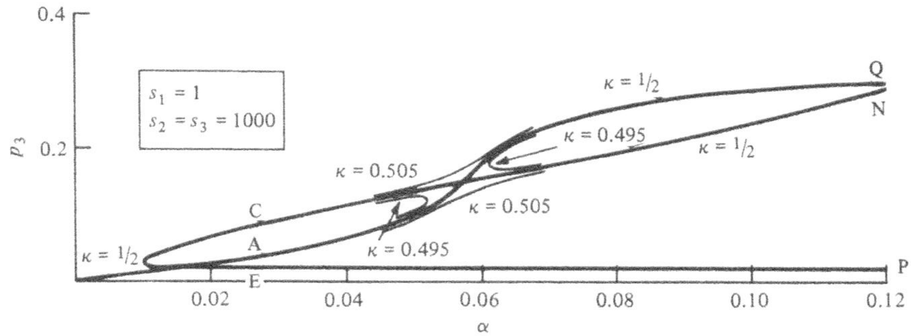

Fig. 11-10. Plot of $p_3(\alpha)$, as in Fig. 11-8 for p_1.

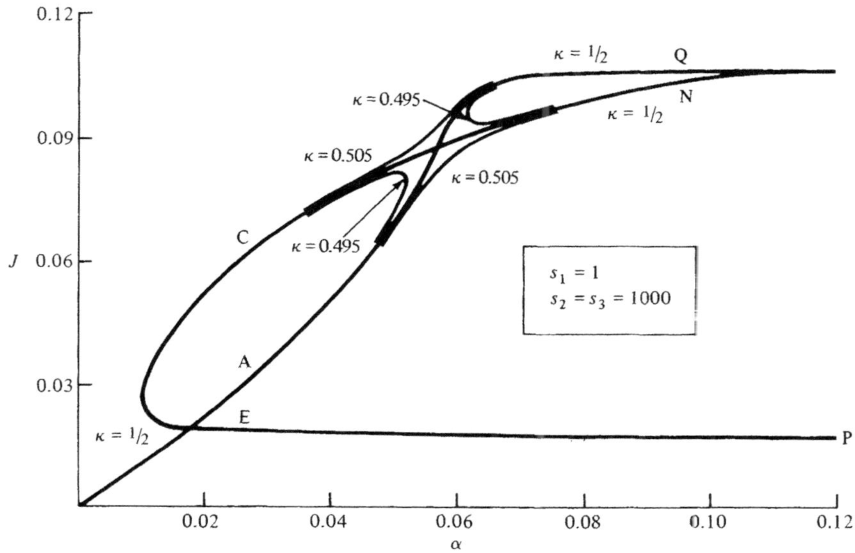

Fig. 11-11. Plot of $J(\alpha)$, as in Fig. 11-8 for p_1.

this case are at lower left (state 2) and upper right (state 3) of Fig. 11-7.

The flux J for $M = 10$ and 20 and segments of curves for $M = 30$ and 50 are shown in Fig. 11-14. The resemblance to the closed BW flux in Fig. 11-11 (on the path OAEPN) is obvious. The same is true of the Monte Carlo 10×10 flux in Fig. 10-28. Also included in Fig. 11-14 is the unperturbed flux for enzyme molecules with no interactions ($s_1 = s_2 = s_3 = 1$; $\kappa = 0.5$): $J_o = \alpha/(1 + 3\alpha)$.

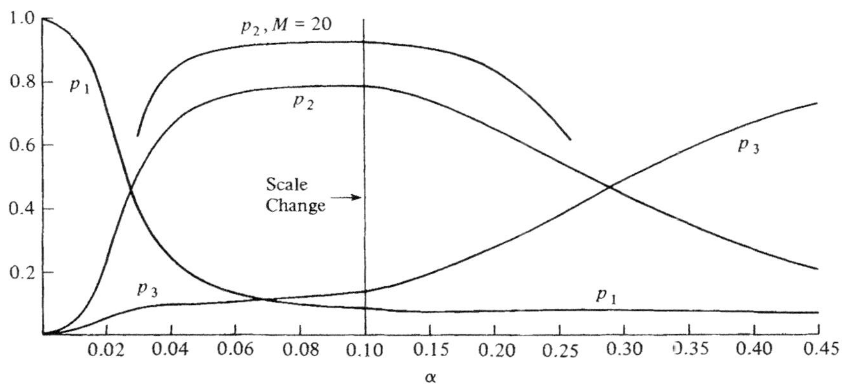

Fig. 11-12. Calculated mean state probabilities $p_1(\alpha)$, $p_2(\alpha)$, and $p_3(\alpha)$ for an open BW system with $M = 10$, $s_1 = 1$, $s_2 = s_3 = 1000$, and $\kappa = 0.5$. A partial $p_2(\alpha)$ for $M = 20$ is included.

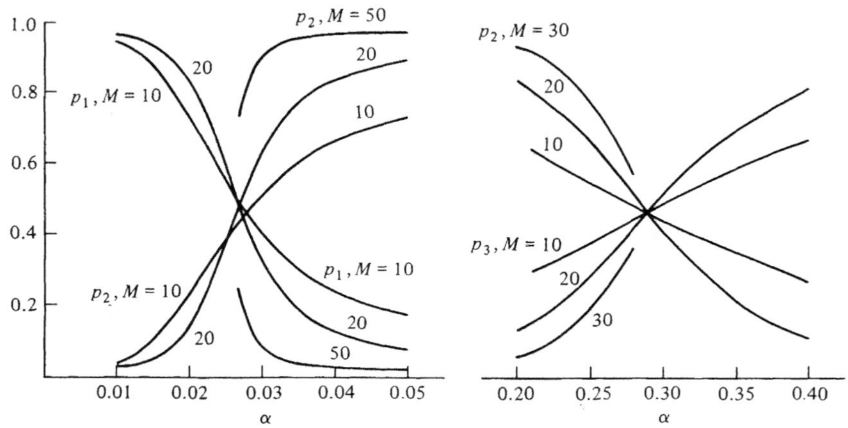

Fig. 11-13. Further details (see Fig. 11-12) on the p_1, p_2 intersection (*left*) and on the p_2, p_3 intersection (*right*), including M dependence.

Naturally, if one could push the open BW calculations to $M \to \infty$, one would expect exact agreement between open BW and closed BW results, except for the fact that vertical cuts in the former case would replace loops in the latter case. In this example, the BW approximation provides a good *qualitative* picture of the behavior of the exact Monte Carlo system (remember, also, that $M = \infty$ in BW, whereas $M = 100$ in the Monte Carlo system).

50. Steady-State Phase Transitions Among Biochemical Cycles

Phase transitions in Ising and related steady-state systems conventionally involve a sudden switch from the dominance of one of two discrete states to the other. However, some steady-state enzyme-lattice systems may also ex-

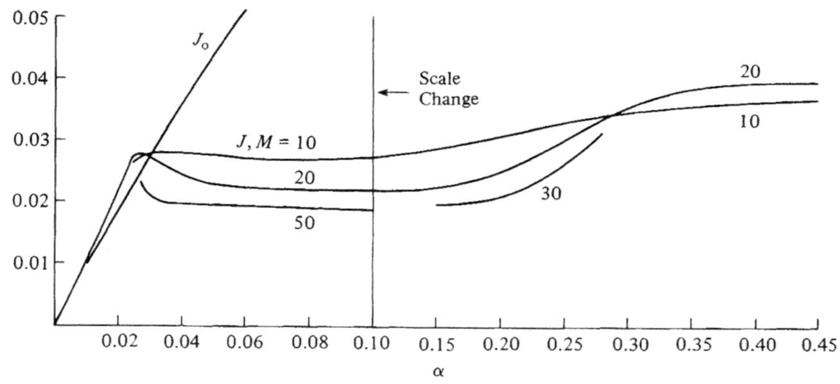

Fig. 11-14. Calculated flux per molecule, J, corresponding to Figs. 11-12 and 11-13. J_o is the unperturbed flux for comparison.

hibit a higher-order type of transition not possible at equilibrium.[7] This may occur if the kinetic diagram of the individual enzyme molecules of the lattice contains more than one cycle. In such cases the phase transition may bring about not only a sudden change in the dominant state or states but also a *sudden change in the dominant cycle or cycles*. Thus the predominant steady-state biochemistry taking place in the lattice may switch precipitously as a result of the phase transition. In the numerical example below, there is active transport of a ligand L (driven by ATP, say) after a phase transition but not before.

The simplest diagram that has more than one cycle and that allows coupling between two different thermodynamic forces is a diagram with four states and three cycles, as shown in Fig. 11-15(a). We use this diagram here because our very limited purpose in this section[7] is simply to illustrate, with one hypothetical example, the general phenomenon described in the preceding paragraph. It remains to be seen, in the future, whether any natural or artificial steady-state enzyme-lattice systems exhibit this type of behavior.

Our discussion above has been in terms of a sharp phase transition. Of course a cooperative but nonprecipitous transition among dominant cycles would also be of considerable interest. A gradual transition could occur either in a lattice with subcritical interaction free energies (see below) or in a small aggregate or complex (e.g., of from two to six subunits). In the latter (finite) case, sharp behavior would not be observed no matter how strong the interactions.

The Model and Some Analytical Properties

Figure 11-15(a) shows the four-state kinetic diagram we use. There are three cycles [Fig. 11-15(b)]; we take the counterclockwise direction as positive in all cycles. Without specifying the details of a biochemical mechanism, let us suppose for concreteness that in the positive direction in each case: cycle *A* transports a ligand L across a membrane from inside to outside; cycle *B*

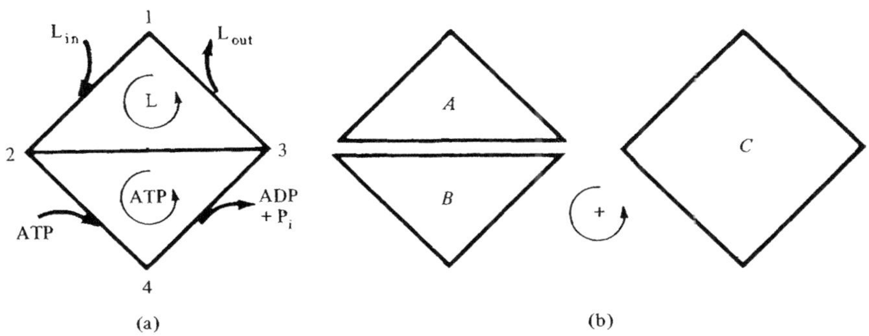

Fig. 11-15. (a) Four-state kinetic diagram for possible active transport of ligand L by ATP. (b) The three cycles that belong to this diagram.

hydrolyzes ATP; and cycle C does both of these. Our primary interest is in cases where the cycle flux J_A and thermodynamic force X_A (for L) are both negative, J_B and X_B (for ATP) are both positive, and J_C and $X_A + X_B$ (the net force in cycle C) are also both positive. Thus, the "downhill" direction for transport of L is out → in (negative direction), but the ATP force X_B is large enough to overcome the negative X_A in cycle C so that L is transported "uphill" (in → out) in cycle C.

The operational (total) flux in L is $J_L = J_A + J_C$, while the operational flux in ATP is $J_T = J_B + J_C$. If J_L is positive, there is net flux of L in → out, against its electrochemical potential gradient (i.e., there is active transport of L by ATP).

We use the BW approximation to take care of nearest-neighbor interactions in the lattice. The rate constants chosen are shown in Fig. 11-16. Here $\alpha < 1$, β is very small, γ is proportional to the ATP concentration (which can be varied), and u is an interaction parameter. Unity is used as a reference rate constant. To keep the mathematics as simple as possible, we assume one kind of interaction only: w_{44}, the free energy of interaction between neighboring molecules both in state 4, is taken as zero or negative relative to all other pair interactions (w_{ij}). The unperturbed rate constants in Fig. 11-16 are 1, α, β, and γ. The perturbed rate constants shown in Fig. 11-16 follow from the BW approximation and the arbitrary choices $f_{24} = \frac{1}{2}$ and $f_{43} = 0$. Thus

$$\frac{\alpha_{24e}}{\alpha_{42e}} = \frac{\alpha_{24}}{\alpha_{42}}\left(\frac{y_{44}}{y_{ij}}\right)^{zp_4} = \frac{\gamma(y_{44}/y_{ij})^{zp_4/2}}{\beta(y_{44}/y_{ij})^{-zp_4/2}} = \frac{\gamma u}{\beta u^{-1}}, \tag{50.1}$$

where

$$u \equiv s^{p_4}, \qquad s \equiv (y_{44}/y_{ij})^{z/2}, \tag{50.2}$$

and z is the nearest-neighbor number of the lattice. The final expressions in Eq. (50.1) are the perturbed rate constants that appear in Fig. 11-16 for $2 \rightleftarrows 4$. Similarly, for $4 \rightleftarrows 3$,

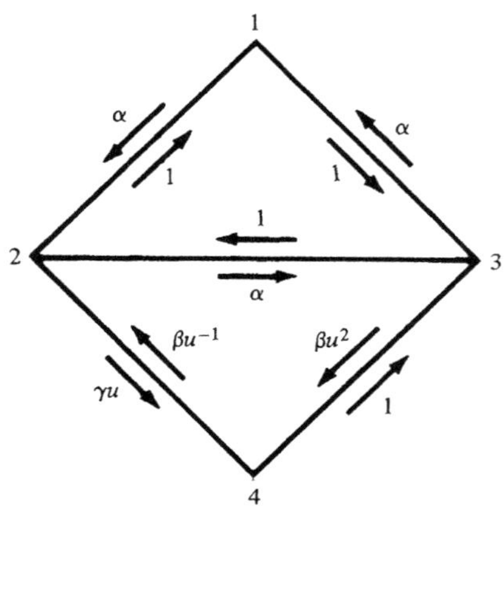

Fig. 11-16. Rate constants of the diagram, used in the example. See text for details.

$$\frac{\alpha_{43e}}{\alpha_{34e}} = \frac{\alpha_{43}}{\alpha_{34}}\left(\frac{y_{44}}{y_{ij}}\right)^{-zp_4} = \frac{1}{\beta(y_{44}/y_{ij})^{zp_4}} = \frac{1}{\beta u^2}. \tag{50.3}$$

The perturbed rate constants for $2 \rightleftarrows 4$ and $4 \rightleftarrows 3$ themselves depend (through u) on the state probability p_4 (in the BW approximation).

This model has the virtue of simplicity, but the efficiency of free-energy transduction $(-J_L X_L/J_T X_T)$ is small. More realistic models would include other interactions besides those of type 44.

Using the diagram method and Fig. 11-16, it is easy to write each of the four steady-state p_i as a sum of eight terms divided by Σ, where Σ is the sum of all 32 terms. Also, the cycle fluxes and forces are

$$J_A = (\alpha^3 - 1)(1 + \beta u^{-1})\,\Sigma^{-1}, \tag{50.4}$$

$$J_B = (\gamma u - \alpha\beta^2 u)(1 + \alpha)\,\Sigma^{-1}, \tag{50.5}$$

$$J_C = (\gamma u \alpha^2 - \beta^2 u)\,\Sigma^{-1}, \tag{50.6}$$

$$e^{X_L/kT} = \alpha^3, \qquad e^{X_T/kT} = \gamma/\alpha\beta^2. \tag{50.7}$$

The forces are independent of s. Because γ and α are of order unity (see below) and $e^{X_T/kT}$ is of order 10^{10}, β is of order 10^{-5}. In our explicit derivation of the p_i and the fluxes below, we therefore take $\beta = 0$ for simplicity.

With $\beta = 0$, we find

$$\begin{aligned} p_1 &= (a + \gamma u\alpha)\,\Sigma^{-1}, \qquad & p_2 &= a\Sigma^{-1}, \\ p_3 &= [a + \gamma u(1 + \alpha)]\,\Sigma^{-1}, \qquad & p_4 &= a\gamma u\Sigma^{-1}, \end{aligned} \tag{50.8}$$

$$\Sigma = 3a + \gamma u b,$$

$$a \equiv 1 + \alpha + \alpha^2, \qquad b \equiv (1 + \alpha)(2 + \alpha),$$

and

$$J_A = (\alpha^3 - 1)\,\Sigma^{-1}, \qquad J_B = \gamma u(1 + \alpha)\,\Sigma^{-1}, \qquad J_C = \gamma u\alpha^2\,\Sigma^{-1},$$

$$J_L = J_A + J_C, \qquad J_T = J_B + J_C. \tag{50.9}$$

Note that $J_T = p_4$. The ps and Js are regarded here as functions of γ (or ATP concentration), for specified s and α.

In numerical work, we use

$$\gamma = 3ap_4/(a - bp_4)u, \tag{50.10}$$

which follows from the equation for p_4 above, to obtain $\gamma(p_4)$ (rather than vice versa). The other quantities in Eqs. (50.8) and (50.9) are then calculated at each γ value.

Numerical example (with $\beta = 0$)

In this example, we choose α so that $J_L^{(c)} = 0$, where c refers to the critical point. This value of α is 0.68740. The critical value of s is then $s_c = 4438.0$.

If we take $z = 6$ (two-dimensional hexagonal lattice) in Eq. (50.2), the critical value of $(w_{ij} - w_{44})/kT$ is 2.80. Figure 11-17 shows the four steady-state ps as functions of γ for $s = s_c$. That is, these are all critical curves. The critical value of γ is $\gamma_c = 0.19338$. Also, $p_4^{(c)} = 0.23815$. Before the transition, p_4 is small because γ is small. However, state 4 has the largest probability just after the transition because of the attractive 44 interactions. After the transition, p_2 is small because the rate constant γu out of state 2 (Fig. 11-16) is relatively large.

Figure 11-18 (solid curves) shows the five critical ($s = s_c$) flux curves (recall that $J_T = p_4$). With the choice $J_L^{(c)} = 0$ (see above), the flux J_L (transport of L) is negative ("downhill", out → in) before the transition but is positive ("uphill") after the transition. That is, as the ATP concentration is increased, the onset of active transport of L by ATP is sudden; the active transport exists after the transition but not before.

Cycle A is the dominant cycle at small γ, while cycles B and C dominate at large γ. This is true even in the absence of interactions ($s = 1$), but the change in cycle status occurs at relatively small values of γ and is cooperative if s is large. Thus the midpoint of the transition (e.g., $J_L = 0$) occurs at $\gamma = 0.168$ for $s = 8000$, at $\gamma = 0.193$ for $s = s_c = 4438$, at $\gamma = 0.234$ for $s = 2000$, and at $\gamma = 1.429$ for $s = 1$.

The dotted J_T ($= p_4$) curves in Fig. 11-18 illustrate the effect of varying s. Exactly analogous curves obtain for the other ps and Js. The transition is subcritical for $s = 2000$ (or for any $s < s_c$). A loop is found in the $s = 8000$ case (or for any $s > s_c$). Because of the symmetry of this model, the vertical cut (stable path) across the loop occurs at $\gamma = 0.168$.

In summary, in this steady-state model with a multicycle diagram, the phase transition involves not only a sudden change in the relative importance of the different states (Fig. 11-17) but also a sudden change in the relative

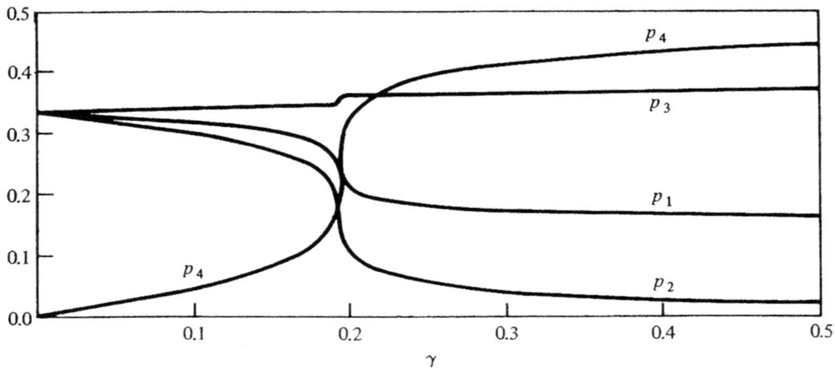

Fig. 11-17. The four steady-state state probabilities as functions of γ, in a numerical example. These are all critical curves ($s = 4438$).

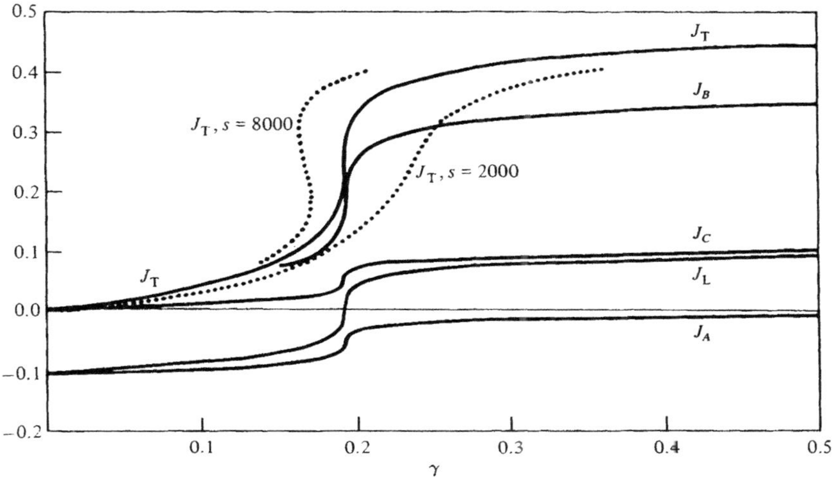

Fig. 11-18. The five critical flux curves corresponding to Fig. 11-17. Dotted curves give J_T ($=p_4$) for s = 2000 and 8000.

importance of the different cycles (Fig. 11-18). The latter effect implies, in turn, a sudden switch in the dominant biochemistry (here, the onset of active transport).

References

1. Hill, T. L. (1960) *Statistical Thermodynamics* (Addison-Wesley, Reading, MA).
2. Hill, T. L. (1977) Proc. Natl. Acad. Sci. USA **74,** 4111.
3. Hill, T. L. and Stein, L. (1978) J. Chem. Phys. **69,** 1139.
4. Hill, T. L. and Chen, Y. (1978) Proc. Natl. Acad. Sci. USA **75,** 3015.
5. Hill, T. L. and Chen, Y. (1981) Proc. Natl. Acad. Sci. USA **78,** 4.
6. Hill, T. L. and Stein, L. (1979) Proc. Natl. Acad. Sci. USA **76,** 1014.
7. Hill, T. L. (1979) Proc. Natl. Acad. Sci. USA **76,** 714.

Index